Stochastic Processes and Random Vibrations

Stochastic Processes and Random Vibrations

Theory and Practice

Júlíus Sólnes

University of Iceland, Reykjavík

JOHN WILEY & SONS

Chichester · New York · Weinheim · Brisbane · Singapore · Toronto

National 01243 779777
International (+44) 1243 779777
e-mail (for orders and customer service enquiries): cs-books@wiley.co.uk
Visit our Home Page on http://www.wiley.co.uk
or http://www.wiley.com

Other Wiley Editorial Offices

John Wiley & Sons, Inc., 605 Third Avenue,
New York, NY 10158-0012, USA

Jacaranda Wiley Ltd, 33 Park Road, Milton,
Queensland 4064, Australia

John Wiley & Sons (Canada) Ltd, 22 Worcester Road,
Rexdale, Ontario M9W 1L1, Canada

John Wiley & Sons (Asia) Pte Ltd, 2 Clementi Loop #02-01,
Jin Xing Distripark, Singapore 129809

VCH Verlagsgesellschaft mbH, Pappelallee 3,
D-69469 Weinheim, Germany

Library of Congress Cataloging-in-Publication Data

Sólnes, Júlíus.
Stochastic processes and random vibrations : theory and practice / J. Sólnes.
p. cm.
Includes bibliographical references and index.
ISBN 0-471-97191-X (cloth).—ISBN 0-471-97192-8 (pbk.)
1. Random vibration. 2. Stochastic processes. I. Title.
TA355.S567 1997
624.1'7—dc21 96-49092
CIP

British Library Cataloguing in Publication Data

A catalogue record for this book is available from the British Library

ISBN 0-471-97191-X; 0-471-97192-8 (pbk.)

Typeset in 10/12pt Times from the author's disks by Thomson Press (India) Ltd, New Delhi

Contents

Preface

The material which is contained in this book is the result of studies and research work that has been going on since early 1960s when I first became acquainted with engineering seismology and the behaviour of structures subjected to earthquake motion while studying at the International Institute of Earthquake Engineering and Seismology in Tokyo. There I was introduced to the work of K. Kanai and H. Tajimi and other renowned Japanese scientists in this field. Inspired by the many outstanding scientists, who had started to apply probabilistic methods in interpreting earthquake motions and the earthquake hazard, I tried to incorporate some of their ideas into my PhD thesis, which had to do with structural vibrations induced by earthquake motions. This work was encouraged by the late Professor K.W. Johansen and the Technical University of Denmark in Copenhagen (DTU), who supervised my doctoral studies there. Later, as a research engineer and associate professor at DTU I came to study another aspect of random phenomena in civil engineering, namely the random nature of atmospheric turbulence and wind forces acting on bluff structures. Following the pioneer work of Alan G. Davenport in the early 1960s this could only be handled properly by applying the theory of random processes. This led to studies of the fundamental theory of stochastic processes and related topics. Therefore, in the early 1970s, I published a set of lecture notes on random vibration theory and stochastic processes that was used in the courses given to civil engineering students at DTU at the time. Analysis of structural reliability and application of probability theory to the fundamental design of engineered structures, treating the loads and the material strength as random variables, also became a new field of investigation. This avenue followed a marked trend in advanced structural design philosophy in those years, where classical deterministic methods gave way to probabilistic methods, which soon became the basis of modern structural design codes.

In the early 1970s, I was asked to give a course on random vibrations and stochastic processes at the Institute of Engineering Seismology and Earthquake Engineering at the Kiril ii Methodii University in Skopje, Yugoslavia, for which I augmented and altered the material in my lecture notes. At this time, I had moved from Copenhagen to Iceland, where I had been appointed a professor of civil engineering at the newly established engineering school of the University of Iceland in Reykjavík. Busy with work that had to do with designing the curriculum in a new engineering school and organizing courses in most fields of structural design, there was little time to continue the very specialized studies in the field of random vibrations and stochastic processes. However, in 1979, I had the opportunity to work with Professor Stephen H. Crandall at MIT in Cambridge, Massachusetts, one of the leading authorities in the field of random vibration theory and who has done so much to advance engineering insight into this important subject. Needless to say, my stay as a visiting professor at MIT and collaborating with Professor Crandall was a source of great inspiration, and I was able to increase and add new and a valuable material to my already growing notes. I would like to take this opportunity to thank Professor Crandall for kindly allowing me to use some of his old examples from his courses on random vibration. Almost immediately after my visit to MIT, I became a DAAD scholar at the Technical University of Karlsruhe in West Germany, where I stayed at the Institute of concrete and applied mechanics, collaborating with the late Professor P. Mueller, who was one of the leading German authorities on earthquake resistant design of structures. I was again asked to give a lecture course on random vibration and stochastic processes and had ample time to greatly improve my work and incorporate a lot of new material and examples.

During the 1980s, I was primarily occupied with administrative work in the university. I also had a brief brush with politics that put a severe strain on my research effort and academic work throughout the last half of the 1980s. It can be said that I was given a new opportunity when I was awarded a full sabbatical year during 1991/92 after leaving the world of politics, having been a member of parliament and the first Icelandic Minister of the Environment. I had the good fortune to be invited to stay as a visiting professor at the Institute of Geophysics of the National Autonomous University of Mexico in Mexico City (UNAM). This visit was instigated by Dr Ingvar Emilsson of Departmento de limnología y ciencias del mar, a former UNESCO expert, who was instrumental in bringing me to Mexico and saw to that my stay evolved without problems. I also offer my thanks to Dr Gerardo Suarez Reynoso, the Director at the time of the Institute of Geophysics, who invited me to stay for almost a year in his institute. He also organized the necessary support from the National Science Foundation of Mexico, CONACYT, which made the stay possible.

Dr Cinna Lomnitz and Dr Krishna Singh of the Instituto de geofisica willingly explained to me the more sophisticated aspects of seismological research, which I was able to incorporate into my lecture notes for a new course on the theory of stochastic processes. I also collaborated with many other fine scientists in UNAM, which gave me a lot of new ideas and possibilities for application of stochastic processes in geophysics and related subjects. The 'students' in my course, some of whom were mature scientists themselves, were also a great source of inspiration. For instance, Dr Miquel de Icaza Herrera from the Instituto de física, who attended my course and sometimes made me feel like the egg that wanted to teach the hen, helped me unravel the mysteries of Kolmogorov's paper on fragmentation. Sr Ricardo Ruiz of Centro de Instrumentos, departmanto de acústica, was another stout ally in the field of stochastics, and persuaded me to give another course in his institute, which helped me put the final touches to the material. Dr Vladimir Kostoglov, formerly of the Institute of Physics of the Interior of the Earth in Moscow, my roommate in the office for visiting professors at Geofísica, with whom I had many interesting conversions about all kinds of scientific and wordly topics, introduced me to the latest computer techniques and word processing possibilities, that greatly simplified my work. The Russian club chaired by the late Dr Lautaro Ponce, which met every afternoon in our room for tea, was a continuous source of merriment and scientific discussions. However, it was Dr Lommitz who claimed that my loose and disorderly lecture notes were worthy of publication. He encouraged me to have the notes published and offered his help and advice. It was thus decided to prepare first an unrefined limited edition of the manuscript as an internal institute report. Dr Rosa Maria Proel undertook to have my notes published as a report under the auspices of the postgraduate section of the Institute.

Without the kind assistance of Dr Proel, Sr Gerardo Zentento and many others of the Institute, this book might never have been published. Finally, I would like to mention my partners and collaborators in Iceland, Mr Ásmundur Ásmundsson, civil engineer, and Mr Sæmundur Jónsson, mechanical engineer. Mr Jónsson flew all the way from Reykjavík to Mexico City to assist with the computer graphics and taught me how to fit complicated mathematical curves into a pretty graphics background. When the computer methods failed, my old friend and schoolmate from Tokyo, Sr Carlos Correa, put his consulting engineering company at my disposal, where some of the more difficult figures were expertly drawn by hand. In this manner, the first printed draft of a manuscript that was later to become a fully fledged textbook came to light.

During the years to follow, the manuscript was being refined from time to time as much as heavy administrative duties allowed. A short stay at the Departmento de geofísica in the University of Chile in Santiago with Dr Lautaro Ponce, produced a lot of new material and modifications of the text. Professor Ottó Björnsson of the department of mathematics at the University of Iceland undertook to read through the first chapter on probability theory and offer critical remarks. This chapter was therefore completely rewritten with the kind assistance of Professor Björnsson. I would like to thank him and the many other colleagues from the Faculties of Engineering and Natural Sciences of the University of Iceland, who gave valuable advice and help with the various chapters. The final touches were then applied during a sabbatical leave at the University of Central Florida in Orlando last year. Finally, the University of Iceland, which has been my alma mater for so many years, has given me the facilities and offered me the opportunity to carry out this work.

The material of this book has been arranged in the following manner. In the first chapter, a short overview of the theory of probability is presented, covering only the essentials that are necessary for understanding the remaining material of the book. In the second chapter, the general theory of stochastic processes is presented, with emphasis on time series analysis, which has become so important for the treatment of the environmental and engineering processes, and also for interpretation of the response of engineered structures. In particular, wind loading processes are studied with examples of application to the analysis of bluff buildings. In the third chapter, a general outline of random vibration and systems analysis is offered, where the emphasis is on analysing the system response to random excitations in stochastic terms. The fourth chapter deals with extreme conditions such as distribution of the largest response peaks, probabilities of exceedance of certain limits and failure problems in connection with overloading and fatigue. A short overview of structural reliability and probability based design of structures is also included. In the fifth chapter, random vibrations of more complicated structural vibratory systems are treated, where examples of the earthquake response of tall multi-storey structures and wind loading of tall towers are offered. Then, in the sixth chapter, certain well-known stochastic processes are treated in more detail. The Gaussian and Poisson processes are covered extensively with application examples concerning the generation of artificial earthquake motion processes. Markov processes, Martingales and non-Markovian processes are also briefly discussed. The seventh chapter offers a presentation or an overview of Fourier transform analysis and data processing with emphasis of digital signal processing techniques. The eight and last chapter contains certain applications of probability theory in earthquake engineering with an emphasis on earthquake hazard and seismic risk analysis.

The subject material which is contained in these eight chapters is of course not original, not purports to show the latest advances in the field of stochastics. It is intended to offer the student an introduction to the world of uncertainties, and give any reader who wants to become familiar with problems of random nature and their interpretative solutions a quick and easy way to do so. This book is therefore a compilation of results and thoughts put forward by the many fine and outstanding scientists that have been involved in this field. They have made their contributions freely to advance both the mathematical part and the practical applications of the theory of stochastic processes and other random phenomena; they have thus advanced human knowledge and made modern communications technology possible, which is the basis of our everyday convenience. They have done so without expectation of being rewarded or remembered, perhaps driven only by the thirst for knowledge and the academic longing to solve puzzles. They have done so without the self-serving motivation that is often the driving force in the world of commerce, where their thoughts, ideas and knowledge have come to be utilized. However, their job is not entirely forgotten, since as long as there are students and seekers of knowledge, they will rely upon and travel the road paved by those scientists that came before. Their renown and reputation will at least be part of a heritage, cherished by those who want to follow in their footsteps, or as it is said so aptly in the ancient Hávamál from the Edda of Snorri Sturluson, the Icelandic medieval author and politician,

Deyr fé,
deyja frændr
deyr sjalfr et sama
en orðstírr
deyr aldrigi
hveim sér góðan getr

or in a poor English translation

Kine dies, kinsmen die
oneself dies the same
But word of fame for deeds well done
never dies

JÚLÍUS SÓLNES
University of Iceland, Reykjavík

1 Fundamentals of Probability Calculus with Applications

In this chapter, the fundamental concepts and definitions in probability calculus will be covered. The treatment can be neither thorough nor exhaustive because of the enormous material available on the theory of probability and statistical analysis. Here, it is primarily presented to provide a necessary basic tool for the treatment of stochastic processes, random vibration and related random phenomena, which will be the main topics of this book. This fundamental material and topics will be covered only superficially and designed more for a refresher course in probability theory and statistics. As such the material can therefore not supplant a fundamental course in this field. The treatment also presupposes that the reader has a sufficient background in applied mathematics. The presentation of the material is application oriented and many practical examples are shown. Of the numerous applications that are presented, general examples, which have to do with the random behaviour of civil and electromechanical structures, are shown. Also material that is believed to be useful for engineers and other scientists involved in earthquake risk analysis, seismic zoning and evaluation of earthquake loads on buildings, wind climate and wind loads on buildings and structures will be covered. However, the presentation is quite general and can be made use of in many kinds of problems that involve probabilistic measures or require application of the theory of stochastic processes. Therefore, the fundamental concepts necessary for understanding probability calculus are presented to give the necessary background, but without the stringency and rigour appropriate for this important subject. It is hoped that the treatment is sufficiently clear and thorough to give the reader a good overview of the necessary tools that can be applied in whichever field the concepts of probability and statistical analysis are found useful.

Besides describing the main topics of probability calculus such as distributions, conditional probability, marginal distributions, transformation of variables, moment generating functions, etc., a section on application of statistical methods, mostly for analysis of earthquake risk, is presented. However, the main emphasis is to present the necessary basis for the topics that are covered in the subsequent chapters. Application-oriented topics will be treated, however, whenever the occasion arises, within the text. Of course, there are many excellent texts on the theory of probability calculus and statistics, which can be of greater advantage in this context. Texts that treat the material in greater depth, albeit with different kind of emphasis, are for instance found in* [13],

---

* Numbers within brackets refer to various textbooks and scientific papers that are listed in a References and Further Reading list at the end of this book. These works have either been utilized, directly or indirectly, for the preparation of the text, or have been included as suitable for additional reading.

[41], [64], [66], [71], [77], [83], [124], [128], [147], [169], [184] and [214], to name but a few.

1.1 DEFINITION OF THE PROBABILITY CONCEPT

1.1.1 Basic Ideas

Since time immemorial, man must have wondered about the laws of chance. Is Nature predictable or is it completely irregular and unpredictable? Can you in any way foretell the future and make predictions about the results of any kind of action caused by natural or unnatural forces? In early times, such questions were probably thought to be related to witchcraft or religious ceremonies. It was not until the late middle ages that scientists started to address such problems in an orderly and scientific manner. To find out and formulate the basic laws of elementary probability theory did not come easily and it is still best to approach the subject in a way similar to that of the old mathematicians [23].

If an experiment that can have several uncertain results is carried out repeatedly, it is not possible to predict with any precision what will be the outcome of a single experiment. This is the central point of the problem at hand, and one must therefore look for a systematic method or means to evaluate the possibilities of different outcomes of the experiment. For example, what is the possibility of getting a six when throwing a die of homogeneous material of perfectly regular shape? How can one find out that the possibility of such an occurrence is 1/6? Of course there are six different faces of the die and intuition may tell us that the possibilities of throwing any of the numbers one to six are equal. Therefore, the probability of throwing a one is 1/6 and similarly that of throwing a six is 1/6. Today, almost anyone can understand that in a game of dice the number 6 will come up on the average 1/6 times when a particular die is thrown many times. However, about 300 years ago, when the French mathematicians Blaise Pascal and Pierre de Fermat started to speculate about this problem in an exchange of letters, it was not so easy [23].

Study another experiment that can have in total m different outcomes or results, which are all equally likely. An outcome can be successful or not successful according to some *a priori* criteria. If the experiment fails, however, or something goes wrong, the outcome is excluded or can be defined as the zero outcome. In this way, all possible outcomes of the experiment have been accounted for. If only the successful outcomes are of interest, what then is the probability that a single experiment is successful when it is known that there are altogether n successful or favourable outcomes. According to Laplace's definition, the probability of a successful experiment can now be defined as follows:

$$\text{probability of success} = \Pr(\text{‘success’}) = P[\text{‘success’}]$$

or

$$P[\text{“Success”}] = \frac{\text{the number } n \text{ of successful outcomes}}{\text{the number } m \text{ of all possible outcomes}} = \frac{n}{m} \tag{1.1}$$

which is often referred to as the ‘favourable fraction’. The probability that any particular experiment will be a failure, i.e. not successful, can be evaluated in a similar manner as

$$P[\text{'Failure'}] = (m - n)/m = 1 - n/m = 1 - P[\text{'Success'}] \tag{1.2}$$

since there will be $(m - n)$ 'favourable' outcomes for an experiment resulting in a failure.

Random events, which are the outcomes or results of a particular experiment, are defined as subsets of the set of all outcomes. They will from now on be designated with capital letters $A, B, C, \ldots$. The probability p of an event A will therefore be written as follows:

$$P[A] = p \tag{1.3}$$

From Eqs. (1.1) to (1.3), already certain conclusions can be drawn. The probability measure is a number between 0 and 1, since the favourable outcomes can either be 0, a fraction of, or the entire number of possible outcomes. Therefore, to measure the probability of an event of a random experiment, the number of favourable outcomes resulting in the occurrence of this event is sought as a fraction of all possible results.

To attempt a further clarification of the concept of the probability of a random event, really a set of random outcomes of any defined experiment or action, consider for instance that of m possible outcomes, where a favourable result is defined when any particular outcome is within a group of j outcomes with prescribed results. 'Such a combined result or set of outcomes is called a random event'. The probability of a favourable outcome of any random experiment with m possible outcomes, j of them favourable, must therefore be j/m. Again assume that another group of k different outcomes, none of them coinciding with the original j, with prescribed results from the experiment is possible. The probability of any such outcome is similarly equal to k/m. What is then the probability that the outcome of any single experiment falls into either group j or k? For the evaluation, count the possible favourable outcomes, which are $(j + k)$. Therefore the probability is $(j + k)/m$.

The above result can be interpreted in a different manner. Let the two random events that an outcome will belong to either group j or k be denoted by A_j and A_k. The combined event, that is, a result belongs to either group j or k without making any further distinction between the groups, is also an event that can be denoted $A_j + A_k$. The probability of this event has already been evaluated and can be written as follows:

$$P[A_j + A_k] = (j + k)/m = j/m + k/m$$

This can also be written as

$$P[A_j + A_k] = P[A_j] + P[A_k]$$

Therefore, the probability of the combined event is the sum of the probabilities that are associated with groups j and k. If the outcomes of the experiment were to fall into several such mutually exclusive groups, the above result is easily expanded to show that

$$P[A_j + A_k + A_l + \cdots] = P[A_j] + P[A_k] + P[A_l] + \cdots \tag{1.4}$$

Thus the addition theorem of Laplace probability theory has been discovered. It can be interpreted as the either–or theorem, that is, if one is moderate and content with a result that belongs to any one of the different groups j, k, $l, \ldots$, the probability of success is increased by addition.

Example 1.1 Two cards are drawn from a well-shuffled stack of 52 cards, one after another. What is the probability that the second cards is an ace

(a) regardless of the value of the first card;
(b) given that the outcome of the first draw was an ace?

Solution (a) There are 52 possible outcomes when drawing any first card and for each first card there are 51 possibilities drawing any second, which means that there are altogether $52 \cdot 51$ possible outcomes. As for drawing an ace on the second card, there are in all $48 \cdot 4 + 4 \cdot 3$ different favourable outcomes, that is, the number of outcomes not drawing an ace for the first time but on second draw, which is $48 \cdot 4$, plus the number of outcomes drawing an ace for the first and second time, which is $4 \cdot 3$. Therefore,

$$P[\text{"The second is an ace"}] = (48 \cdot 4 + 4 \cdot 3)(52 \cdot 51) = 4/52 = 1/13$$

(b) Given that the first draw produced an ace, there are 51 cards left with three aces, so the possible outcomes drawing an ace on the second try are simply 3. Thus,

$$P[\text{"The second is an ace"}|\text{given that the first draw resulted in an ace}] = 3/51 = 1/17$$

Example 1.2 Two students have an agreement to meet in the university cafeteria between noon and 1 p.m. The one who comes first is obligated to wait for 20 minutes and then leave. If the arrival times of the students are independent and any time between 12:00 and 13:00 is equally likely, what is the probability that a meeting will take place

Solution Let the arrival times of the two students, denoted t_A and t_B, be shown on two orthogonal times axis, Fig. 1.1. When student A comes at time t_A, a meeting will take place if student B comes in the time interval $(-20 + t_A) < t_B < (t_A + 20)$. Favourable combinations for a meeting will therefore fall within the two lines as shown by the shaded area.

$$t_B = t_A + 20$$

$$t_B = t_A - 20$$

If the arrival times are independent and any time is equally likely, then

$$P[\text{"A meeting takes place"}] = (\text{shaded area})/(\text{whole area}) = 5/9$$

It should be noticed that although the calculation of the probability is intuitively clear, it does not fall under the definition of probability given by Eq. (1.1), which presumes a finite number of possible outcomes.

The Laplace definition of probability or the concept of the favourable fraction has been widely used in probability theory. Even if it can be successfully used to solve various probabilistic problems, it has certain logical flaws and lacks the stringency and completeness which is needed to establish a solid basis for the probability concept. Therefore, the axiomatic probability concept introduced by Kolmogorov in 1933 was a very welcome

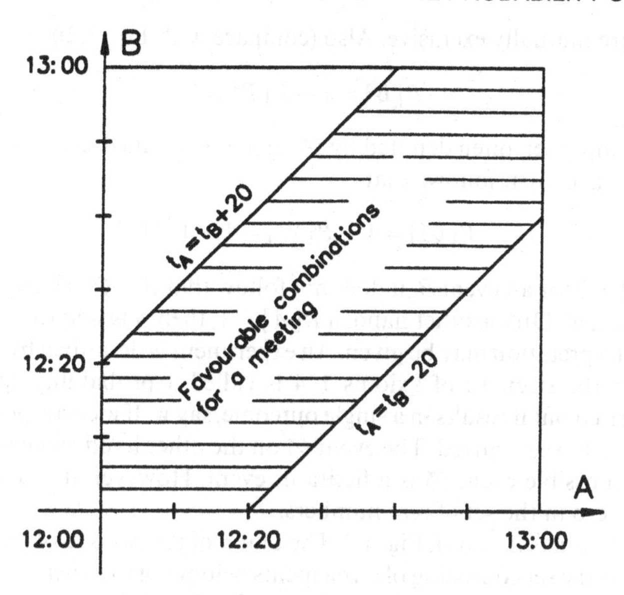

Fig. 1.1 *Random arrival times*

contribution to the development of probability theory [113]. The Kolmogorov axioms, which are very simple yet rigorous and disciplined, can be essentially stated as follows:

1. All possible outcomes of a random experiment are considered to be elements in the outcome set W, also called the sample space.

2. Any random event A associated with the experiment is a subset of W, $(A \in W)$, and is assigned a real number $p = P[A]$, which is interpreted as the probability of the event A. By definition, as has already been estabilshed,

$$0 \leqslant P[A] = p \leqslant 1 \tag{1.5}$$

3. The whole sample space W can also be considered as an event, that is, the certain event, which has the probability equal to one,

$$P[W] = 1 \tag{1.6}$$

4. The probability of a sum of a finite number or at least a countable infinite number of events that are mutually exclusive is equal to the sum of probabilities of each single event, that is,

$$P[A_1 + A_2 + A_3 + \cdots] = P[A_1] + P[A_2] + P[A_3] + \cdots \tag{1.7}$$

Let B be an event, which is a subset of W and denote by B^* the complement of the set B, i.e. $B + B^* = W$, or in other words B^* is the set of all outcomes or elements in W that do not belong to B. B^* is the complementary event of B and from axioms 1–4 it follows that

$$P[W] = 1 = P[B + B^*] = P[B] + P[B^*]$$

since B and B^* are mutually exclusive. Also (compare with Eq. (1.2))

$$P[B] = 1 - P[B^*] \tag{1.8}$$

Now, W^* is the empty set, often denoted by $\varnothing$, and is also called the impossible event. As a special case of Eq. (1.8), it follows that

$$P[\varnothing] = 1 - P[\mathrm{W}] = 1 - 1 = 0$$

Given that $P[B] = 0$ for an event B, it does not follow that $B = \varnothing$. However, if $P[B] = 0$, B is called a null event. On the other hand, if $P[A] = 1$, then A is said to be a certain event. The following interpretation may be given. An experiment is described by a mathematical model, which on the strength of axioms 1–4 is called a probability space. When the experiment is carried out it results in a single outcome, say w. If w corresponds to the event A, then A is said to have occurred. The event $\varnothing$ on the other hand can never occur, hence its name. The impossible event $\varnothing$ is a fictitious event. However, it is a very convenient concept like the zero in the set of real numbers.

Next consider two sets A and B, Fig. 1.2. The union of the two sets A and B, denoted by $A \cup B$, is defined as the set consisting of all elements belonging to either A or B or both. The intersection of the two sets, denoted by $A \cap B$, is defined as the set containing all elements that belong to both A and B. On the other hand, the elements remaining in A when all elements that also belong to B have been subtracted, that is, $A - B$, will be the same as the intersection $A \cap B^*$.

For any A and B the following relations are easily derived,

$$A = (A \cap B^*) \cup (A \cap B) = (A - B) \cup (A \cap B) \qquad \text{and}$$
$$A \cap B = A \cup (B - (A \cap B)) = A \cup (B \cap A^*)$$

where $(A - B)$ and $A \cap B$ are mutually exclusive as are also A and $B \cap A^*$. Obviously the set $(B \cap (A \cap B))$ is the same as $(A \cap B)$. Therefore by Eq. (1.7),

$$P[A] = P[A - B] + P[A \cap B]$$

or

$$P[A - B] = P[A] - P[A \cap B] \tag{1.9}$$

which is the theorem of subtraction. From Fig. 1.2 and Eq. (1.9), it easily follows that

$$\begin{aligned} P[A \cup B] &= P[A] + P[B - (A \cap B)] = P[A] + P[B] - P[B \cap (A \cap B)] \\ &= P[A] + P[B] - P[A \cap B] \end{aligned} \tag{1.10}$$

which is the theorem of addition.

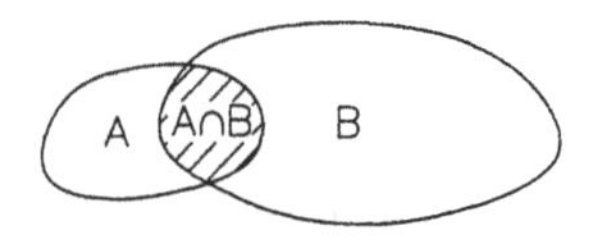

Fig. 1.2 *Two sets A and B*

The addition theorem is easily expanded to cover any finite number of events, e.g. (the reader is encouraged to work out the proof by himself),

$$\begin{aligned} P[A \cup B \cup C \cup D] = {} & P[A] + P[B] + P[C] + P[D] - P[A \cap B] - P[A \cap C] - P[A \cap D] \\ & - P[B \cap C] - P[B \cap D] - P[C \cap D] + P[A \cap B \cap C] + P[A \cap C \cap D] \\ & + P[B \cap C \cap D] + P[A \cap B \cap D] - P[A \cap B \cap C \cap D] \end{aligned} \quad (1.11)$$

If the sets A, B, C, and D are mutually exclusive, the simpler theorem of addition, Eq. (1.7), is obtained. Note that if A and B are exclusive events, then $A \cap B$ is the impossible event, that is, $A \cap B = \emptyset$ and $P[A \cap B] = 0$.

1.1.2 The Conditional Probability

Consider anew the example of drawing an ace from 52 playing cards, Example 1.1. Assume now that there are two different stacks of 52 cards. What is the probability of drawing an ace from the second stack, event B, given that already an ace was drawn from the first one, event A. Here one must take care when counting all the different outcomes. For instance, the drawing of any two cards, one from each stack, can be done in $52 \cdot 52$ different ways. Drawing of two aces from the two stacks can be done in $4 \cdot 4$ different ways. The probability of the random event $A \cap B$, that is, that any two aces, one after another, will be drawn from the two stacks, must therefore be evaluated as

$$P[A \cap B] = (4 \cdot 4)/(52 \cdot 52) = (4/52) \cdot (4/52)$$

Now to this reasoning must be added that the two events A and B obviously occur independently of one another. Drawing an ace from the first stack has nothing to do with drawing or not drawing an ace from the second stack. Therefore,

$$P[A] = P[B] = 4/52$$

and it follows by inference that

$$P[A \cap B] = P[A] \cdot P[B] \quad (1.12)$$

Equation (1.12) is valid for any number of independent events and is referred to as the multiplication theorem for independent events. It should be compared with the addition theorem, Eq. (1.7). It can be interpreted as describing greed. If one wants both this and that at the same time, the chances of success will become less, which is obvious since the result is a multiplication of two numbers less than one, that is, the probabilities $P[A]$ and $P[B]$.

In general, the outcomes of an experiment are not independent. Consider again the set of 52 playing cards. Suppose that a single card is drawn from the pack and is observed to be a heart. What is the probability that it is the ace of hearts? This is called the conditional probability $P[B|A]$, that is, the probability of the outcome B given that (or conditional to the fact that) A is known to have happened. The drawing of any one card from the stack can be taken as one of the 52 independent and equally likely, (1/52), elemental outcomes. The above problem of drawing the ace of hearts, conditional to the fact that a heart has already been drawn, can now be resolved as follows:

$$P[B|A] = (1/13) = (1/52)/(13/52) = P[A \cap B]/P[A]$$

where $(A \cap B)$ is the combined event "drawing the ace of hearts and drawing a heart".

It is now possible to define in general terms the conditional probability of B with respect to A subject to the additional condition that $P[A] \neq 0$

$$P[B|A] = P[A \cap B]/P[A] \tag{1.13}$$

The conditional probability of A relative to B is similarly defined as

$$P[A|B] = P[B \cap A]/P[B]$$

given that $P[B] \neq 0$.

By comparing Eqs. (1.13) and (1.12), the general theorem of multiplication can be derived as

$$P[A \cap B] = P[A] \cdot P[B|A] = P[B] \cdot P[A|B] \tag{1.14}$$

since $A \cap B = B \cap A$. In other words, if two events A and B have the probability of occurrence equal to $P[A]$ and $P[B]$, then the probability of the joint occurrence of both A and B is given by Eq. (1.14).

The multiplication theorem is easily generalized. For instance consider four random events A, B, C and D. Then,

$$P[A \cap B \cap C \cap D] = P[A] \cdot P[B|A] \cdot P[C|A \cap B] \cdot P[D|A \cap B \cap C] \tag{1.15}$$

given that $P[A \cap B \cap C] \neq 0$. If the random events A, B, C and D are mutually independent, then

$$P[B|A] = P[B],\ P[C|A \cap B] = P[C] \text{ and } P[D|A \cap B \cap C] = P[D]$$

and the simple multiplication theorem Eq. (1.12) is obtained:

$$P[A \cap B \cap C \cap D] = P[A] \cdot P[B] \cdot P[C] \cdot P[D] \tag{1.16}$$

It is important to notice that three or more random events can be independent in pairs without being mutually independent as is shown in the following example.

Example 1.3 In Fig. 1.3, a regular tetrahedron is shown with the faces painted with different colours. Face ABC is violet, yellow and white, face OAB is violet, face OBC is yellow and face OCA is white. The tetrahedron is put into a beaker, and then thrown out after the beaker has been shaken. The outcome is defined as the colour of the face on which the tetrahedron comes to rest. The following events are now defined:

F: the outcome is OAB or ABC, i.e. the face contains a violet colour.
G: the outcome is OBC or ABC, i.e. the face contains a yellow colour.
H: the outcome is OCA or ABC, i.e. the face contains a white colour.

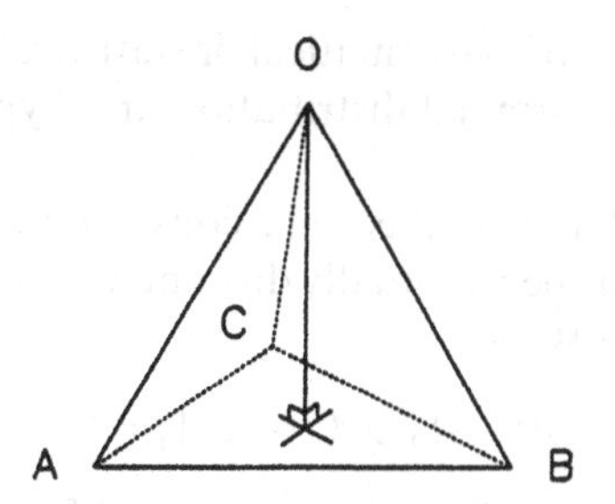

Fig. 1.3 *A regular tetrahedron*

Since the tetrahedron is completely regular as regards the geometry and mass distribution, any face is equally likely to come up as the base on which the tetrahedron comes to a rest with a probability equal to 1/4. Furthermore, the probability of each of the events F, G or H is equal to 1/2 as each event corresponds to two out of four faces. The probability of having all of F, G and H happen, that is, the face contains all colours is expressed by $P[F \cap G \cap H]$. This happens when ABC is the base and has a probability of 1/4. Therefore,

$$P[F \cap G \cap H] = P[\{ABC\}] = 1/4 \neq P[F] \cdot P[G] \cdot P[H] = (1/2)^3 = 1/8$$

which shows that the events F, G and H are not mutually independent. On the other hand, they are independent when combined in pairs, that is,

$$\begin{aligned} P[F \cap G] = P[G \cap H] = P[H \cap F] &= P[\{ABC\}] = 1/4 \\ &= (1/2)^2 = P[F] \cdot P[G] = P[G] \cdot P[H] = P[H] \cdot P[F] \end{aligned}$$

1.1.3 Bayesian Probabilities

By Eq. (1.13), the conditional probability of B given that event A has taken place was defined. In this case, both events A and B were clearly defined and understood. Often one is faced with a situation of having to assess the probability of a certain event with some *a priori* information about the conditions for the event taking place. Such information may be used on some knowledge about the processes being studied, on past experience and even on individual hunches. Even if the exact law of nature governing the event is unknown, it is unusual not to have some notion about the event which could and should be used to assist in a decision. Taking such information into account can be formalized applying the Bayesian principle. The Bayesian principle assigns to each possible action a measure of the consequences of that action in form of a hypothesis or an educated guess formulated in the form of a probability statement. In Bayesian statistics the hypothesis is not tested but modified by the data represented by the event A, thus furnishing a posterior, improved probability distribution. Because of the arbitrariness and guesswork involved in selecting the prior distributions and hypotheses, the use of Bayes' principle has been criticized. However, it has found widespread use in engineering applications, due to perhaps the engineer's ability (and necessity) to stipulate assumptions and come up with solutions when very little information is available. Alternative methods have been suggested to overcome the lack of mathematical or statistical rigour, but they do no more than disguise the fact that the original distribution and hypotheses are subject to some kind of choice [129], [151].

To establish the so-called Bayes' theorem, consider first the following elementary problem. Let $A_1, A_2, A_3, \ldots, A_n$ be n mutually disjoint non-null events, the union of which is the entire probability space W, i.e.

$$A_1 \cup A_2 \cup A_3 \cdots \cup A_n = W$$

The probability $P[B]$ of any event B can be written as $P[B \cap W]$ since the intersection of B and the entire probability space must be equal to B itself. Then

$$\begin{aligned} P[B] &= P[(A_1 \cup A_2 \cup A_3 \cup \cdots \cup A_n) \cap B] \\ &= P[(A_1 \cap B) \cup (A_2 \cap B) \cup (A_3 \cap B) \ldots \cup (A_n \cap B)] \\ &= \sum_{i=1}^{n} P[(A_i \cap B)] = \sum_{i=1}^{n} P[A_i] P[B|A_i] \end{aligned}$$

where the addition theorem for disjoint events and the definition of conditional probabilities, Eqs. (1.13) and (1.14), have been applied.

The conditional probability of event A_i relative to event B can now be written. Using Eq. (1.14),

$$P[A_i|B] = \frac{P[A_i]P[B|A_i]}{\sum_{k=1}^{n} P[A_k]P[B|A_k]}$$

since the denominator is the same as $P[B]$. Bayes' theorem is then stated as follows. Let $\{H_i\}$, $i = 1, \ldots, n$ denote n exhaustive, mutually exclusive hypotheses which may be descriptive of certain behaviour or inherent quality in connection with a certain action or an event A. Then $P[H_i]$ is the prior probability that the hypothesis H_i is true independent of the occurrence of the event A. On the other hand, $P[A|H_i]$ is the conditional probability that event A takes place when it is known that hypothesis H_i is true, and $P[H_j|A]$ is the conditional probability that the hypothesis H_j is true when it is known that event A has occurred, actually the improved or posterior probability that the hypothesis is true. Then

$$P[H_j|A] = \frac{P[H_j]P[A|H_j]}{\sum_{i=1}^{n} P[H_i]P[A|H_i]} \tag{1.17}$$

The obvious advantage of the Bayesian approach lies in the possibility of modifying original predictions through incorporation of new data. A particular hypothesis which maximizes $P[H|A]$ will be preferred at each step. The drawback, however, is that an initial prediction depends to a considerable extent on initial assumptions concerning the prior probabilities $P[H_j]$. Where the knowledge of the mechanism of the process under study is weak, these prior probabilities may be little better than wild guesses.

Example 1.4 It has been found that 5‰ of persons in an entire population W of individuals have cancer. At a cancer clinic, a random sample of N individuals of the population W is tested for cancer. The test is carried out with an accuracy of 95%. Find the conditional probability that a person really has cancer if the test was positive.

The following events can be defined:

A : The test is positive whether the individual has cancer or not
A^*: The test is negative whether the indivitual has cancer or not
C : The individual has cancer whether the test showed it or not
C^*: The individual has no cancer whether the test showed it or not

The conditional probability $P[A|C]$ is the same as the accuracy of the test, which is here equal to 95%. That is, given that a person has cancer, there is 95% probability that the test will reveal that condition. The accuracy of the test $P[A|C]$ and the conditional probability $P[A^*|C^*]$ completely characterize the statistical properties of the test. Here $P[A^*|C^*]$ is also equal to 0.95 and is the probability that the test will result in a negative outcome, given that the person does not have cancer. In general these two characteristics, however, have different values. Therefore, the conditional probability that the test is positive when it is known that the person does not have cancer is $P[A|C^*] = 1 - P[A^*|C^*] = 0.05$. As the frequency of cancer in the population was 5‰, $P[C] = 0.005$, and $P[C^*] = 0.995$. All ingredients for using the Bayesian formula are now in place and choosing $H_1 = C$ and $H_2 = C^*$, $n = 2$,

$$P[C|A] = \frac{P[A|C]\cdot P[C]}{P[A|C]\cdot P[C] + P[A|C^*]\cdot P[C^*]} = \frac{0.95\cdot 0.005}{0.95\cdot 0.005 + 0.05\cdot 0.995} = 0.087$$

so the probability that a person has cancer if the test was positive is about 9%, which does not seem to make much sense as regards the usefulness of the test. Actually this result is quite controversial as pointed out by Parzen, [169]. On the one hand, the cancer diagnostic test is highly accurate, since it will detect cancer in 95% of the cases in which cancer is present. On the other hand, in only about 9% of the cases in which the test gives a positive result and asserts cancer, it is actually true that there is cancer in that individual! The reason for the inefficiency of this enterprise at the cancer clinic, as can be seen by the above fraction, is the scarcity of the disease, $P[C] \ll 1$. Nowadays, the medical profession uses other characteristics to estimate the usefulness of its tests, namely $P[C|A]$ and $P[C^*|A^*]$, which are called the sensitivity and specificity of the test respectively. For efficient test, a high value of these quantities is usually required.

Example 1.5 A contractor assumes the responsibility to carry out and complete satisfactorily a certain project. The buyer hires an independent controller to check out whether the project has been completed satisfactorily according to prescribed standards. The controller will make random inspections and it may be assumed that the probability that one particular inspection reveals the quality of the whole project is 85%, that is there is 15% probability that the controller (or any other controller) will draw erroneous conclusions from that inspection. On the other hand, from previous experience, 92% of the projects carried out by the contractor have been accepted by a controller. What is the probability that the project is satisfactory and will be approved by the controller? What is the probability that the project will be approved if it has been completed in a satisfactory manner?

Define the two events,

A: the project has been satisfactorily completed
B: the controller will accept the work

Note that the above situation is analogous to that of Example 1.4. The controller plays the role of the test. Now the accuracy of the inspection based on the information given may be interpreted as the conditional probability

$$P[A|B]=0.85$$

and the probability of acceptance, based on past experience, is $P[B]=0.92$. Therefore by Eq. (1.14),

$$P[A \cap B]=P[A|B] \cdot P[B]=0.85 \cdot 0.92=0.782$$

which is the probability of having both A and B and answers the first question.

The probability that the project will be approved if it has been completed satisfactorily may be interpreted as the conditional probability $P[B|A]$. By Eq. (1.14) $P[B|A]=P[B]P[A|B]/P[A]$. Bayes' theorem (Eq. (1.17)) can be applied to obtain $P[A]$, which is the only remaining unknown quantity. Using the two hypotheses B and $B^*=1-B$, which are descriptive of whether the project will be accepted or not, the denominator of Eq. (1.17) is

$$\begin{aligned} P[A] &= P[A|B]P[B]+P[A|B^*]P[B^*] \\ &= P[A|B]P[B]+(1-P[A^*|B^*])(1-P[B]) \end{aligned}$$

Obviously, the conditional probability $P[A^*|B^*]$, 'the second characteristic of the test', is needed in order to calculate $P[A]$, where the two complementary events are:

A^*: the project has not been satisfactorily completed
B^*: the controller will not accept the work

As in Example 1.4, assume that the conditional probability $P[A^*|B^*]$ is also equal to 0.85, that is, the probability that the project is not satisfactorily completed given that the work has not been accepted is the same as the probability that the project is satisfactorily completed given that the work has been accepted. Then,

$$P[A]=0.85 \cdot 0.92+(1-0.85)(1-0.92)=0.794$$

which is the 'total' probability in the denominator of Eq. (1.17). Therefore,

$$P[B|A]=P[B]P[A|B]/P[A]=(0.92 \cdot 0.85)/0.794=0.985$$

so there is 98.5% probability that satisfactory work will be accepted.

1.2 RANDOM VARIABLES AND DISTRIBUTION FUNCTIONS

The probability model based on the Kolmogorov axioms, which were introduced in the previous section, can be described by the triplet $(W, \mathscr{W}, P)$. Here W denotes the outcome set or the sample space, $\mathcal{W}$ is a collection of subsets, events E, of W, and P is a mapping of $\mathcal{W}$ into the unit interval of real numbers $[0,1]$ such that to each event $E \in \mathcal{W}$ is assigned a probability $P[E]$ between 0 and 1. If W is very large, an awkward situation may arise that there are subsets of W, which can not be included in the collection $\mathcal{W}$, that is, can not be considered as events with well-defined probabilities. The mathematical term for $\mathcal{W}$ is σ-algebra over W. This situation is treated in more sophisticated texts on the theory of probability and is beyond the scope of this book.

The fundamental concept of a random variable can now be defined. Let X denote a mapping or transformation of W into the set of real numbers $\mathbb{R}$, i.e.

$$X: W \rightarrow \mathbb{R}$$

For any real number c, the symbol $[X \leqslant c]$ shall denote the set of outcomes w for which $X(w) \leqslant c$. The symbols $[X < c]$, $[X = c]$, $[X > c]$ etc. are similarly defined. If for every c, these sets of outcomes w are events, i.e. belong to $\mathcal{W}$, then X is a real-valued one-dimensional random variable. Hence the expression $P[X \leqslant c]$ denotes the probability that the random variable X, corresponding to a given experiment, will assume a value less than or equal to c. If $\{X_i\}$, $i = 1, 2, \ldots, n$ are random variables, then the ordered n-tuple $(X_1, X_2, \ldots, X_n)$ is a mapping of W into $\mathbb{R}^n$ and is called an n-dimensional random vector. Now, suppose E is an event. Then the mapping X defined by

$$X(w) = \begin{cases} 1 & \text{if} \quad w \in E \\ 0 & \text{if} \quad w \in E^* \end{cases}$$

is a random variable, which can be considered to classify the outcomes in two exclusive classes E and E^*. The following example shows a more general classification variable.

Example 1.6 In a zone of devastation after a large earthquake, the damage of houses is being classified by six different damage classes, that is, no damage, very little damage, little damage, considerable damage, severe damage and total damage. A survey of the damage indicates that 15% of the houses suffered no damage, 20% very little damage, 25% little damage, 20% considerable damage, 15% severe damage and 5% of the house were completely destroyed. The damage of each particular house in the zone is therefore a random event with an outcome as described. Assigning a real number to each of the above damage classes, the transformation indicated in Table 1.1 can be established.

Table 1.1 *Earthquake damage ratios*

$w \in W$	$\rightarrow$	$X(w) = x \in \mathbb{R}$	Portion of houses in class (%)
no damage		0	15
very little damage		1	20
little damage		2	25
considerable damage		3	20
severe damage		4	15
total damage		5	5

This defines the random vairable $X: W \to \mathbb{R}$, which can be used to describe the damage caused by the earthquake in probabilistic terms. For instance, the probability that a house will be classified with considerable damage is stated as $P[X=3]=0.20$.

In the following, random variables will be denoted by capital letters, $X, Y, Z, \ldots$. When they have been assigned a number it will be denoted by lower case letters, $a, b, c, \ldots$, (i.e., a specific or a deterministic value). When such random variables can only have a limited number or a countable infinite number of different real values as in the example above, they are called discrete random variables. If no single real number can be assigned to the outcome, rather it has to be described by a range of real numbers, the variables on the other hand are continuous.

Having defined the random variable X, an important function related to the variable can be defined through the mapping $F_X: \mathbb{R} \to [0, 1]$, that is,

$$F_X(x) = P[X \leqslant x] \tag{1.18}$$

This function, which is called the cumulative distribution function or simply the distribution function of the random variable X, obviously gives the probability that the random variable assumes a value less than or equal to a real number x on the axis of real numbers. It can easily be shown to have the following properties where F is written instead of F_X:

$$\begin{aligned} &F(a) - F(b) = P[a < X \leqslant b] \geqslant 0, \; a \leqslant b \\ &F(a) \leqslant F(b) \text{ (non-decreasing)} \\ &F(-\infty) = 0, \; F(+\infty) = 1 \end{aligned} \tag{1.19}$$

In Eqs. (1.19), $F(-\infty) = \lim F(x)$, $x \to -\infty$, and $F(\infty) = \lim_{x \to \infty} F(x)$, and the probabilities simply represent the intuitive result that one can not have a number less than $-\infty$ and a number less than $+\infty$ is certain.

The distribution function as shown in Fig. 1.4 is a non-decreasing function, increasing from 0 to 1, but can have a countable set of discontinuities. The point $x = a$ is a discontinuity point, and the discontinuity or jump at a is $P[X = a]$. On the other hand, F is continuous at $x = b$ so $P[X = b] = 0$. In a similar manner, the distribution function of an n-dimensional random vector $\{X_i\}$, $i = 1, 2, \ldots, n$, defined by the mapping $F_X: \mathbb{R}^n \to [0, 1]$, describing more complex situations where the outcome of a random event is represented

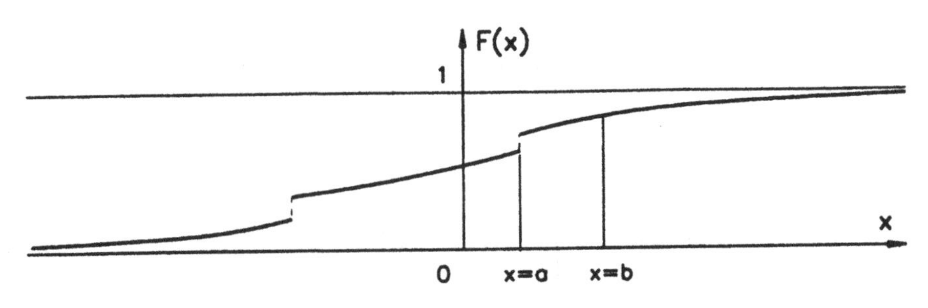

Fig. 1.4 *The cumulative distribution function*

by n different real numbers, is given by

$$F_{X_1,\ldots,X_n}(x_1,\ldots,x_n) = P[(X_1 \leqslant x_1) \cap \cdots \cap (X_n \leqslant x_n)] \tag{1.20}$$

The probabilistic properties of a random variable are those and only those which can be characterized by its distribution functions. Random variables are therefore classified according to the properties of their distribution functions. The class of absolutely continuous random variables X includes those which have distribution functions that can be represented by the Riemann integral

$$F(x) = \int_{-\infty}^{x} p(u)\mathrm{d}u \tag{1.21}$$

in which $p(u)$ is a certain non-negative function. In that case, $F(x)$ is a continuous function and has the derivative

$$p(x) = \frac{\mathrm{d}F(x)}{\mathrm{d}x} \tag{1.22}$$

if $p(x)$ is continuous at x. To obtain the whole class of continuous random variables, the Riemann integral has to be extended to that of Lebesgue. A knowledge of the Lebesgue integral, however, will not be assumed as it will not be used in the following text. The second main class consists of discrete random variables, which have distribution functions $F(x)$ that can be represented by a step function with a discontinuity or a jump equal to the probability $P[X = x_i]$ at most countably many points x_i, $i = 1, 2, \ldots$. The discrete distribution functions will be discussed in more detail in Section 1.2.2.

1.2.1 Probability Density of Continuous Random Variables

By Eqs. (1.21), the probability that an absolutely continuous random variable X assumes a specific value a is equal to zero, which makes the treatment fundamentally different from that of discrete random variables, which can assume a real number with a probability that need not be 0. The integrand function $p(x)$ is called the probability density and can look like the example shown in Fig. 1.5. From Eqs. (1.18) and (1.21), the probability $P[X \leqslant x]$ is obviously equal to the area under the probability density curve up to the point x, that is,

$$P[X \leqslant x] = F(x) = \int_{-\infty}^{x} p(u)\mathrm{d}u \tag{1.21}$$

Furthermore, since $F(+\infty) = 1$ by the last of Eqs. (1.19)

$$1 = \int_{-\infty}^{\infty} p(x)\mathrm{d}x \tag{1.23}$$

that is, the area under the entire density curve must be equal to one.

By the first of Eqs. (1.19), the probability that a continuous random variable assumes a specific number a is equal to zero. On the other hand, the probability that the variable X assumes a value in a narrow interval $[x, x, + \Delta x]$, where $\Delta x > 0$ is an arbitrarily small

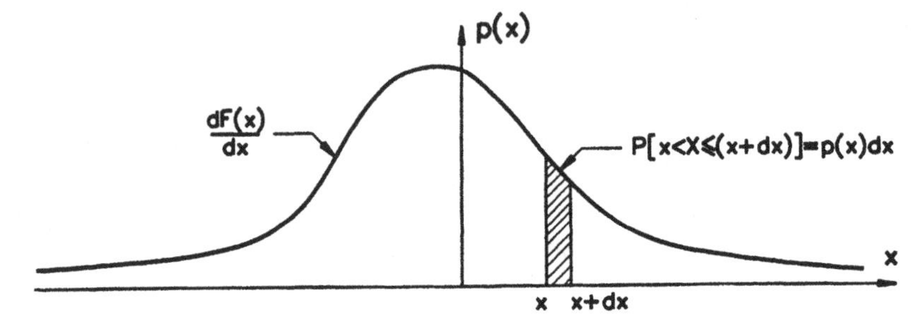

Fig. 1.5 *The probability density function p(x)*

number, can be computed as

$$P[x < X \leqslant x + \Delta x] = F(x + \Delta x) - F(x) = \int_{-\infty}^{x+\Delta x} p(u)\mathrm{d}u - \int_{-\infty}^{x} p(u)\mathrm{d}u$$
$$= \int_{x}^{x+\Delta x} p(u)\mathrm{d}u \cong p(x)\Delta x \qquad (1.24)$$

that is, the probability is represented by the shaded column in Fig. 1.5.

By Eq. (1.20), multi-dimensional probability distributions were introduced. As a natural extension of the probability density given by Eq. (1.21), multi-dimensional probability densities are defined in the same manner. For instance, consider the two random variables (X, Y). By Eqs. (1.20) and (1.22), the probability density function is given by

$$p_{XY}(u, v) = \frac{\partial^2 F_{XY}(u, v)}{\partial u\, \partial v} \qquad (1.25)$$

Example 1.7 Investigate whether the function

$$p(x_1, x_2) = \frac{1}{2\pi}\frac{1}{a^2 + x_1^2 + x_2^2}$$

where a is a real constant, is an admissible two-dimensional probability density for two random variables (X_1, X_2) that are distributed over the entire (x_1, x_2) plane.

Solution

$$\int_{-\infty}^{\infty}\int_{-\infty}^{\infty} p(x_1, x_2)\mathrm{d}x_1\,\mathrm{d}x_2 = 1 \qquad \text{(the normalization test)}$$

$$\frac{1}{2\pi}\int_{-\infty}^{\infty}\int_{-\infty}^{\infty}\frac{\mathrm{d}x_1\,\mathrm{d}x_2}{a^2 + x_1^2 + x_2^2} = \frac{1}{2\pi}\int_{-\infty}^{\infty}\frac{\pi}{\sqrt{a^2 + x_1^2}}\mathrm{d}x_1$$
$$= \frac{1}{2}\cdot 2\cdot \lim_{u\to\infty}\int_{0}^{u}\frac{\mathrm{d}x_1}{\sqrt{a^2 + x_1^2}} = \lim_{u\to\infty} \operatorname{arcsinh}\left(\frac{\mathrm{u}}{\mathrm{a}}\right) = \infty!!$$

The density function is not admissible, since it does not satisfy the normalization test.

1.2.2 Probability Density of Discrete Random Variables

In Example (1.6), the probability distribution for a discrete variable X was shown to be represented by a certain probability function, which will be called a density function.

$$p_X(x) = P(X = x) \tag{1.26}$$

that is, the probability that the discrete variable X assumes a value x is equal to the density function $p_X(x)$. According to the Kolmogorov axioms, the probability density function must have the following properties:

$$\begin{aligned} & p_X(x_i) \geqslant 0 \qquad \text{for all } x_i \\ & \sum_i p_X(x_i) = 1 \end{aligned} \tag{1.27}$$

where all possible values x_i that can be assumed by X are taken into account. The probability distribution function for a random variable X has already been defined according to Eq. (1.19). The probability $P[X \leqslant x]$ that the discrete variable X will not exceed a certain value x is therefore given by

$$F_X(x) = \sum_j p_X(x_j) \qquad \text{for all } x_j \leqslant x \tag{1.28}$$

where $F_X(x)$ is the probability distribution function. If the value x is allowed to approach a very large positive number, the probability tends to zero, and if x is allowed to approach a very large positive number, the probability tends to one in accordance to Eq. (1.19), that is,

$$F_X(-\infty) = 0,\ F_X(\infty) = 1,\ 0 \leqslant F_X(x) \leqslant 1 \tag{1.29}$$

Example 1.8 A seismological institute buys and installs three strong motion acceleration meters in three different places. Based on past experience, there is 20% chance that any instrument will malfunction during the first six months after installation. Therefore, the probability p that any one of the three instruments will function properly without trouble during the first six months in use is 0.80. Also, the behaviour of each instrument is independent of the other two. Given that the random variable X is the number of instruments that operates without malfunction the first six months, find the probability density and the probability distribution for X.

Solution The random variable X can obviously assume the following values:

0: all instruments do malfunction
1: two instruments do malfunction, one does not
2: one instrument malfunctions, two do not
3: all instruments function properly.

The probability density is obtained by using the multiplication and addition theorems (Eqs. (1.16) and (1.7)),

$$p_X(0) = P[X = 0] = (1 - p)^3 = 0.008$$

$$p_X(1) = P[X = 1] = p(1 - p)^2 + (1 - p)p(1 - p) + (1 - p)^2 p = 3p(1 - p)^2 = 0.096$$

$$p_X(2) = P[X = 2] = p{\cdot}p(1 - p) + p(1 - p)p + (1 - p)p{\cdot}p = 3p^2(1 - p) = 0.384$$

$$p_X(3) = P[X = 3] = p^3 = 0.512$$

For all other values of x, $p_X(x) = 0$. The probability distribution function can now be computed using Eq. (1.28):

$$F_X(0) = 0.008$$

$$F_X(1) = 0.008 + 0.096 = 0.104$$

$$F_X(2) = 0.008 + 0.096 + 0.384 = 0.488$$

$$F_X(3) = 0.008 + 0.096 + 0.384 + 0.512 = 1.000$$

The above probabilities and the values of the distribution function are shown in Fig. 1.6. The graph of $p_X(x)$ is a spike diagram. Note that the graph of $F_X(x)$ is a step function.

It is noteworthy that the above probabilities follow the so-called *binomial law* [169]. The binomial law is usually associated with a so-called *Bernoulli trial*, which is the name given to an experiment that has only two possible outcomes. Consider an experiment that is carried out repeatedly and the outcome from each run can only be described as either a success S or failure F. The probability of a successful outcome will be denoted by p and the probability of failure by q. Therefore,

$$P[S] = p, P[F] = q, p \geqslant 0, q \geqslant 0 \qquad \text{and} \qquad p + q = 1 \tag{1.30}$$

If n independent Bernoulli trials are carried out, what is then the probability of having k successful outcomes? The number of ways k successes can occur in n independent trials, is the same as the number of ways k numbers can be selected out of n numbers, that is, $\binom{n}{k}$, which is the number of descriptions containing exctly k successes and $n - k$ failures. Using the multiplication combining probabilities, Eq. (1.16), each such description has the

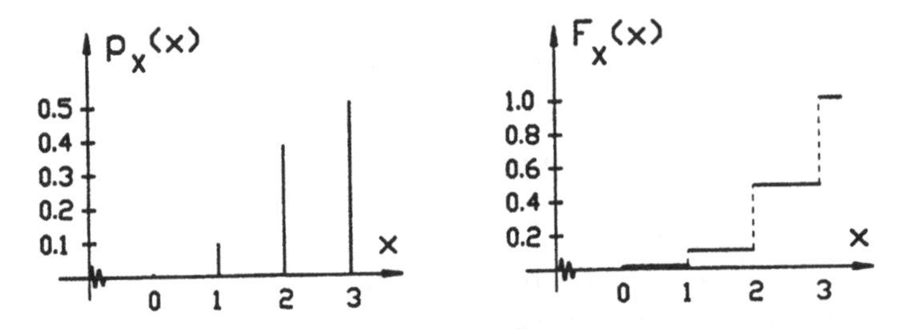

Fig. 1.6 *The probability density and the distribution function*

probability $p^k q^{n-k}$ and the probability that n independent trials will have k successes, denoted by $b(k; n, p)$, is then given by

$$b(k; n, p) = \binom{n}{k} p^k q^{n-k} \tag{1.31}$$

Obviously,

$$(p+q)^n = \sum_{k=0}^{n} \binom{n}{k} p^k q^{n-k} = 1 \tag{1.32}$$

which is the usual binomial theorem, hence the name.

In the following text, the main emphasis will be placed on continuous variables. It will therefore be of convenience to be able to treat all variables as continuous variables in some generalized sense. This can be done by introducing *Dirac's delta function* (see Chapter 7, Example 7.4), that is,

$$\left.\begin{aligned} &\delta(x-a) = \begin{cases} \infty, & x = a \\ 0, & x \neq a \end{cases} \\ &\int_{-\infty}^{\infty} \delta(x-a)\mathrm{d}x = 1 \\ &\int_{-\infty}^{\infty} f(x)\delta(x-a)\mathrm{d}x = f(a) \end{aligned}\right\} \tag{1.33}$$

Analogous to the treatment of discrete and continuous Fourier transforms (Chapter 7), the discrete probability density is considered as a series of delta functions or spikes and can, through Eq. (1.33), be treated as if it were a continuous function. For instance, let $P[X = a] = p$, then

$$p(x) = p\delta(x-a) \qquad \text{and } P[X = a] = \int_{-\infty}^{\infty} p(x)\mathrm{d}x = p\int_{-\infty}^{\infty} \delta(x-a)\mathrm{d}x = p$$

1.2.3 Statistical Moments and Frequently Used Parameters

Suppose X is an absolutely continuous random variable with the distribution function (Eq. 1.21), that is,

$$P[X \leqslant x] = F(x) = \int_{-\infty}^{x} p(u)\mathrm{d}u \tag{1.21}$$

Consider the functional relationship $y = r(x)$, where $r(x)$ is a real-valued function. Now suppose that $Y = r(X)$ is a random variable, which is produced through the functional transformation $r(X)$. Then the mean value of Y, denoted by μ_Y, is defined as follows:

$$\mu_Y = \int_{-\infty}^{\infty} r(x)p(x)\mathrm{d}x \tag{1.34}$$

if

$$\int_{-\infty}^{\infty} |r(x)|p(x)\mathrm{d}x < +\infty$$

In particular, if $r(x) = x$ or $Y = X$, then the mean value of the random variable X is given by

$$\mu_X = \int_{-\infty}^{\infty} xp(x)\mathrm{d}x \tag{1.35}$$

If X is a discrete random variable with a probability density function $p_X(x_i) = p_i\delta(x - x_i)$, $i = 1, 2, 3, \ldots$, then by Eqs. 1.33 and 1.35,

$$\mu_X = \sum_i x_i p_X(x_i) \tag{1.36}$$

provided that

$$\sum_i |x_i| p_X(x_i) < +\infty$$

The mean value of a random variable is also referred to as its mathematical expectation or its expected value denoted by $E[X]$. The expectation is the value the random variable is expected to assume on the average. It is one of the fundamental characteristics of the probability distribution of a random variable. The expectation has certain elementary properties, which will be useful in the following. For instance, by Eqs. (1.35) and (1.36),

$$E[X] \geqslant 0 \qquad \text{for } X \geqslant 0$$

and

$$|E[X]| \leqslant E[|X|] \tag{1.37}$$

Let X and Y denote two random variables. Then it can be proven that the linear combination $aX + bY$ is also a random variable for any real constant a and b. In mathematicial terms, the class of all random variables of the same dimension is a linear vector space. Not every random variable has a finite mathematical expectation, but if both X and Y do, then the following relation between their expectations holds:

$$E[aX + bY] = aE[X] + E[Y] \tag{1.38}$$

which can intuitively be deducted from Eq. (1.34) by putting $r(x) = r(ax + by) = ar(x) + br(y)$. Thus it has been shown that the expectation $E[*]$ is a linear functional.

Example 1.9 An experiment is carried out, which consists of measuring n sample values for the two random variables Y and X having the functional relationship $Y = r(X)$. Thus a collection of sample values x_i, $i = 1, 2, 3, \ldots, n$, produces a collection of sample values $r(x_i)$, $i = 1, 2, 3, \ldots, n$, for the variable Y. The average value of the sample values for Y is given by

$$\bar{Y} = \sum_{i=1}^{n} r(x_i)\frac{1}{n}$$

where a bar over the character indicates an average value. As shown in all standard textbooks on the theory of statistics, the average value of n samples of a random variable gives a reasonably accurate estimate of the theoretical mean value of Y. In fact, for $n \to \infty$, the average sample value converges to the mean value of the expectation, i.e. $\mu_Y = E[Y] = \lim_{n\to\infty} \bar{Y}$.

If a collection of samples x_i is not available, the sample values $r(x_i)$ can be related to the probability that X assumes the value x_i. Given the probability density function $p(x)$, the probability that X assumes the value x_i can be evaluated as $p(x_i)\Delta x_i$, using Eq. (1.24). The value $r(x_i)$ will therefore be encountered with the probability $p(x_i)\Delta x_i$ or the weighted value is $r(x_i)p(x_i)\Delta x_i$. Thus the weighted mean value of Y is given by the following relation:

$$\mu_Y^* = \lim_{n\to\infty} \left[\frac{\sum_{i=1}^{n} r(x_i)p(x_i)\Delta x}{\sum_{i=1}^{n} p(x_i)\Delta x_i} \right]$$

By going to the limit, i.e. letting $\Delta x_i \to 0$ as $n \to \infty$, the weighted mean value μ_Y^*, is given by

$$\mu_Y^* = \int_{-\infty}^{\infty} r(x)p(x)\mathrm{d}x$$

since $\int_{-\infty}^{\infty} p(x)\mathrm{d}x = 1$ (cf. Eq. (1.23)). Thus the weighted mean value of a random variable is equivalent to the mean value or the expected value $E[Y]$ in the limit.

Example 1.10 Let X denote a random variable with a cumulative distribution function $F(x)$. Then an illustrative interpretation of the mean value (Eqs. (1.34) and (1.35)) can be given by imagining the area under the density curve in Fig. 1.7 to be covered with dense material with a uniform thickness, that is, the probability density is to be modelled as a mass density. The total mass under the probability density curve must be equal to 1 (Eq. (1.23)). Now the abscissa of the centre of gravity of the mass underneath the curve is obtained by equating the static moments of the mass with respect to the vertical axis, Fig. 1.7. Denoting the abscissa by μ_X,

$$\mu_X \cdot 1 = \int_{-\infty}^{\infty} xp(x)\mathrm{d}x = \mu_X$$

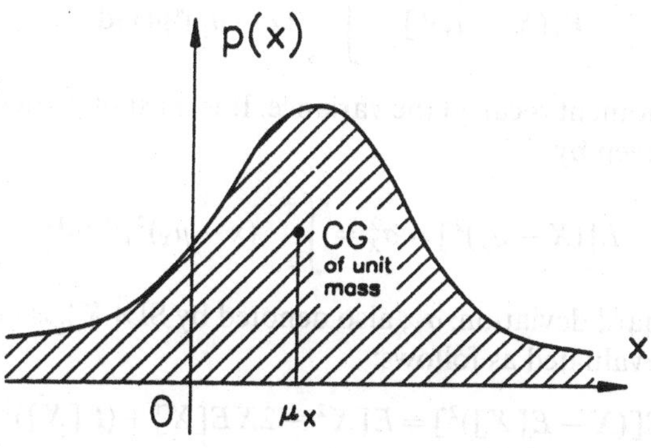

Fig. 1.7 *The centre of gravity of the probability density*

That is, the mean value of the random variable X corresponds to the abscissa of the centre of gravity of the area under the probability density curve and is the same as the static moment of the unit mass about the vertical axis.

Another more precise interpretation is to imagine that the real axis is a thin stiff wire having a unit mass distributed along the axis in accordance with the probability density. Let μ_X denote the mass centre of the string and subdivide the axis into small disjoint intervals $x_i < x \leqslant x_i + \Delta x_i$, $i = 0, \pm 1, \pm 2, \pm 3, \ldots$. By Eq. (1.21), the mass of the piece of string in the interval $x_i < x \leqslant x_i + \Delta x_i$ is $F(x_i + \Delta x_i) - F(x_i)$. The mass centre of the entire string is then given by

$$\mu_X \cdot 1 = \sum_{i=-\infty}^{\infty} x_i (F(x_i + \Delta x_i) - F(X_i)) = \sum_{i=-\infty}^{\infty} x_i \Delta F(x_i)$$

Going to the limit, i.e. max $\{\Delta x_i\} \to 0$, the above approximation becomes exact and

$$E[X] = \mu_X = \lim_{\max \Delta x_i \to 0} \sum_{i=-\infty}^{\infty} x_i \Delta F(x_i) = \int_{-\infty}^{\infty} x \mathrm{d}F(x) \tag{1.39}$$

In the general case, the random variable X is said to have a finite mean value or expectation if the limit in Eq. (1.39) exists as a finite real number.

Using the above analogy, Example 1.10, the following quantities, defined by the function $r(x) = x^n$, are called the moments of the random variable X:

$$E[X^n] = \int_{-\infty}^{\infty} x^n p(x) \mathrm{d}x \tag{1.40}$$

is the nth moment of X. The two first moments are expecially useful, the first moment, that is, the mean value or the expectation $E[X]$, and the second moment

$$E[X^2] = \overline{X^2} = \int_{-\infty}^{\infty} x^2 p(x) \mathrm{d}x \tag{1.41}$$

or the quadratic mean.

The central moments are derived by referring all moments to the mean value, that is, $r(x) = (x - \mu_X)^n$.

$$E[(X - \mu_X)^n] = \int_{-\infty}^{\infty} (x - \mu_X)^n p(x) \mathrm{d}x \tag{1.42}$$

The second central moment is called the variance. It is most often denoted by σ_X^2, but also by Var $[X]$, and is given by

$$E[(X - \mu_X)^2] = \sigma_X^2 = \int_{-\infty}^{\infty} (x - \mu_X)^2 p(x) \mathrm{d}x \tag{1.43}$$

from which the standard deviation, σ_X, also denoted by SD$[X]$, is derived. By Eq. (1.38), the variance can be evaluated as follows:

$$\begin{aligned} \sigma_X^2 &= \mathrm{Var}[X] = E[(X - E[X])^2] = E[X^2 - 2XE[X] + (E[X])^2] \\ &= E[X^2] - 2E[X]E[E[X]] + E[(E[X])^2 = E[X^2] - (E[X])^2 \end{aligned} \tag{1.44}$$

since the expected value of a constant, in this case the mean value $E[X]$, is the same as the constant. This last remark is of major importance, since it implies that $\sigma_X^2 = 0$ if and only if X is a constant with probability one, i.e. $P[X = \mu_X] = 1$.

Another important parameter that can be very useful in interpreting statistical phenomena is the coefficient of variation. It is simply defined as the standard deviation divided by the mean, or

$$V_X = \mathrm{SD}[X]/E[X] = \sigma_X/\mu_X \tag{1.45}$$

The coefficient of variation can be said to describe the relative variability of a random variable about its mean. If the coefficient is small, there is little variation and the random variable is close to being a constant (for which the coefficient must be zero). A large value means great scattering about the mean, i.e. increased randomness of the data represented by the random variable.

Example 1.11 A random variable U has the distribution function $\Phi(u)$ given by the probability density

$$\varphi(u) = \frac{1}{\sqrt{2\pi}} \exp\left[-\frac{u^2}{2}\right] \tag{1.46}$$

Find the expected value, the quadratic mean and the standard deviation of U.

From Eq. (1.35), the expected value $E[U] = 0$, since the density $\varphi(u)$ is symmetric about zero. By Eqs. (1.41) and (1.42), the quadratic mean is equal to the variance, which can be evaluated through integration by parts of Eq. (1.23), that is,

$$1 = \frac{1}{\sqrt{2\pi}} \int_{-\infty}^{\infty} e^{-u^2/2}\,du = \frac{1}{\sqrt{2\pi}} (ue^{-u^2/2})_{-\infty}^{\infty} + \frac{1}{\sqrt{2\pi}} \int_{-\infty}^{\infty} u^2 e^{-u^2/2}\,du$$

$$= \frac{1}{\sqrt{2\pi}} \int_{-\infty}^{\infty} u^2 e^{-u^2/2}\,du = \sigma_U^2$$

and the variance as well as the standard deviation is equal to one. It is worth while to make a coordinate transformation, whereby a new random variable $X = aU + b$ is defined, $(a \neq 0)$. Then, by Eq. (1.38),

$$E[X] = aE[U] + b = b = \mu_X$$

and

$$\mathrm{Var}[X] = E[X^2] - \mu_X^2 = E[a^2U^2 + 2abU + b^2] - \mu_X^2 = a^2 = \sigma_X^2$$

The distribution function of X is given by

$$P[X \leqslant x] = P[aU + b \leqslant x] = P\left[U \leqslant x - \frac{b}{a}\right] = \Phi\left(\frac{x - \mu_X}{\sigma_X}\right)$$

Hence X has the probability density

$$p(x) = \frac{\mathrm{d}}{\mathrm{d}x}\Phi\left(\frac{x-\mu_X}{\sigma_X}\right) = \frac{1}{\sigma_X}\varphi\left(\frac{x-\mu_X}{\varphi_X}\right) = \frac{1}{\sigma_X\sqrt{2\pi}}\exp\left[-\frac{(x-\mu_X)^2}{2\sigma_X^2}\right] \tag{1.47}$$

for $-\infty < x < \infty$.

The above distribution, Eqs. (1.46) and (1.47), is the normal distribution. It is a fundamental distribution of probability theory, and is also often referred to as the Gaussian distribution after the German mathematician Gauss. It is completely defined through its two first moments μ_X and σ_X^2.

The mean value of the variable U that was used in Example 1.11 is equal to zero and its variance equal to 1. Such a variable is called a standardized variable, in this case a standardized normal variable. Through the cumulative distribution function of such a variable, that is,

$$P[U \leqslant u_p] = \Phi(u_p) = p \tag{1.48}$$

the so-called fractile values u_p, or the p-fractile, can be determined (the probability that the random variable has a value that is equal to or less than the fraction u_p is p). The p-fractile, sometimes also called the p-percentile or the p-quantile, is then given by the inverse of the distribution function or

$$u_p = \Phi^{-1}(p) \tag{1.49}$$

Tabulated inverse values of the most common distributions, $F^1(p) = x_p$, can be found in a number of textbooks and tables of mathematics, [66], [83], [124], [214] and [232].

If the random variable X has the cumulative distribution function $\Phi((x-\mu)/\sigma)$, i.e. $X = \sigma U + \mu$, and u_p denotes the p-fractile of the standardized variable U, then the p-fractile of the variable X is simply given by

$$x_p = \sigma u_p + \mu = \mu(1 + V u_p) \tag{1.50}$$

where V is the coefficient of variation Eq. (1.45).

The fractile values of random variables are useful tools in statistical evaluations. For instance, let u_{75} be the 75% fractile, that is, $p = 0.75$, (sometimes called the upper 25% fractile), and u_{25} the lower 25% fractile of the distribution $\Phi(u)$ (u_{25} and u_{75} are also referred to as the first and second quartiles). Then

$$u_{75} = \Phi^{-1}(0.75) \text{ and } u_{25} = \Phi^{-1}(0.25)$$

The difference $(u_{75} - u_{25})$ is called the inter-quartile range and can give information about the spread of the distribution. Lower and upper fractile fractile values are shown in Fig. 1.8.

Example 1.12 A steel manufacturer guarantees that the upper yield level of a certain structural steel is equal to 240 MPa (N/mm^2), which is the normal level for most steel structures. Upon inquiry, it is made known that this level corresponds to the 1% fractile value, that is, there is about 1% chance that the steel will have an upper yield level less than 240 MPa and the production is subject to a quality variation, i.e. the coefficient of variation is 0.09. What is the mean value or the expected value of the upper yield level of the steel?

Solution The upper yield limit of structural steel can be interpreted as a random variable Y. The so-called 1% fractile, $y_{1\%}$, see Fig. 1.8, has been given as 240 MPa. If the distribution of the random variable is not well known or is unknown, the small fractile

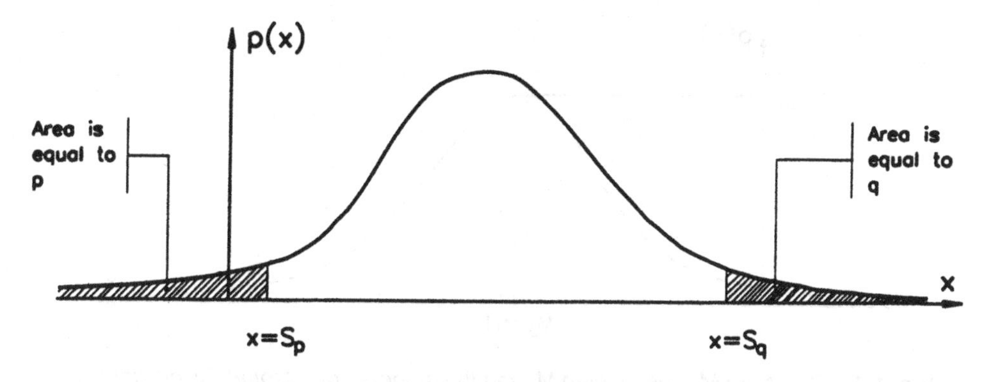

Fig. 1.8 *The p-fractile and the upper q-fractile*

values are often selected from the normal distribution. When the shapes of the various distributions are studied, the end tails are often like those of normal distributions, a fact that can be useful. Tabulated values of the standardized normal distribution show that the 1% fractile corresponding to the probability 0.01 is equivalent to −2.33. Then by Eq. (1.50),

$$y_{1\%} = \mu_Y(1 + V_Y u_{1\%})$$

or

$$240 = \mu_Y(1 + 0.09(-2.33)), \qquad \mu_Y = 304 \text{ MPa}$$

Example 1.13 The three Ms There are three important parameters that are of significance for characterizing probability distributions. The first one is the mean value, M_1, which has already been defined. Another important parameter is the median of the distribution, M_2, which is defined as the value of the random variable X which it takes with a probability equal to one half, that is

$$P[X \leqslant M_2] = \int_{-\infty}^{M_2} p(x)\mathrm{d}x = \int_{M_2}^{\infty} p(x)\mathrm{d}x = \tfrac{1}{2} \tag{1.51}$$

The median describes the centre of the distribution. Finally the third parameter is the mode of the distribution, M_3, which corresponds to the value x_{max} at which the density function has its maximum, Fig. 1.9. Therefore, if the maximum of $p(x)$ exists, the mode is the most likely value of the random variable, i.e. the probability that X is close to M_3 is maximum according to Eq. (1.24). Finally two interesting parameters can be mentioned, namely the skewness and the kurtosis. The coefficient of skewness is simply the third central moment of a random variable divided by the cube of the standard deviation, thus

$$\alpha_3 = E[(X - u_X)^3]/(\sigma_X)^3 \tag{1.52}$$

The skewness, as the name indicates, describes the shape of the density function. If the skewness is zero, the density function is symmetric about the mean value, otherwise the distribution will be skewed to the left or to right of the mean value according to the sign of

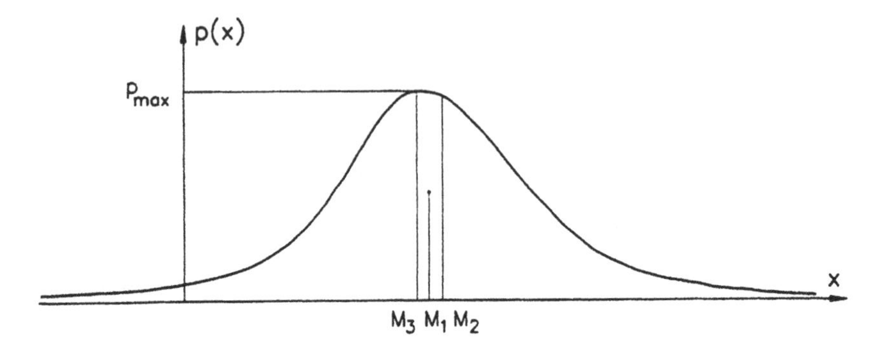

Fig. 1.9 *The mean M_1, the median M_2 and the mode M_3 of a probability distribution*

α_3. If $\alpha_3 > 0$, the distribution has a longer tail to the right, whereas if $\alpha_3 < 0$, the left tail is more prominent.

The coefficient of kurtosis is defined as the fourth central moment divided by the fourth power of the standard deviation, that is,

$$\alpha_4 = E[(X - \mu_X)^4]/(\sigma_X)^4 \tag{1.53}$$

Kurtosis is the degree of peakedness of the density function, usually with reference to the symmetrical density of the normal distribution, which has a coefficient of kurtosis equal to 3 (cf. Eq. (1.46)). A distribution having a high narrow peak, for instance $\alpha_4 > 10$, is called leptokurtic, while a flat-topped density, $\alpha_4 < 1$, is called platykurtic. Distributions with values of α_4 between, say, 1 and 10 are mesokurtic.

1.2.4 Marginal and Conditional Distributions

For the desription of multi-dimensional distributions involving many random variables, it is often required to elucidate the probabilities for one of them regardless of the values or conditions of the other. Consider for instance the two-dimensional distribution for the random variables (X, Y), given by

$$F(x, y) = \int_{-\infty}^{x} \int_{-\infty}^{y} p(u, v)\mathrm{d}u\,\mathrm{d}v, \qquad x, y \in \mathbb{R} \tag{1.54}$$

It can be asked, what is the probability that X has a value less than or equal to u while Y may assume any value within its range of definition $(-\infty < y < +\infty)$? Then by Eq. (1.54),

$$P[[X \leqslant u] \cap [-\infty < Y < \infty]] = \int_{-\infty}^{u} \int_{-\infty}^{\infty} p(x, y)\mathrm{d}x\,\mathrm{d}y = \int_{-\infty}^{u} \left(\int_{-\infty}^{\infty} p(x, y)\mathrm{d}y \right) \mathrm{d}x$$

$$= \int_{-\infty}^{u} f_1(x)\mathrm{d}x \tag{1.55}$$

in which the new function $f_1(x)$ is called the one-dimensional marginal probability density of the variable X. In the same manner, the probability density of the marginal distribution for Y is

$$f_2(y) = \int_{-\infty}^{\infty} p(x, y)\mathrm{d}x \tag{1.56}$$

Clearly by Eq. (1.50), $f_1(x)$ and $f_2(x)$ are the usual probability densities for the variables X and Y respectively.

In general, any function that is obtained through integration of one or more of the variables in a multi-dimensional distribution is a marginal distribution. For instance,

$$p(x_1, x_2) = \int_{-\infty}^{\infty}\int_{-\infty}^{\infty} p(x_1, x_2, x_3, x_4)\mathrm{d}x_3\,\mathrm{d}x_4 \tag{1.57}$$

is a two-dimensional marginal probability density.

By Eq. (1.12) the conditional probability $P[A|B]$ was defined. Consider next conditional distributions. Let $p(x, y)$ be the density of the random vector (X, Y). For any value y and for any small interval $(x, x+h]$, the conditional probability that Y assumes a value less than or equal to y, given that X has a value within the given interval, can be written as

$$P[[Y \leqslant y]|[x < X \leqslant x+h]] = \frac{P[[Y \leqslant y] \cap [x < X \leqslant x+h]]}{P[x < X \leqslant x+h]}$$

$$= \frac{\displaystyle\int_{x}^{x+h}\int_{-\infty}^{y} p(u, v)\mathrm{d}u\,\mathrm{d}v}{\displaystyle\int_{x}^{x+h}\int_{-\infty}^{\infty} p(u, v)\mathrm{d}u\,\mathrm{d}v} \tag{1.58}$$

This expression is used for determining the distribution function $F(y|x)$ for the random variable Y under the condition that the random variable X lies in the interval $(x, x+h]$. Difficulties arise when the conditional probability $P[[Y < y]|[X = x]]$ is to be evaluated since $P[X = x] = 0$. One has to assume that going to the limit, $h \to 0$, the limit value of the expression, Eq. (1.58), exists. Then,

$$F(y|x) = \lim_{h\to 0} P[[Y \leqslant y]|[x < X \leqslant x+h]] = \frac{\displaystyle\int_{-\infty}^{y} p(x, v)\mathrm{d}v}{f_1(x)} \tag{1.59}$$

where $f_1(x)$ is the marginal density (cf. Eq. (1.56)). The conditional density $p(y|x)$ is now easily obtained by differentiation of Eq. (1.59) with respect to y, giving the result

$$p(y|x) = p(x, y)/f_1(x) \tag{1.60}$$

Similarly, the distribution function and the density of the variable X under the condition that $Y = y$, can be derived. Conditional expected values like $E[Y|X = x_1] = E[Y|x_1]$ can also be obtained. For instance,

$$E[Y|x_1] = \int_{-\infty}^{\infty} y p(y|x_1)\mathrm{d}y \tag{1.61}$$

If the value x_1 in Eq. (1.61) is replaced by the variable X itself, a function $g(X)$ is obtained which has to be interpreted as a new random variable Z as follows:

$$Z = g(X) = E[Y|X] = \int_{-\infty}^{\infty} yp(y|X)\mathrm{d}y \tag{1.62}$$

The expected value of Z, $E[Z]$, is then obtained in accordance with Eqs. (1.34) and (1.60), that is,

$$E[Z] = E[g(X)] = E[E[Y|X]] = \int_{-\infty}^{\infty} g(x)f_1(x)\mathrm{d}x = \int_{-\infty}^{\infty}\left(\int_{-\infty}^{\infty} yp(y|x)\mathrm{d}y\right)f_1(x)\mathrm{d}x$$

or

$$\begin{aligned} E[E[Y|X]] &= \int_{-\infty}^{\infty}\int_{-\infty}^{\infty} yp(x,y)\mathrm{d}x\,\mathrm{d}y = \int_{-\infty}^{\infty} y\left(\int_{-\infty}^{\infty} p(x,y)\mathrm{d}x\right)\mathrm{d}y \\ &= \int_{-\infty}^{\infty} yf_2(y)\mathrm{d}y = E[Y] \end{aligned} \tag{1.63}$$

Similarly, $E[E[X|Y]] = E[X]$, a quite pleasant if surprising result, which is sometimes called the law of total probability.

1.2.5 Dependent and Independent Variables

Let X and Y be two independent random variables with distribution functions $H(x)$ and $G(y)$ respectively. By Eq. (1.16), the probabilities of all events that are associated with the two variables must be in accordance with the simple law of multiplication, that is,

$$P[[X \leqslant x] \cap [Y \leqslant y]] = P[X \leqslant x]P[Y \leqslant y]$$

or

$$F(x, y) = H(x)G(y) \tag{1.64}$$

where $F(x, y)$ is the distribution function of the random vector (X, Y). By differentiation of Eq. (1.64), one obtains

$$\frac{\partial^2 F(x,y)}{\partial x \partial y} = p(x,y) = \frac{\partial H(x)}{\partial x}\cdot\frac{\partial G(y)}{\partial y} = h(x)g(y) \tag{1.65}$$

where $h(x)$ and $g(y)$ are the probability density functions of X and Y respectively. If, on the other hand, it is given that Eq. (1.65) is satisfied by two random variables, then

$$F(x,y) = \int_{-\infty}^{x}\int_{-\infty}^{y} p(u,v)\mathrm{d}u\,\mathrm{d}v = \int_{-\infty}^{x} h(u)\mathrm{d}u \int_{-\infty}^{y} g(v)\mathrm{d}v = H(x)G(y)$$

and the random variables X and Y are independent. Thus the distribution and density functions for two independent random variables X and Y are functions of each basic variable only.

From Eq. (1.55), obviously the two functions $h(x)$ and $g(y)$ are the previously defined marginal density function $f_1(x)$ and $f_2(y)$, and from Eq. (1.59)

$$\left.\begin{aligned} F(y|x) &= \frac{\int_{-\infty}^{y} f_1(x)f_2(u)du}{f_1(x)} = \int_{-\infty}^{y} f_2(u)du = G(y) \\ F(x|y) &= H(y) \end{aligned}\right\} \tag{1.66}$$

If the two random variables X and Y are independent, what is the expected value of the product XY of the two variables?

$$E[XY] = \int_{-\infty}^{\infty}\int_{-\infty}^{\infty} p(x,y)xydxdy = \int_{-\infty}^{\infty} xf_1(x)dx \int_{-\infty}^{\infty} yf_2(y)dy = E[X]E[Y] \tag{1.67}$$

The above relation $E[XY] = E[X]E[Y]$ holds in general for independent variables with finite mean values.

Example 1.14 Two independent random variables X_1 and X_2 are identically distributed as indicated by the density function shown in Fig. 1.10. Find the probability density function for the random variable $Z = X_1 + X_2$.

Solution The pair (X_1, X_2) is uniformly distributed over an unit rectangle in the (x_1, x_2) plane. The line $z = x_1 + x_2$ will intercept both axes x_1 and x_2 at equal lengths from zero, Fig. 1.11. The distribution function $F_Z(z)$ must therefore be given by the following relation:

$P[Z < z] = P[Z < (x_1 + x_2)] = F_Z(z) =$ "the area under the line $z = x_1 + x_2$ within the rectangle"

Thus,

$F_Z(z) = z^2/2$, for $0 < z \leqslant 1$, and $F_Z(z) = z^2/2 - 2((z-1)^2/2)$, for $1 < z \leqslant 2$ as shown in Fig. 1.12

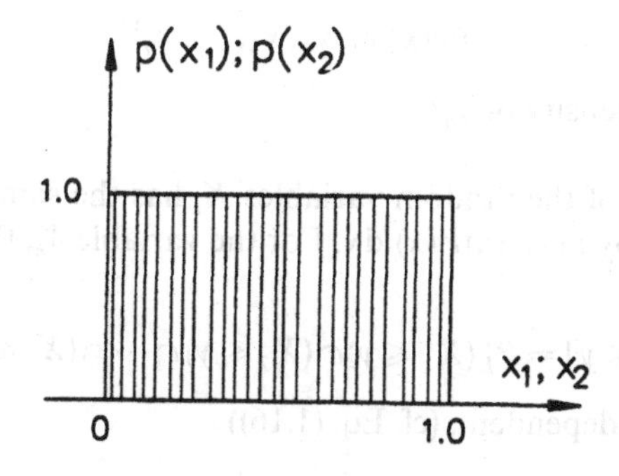

Fig. 1.10 *The density functon for the uniform distribution on [0, 1]*

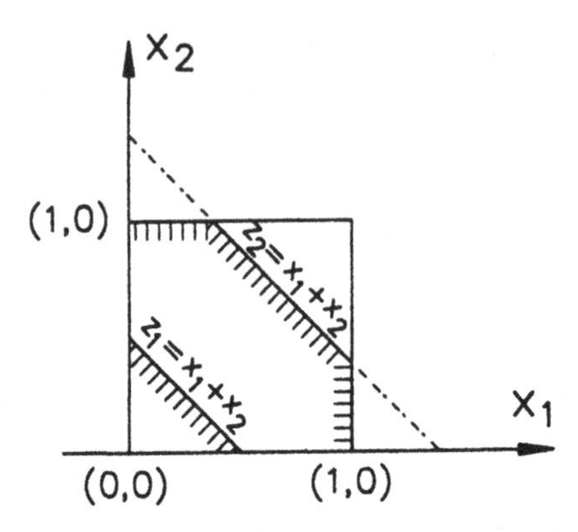

Fig. 1.11 *The probability mass in the domain of $F_z(z)$ is uniform over the unit square*

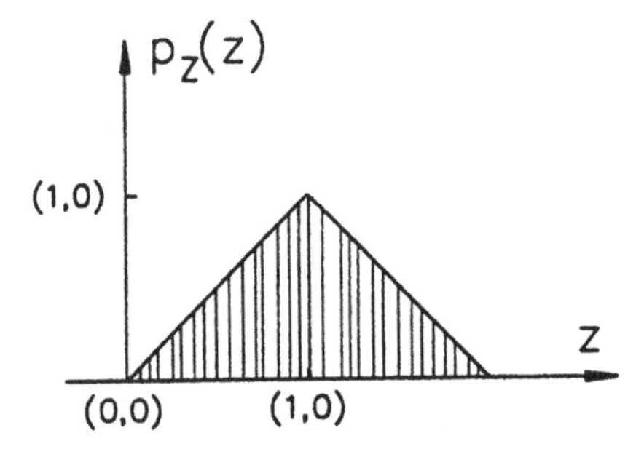

Fig. 1.12 *The probability density $p(z)$; a triangular distribution*

Example 1.15 Consider n random variables, $X_i, i = 1, 2, \ldots, n$, that are mutually independent and identically distributed (IID), sharing the common probability density $p(x)$. This is often the case for series of outcomes from a single experiment repeated under identical conditions. The numerical value assigned to each variable for one particular outcome is the observed value of the variable, sometimes called a sample value or just a sample. Now consider a new random variable Y_n with observed values y, which are the maximum values of the n observed values $x_1, x_2, x_3, \ldots, x_n$, i.e.

$$y = \max\{x_1, x_2, x_3, \ldots, x_n\}$$

What is the probability density of Y_n?

Solution Any of the random variables X_i has the same distribution function $F(x) = P[X \leqslant x]$, whereby $p(x) = \mathrm{d}F(x)/\mathrm{d}x$. For the variable Y_n, the probability is evaluated as follows:

$$P[Y_n \leqslant y] = P[(X_1 \leqslant y) \cap (X_2 \leqslant y) \cap \cdots \cap (X_n \leqslant y)]$$

Since the variables are independent, (cf. Eq. (1.16))

$$P[Y_n \leqslant y] = P[X_1 \leqslant y] P[X_2 \leqslant y] \cdots P[X_n \leqslant y]$$

As the variables have identical distributions and densities,

$$P[Y_n \leqslant y] = Q(y) = (P[X \leqslant y])^n = (F(y))^n \tag{1.68}$$

or

$$q(y) = \frac{\mathrm{d}Q(y)}{\mathrm{d}y} = n(F(y))^{n-1}\frac{\mathrm{d}F(y)}{\mathrm{d}y} = n(F(y))^{n-1}p(y) \tag{1.69}$$

The above relations are well known from reliability theory and the theory of extreme values. They are very useful when dealing with maximum values. For instance, the largest annual wind speeds encountered at a certain location, the largest earthquake magnitudes to be expected at a certain source and countless other applications. An overview of extreme value distributions will be given in Section 1.2.12.

Next consider the case when the random variables are not necessarily independent. Formally, the joint moments of the random variables X and Y can be obtained through integration as follows:

$$m_{ij} = E[X^i Y^j] = \int_{-\infty}^{\infty}\int_{-\infty}^{\infty} x^i y^j p(x, y)\mathrm{d}x\,\mathrm{d}y \tag{1.70}$$

The central moments are defined in a manner similar to that for a single variable. They will be denoted by μ_{ij}, i.e.

$$\mu_{ij} = E[(x - \mu_X)^i (y - \mu_Y)^j] \tag{1.71}$$

where the simple expressions $\mu_X = E[X]$ and $\mu_Y = E[Y]$ have been used for the first moments instead of m_{10} and m_{01}. The central moment μ_{11} is called the covariance of the two variables and is also written as $\mathrm{Cov}[X, Y]$. Also the denotation Γ_{XY} is frequently encountered in the literature. So

$$\mathrm{Cov}[X, Y] = \Gamma_{XY} = E[(X - \mu_X)(Y - \mu_Y)] = E[XY] - \mu_X \cdot \mu_Y \tag{1.72}$$

When two random variables are independent, one has

$$\mathrm{Cov}[X, Y] = 0$$

since $E[XY] = E[X]E[Y] = \mu_X \cdot \mu_Y$. The reverse statement, however, need not be true, that is, if the covariance of two random variables is zero, it does not necessarily imply that the variables are independent.

Finally, the correlation coefficient ϱ should be mentioned. If the standard deviations, σ_X and σ_Y, of the two variables are not equal to zero, a very important parameter, the correlation coefficient, is defined as follows:

$$\varrho_{XY} = \frac{E[(X - \mu_X)(Y - \mu_Y)]}{\sqrt{E[(X - \mu_X)^2]E[(Y - \mu_Y)^2]}} = \frac{\Gamma_{XY}}{\sigma_X \sigma_Y} \tag{1.73}$$

Example 1.16 Show that the correlation coefficient ϱ_{XY} satisfies the following inequalities:

$$-1 \leqslant \varrho_{XY} \leqslant 1 \tag{1.74}$$

Solution Form the positive expression $(u(X-\mu_X)+(Y-\mu_Y))^2 \geqslant 0$ where u is an arbitrary real number. From Eqs. (1.37) and (1.38), the expection is also positive and can be evaluated as follows:

$$E[(u(X-\mu_X)+(Y-\mu_Y))^2] = u^2 E[(X-\mu_X)^2] + E[(Y-\mu_Y)^2] + 2uE[\{(X-\mu_X)(Y-\mu_Y)\}] \geqslant 0$$

or

$$u^2\sigma_X^2 + 2u\Gamma_{XY} + \sigma_Y^2 = f(u) \geqslant 0 \quad \text{for all } u \in \mathbb{R}$$

Since $f(u)$ is a parable, never crossing the u-axis, it follows that the equation $f(u)=0$ can not have two distinctive real roots. Therefore the discriminant D of the equation

$$u^2\sigma_X^2 + 2u\Gamma_{XY} + \sigma_Y^2 = 0$$

must be non-positive or

$$D = 4\Gamma_{XY}^2 - 4\sigma_X^2\sigma_Y^2 \leqslant 0$$

whereby

$$|\Gamma_{XY}| \leqslant \sigma_X\sigma_Y \rightarrow |\Gamma_{XY}/\sigma_X\sigma_Y| \leqslant 1$$

If $u=u_0$ is a double real root of $f(u)$, then the correlation coefficient $\varrho_{XY} = \pm 1$. Actually this means that the variance of the variable u_0X+Y is zero, which can happen if and only if u_0X+Y is a constant equal to $u_0\mu_X\mu_Y$ with probability one. In this case, $Y = \mu_Y - u_0(X-\mu_X)$ and $\varrho_{XY} = -u_0/|u_0|$.

The value of the correlation coefficient reveals the interrelationship between two random variables X and Y. If the correlation coefficient is equal to one, a perfect 'in phase' condition exists, that is, the value of X is always equal to that of Y and both variables have the same sign. If the correlation coefficient is equal to -1, a perfect 'out of phase' condition exists, that is, the value of X is always opposite in sign to that of Y but the magnitudes are equal. For values of ϱ_{XY} in between -1 and 1, the interrelationship of course is weaker.

Example 1.17 Consider the two-dimensional probability density

$$p(x_1,x_2) = \frac{1}{2\pi\sqrt{3}}\exp\{-\tfrac{1}{3}(x_1^2 + x_1x_2 + x_2^2)\}, \qquad (x_1,x_2)\in\mathbb{R}^2$$

for the two random variables X_1 and X_2.

(1) are X_1 and X_2 statistically independent?
(2) find $f_1(x_1)$ and $f_2(x_2)$
(3) determine $p(x_1|x_2)$ and find $E[X_1|x_2]$
(4) find the correlation coefficient ϱ_{12}

Solution (1) The expression $x_1^2 + x_1x_2 + x_2^2$ shows that the function $p(x_1x_2)$ can not be factored into a product of two functions $f(x_1)$ and $g(x_2)$. Therefore, the variables are not independent.

(2)
$$p(x_1, x_2) = \frac{1}{2\pi\sqrt{3}} \exp\{-\tfrac{1}{3}(x_1^2 + x_1 x_2 + x_2^2)\}$$

$$= \frac{1}{2\pi\sqrt{3}} \exp\left(-\frac{x_1^2}{4}\right) \exp\left(-\frac{(x_2 + \frac{1}{2}x_1)^2}{3}\right)$$

$$f_1(x_1) = \int_{-\infty}^{\infty} p(x_1, x_2) \mathrm{d}x_2 = \frac{1}{2\sqrt{\pi}} \exp\left(-\frac{x_1^2}{4}\right)$$

since

$$\int_{-\infty}^{\infty} \exp\left(-\frac{(x_2 - \mu)^2}{3}\right) \mathrm{d}x_2 = \sqrt{\frac{3}{2}} \cdot \sqrt{2\pi}$$

Also, because of symmetry

$$f_2(x_2) = \frac{1}{2\sqrt{\pi}} \exp\left(-\frac{x_2^2}{4}\right)$$

Note that $E[X_1] = E[X_2] = 0$, $\sigma_1^2 = \sigma_2^2 = 2$.

(3) The conditional distribution is

$$p(x_1 | x_2) = \frac{p(x_1, x_2)}{f_2(x_2)} = \frac{1}{\sqrt{3\pi}} \exp(-\tfrac{1}{3}(x_1 + \tfrac{1}{2}x_2)^2)$$

and the conditional expected value is

$$E[X_1 | x_2] = \int_{-\infty}^{\infty} x_1 p(x_1 | x_2) \mathrm{d}x_1 = \frac{1}{\sqrt{\frac{3}{2}} \cdot \sqrt{2\pi}} \int_{-\infty}^{\infty} u \exp\left(-\frac{(u + \frac{1}{2}x_2)^2}{2(\sqrt{\frac{3}{2}})^2}\right) \mathrm{d}u = -\tfrac{1}{2}x_2$$

which means that, on the average, the value of X_1 is minus one half of the value of X_2.

(4)
$$E[X_1 X_2] = \frac{1}{2\pi\sqrt{3}} \int_{-\infty}^{\infty} \left(\int_{-\infty}^{\infty} uv \cdot \exp\left(-\frac{u^2}{4}\right) \exp\left(-\frac{(v + u/2)^2}{3}\right) \mathrm{d}v \right) \mathrm{d}u$$

$$= \frac{1}{2\sqrt{\pi}} \int_{-\infty}^{\infty} u\left(-\frac{u}{2}\right) \exp\left(-\frac{u^2}{4}\right) \mathrm{d}u$$

$$= -\frac{1}{\sqrt{2}\sqrt{2\pi}} \frac{1}{2} \int_{-\infty}^{\infty} u^2 \exp\left(-\frac{u^2}{2(\sqrt{2})^2}\right) \mathrm{d}u = -1$$

and

$$\varrho_{12} = \frac{\Gamma_{12}}{\sigma_1 \sigma_2} = \frac{E[X_1 X_2] - \mu_1 \mu_2}{\sigma_1 \sigma_2} = -\tfrac{1}{2}$$

which can be compared with the previous result (3).

Example 1.18 A random variable Z is formed as the product of three independent random variables X_1, X_2 and X_3. Find an expression that gives the coefficient of variation of Z in terms of the coefficients of variation of the three variables.

Solution

$$Z = X_1 X_2 X_3$$

and

$$E[Z] = E[X_1 X_2 X_3] = \mu_Z = E[X_1]E[X_2]E[X_3] = \mu_1 \mu_2 \mu_3$$

$$\begin{aligned}
\mathrm{Var}[Z] = \sigma_Z^2 = \mathrm{Var}[X_1 X_2 X_3] &= E[X_1^2 X_2^2 X_3^2] - \mu_1^2 \mu_2^2 \mu_3^2 \\
&= E[X_1^2]E[X_2^2]E[X_3^2] - \mu_1^2 \mu_2^2 \mu_3^2 \\
&= (\sigma_1^2 + \mu_1^2)(\sigma_2^2 + \mu_2^2)(\sigma_3^2 + \mu_3^2) - \mu_1^2 \mu_2^2 \mu_3^2 \\
&= \sigma_1^2 \sigma_3^2 \mu_2^2 + \sigma_2^2 \sigma_3^2 \mu_1^2 + \sigma_1^2 \sigma_2^2 \mu_3^2 \\
&\quad + \sigma_1^2 \sigma_2^2 \sigma_3^2 + \sigma_1^2 \mu_2^2 \mu_3^2 + \sigma_2^2 \mu_1^2 \mu_3^2 + \sigma_3^2 \mu_1^2 \mu_2^2
\end{aligned}$$

Dividing by $\mu_Z^2 = \mu_1^2 \mu_2^2 \mu_3^2$ on both sides, noting that $V = \sigma/\mu$, the following expression is obtained:

$$V_Z^2 = V_1^2 + V_2^2 + V_3^2 + V_1^2 + V_2^2 + V_2^2 + V_3^2 + V_1^2 V_3^2 + V_1^2 V_2^2 V_3^2 \tag{1.75}$$

If one of the three variables is deterministic, that is, a constant, for instance X_3, then $\mu_3 = X_3$ and $\sigma_3 = V_3 = 0$, so

$$V_Z^2 = V_1^2 + V_2^2 + V_1^2 V_2^2$$

Example 1.19 In continuation of Example 1.15, consider again the maximum value of a collection of sample values which are the result of a certain experiment or an action, described by random variables X_i, $i = 1, 2, \ldots, n + m$. Y_n and Y_m are now defined as two random variables the sample values of which are the maximum values of two groups of sample values, viz. X_i, $i = 1, 2, \ldots, n$ and X_j, $j = n + 1, n + 2, \ldots, n + m$. All the random variables X_i and X_j are mutually independent and identically distributed, IID, as before. Find the probability that $Y_m \leqslant Y_n$.

Solution

$$y_n = \max\{x_1, x_2, x_3, \ldots, x_n\} \qquad \text{and} \qquad y_m = \max\{x_{n+1}, x_{n+2}, x_{n+3}, \ldots, x_{n+m}\}$$

As was already established, (Example 1.15, Eq. (1.68)),

$$P[Y_m \leqslant y_m] = (F(y_m))^m = H(y_m)$$

The two random variables Y_n and Y_m can now be treated as a pair with the probability density $p(y_n, y_m)$ defined in the (y_n, y_m) plane, Fig. 1.13. From the figure, it is obvious that

$$P[Y_m < Y_n] = \iint_{A^*} p(y_m, y_n) \mathrm{d}y_n \mathrm{d}y_m$$

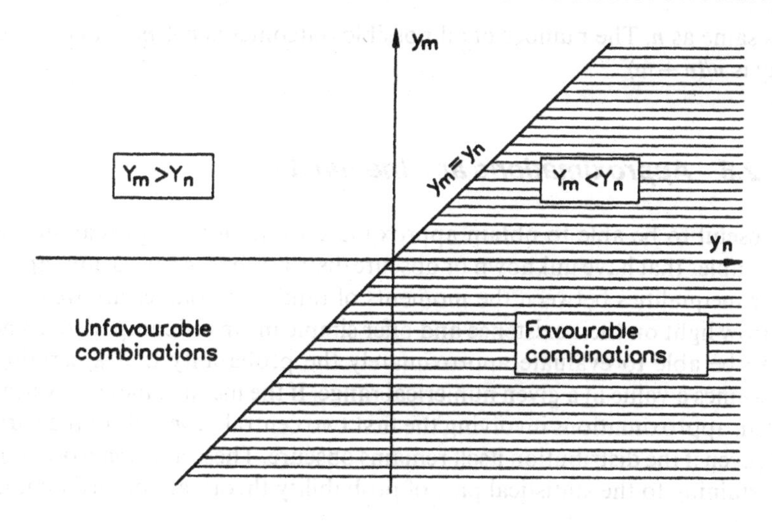

Fig. 1.13 *The* (y_n, y_m) *plane*

Since Y_m and Y_n are independent, $p(y_n, y_m) = q(y_n)h(y_m)$, and

$$P[Y_m < Y_n] = \int_{-\infty}^{\infty}\int_{-\infty}^{y_n} q(y_n)h(v)\mathrm{d}y_n\,\mathrm{d}v = \int_{-\infty}^{\infty} q(y_n)\left[\int_{-\infty}^{y_n} h(v)\mathrm{d}v\right]\mathrm{d}y_n$$

$$= \int_{-\infty}^{\infty} q(y_n)H(y_n)\mathrm{d}y_n = \int_{-\infty}^{\infty} n[F(y_n)^{n-1}][F(y_n)]^m \cdot \frac{\mathrm{d}F}{\mathrm{d}y_n}\cdot \mathrm{d}y_1$$

$$= \frac{n}{n+m}[(F(y_n)^{n+m})]_{-\infty}^{\infty} = \frac{n}{n+m}$$

This result can indeed be obtained by using simpler means. Consider a collection of sample values x_i, $i = 1, 2, \ldots, m$ whereby the two sample values y_n and y_m can be studied at the same time. The generation of sample values can be interpreted as the assigning of real numbers to x_i at each of the $n + m$ stations, Fig. 1.14. All numbers x_i are derived from the same base distribution and are mutually independent. The largest value of the entire collection will be found at any of the $n + m$ stations with equal likelihood, that is, the probability of appearance of the largest value at any of the $n + m$ stations is uniformly distributed. The favourable outcomes $Y_m < Y_n$, that is, that the largest value belongs to the

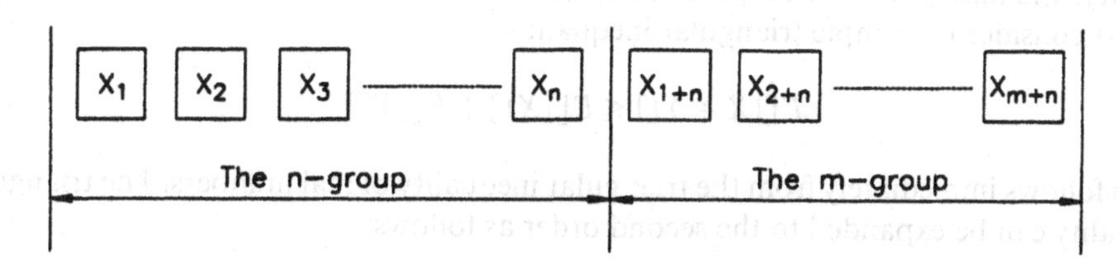

Fig. 1.14 *The* $n + m$ *stations for X*

group n, is same as n. The number of all possible outcomes is $n + m$ of and therefore the probability is $n/(n + m)$.

1.2.6 *Approximations and Inequalities*

It is often useful to be able to obtain approximate values for the probability of certain random variables that have unknown or obscure distributions. Also, various approximate relations or inequalities between the moments of random variables are well established and can throw light on their existence and offer simple means of comparison. Then it can be useful to be able to evaluate approximately the probability that a certain random variable assumes a value in a given numerical range. If the mean values and variances can be evaluated, approximations involving the first two central moments of the variable can be derived, even if the distribution itself remains obscure. These and some other questions that are pertaining to the statistical part of probability theory are briefly summarized in the following.

Let X and Y be two random variables with finite second moments, i.e. m_{02}, m_{11} and m_{20} are finite. Consider the random variable $(uX + Y)^2$ where u is an arbitrary real number. Since $(uX + Y)^2 \geqslant 0$, its expectation is also larger than or equal to zero according to Eq. (1.37). Using Eq. (1.38), the expectation can be evaluated as

$$E[(uX + Y)^2] = u^2 E[X^2] + 2uE[XY] + E[Y^2]$$
$$= u^2 m_{02} + 2um_{11} + m_{20} = f(u) \geqslant 0 \qquad \text{for all } u$$

whereby the positive expression inside the expectation on the right-hand side has been turned into a quadratic in the real variable u, which is non-negative for all u. Using the same argument as in Example 1.16, the discriminant of the quadratic equation must be non-positive, that is,

$$4u^2 m_{11}^2 - 4u^2 m_{20} m_{02} \leqslant 0$$

or

$$(E[XY])^2 \leqslant E[X^2]E[Y^2] \tag{1.76}$$

This inequality is known as the Schwartz inequality. The two variables can be replaced by their absolute values, $|X|$ and $|Y|$, without altering the inequality nor the following derived expressions. If $u = u_0$ is a double root of $f(u)$, then and only then the inequality becomes an equality, and $u_0 X + Y = 0$ with probability one.

Next consider the simple triangular inequality,

$$E[|X + Y|] \leqslant E[|X|] + E[|Y|] \tag{1.77}$$

which follows immediately from the triangular inequality of real numbers. The triangular inequality can be expanded to the second order as follows:

$$|X + Y|^2 \leqslant |X|^2 + |Y|^2 + 2|X||Y|$$

and

$$(|X|-|Y|)^2 \geqslant 0, \qquad \text{whereby} \qquad |X|^2+|Y|^2 \geqslant 2|X||Y|$$

Combining these two expressions and taking the expectation,

$$E[|X+Y|^2] \leqslant 2E[|X|^2]+2E[|Y|^2] \tag{1.78}$$

From the inequalities, Eqs. (1.78) and (1.76), the existence of the second-order moments of the two random variables X and Y guarantees the existence of the second moment of the sum variable $X+Y$ and the multiplication variable XY. This statement can also be expanded to higher-order moments. Finally, it may be concluded that for the two variables X and Y with finite second-order moments, the two mean values, that is, the first-order moments exist, and hence the existence of the variance and the covariance of the two variables is also guaranteed.

If the second variable is identically equal to one, $Y \equiv 1$, the inequality, Eq. (1.76) becomes

$$E[X] \leqslant \sqrt{E[X^2]}$$

which shows that if the second moment is finite, $E[X^2] \leqslant \infty$, then the first moment, the mean, is also finite. This statement can be extended to the higher moments. If the moment of order n is finite, then all moments of the lower order must also be finite. That is,

$$\sqrt[n]{E[|X|^m]} \leqslant \sqrt[n]{E[X^n]}, \qquad m=1,2,\ldots,n-1 \tag{1.79}$$

if $E[X^n] \leqslant +\infty$. Thus the existence of the nth moment guarantees the existence of all lower moments.

Finally, consider the basic inequality for a random variable X with an unknown distribution function $F(x)$. Let the real function $h(x) \geqslant 0$ for all values of x and $h(x) \geqslant b \geqslant 0$ (b is a positive constant) for all values of x in a certain set of values A. Then,

$$P[X \in A] \leqslant E(h(X)]/b \tag{1.80}$$

This inequality can be proved as follows:

$$E[h(X)] = \int_A h(x)\mathrm{d}F(x) + \int_{A^*} h(x)\mathrm{d}F(x)$$

where A^* is the complementary set to A. If the last integral is dropped, the right-hand side and hence the expectation becomes smaller; also if $h(x)$ is replaced by b, which is less than or equal to $h(x)$, it becomes smaller again. Therefore,

$$E[h(X)] \geqslant b\int_A F(x) = bP[X \in A]$$

and by dividing by b, the inequality Eq. (1.80) is obtained. Now choose the function $h(x)=(x-\mu_X)^2$, which is a non-negative function, and let A denote the set of values for x such that $|x-\mu_X| \geqslant c$. Then,

$$P[|X-\mu_X| \geqslant c] \leqslant E[(X-\mu_X)^2]/c^2 = \sigma_X^2/c^2$$

or by rearranging the terms and choosing $c = a\sigma_X$, $a \geqslant 0$,

$$P[(\mu_X - a\sigma_X) \leqslant X \leqslant (\mu_X - a\sigma_X)] \geqslant 1 - (1/a^2) \tag{1.81}$$

which is known as the Chebyshev inequality. The Chebyshev inequality is a useful tool for the purpose of evaluating the probabilities of a random variable with an unknown distribution when the mean value and the variance are known. It also shows how the variance or standard deviation measures the dispersion of the probability mass about the mean value.

Example 1.20 In Example 1.12, it was found that the mean value μ_Y of the upper yield level of certain structural steel was equal to 304 MPa and the quality variation during production corresponded to a coefficient of variation equal to 0.09. Find the approximate probability that the upper yield level assumes a value that is less than or equal to 250 MPa.

Solution The difference between the mean value and the selected value is 54 MPa. The standard deviation of the yield level is given as $0.09 \times 304 = 27.36$ MPa such that the difference corresponds to 1.97 standard deviations. Using the Chebyshev inequality of the form given by Eq. (1.81), the following relation is obtained:

$$P[(304 - 54) \leqslant Y \leqslant (304 + 54)] \geqslant 1 - (1/1.97^2) = 0.743$$

So the probability that the upper yield level lies outside the range of values, within about two standard deviations to each side of the mean, is approximately 25%. That is,

$$P[(Y \geqslant 250) \cup (Y \leqslant 358)] = 1 - 0.743 = 0.257$$

By assuming that the distribution of the upper yield level is symmetrical about the mean, the approximate probability that the yield level will be less than or equal to 250 MPa is

$$P[Y \leqslant 250] = 0.257/2 = 0.1285 \text{ or almost } 13\%$$

1.2.7 The Transformation of Multi-dimensional Random Variables

In many cases, one has to deal with distributions of random variables that are functions of other basic random variables, that is, they are derived through transformation of the basic variables. As an example let (X, Y) denote a two-dimensional random vector or point in $\mathbb{R}^2$, the (x, y) plane, which through the functional transformation

$$\begin{aligned} U &= f(X, Y) \\ V &= g(X, Y) \end{aligned} \tag{1.82}$$

produces another random vector (U, V). If the probability density $p(x, y)$ of (X, Y) is given, the density $q(u, v)$ of (U, V) is sought.

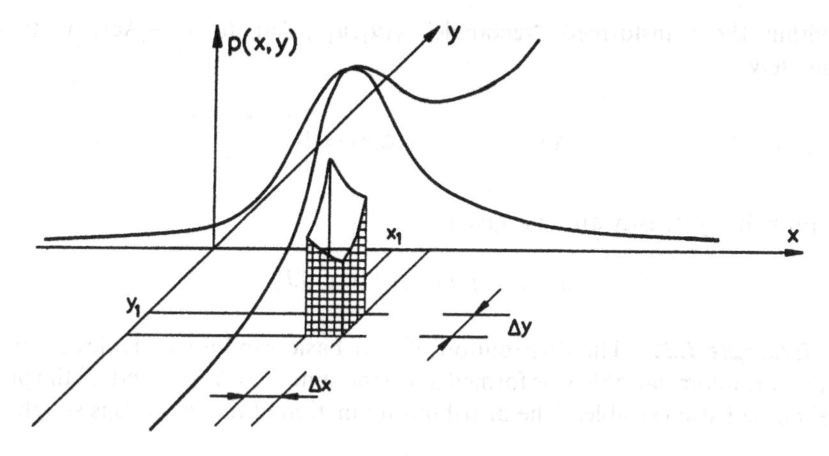

Fig. 1.15 *The probability density of (X, Y)*

The probability that (X, Y) has a value within the rectangular square,

$$\{(x_1 < X \leqslant x_1 + \Delta x_1), (y_1 < Y \leqslant y_1 + \Delta y_1)\}$$

i.e. has a value that is close to (x_1, y_1), is represented by the integral

$$\int_{x_1}^{x_1+\Delta x} \int_{y_1}^{y_1+\Delta y} p(x, y) \mathrm{d}x \, \mathrm{d}y$$

which corresponds to the volume of the column shown in Fig. 1.15. Through the presumed one to one coordinate transformation

$$u = f(x, y) \leftrightarrow x = h(u, v)$$

$$v = g(x, y) \leftrightarrow y = k(u, v)$$

the volume of a corresponding column in the (u, v) plane is given by

$$\int_{u_1}^{u_1+\Delta u} \int_{v_1}^{v_1+\Delta v} q(u, v) \mathrm{d}u \, \mathrm{d}v = \int_{u_1}^{u_1+\Delta u} \int_{v_1}^{v_1+\Delta v} p(h(u, v), k(u, v)) |J| \mathrm{d}u \, \mathrm{d}v$$

where $|J|$ denotes the absolute value of the functional determinant, the Jacobian, of the inverse transformation, Eq. (1.83), that is,

$$J = \begin{vmatrix} \dfrac{\partial h}{\partial u} & \dfrac{\partial h}{\partial v} \\ \dfrac{\partial k}{\partial u} & \dfrac{\partial k}{\partial v} \end{vmatrix} \tag{1.84}$$

where by supposition, the inverse transformation functions $h(u, v)$ and $k(u, v)$ possess continuous first-order partial derivatives. The probability that the pair (U, V) has a value within the transformed "rectangle" $\{(u_1, u_1 + \Delta u), (v_1, v_1 + \Delta v)\}$ is therefore approximately

$$P[(u_1 < U \leqslant u_1 + \Delta u), (v_1 < V \leqslant v_1 + \Delta v)] = \int_{u_1}^{u_1+\Delta u} \int_{v_1}^{v_1+\Delta v} q(u, v) du\, dv$$

and the probability density $q(u, v)$ is given as

$$q(u, v) = p(h(u, v), k(u, v))|J| \tag{1.85}$$

Example 1.21 The distribution of two basic random variables X and Y is given. A new random variable U is formed as (1) the sum, $U = X + Y$, and (2) the product, $U = XY$, of the basic variables. The distribution function of the two U's is sought.

Solution By introducing an auxiliary variable $V = X$, the problem can be treated as a transformation exercise. Therefore, let

(1) $$U = X + Y, V = X$$

Then,

$$x = h(u, v) = v \text{ and } y = k(u, v) = u - v$$

and

$$\mathbf{J} = \begin{vmatrix} 0 & 1 \\ 1 & -1 \end{vmatrix} = -1, \qquad |\mathbf{J}| = 1$$

whereby the probability density of the two variables U and V is equal to $p(v, u - v)$. The density $\psi(u)$ of the variable $U = X + Y$ must then be obtained as the marginal density from the two-dimensional distribution for the random vector (U, V) by integration over v from $-\infty$ to ∞.

$$\psi(u) = \int_{-\infty}^{\infty} p(v, u - v) dv = \int_{-\infty}^{\infty} p(x, u - x) dx$$

If the two basic variables X and Y were independent, then $p(x, y) = f_1(x) f_2(y)$ and

$$\psi(u) = \int_{-\infty}^{\infty} f_1(x) f_2(u - x) dx$$

(2) $$U = XY, V = X, x = h(u, v) = v, y = k(u, v) = u/v$$

$$\mathbf{J} = \begin{vmatrix} 0 & 1 \\ \frac{1}{v} & -\frac{u}{v^2} \end{vmatrix} = -\frac{1}{v}, \qquad |\mathbf{J}| = \frac{1}{|v|}$$

and

$$q(u, v) = p\left(v, \frac{1}{v}\right) \frac{1}{|v|}$$

The density of U is again obtained by integration over v, that is,

$$\varphi(u) = \int_{-\infty}^{\infty} p\left(x, \frac{u}{x}\right)\frac{1}{|x|}\mathrm{d}x$$

Notice that $x = 0$ must be excluded as it represents the y-axis in the xy-plane, which will be transformed into the u-axis in the uv-plane where $v = 0$. Since both these axes support the probability mass zero, this does not affect the calculation.

Example 1.22 Let Y be a random variable that is constructed as the weighted sum of n random variables X_i, $i = 1, 2, \ldots, n$. Thus

$$Y = \sum_{i=1}^{n} a_i X$$

where the constants a_i, the weights, have real values. Find the expected value and the variance of Y.

Solution To obtain the moments of $Y$by applying the method of transformed variables with multi-dimensional distributions would seem to be an insurmountable amount of work according to the above exercise. It is therefore more advantageous to obtain the moments directly from their definition. Thus

$$\begin{aligned} E[Y] &= E[a_1X_1 + a_2X_2 + a_3X_3 + \cdots a_nX_n)] \\ &= a_1E[X_1] + a_2E[X_2] + a_3E[X_3] + \cdots + a_nE[X_n] \end{aligned}$$

or

$$E[Y] = \sum_{i=1}^{n} a_i E[X_i] \tag{1.86}$$

since the operator $E[*]$ is linear-additive. The second moment, the variance, is obtained as follows:

$$\begin{aligned} \mathrm{Var}[Y] = E[Y - E[Y])^2] &= E\left[\left(\sum_{i=1}^{n} a_i X_i - \sum_{i=1}^{n} a_i E[X_i]\right)^2\right] = E\left[\left(\sum_{i=1}^{n} a_i (X_i - E[X_i])\right)\right] \\ &= E\left[\sum_{i=1}^{n}\sum_{j=1}^{n} a_i a_j (X_i - E[X_i])(X_j - E[X_j])\right] = \sum_{i=1}^{n}\sum_{j=1}^{n} a_i a_j \mathrm{Cov}[X_i, X_j] \\ &= \sum_{i=1}^{n} a_1^2 \mathrm{Var}[X_i] + \sum_{i \neq j}\sum a_i a_j \mathrm{Cov}[X_i, X_j] \end{aligned} \tag{1.87a}$$

or, by introducing the correlation coefficient and the standard deviations,

$$\sigma_Y^2 = \sum_{i=1}^{n}\sum_{j=1}^{n} a_i a_j \rho_{ij} \sigma_{X_i} \sigma_{X_j} \tag{1.87b}$$

If the random variables $\{X_i\}$ are pairwise uncorrelated, $\text{Cov}[X_i, X_j] = 0$ and

$$\text{Var}[Y] = \sum_{i=1}^{n} a_i^2 \,\text{Var}[X_i] \tag{1.87c}$$

An important special case of Eq. (1.87c) is when the X_i variables are mutually independent.

1.2.8 Approximate Moments and Distributions

When dealing with random variables that have complicated functional relationships with the basic variables, it is often possible to approximately evaluate their statistical properties, sufficiently accurately for most practical problems.

Consider a random variable Y that is given as a reasonably well-behaved function $Y = g(X)$ of the random variable X, which has a coefficient of variation V_X that is not too large. Then the difference between X and its mean value will not be too large either, and by expanding $g(x)$ into a Taylor series about the mean value μ_X, the following approximation is obtained:

$$g(X) \simeq g(\mu_X) + (X - \mu_X)\, g'(\mu_X) + \frac{(X - \mu_X)^2}{2}\, g''(\mu_X) \tag{1.88}$$

in which the derivatives are denoted by

$$g'(\mu_X) = \left.\frac{\mathrm{d}g(x)}{\mathrm{d}x}\right|_{X=\mu_x} \quad \text{and} \quad g''(\mu_X) = \left.\frac{\mathrm{d}^2 g(x)}{\mathrm{d}x^2}\right|_{X=\mu_x}$$

Taking the expected value of both sides of the above equation, the following relation is found:

$$E[Y] = E[g(X)] \simeq g(E[X]) + \tfrac{1}{2}\text{Var}[X]\, g''(\mu_X)$$

or

$$\mu_Y \simeq g(\mu_X) + \tfrac{1}{2}\sigma_X^2\, g''(\mu_X) \tag{1.89}$$

The evaluation of the variance of both sides of Eq. (1.88) is somewhat more cumbersome. By applying Eq. (1.87a) in conjunction with Eq. (1.88), the following expression is obtained:

$$\begin{aligned}\text{Var}[Y] &\simeq (g'(\mu_X)^2\, \text{Var}[X] + g'(\mu_X)\, g''(\mu_X)\, E[(X - \mu_X)^3] \\ &\quad + \tfrac{1}{4}(g''(\mu_X))^2\, \text{Var}[(X - \mu_X)^2] \\ &= \sigma_X^2 \{(g'(\mu_X)^2 - 2\mu_X g'(\mu_X)\, g''(\mu_X) + \mu_X^2 (g''(\mu_X))^2\} \\ &\quad + \tfrac{1}{4}(g''(\mu_X)^2\, \text{Var}[X^2] + g'(\mu_X)\, g''(\mu_X)\, \text{Cov}[X, X^2]\end{aligned}$$

or

$$\sigma_Y^2 \simeq \sigma_X^2 \{(g'(\mu_X))^2 - 2\mu_X g'(\mu_X) g''(\mu_X) + \mu_X^2 (g''(\mu_X))^2\} \tag{1.90}$$

by neglecting the last two higher-order terms on the assumption that the coefficient of variation V_X is small. For instance, if $|V_X| \ll 1$,

$$\text{Var}[(X/\mu_X)^2] \ll \text{Var}[X/\mu_X] \qquad \text{or} \qquad \text{Var}[X^2] \ll \mu_2^X \text{Var}[X] = \mu_X^2 \sigma_X^2$$

Most often, only the first two elements of the expansion in the right hand side of Eq. (1.88) are needed to obtain a reasonably accurate value for the mean, and for the variance only the first element is needed. Thus,

$$\begin{aligned} E[Y] &= E[g(X)] \approx g(E[X]) \\ \text{Var}[Y] &= \text{Var}[g(X)] \approx \text{Var}[X](g'(\mu_X))^2 \end{aligned} \tag{1.91}$$

Similar approximations are also possible for multidimensional functional relationship. For instance, let

$$Y = g(X_1, X_2, X_3, \ldots, X_n)$$

The second-order approximation for the mean is obtained by

$$E[Y] \simeq g(\mu_1, \mu_2, \ldots, \mu_n) + \frac{1}{2} \sum_{i=1}^{n} \sum_{j=1}^{n} \left(\frac{\partial^2 g}{\partial x_i \partial x_j} \right)_{\mu} \text{Cov}[X_i X_j] \tag{1.92}$$

and noting that

$$\left(\frac{\partial^2 g}{\partial x_i \partial x_j} \right)_{\mu} = \left. \frac{\partial^2 g}{\partial x_i \partial x_j} \right|_{\{x_i = \mu_i\}}$$

The first-order approximation for the variance is

$$\text{Var}[Y] \simeq \sum_{i=1}^{n} \sum_{j=1}^{n} \left. \frac{\partial g}{\partial x_i} \right|_{\mu} \left. \frac{\partial g}{\partial x_j} \right|_{\mu} \text{Cov}[X_i X_j] \tag{1.93}$$

When the random variables $\{X_i\}$ are mutually independent, the following simpler expressions are obtained:

$$E[Y] \simeq g(\mu_1, \mu_2, \ldots, \mu_n) + \frac{1}{2} \sum_{i=1}^{n} \sum_{j=1}^{n} \left(\frac{\partial^2 g}{\partial x_i^2} \right)_{\mu_i} \sigma_i^2 \tag{1.94}$$

and

$$\text{Var}[Y] \simeq \sum_{i=1}^{n} \left(\frac{\partial g}{\partial x_i} \right)_{\mu_i}^2 \sigma_i^2 \tag{1.95}$$

where μ_i and σ_i^2 are the mean and variance of the random variable X_i.

Finally, by series expansion, (cf. Eq. (1.88)), a simple approximation for the distribution of Y can be obtained. By conserving the first two elements of the right-hand side of Eq. (1.88), the following expression for the mean is obtained:

$$Y \simeq g(\mu_X) - \mu_X g'(\mu_X) + x g'(\mu_X)$$

which is of the form $Y = a + bX$. Then if X has the density function $p(x)$,

$$q(y) = p\left(\frac{y-a}{b}\right)\left|\frac{1}{b}\right|$$

or

$$q(y) = \left|\frac{1}{(\mathrm{d}g/\mathrm{d}x)_{x=\mu_X}}\right| p\left\{\frac{y - g(\mu_X) + \mu_X(\mathrm{d}g/\mathrm{d}x)_{x=\mu_X}}{(\mathrm{d}g/\mathrm{d}x)_{x=\mu_X}}\right\} \tag{1.96}$$

Example 1.23 In Fig. 1.16, the internal stress distribution in a cross section of a reinforced concrete beam in pure bending is shown, just before breaking. Now let the ultimate breaking stress of the concrete correspond to the random variable C with $\mu_C = 28$ MPa and a coefficient of variation $V_C = 0.25$, and the yield stress of the reinforcing steel bars correspond to the random variable S with $\mu_S = 308$ MPa and $V_S = 0.1$. The cross section of the beam has a so-called balanced design, that is, in the ultimate state, the concrete and the steel will reach their ultimate limits simultaneously. The mean value and the variance of the breaking moment of the beam will be determined, given that the steel reinforcement cross section area is $A_s = 942\,\text{mm}^2$.

Solution The ultimate breaking moment of an RC beam with a balanced design can be expressed as follows (cf. Fig. 1.16):

$$M = SA_s\left(h - \gamma\frac{SA_s}{cb}\right), \qquad \gamma = \frac{\beta}{\alpha} \sim 0.5$$

thus $M = g(S, C)$ is a function of the two random variables S and C that must be independent due to the nature of the problem, i.e. the strength of steel has nothing to do with the strength of concrete.

First, the partial derivatives must be evaluated:

$$g(s,c) = 5.28 \times 10^5 \times s - 1400s^2/c \text{ Nmm}$$

$$g'_s = 5.28 \times 10^5 - 2800s/c,\ g''_s = -2800/c$$

$$g'_c = 1400s^2/c^2, \qquad g''_{cc} = -2800s^2/c^3$$

$$(g''_{ss})^2 = 10^4 \text{ for } c = \mu_C,\ (g''_{cc})^2 = 11^4 \times 10^4 \qquad \text{for} \quad c = \mu_C \quad \text{and} \quad s = \mu_S$$

By the expressions Eqs. (1.94) and (1.95), it is now possible to obtain the necessary information about the ultimate breaking moment:

$$g(\mu_S, \mu_C) = 5.28 \times 10^5 \times 308 - 1400 \times 308^2/28$$

$$= 162.6 \times 10^6 - 4.7 \times 10^6 = 157.9 \times 10^6 \text{ Nmm}$$

$$\frac{1}{2}\sum_{i=1}^{2}(g''_{x_ix_i})\mu_i\sigma_i^2 = -(2800/28 \times (0.1 \times 308)^2$$

$$+ 2800 \times 308^2/28^3 \times (0.25 \times 28)^2)/2 = -0.34 \times 10^6$$

$$\mu_M = 157.9 - 0.34 = 157.6 \text{ KNm}$$

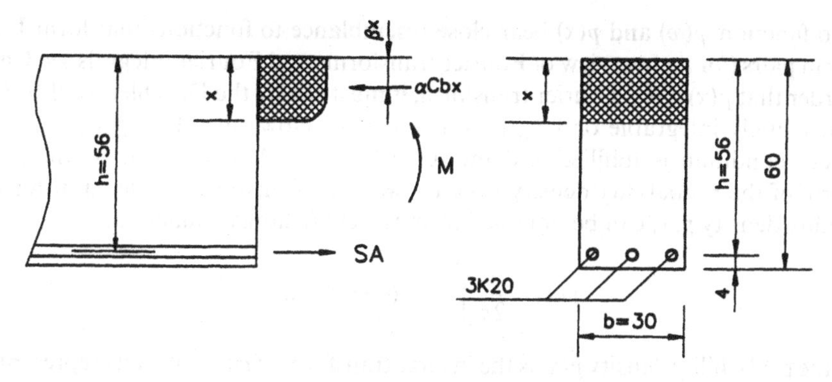

Fig. 1.16 *The ultimate breaking state of a RC beam*

The variance of the breaking moment is obtained from Eq. (1.95):

$$\sigma_M^2 = \sum_{i=1}^{2} (g'_{x_i})^2_{\mu_i} \sigma_i^2 = [(5.28 \times 10^5 - 2800 \times 308/28)^2 (0.1 \times 308)^2$$

$$+ (1400 \times 308^2/28^2)^2 (0.25 \times 28)^2] = (15.35 \times 10^6)^2 \,(\text{Nmm})^2$$

and the co-efficient of variation $V_M = \sigma_M/\mu_M = 15.35/157.6 = 0.0973$. Therefore the fundamental assumption for the approximation is fulfilled, that is, the coefficient of variation is sufficiently small, ($V \leqslant 0.1$).

If a reliable design for the moment is defined as the mean value minus two standard deviations, $M_D = 157.6 - 2 \times 15.35 = 126.9$ KNm, the probability of failure can be computed as $P[M \leqslant M_D] = p_f$. Assuming that the moment is a normally distributed variable, the probability of failure can be obtained from the standardized variable $U = (M - \mu_M)/\sigma_M$, (cf. Ex. 1.12), that is,

$$p_f = P[U \leqslant (-2)] = 1 - P[U \leqslant 2] = 1 - 0.9772 = 0.0228$$

In other words, if 1000 beams according to the above conditions were to be made, 23 of them could be expected to be a failure.

1.2.9 Characteristic Functions

The concept of characteristic functions of a random variable can be very useful in many applications. The characteristic function of the random variable X is defined as the function

$$\varphi(\omega) = E[e^{i\omega X}]$$

in which ω is a real variable and i is the imaginary unit. The expected value of the function $\exp(i\omega X)$ is obtained in the usual manner by integration:

$$\varphi(\omega) = E[e^{i\omega x}] = \int_{-\infty}^{\infty} e^{i\omega X} p(x) \mathrm{d}x \tag{1.97}$$

The two function $\varphi(\omega)$ and $p(x)$ bear close resemblance to functions that form Fourier transform pairs (for an overview of Fourier transforms and Fourier integrals see Chapter 7). In order that $p(x)$ has a Fourier transform, it must satisfy the Dirichlet condition, that is, be absolutely integrable or $\int_{-\infty}^{\infty}|p(x)|\mathrm{d}x = c < \infty$. Now since $\int_{-\infty}^{\infty}|p(x)|\mathrm{d}x = 1 < \infty$, Dirichlet's condition is fulfilled and the function $\varphi(-\omega)$ is the same as the Fourier transform of the probability density $p(x)$. By way of the inverse Fourier transform, the probability density $p(x)$ can be obtained from the characteristic function:

$$p(x) = \frac{1}{2\pi}\int_{-\infty}^{\infty} \varphi(\omega)e^{-i\omega X}\mathrm{d}\omega \tag{1.98}$$

that is, the probability density $p(x)$ is the inverse transform of $\varphi(-\omega)$, or its representation in the x-domain.

If the random variable X has a finite mean value, then Eq. (1.97) can be differentiated with respect to ω and

$$\varphi'(\omega) = E[(iX)e^{i\omega X}] = \int_{-\infty}^{\infty} (ix)e^{i\omega x}p(x)\mathrm{d}x$$

If $\omega = 0$, then from the above relation,

$$\varphi'(0) = iE[X] = i\int_{-\infty}^{\infty} xp(x)\mathrm{d}x$$

and

$$E[X] = \frac{1}{i}\varphi'(0)$$

If the kth moment of X is finite, then through repeated differentiation of Eq. (1.97), one finally obtains for the kth derivative,

$$\varphi^{(k)}(\omega) = E[i^k X^k e^{i\omega X}] \tag{1.99}$$

or with $\omega = 0$

$$E[X^k] = \frac{1}{i^k}\varphi^{(k)}(0) \tag{1.100}$$

Thus the kth moment of a random variable has the same magnitude as the kth derivative of its characteristic function at the point $\omega = 0$, and a sign corresponding to i^k.

Example 1.24 The probability density of the Gaussian distribution is

$$p(x) = \frac{1}{\sigma\sqrt{2\pi}}\exp\left\{-\frac{(x-\mu)^2}{2\sigma^2}\right\}$$

Find the moments of the distribution.

Solution Firstly, the variable X will be standardized, that is,

$$U = (X-\mu)/\sigma, \qquad X = \sigma U + \mu$$

The characteristic functions of U and X are then related through the expression

$$\varphi_X(\omega) = E[e^{(\sigma U+\mu)i\omega}] = E[e^{\sigma U i\omega}]e^{i\mu\omega} = e^{i\mu\omega}\varphi_U(\omega)$$

As $p(u) = \exp(-u^2/2)/\sqrt{(2\pi)}$, the characteristic function $\varphi_U(\omega)$ is easily obtained:

$$\varphi_U(\omega) = \frac{1}{\sqrt{2\pi}}\int_{-\infty}^{\infty} e^{i\omega\mu}e^{-u^2/2}du = \frac{1}{\sqrt{2\pi}}\int_{-\infty}^{\infty} e^{-(u-i\omega)^2/2}e^{-\omega^2/2}\,du$$

$$= e^{-\omega^2/2}\frac{1}{\sqrt{2\pi}}\int_{-\infty}^{\infty} e^{-(u-i\omega)^2/2}\,du = e^{-\omega^2/2}$$

Therefore,

$$\varphi_X(\omega) = e^{i\mu\omega}e^{-\sigma^2\omega^2/2} = \exp\left(i\mu\omega - \frac{\sigma^2\omega^2}{2}\right) \tag{1.101}$$

It can be observed that all moments of the normal distribution are functions of μ and σ only. That is, the normal distribution is completely characterized by its mean and variance.

The log-characteristic function should also be mentioned:

$$\psi(\omega) = \log_e \varphi(\omega) \tag{1.102}$$

It can be used for obtaining the so-called semi-invariants or cumulants κ_s, first introduced by the Danish mathematician T. N. Thiele (see, for instance, Cramer [41]). Indeed,

$$\kappa_s = \frac{1}{i^s}\left[\frac{d^s\psi(\omega)}{d\omega^s}\right]_{\omega=0} \tag{1.103}$$

The semi-invariants of the random variable X are thus related to the central moments as follows:

$$\begin{aligned}\kappa_1 &= m_1 = \mu_X = \mu_1\\ \kappa_2 &= \mu_2 = \sigma_X^2\\ \kappa_3 &= \mu_3\\ \kappa_4 &= \mu_4 - 3\mu_2^2\end{aligned} \tag{1.104}$$

Example 1.25 From the characteristic function $\varphi(\omega)$, an other important function, the moment generating function, can be obtained by changing the argument ω to it where i is the imaginary unit. Thus a real function $M(t)$ is obtained, i.e.

$$M(t) = \varphi(it) = E[e^{-tX}] \tag{1.105}$$

Using the Taylor series expansion of the exponential function,

$$e^{-tx} = 1 - tx + (tx)^2/2! - (tx)^3/3! + (tx)^4/4! - \cdots$$

the moment generating function is expanded as follows if the first n moments are finite:

$$M(t) = \sum_{j=0}^{n} \frac{(-t)^j}{j!} \int_{-\infty}^{\infty} x^j p(x) \mathrm{d}x + O(t^{n+1}) = \sum_{j=0}^{n} \frac{(-t)^j}{j!} E[X^j] + O(t^{n+1}) \quad (1.106)$$

Thus, the jth moment of the random variable X is the coefficient of $t^j/j!$ in Eq. (1.106)

$$E[X^j] = M^{(j)}(0) \quad (1.107)$$

The characteristic function itself can be expanded in the same manner (cf. Eq. (1.100)),

$$\varphi(\omega) = 1 + E[X](\mathrm{i}\omega) + E[X^2]\frac{(\mathrm{i}\omega)^2}{2!} + E[X^3]\frac{(\mathrm{i}\omega)^3}{3!} + E[X^4]\frac{(\mathrm{i}\omega)^4}{4!} + O(\omega^5) \quad (1.108)$$

Example 1.26 The Poisson distribution for a discrete random variable X can be very useful when dealing with random arrival times, for instance for accidents occurring at a certain road sections, earthquake waves arriving at a specific site etc. The discrete variable X has a Poisson distribution, that is, it can assume the values $x_k = k$, where the k's are non-negative integers, with the probability

$$P[X = k] = \frac{\lambda^k}{k!}\mathrm{e}^{-\lambda} \quad (1.109)$$

in which λ is a positive real constant. Obtain the characteristic function $\varphi(\omega)$ and the semi-invariants of the Poisson distribution.

Solution By Eqs. (1.100) and (1.33),

$$\varphi(\omega) = E[\mathrm{e}^{\mathrm{i}\omega X}] = \int_{-\infty}^{\infty} \mathrm{e}^{\mathrm{i}\omega x} \left\{ \sum_{k=0}^{\infty} \frac{\lambda^k}{k!} \mathrm{e}^{-\lambda} \delta(x-k) \right\} \mathrm{d}x = \sum_{k=0}^{\infty} \mathrm{e}^{\mathrm{i}\omega k} \mathrm{e}^{-\lambda} \frac{\lambda^k}{k!}$$

so

$$\varphi(\omega) = \mathrm{e}^{-\lambda} \sum_{k=0}^{\infty} \frac{(\lambda \mathrm{e}^{\mathrm{i}\omega})^k}{k!} = \mathrm{e}^{\lambda \mathrm{e}^{\mathrm{i}\omega}} \quad (1.110)$$

Then

$$\psi(\omega) = \log_e \varphi(\omega) = \lambda(\mathrm{e}^{\mathrm{i}\omega} - 1) = \lambda\left(\sum_{k=0}^{\infty} \frac{(\mathrm{i}\omega)^k}{k!} - \right) = \sum_{k=1}^{\infty} \lambda \mathrm{i}^k \frac{\omega^k}{k!}$$

Also by Eq. (1.103),

$$\lambda = \frac{1}{\mathrm{i}^k} \left[\frac{\mathrm{d}^k \psi(\omega)}{\mathrm{d}\omega^k} \right]_{\omega=0} = \kappa_k, \qquad (k = 1, 2, 3, \ldots)$$

so the three first central moments of the Poisson distribution are equal to λ, cf. Eq. (1.104).

The characteristic function Eq. (1.97) can easily be extended to describe more than one random variable. Let X, Y and Z be three random variables. Then the joint characteristic function is given by

$$\varphi_{XYZ}(\omega_1, \omega_2, \omega_3) = E[e^{i(\omega_1 X + \omega_2 Y + \omega_3 Z)}] \tag{1.111}$$

and the log-characteristic function is given by

$$\psi_{XYZ}(\omega_1, \omega_2, \omega_3) = \log_e \varphi_{XYZ}(\omega_1, \omega_2, \omega_3) \tag{1.112}$$

The joint characteristic function can be obtained from integration of the joint density in the usual manner, Eq. (1.98):

$$\varphi_{XYZ}(\omega_1, \omega_2, \omega_3) = \int_{-\infty}^{\infty}\int_{-\infty}^{\infty}\int_{-\infty}^{\infty} e^{i(\omega_1 x + \omega_2 y + \omega_3 z)} p_{XYZ}(x,y,z) dx\, dy\, dz \tag{1.113}$$

Thus the joint characteristic function is, but for the sign of the power in the exponential, the triple Fourier transform of the joint density function. The inverse relation is obtained in the same way as for the one-dimensional characteristic function, Eq. (1.99):

$$p_{XYZ}(x,y,z) = \frac{1}{(2\pi)^3}\int_{-\infty}^{\infty}\int_{-\infty}^{\infty}\int_{-\infty}^{\infty} e^{-i(\omega_1 x + \omega_2 y + \omega_3 z)} \varphi_{XYZ}(\omega_1,\omega_2,\omega_3) d\omega_1\, d\omega_2\, d\omega_3 \tag{1.114}$$

The marginal characteristic functions can be defined and expressed in the same manner as marginal densities, Eq. (1.55). For instance, putting $\omega_2 = \omega_3 = 0$ in Eq. (1.113), the one-dimensional characteristic function is obtained through double integration, i.e.

$$\varphi_X(\omega) = E[e^{i\omega X}] = \varphi_{XYZ}(\omega,0,0) \tag{1.115}$$

Now, consider the sum variable $Z = aX + bY$. The characteristic function of Z is

$$\varphi_Z(\omega) = E[e^{i\omega Z}] = E[e^{i(a\omega X + b\omega Y)}]$$

The last exponential is the same as in Eq. (1.111), where ω_1 and ω_2 have been replaced by $a\omega$ and $b\omega$. Therefore,

$$\varphi_Z(\omega) = \varphi_{XY}(a\omega, b\omega)$$

If the two variables X and Y are independent, then

$$E[e^{i(\omega_1 x + \omega_2 y)}] = E[e^{i\omega_1 x}]\, E[e^{i\omega_2 y}]$$

and

$$\varphi_{XY}(\omega_1,\omega_2) = \varphi_X(\omega_1)\cdot\varphi_Y(\omega_2) \tag{1.116}$$

Using the inverse transform, Eq. (1.114), it is easy to show that the converse statement is true.

Example 1.27 The two discrete random variables X and Y are independent and Poisson-distributed, Example 1.26, with the parameters α and β respectively. What is the distribution of the sum variable $Z = X + Y$?

Solution From Example 1.24, Eq. (1.110),

$$\varphi_X(\omega) = e^{-\alpha} e^{\alpha e^{i\omega}} \quad \text{and} \quad \varphi_Y(\omega) = e^{-\beta} e^{\beta e^{i\omega}}$$

By Eq. (1.116),

$$\varphi_Z(\omega) = e^{-\alpha} e^{\alpha e^{i\omega}} e^{-\beta} e^{\beta e^{i\omega}} = e^{-(\alpha+\beta)} e^{(\alpha+\beta) e^{i\omega}}$$

so the sum variable Z is also Poisson-distributed with a parameter $(\alpha + \beta)$. It can be proven that the converse statement is also true.

1.2.10 Laws of Large Numbers

In many cases the result or outcome of an action, which is repeated a great number of times, will gradually tend to a pattern that can be predicted at least intuitively. For instance, tossing a coin many times, the heads will come up in 50% of the tosses on the average. In other words, the proportion of heads and tails will tend to 1/2 as the number of tosses becomes sufficiently large. As another example consider a basket that contains 100 yellow and 300 white golf balls. If a ball is taken out at random, there is no obvious pattern to be noticed. However, when a large number of balls have been taken out of the basket, one would expect the ratio of yellow to white golf balls thus extracted to be 1/3. It can be useful to dress this simple probabilistic model into a theorem that reflects this behaviour. Actually, it is this theorem, which is called the law of large numbers, that makes probability theory useful as a basic tool for statistical evaluations.

Let X_i, $i = 1,2,\ldots,n$ be independent random variables, all with the mean value μ and variance σ^2. By forming the statistical average, a new random variable Y is formed:

$$Y = \frac{1}{n} \sum_{i=1}^{n} X$$

The mean value of Y is easily obtained,

$$E[Y] = \frac{1}{n} \sum_{i=1}^{n} E[X_i] = \mu$$

and the variance

$$\text{Var}[Y] = \frac{1}{n^2} \sum_{i=1}^{n} \text{Var}[X_i] = \frac{\sigma^2}{n}$$

The law of large numbers may then be stated as follows. For any prescribed constant $\delta > 0$, no matter how small,

$$\lim_{n \to \infty} P[|Y - \mu| > \delta] = 0$$

that is, the probability that Y and the mean μ remain apart by as much as δ tends to zero for a sufficiently large number of variables. The proof is easily carried out by using the Chebyshev inequality, Eq. (1.81), that is,

$$P[|Y - E[Y]| > \delta] \leqslant \frac{\text{Var}[Y]}{\delta^2} = \frac{\sigma^2}{n\delta^2}$$

so

$$\lim_{n\to\infty} P[|Y-\mu|>\delta]=0$$

The law of large numbers provides a theoretical counterpart of the interpretation of probability as the favourable fraction that was discussed in Section 1.1.1, Eq. (1.1). For this purpose, consider first a discrete random variable X that can assume only two different values with probabilities p and $q=1-p$ and is therefore a Bernoulli type variable, Eq. (1.30). Without loss of generality, the two values can be taken as 0 and 1. Its probability distribution can be conveniently expressed by the following formula:

$$P[X=k]=p^k(1-p)^{1-k}, \qquad \text{for} \quad k=0 \quad \text{or} \quad 1$$

Its probability density function can be written using the delta function, Eq. (1.33):

$$p(x)=(1-p)\delta(x)+p\delta(x-1)$$

The expectation of any function $g(X)$ of a Bernoulli variable is then easily obtained by integration, that is,

$$E[g(X)]=g(1)p+g(0)(1-p)$$

For instance, the kth moment $E[X^k]=p$, so $\mu=p$. Now if a large sequence of n Bernoulli trials is carried out with j successes, (e.g. $X=1$ is a success), the law of large numbers states that

$$\lim_{n\to\infty} P\left[\left(\frac{j}{n}-p\right)>\delta\right]=0 \tag{1.118}$$

That is the probability, that the favourable fraction j/n is different from the actual probability p by more than a small prescribed number δ, tends to zero for large number of trials n.

Sometimes a distinction is made whether the convergence implied by Eq. (1.118) is strong or weak. The weak law of large numbers as stated above is a special case of Khinchine's theorem, which states that a random variable Y, which is the average value of the n random IID variables $\{X_i\}$, converges to μ in probability even if the variances, $\text{Var}[X_i]$, are infinite, that is,

$$\lim_{n\to\infty} P[|Y-\mu|>\delta]\to 0 \qquad \text{for all } \delta>0 \tag{1.119}$$

The strong law of large numbers on the other hand implies that the covergence is stronger as the name indicates, that is,

$$\lim_{n\to\infty} P[Y\to\mu]=1 \tag{1.120}$$

which states that Yconverges to μ with probability one.

The weak law of large numbers can be proved by applying the characteristic function of the common variable X. By Eq. (1.108), the two first terms of the expansion of the characteristic function are given by

$$\varphi(\omega) = 1 + \mathrm{i}\mu\omega + o(\omega)$$

The first derivative is $\mathrm{i}\mu$ and is also continuous at $\omega = 0$. The log-characteristic function Eq. (1.103) is $\psi(\omega) = \log_e \varphi(\omega)$ and its first derivative by Eq. (1.104) is $\mathrm{i}\mu$. Now let $\varphi^*(\omega)$ be the characteristic function of the average value Y. It can be interpreted by studying the log-characteristic function $\psi^*(\omega) = \log_e \varphi^*(\omega)$,

$$\psi^*(\omega) = \log_e \varphi^*(\omega) = \log_e E\left[\exp\left(\mathrm{i}\omega \frac{(X_1 + X_2 + X_3 + \cdots)}{n}\right)\right] = \log_e \left[\varphi\left(\frac{\omega}{n}\right)\right]^n$$

$$= n\psi(\omega/n) = \omega \frac{\psi(\omega/n)}{\omega/n} = \omega \frac{(\psi(\omega/n) - \psi(0))}{(\omega/n - 0)}$$

In the limit, as n goes to infinity, the last fraction becomes the first derivative of the log-characteristic function taken at $\omega = 0$, i.e. $\mathrm{i}\mu$, so the whole exercise yields the value $\mathrm{i}\mu\omega$. As the logarithm function is continuous, the reverse function is obtained as

$$\varphi^*(\omega) = \mathrm{e}^{\mathrm{i}\mu\omega}$$

which is the characteristic function of a random variable that is a constant, equal to μ! Therefore, the characteristic function of the random variable Y degenerates to the form where Y is equal to μ, such that Y tends to assume this value in the limit.

1.2.11 The Central Limit Theorem

Whereas the law of large numbers stated that the average value of a number of IID variables X_i, $i = 1,2,\ldots,n$, tends to become a constant for large values of n, this does not mean that the sum of the variables approaches a constant multiple of n in the limit. On the contrary, the sum $S_n = X_1 + X_2 + X_3 + \cdots + X_n$ shows increasing variability as the number n grows larger and the spread of the distribution of S_n becomes larger and larger. The mean value of S_n is obviously equal to $n\mu$ and the variance is $n\sigma^2$. Thus for finite values of μ and σ, as n becomes a very large number, the centre of the distribution of S_n moves off to infinity with a very large spread.

The above relation about the variable S_n may seem disappointing. However, it can nevertheless be useful to have some idea about the shape of the distribution of S_n. Actually, it can be shown that S_n has a universal limiting shape, regardless of the distribution of the summands X_i, subject to reasonable restrictions. This remarkable feature of S_n is the kernel of the central limit theorem, which can be stated as follows. Let X_i, $i = 1,2,\ldots,n$, be a sequence of independent and identically distributed random variables. The mean and variance of each of the variables X_i is μ and σ^2. Denote by S_n the sum of the variables. Then, the standardized variable,

$$Z_n = \frac{S_n - n\mu}{\sigma\sqrt{n}}$$

will be asymptotically normally distributed, that is, as n goes to infinity, the distribution function of Z_n approaches the standardized normal distribution, $N(z)$ with mean 0 and variance 1. Thus

$$\lim_{n\to\infty} P\left[Z_n = \frac{S_n - n\mu}{\sigma\sqrt{n}} < z\right] = \frac{1}{\sqrt{2\pi}}\int_{-\infty}^{z} e^{-u^2/2}du \tag{1.121a}$$

for any value of z. Hence, approximately for large n,

$$P[S_n < y] = N\left(\frac{y - n\mu}{\sigma\sqrt{n}}\right) \tag{1.121b}$$

To prove this important theorem, it is assumed that the standard deviation σ of the common basic variable X_i is finite and not equal to zero. Then, the characteristic function of the distribution of $X_i - \mu$ can be expressed as follows (Eq. (1.108)):

$$\varphi_{X_i-\mu}(\omega) = 1 - \sigma^2\frac{\omega^2}{2!} + o(\omega^2)$$

Now form the variable,

$$Y_n = \frac{X_i - \mu}{\sigma\sqrt{n}}$$

The characteristic function of $Z_n = Y_1 + Y_2 + Y_3 + \ldots + Y_n$ is obtained from the characteristic function of the Y_i:

$$\varphi_{Y_i}(\omega) = \varphi_{X_i-\mu}\left(\frac{\omega}{\sigma\sqrt{n}}\right) = 1 - \frac{\omega^2}{2n} + o\left(\frac{\omega^2}{n}\right)$$

and by raising it to the nth power,

$$\varphi_{Z_n}(\omega) = \left[1 - \frac{\omega^2}{2n} + o\left(\frac{\omega^2}{n}\right)\right]^n = \left[1 - \frac{\omega^2/2 + no(\omega^2/n)}{n}\right]^n$$

Now recalling that $e^{-x} = \lim(1 - x/n)^n$ for $n \to \infty$, also noting that $no(t^2/n)$ becomes zero in the limit for an arbitrary value of ω, the characteristic function for Z_n is obtained by going to the limit, i.e.

$$\varphi_{Z_n}(\omega) = \lim_{n\to\infty}\left[1 - \frac{\omega^2/2 + no(\omega^2/n)}{n}\right]^n = e^{-\omega^2/2}$$

which is the characteristic function of a standardized normal variable, cf. Eq. (1.102). Since the characteristic function of a random variable is unique this implies that in the limit, Z_n has the standardized normal distribution.

The central limit theorem has been proved for a large number of random contributions that are identically distributed. This condition can be relaxed and the central limit theorem is also valid for a large sum of small random contributions that are independent but can have different probability laws. As pointed out by Kolmogorov, [5], perhaps the greatest service rendered by the classical Russian school of mathematics is the work done by Chebyshev and his students, Markov, Bernstein and Lyapunov to formulate, expand and generalize the conditions for the laws of large numbers and the central limit theorem. Under fairly general conditions, Lyapunov and Bernstein showed that for an arithmetic mean of random variables,

$$Y = (X_1 + X_2 + \cdots + X_n)/n \tag{1.122}$$

where the X_i are random variables with bounded individual variances, not necessarily independent and not necessarily with identical probability distributions,

$$\lim_{n\to\infty} P[t_1\sigma_Y < (Y - E[Y]) < t_2\sigma_Y] = \frac{1}{\sqrt{2\pi}}\int_{t_1}^{t_2} e^{-t^2/2}\,dt = \tfrac{1}{2}\mathrm{Erf}(t_2 - t_1) \tag{1.123}$$

where the error function,

$$\mathrm{Erf}(x) = \frac{\sqrt{2}}{\sqrt{\pi}}\int_0^x e^{-t^2/2}\,dt \tag{1.124}$$

which is sometimes used in the literature to represent the normal distribution, has been introduced for future reference. Finally, it can be useful to state the conditions for the above result, Eq. (1.121a), for independent variables with different probability distributions. In this form, the theorem is often referred to as Lyapunov's theorem, [71]. If for a sequence of mutually independent random variables $X_1, X_2, \ldots, X_k, \ldots$, a positive constant δ can be found such that for $n \to \infty$

$$\lim_{n\to\infty} \frac{1}{B_n^{2+\delta}} \sum_{k=1}^{n} E[|X_k - a_k|^{2+\delta}] = 0, \tag{1.125}$$

where

$$a_k = E[X_k], \qquad b_k^2 = \mathrm{Var}[X_k] \text{ and } B_n^2 = \sum_{k=1}^{n} b_k^2 = \mathrm{Var}\left[\sum_{k=1}^{n} X_k\right] \tag{1.126}$$

then as $n \to \infty$

$$\lim_{n\to\infty} P\left[\frac{1}{B_n}\sum_{k=1}^{n}(X_k - a_k) \leqslant x\right] = \frac{1}{\sqrt{2\pi}}\int_{-\infty}^{x} e^{-u^2/2}\,du \tag{1.127}$$

Example 1.28 The sum of n real numbers is calculated by rounding each number off to the nearest integer in the sum. Given that the round-off error for each number is uniformly distributed over the interval $[-0.5, 0.5]$, find the probability distribution for the round-off error for the sum itself.

Solution Letting X_i denote the random value of the round-off error, the mean value is obviously equal to zero as the density is equal to one in the interval $[-0.5, 0.5]$. The variance of the round-off error on the other hand is given by the integral,

$$\sigma^2 = \int_{-0.5}^{0.5} x^2 p(x)\,dx = \left[\frac{x^3}{3}\right]_{-0.5}^{0.5} = \frac{1}{12}$$

The probability that the sum of the round-off errors, denoted by S_n, is less than any given number approaches the normal distribution for large value of n, cf. Eq. (1.121b). Hence,

$$P[S_n < y] \simeq N\left(2y\sqrt{\frac{3}{n}}\right)$$

Example 1.29 Kolmogorov's law of fragmentation Many physical problems have to do with the fragmentation of a larger piece of material into many small pieces and also the size of grains or particles in a large collection. A typical example is the size of gold particles in a large sample of gold sand. Many studies of such problems have shown that the logarithm of the dimension of the grains or fragmented pieces from a large sample will approximately follow the normal distribution, that is, if a typical dimension of the particle is called $D = e^Y$, $Y = \log_e D$ is normally distributed. That is

$$p_Y(y) = \frac{1}{\sigma_Y\sqrt{2\pi}}\exp\left[-\frac{(y-\mu_Y]^2}{2\sigma_Y^2}\right]$$

whereby the lognormal distribution has the following probability density:

$$p_D(d) = \frac{1}{d\sigma_Y\sqrt{2\pi}}\exp\left[-\frac{(\log_e d-\mu_Y]^2}{2\sigma_Y^2}\right], \quad d > 0 \tag{1.128}$$

Using the moment generating function for $Y = \log_e D$, Eqs. (1.102) and (1.105),

$$E[D^k] = E(e^{kY}] = M(k) = \exp(k\mu_Y + \sigma_Y^2 k^2/2)$$

the mean value, $k = 1$, is

$$\mu_D = \exp(\mu_Y + \tfrac{1}{2}\sigma_Y^2) \tag{1.129}$$

and the variance

$$\sigma_D^2 = E[D^2] - (E[D])^2 = (\exp(\sigma_Y^2) - 1)\exp(2\mu_Y + \sigma_Y^2) \tag{1.130}$$

The median of D, Ex. 1.13, is easily obtained by noting that

$$P[X \leqslant M_2] = P[e^Y \leqslant M_2] = P[Y \leqslant \log_e M_2] = \tfrac{1}{2}$$

whereby $\log_e M_2 = \mu_Y$, so $M_2 = \exp(\mu_Y)$.

In a short paper, Kolmogorov [111] (reproduced by permission of Prof. Dr. A.N. Shiryaev), presented and formalized this interesting probabilistic behaviour, which may be referenced as Kolmogorov's law of fragmentation. This work has perhaps not received the attention that it deserves. In the following, a translation that follows loosely the German text is presented. Since the paper is very short and compact without many explanations, further explanations by the author and Dr Miguel de Icaza have been added, [88].

Following Kolmogorov, let $N(t)$ be the total number of particles at integer times t, $t = 0, 1, ,2, 3, \ldots$, due to fragmentation from a larger piece of material (rock, gold etc.). Further, $N(r, t)$ is the total number of particles that at time t have a typical dimension (diameter, volume, weight etc.), $\varrho \leqslant r$, that is, the fragmentation of a piece of material of size r can only yield particles that are smaller. Now assume that between times t and $t+1$, the probability that one single particle of size r is fragmented into n smaller particles is p_n. Introduce the relative ratio of the dimension of each of the n fragmented particles to the dimension of the parent piece r as $k_i = r_i/r$. Obviously, k_i is a random variable K_i and the joint probability distribution can be written as follows:

$$F_n(a_1, a_2, \ldots, a_n) = P[K_1 \leqslant a_1, K_2 \leqslant a_2, \ldots, K_n \leqslant a_n] \qquad (1.131)$$

where the n fragmented particles have been numbered according to increasing dimensions, $r_1 \leqslant r_2 \leqslant r_3 \leqslant \cdots \leqslant r_n$, so the distribution Eq. (1.131) is only defined for $0 \leqslant a_1 \leqslant a_2 \leqslant \cdots \leqslant a_n \leqslant 1$. Now, the average number of particles of dimension $\varrho \leqslant kr$, called $Q(k)$, that were produced by one single particle of an arbitrary dimension r between times t and $t+1$ is obviously

$$Q(k) = \sum_{n=1}^{\infty} p_n\{F_n(k, 1, 1, \ldots, 1, 1) + F_n(k, k, 1, \ldots, 1, 1)$$
$$\cdots + F_n(k, k, k, \ldots, k, 1) + F_n(k, k, k, \ldots, k, k)\} \qquad (1.132)$$

where each term in the sum is the probability p_n times the probabilities that the dimension kr of each particle belongs to the ordered groups. This last statement, Eq. (1.132), is perhaps not as clear to everybody as it was to Kolmogorov. The following explanation may be given. In addition to the above event that a single particle of dimension r is broken up between times t and $t+1$, define the subevents: (a) J, which is the event that precisely j particles were produced having dimensions less than or equal to k, and (b) J^*, the event that at least j, $(\geqslant j)$, particles were produced, having $K \leqslant k$, in both cases given that in all n particles were produced. Obviously the two events, J and $(J+1)^*$ are disjoint events and

$$J + (J+1)^* = J^*$$

Therefore, the conditonal probability of the event J, called q_i, is given by

$$P[J/n] = q_j = P[J^*/n] - P[(J+1)^*/n] \qquad (1.133)$$

The probability q_j is then given by Eq. (1.133) and the distribution Eq. (1.131), that is,

$$q_i = F_n(\{k, k, \ldots, k\}_i 1, 1, \ldots, 1) - F_n(\{k, k, \ldots, k\}_{i+1} 1, 1, \ldots, 1) \qquad (1.134)$$

where $F_n(\{k, k, \ldots, k\}_{i+1} 1, 1, \ldots, 1)$ is defined as zero if $i = n$. It is now possible to write up $Q(k)$ as follows

$$Q(k) = \sum_{n=1}^{\infty} p_n E[J|n] \qquad (1.135)$$

where

$$E[J|n] = \sum_{i=1}^{n} i q_i = \sum_{i=1}^{n} \sum_{j=1}^{n} q_j = \sum_{i=1}^{n} F_n(\{k, k, \ldots, k\}_i, 1, 1, \ldots, 1) \qquad (1.136)$$

is the conditional mean value of the number of particles with $K \leqslant k$, conditional that exactly n particles were produced, and q_i is the probability to have exactly i particles for which $K \leqslant k$, Eq. (1.134). By introducing this result, i.e. the last sum, Eq. (1.136), into Eq. (1.135), the original Kolmogorov equation, Eq. (1.132), is obtained.

The following reasonable assumptions are now made:

(a) the probabilities p_n and the distribution F_n are not dependent on the absolute dimensions of the particles, nor on the history of the fragmentation, that is, how a particle of any size at an earlier stage was broken up, nor on the destiny of other particles.
(b) the expectation $q(1)$, that is, the average number of particles produced by fragmentation of one single particle of size r, during the time interval $[t, t+1]$ without regard to size of the fragmented particles, $\varrho \leqslant r$, is both limited and larger than one.
(c) the integral $\int_0^1 |\log_e k|^3 \, dQ(k)$ is finite.
(d) at the beginning, $t = 0$, a certain number $n(0)$ of particles is at hand, which can have an arbitrary distribution of size $N(r, 0)$.

Under these assumptions, the expectation or the average number $N(t)$ of particles at time t is

$$E[N(t)] = n(0)(Q(1))^t \tag{1.137}$$

Now, form the ratio

$$\frac{E[N(\mathrm{e}^x, t)]}{E[N(t)]} = \frac{E[N(\mathrm{e}^x, t)]}{n(0)(Q(1))^t} = T(x, t) \tag{1.138}$$

where the function $N(\mathrm{e}^x, t)$ represents the number of particles at time t that have the dimension $\varrho \leqslant r = \mathrm{e}^x$ with x as a real variable. (As pointed out by Kolmogorov, it is completely irrelevant what the dimension ϱ really is.)

The real task is now to estimate the function $T(x, t)$. From the assumptions (a) and (b), it follows that

$$E[N(r, t+1)] = \int_0^1 E\left[N\left(\frac{r}{k}, t\right)\right] \mathrm{d}Q(k) \tag{1.139}$$

that is the average number of particles at time $t+1$, $E[N(r, t+1)]$, is obtained by summing over all particles with a dimension $\varrho \leqslant kr$, that are produced from the available particles at time t, letting k run through all values between 0 and 1. Equation (1.139) needs further explanation since it is far from being obvious. Take any partition at time t, $\{k_1, k_2, \ldots, k_m\}$ such that $0 < L \leqslant k_1 \leqslant k_2 \leqslant k_3 \leqslant \cdots \leqslant k_m = 1$, L being a very small number. $E[N(r/k_{s+1}, t)]$ is the average number of particles with $\varrho \leqslant r/k_{s+1}$. However, these particles are also counted or included in the number $E[N(r/k_s, t)]$. Therefore, the particles $E[N(r, t)]$ are counted or included in all the numbers $E[N(r/k_s, t)]$, $s = 1, 2, 3, \ldots, m$.

Now from $E[N(r, t)]$,

$$E[N(r, t)]Q(1) = E[N(r/k_m), t)]Q(k_m)$$

particles are produced between t and $t+1$.

From $E[N(r/k_{m-1}, t) - N(r/k_m, t)]$,

$$E[N(r/k_{m-1}, t) - N(r/k_m, t)]Q(k_{m-1})$$

particles are produced. From $E[N(r/k_1, t) - N(r/k_2, t)]$,

$$E[N(r/k_1, t) - N(r/k_2, t)]Q(k_1)$$

particles are produced. Summing up,

$$E\left[N\left(\frac{r}{k_m}, t\right)\right]Q(k_m) + \sum_{i=m}^{2} E\left[N\left(\frac{r}{k_{i-1}}, t\right) - N\left(\frac{r}{k_i}, t\right)\right]Q(k_{i-1})$$

By making the partition finer and finer, the above sum tends to the integral, $L \to 0+$,

$$E[N(r,t)]Q(1) + \int_1^L \mathrm{d}E\left[N\left(\frac{r}{x}, t\right)\right]Q(x)$$

$$= +E[N(r,t)]Q(1) + E\left[N\left(\frac{r}{L}, t\right)\right]Q(L) - E[N(r,t)]Q(1)$$

$$+ \int_L^1 E\left[N\left(\frac{r}{x}, t\right)\right]\mathrm{d}Q(x) \to \int_0^1 E\left[N\left(\frac{r}{x}, t\right)\right]\mathrm{d}Q(x)$$

whereby Eq. (1.139) is obtained.

Now put

$$Q(k) = Q(1)S(\xi), \xi = \log_e k, \qquad 0 < k < 1 \quad \text{and} \quad -\infty < \xi < 0 \tag{1.140}$$

Then by Eqs. (1.138) and (1.139),

$$T(x, t+1) = \int_{-\infty}^{0} T(x - \xi, t)\mathrm{d}S(\xi) \tag{1.141}$$

By Eqs. (1.138) and (1.140), the two functions, $S(x)$ and $T(x, 0) = N(e^x, 0)/n(0)$ fulfil all the conditions of probability distribution functions. If $x \to -\infty$, $k \to 0$, and both $S(x)$ and $T(x, 0)$ tend to zero. If $k \to 1$, $x \to 0$ and both functions tend to one. If $x > 0$, $S(x) = S(0) = 1$ and the upper limit 0 in integrals involving $S(x)$ can be replaced by ∞. The two functions, therefore, have the same character as ordinary probability distribution functions, even if they should not be considered as such in the strictest sense. From the recurrence equation, Eq. (1.141), the same applies to the function $T(x, t)$ for any integer $t > 0$.

From the assumption (c), it follows that the integral $\int_{-\infty}^{0} |x|^3 \,\mathrm{d}S(x)$ is finite. It is now possible to apply Lyapunov's theorem, Eq. (1.127), which allows the interpretation that for $t \to \infty$,

$$\lim_{t \to \infty} T(x, t) = \frac{1}{B\sqrt{2\pi t}} \int_{-\infty}^{x} e^{-(\xi - At)^2/2B^2 t} \,\mathrm{d}\xi \tag{1.142}$$

in which

$$A = \int_{-\infty}^{0} x\,\mathrm{d}S(x) \quad \text{and} \quad B^2 = \int_{-\infty}^{0} (x - A)^2 \mathrm{d}S(x) \tag{1.143}$$

Through Eq. (1.133), these integrals can be replaced by

$$A = \frac{1}{Q(1)} \int_0^1 \log_e k \, dQ(k) \quad \text{and} \quad B^2 = \frac{1}{Q(1)} \int_0^1 (\log_e k - A)^2 dQ(k) \tag{1.144}$$

in which A and B^2 represent the mean value and the variance of the variable $\log_e K$, which by Eq. (1.142) is normally distributed, so K is lognormal.

As in many of the above derivations, the final result Eq. (1.132) is far from being obvious. An explanation may be sought as follows. The recurrence equation Eq. (1.141) is a kind of a Stieltjes integral for two distributon functions, [71]. In fact, consider two independent random variables U and V and their sum $W = U + V$, (cf. Example 1.21). Then

$$P[W \leqslant w] = F_W(w) = \int_{-\infty}^{\infty} F_U(w - x) dF_V(x) \tag{1.145}$$

If $S(x)$ and $T(x, t)$ are considered to be two such distribution functions, that is $T(x, 0)$ is the distribution function of the variable U and $S(x)$ of V, then by Eqs. (1.141) and (1.145), $T(x, 1)$ is the distribution function of the variable $U + V_1$, $T(x, 2)$ is the distribution function of the variable $U + V_1 + V_2$ and $T(x, n)$ is the distributon function of the variable $U + V_1 + V_2 + \cdots + V_n$. Introducing the mean values and the variances of the individual variables and the sum divided by n (n tends to a large number),

$$A_k = A = E[V_k] \quad \text{and} \quad A_n/n = E[U + V_1 + V_2 + \cdots + V_n]/n = E[U]/n + A \to A$$

and

$$B_k^2 = B^2 = \mathrm{Var}[V_k] \quad \text{and} \quad B_n^2/n = \mathrm{Var}[U + V_1 + V_2 + \cdots + V_n]/n$$
$$= \mathrm{Var}[U]/n + B^2 \to B^2$$

where the A and B are given by the above integrals Eq. (1.143). By selecting the arbitrary number δ in Lyapunov's condition Eq. (1.125) equal to one, it suffices to show that

$$E[/V_k - A_k/^3] \leqslant \alpha < \infty$$

where α is a positive bounded number. From the condition (c) it follows that

$$E[|V_k - A_k|^3] = \int_{-\infty}^{0} |x - A|^3 dS(x) \leqslant \int_{-\infty}^{0} |u|^3 dS(u) = \alpha < \infty$$

so

$$\lim_{n \to \infty} \frac{1}{B_n^3} \sum_{k=1}^{n} E[|X_k - a_k|^3] \leqslant \frac{n\alpha}{(\mathrm{Var}[U] + n\,\mathrm{Var}[V])^{3/2}} \approx \frac{1}{n^{1/2}} \to 0$$

the conditions for Lyapunov's theorem are fulfilled, and it has been shown that $T(x, t)$ has the distribution Eq. (1.142).

The Kolmogorov law of fragmentation has proved very useful in many different and often unrelated problems. For instance, the law of fragmentation has been used for the study of stress release in earthquake fracture zones, and in soil mechanics studies, [129], [131].

1.2.12 Distributions of Extreme Values

In Examples 1.9 and 1.18, the probability distribution of the largest value of a collection of sample values was obtained, Eq. (1.37). The distributions of extreme values from a population or series of measurements or observation of certain physical quantities show up in many kinds of applications, dealing with the assessment or evaluation of the maxima of these quantites. Maximum annual wind speeds at a given location, maximum earthquake magnitudes to be expected in certain areas, the lowest values of strength to be expected in a material, to name but a few types of problems that have to be addressed through the probability distributions of extreme values of the quantities involved. Due to the importance of this particular field of application of probability theory to engineers and other scientists, a brief overview of the distributions of extreme values will be presented, following the classical approach, [57], [73], [87], [176] and [239].

By Eq. (1.67) it was shown that the probability distribution of the largest or extreme value Y_n from a collection of n random variables X_i, $i = 1, 2, \ldots, n$ that are independent and identically distributed with the parent distribution $F(x)$, is found by the nth power of the parent distribution, that is

$$y = \max\{x_1, x_2, \ldots, x_n\}$$

$$P[Y_n \leqslant y] = Q_n(y) = (F(y))^n \tag{1.68}$$

Under certain assumptions, it can be shown that as n grows very large, $n \to \infty$, $Q_n(y)$ approaches asymptotically a distribution that only depends on few characteristic features of the parent distribution $F(x)$. An asymptotic extreme value distribution, therefore, has the common property that its mean value and variance may depend on n but its form will be independent of n.

It is customary to introduce the so-called attraction coefficients α_n and β_n to facilitate the above implied convergence such that

$$(F(y))^n = Q_n\left(\frac{y - \alpha_n}{\beta_n}\right) \tag{1.146}$$

$$\lim_{n \to \infty} (F(y))^n = Q(y) \tag{1.147}$$

The first equation Eq. (1.146) can be shown to have, in general, three different solutions or types of limiting distribution functions for large values of n that are governed by some simple fundamental characteristics of the parent distribution $F(x)$. (1) If the random variable X is unlimited and $(1 - F(x))$ is asymptotically equivalent to the exponential function for large x, an asymptotic extreme value distribution of the first kind is obtained. (2) If the random variable X has a lower limit such that $(X - \alpha_1)$ can only have positive values, and $(1 - F(x))$ is asymptotically equivalent to cx^{-k} for large x, an asymptotic extreme value distribution of the second kind is obtained. (3) The asymptotic extreme value distribution of the third kind is obtained when $(X - \alpha_1)$ can only assume negative values and $(1 - F(x))$ is asymptotically equivalent to $c(\alpha_1 - x)^k$ for values of x close to the limit α_1. This surprising result, that regardless of the diverse parent distributions that can be encountered in various applications, only three distinct extreme value distributions emerge from the limiting process, is truly remarkable.

Table 1.2 *Properties of asymptotic extreme value distributions (After Dyrbye et al. [57]), reproduced by permission of the authors*

Type	$Q_i(x)$	α_n	β_n	$\mu_n - \alpha$	σ_n^2
I	$\exp[-e^{-x}]$	$\alpha_1 + \beta_1 \log_e n$	β_1	$\beta_n \gamma^*$	$\beta_n^2 \pi^2/6$
II	$\exp[-x^{-k}]$	α_1	$\beta_1 n^{1/k}$	$\beta_n \Gamma(1-1/k)$	$\beta_n^2[\Gamma(1-2/k)$ $-\Gamma^2(1-1/k)]$
III	$\exp[-(-x)^k]\alpha_1$	$\beta_1 n^{-1/k}$	$-\beta_n \Gamma(1-1/k)$	$\beta_n^2[\Gamma(1+2/k)$ $-\Gamma^2(1+1/k)]$	

In Table 1.2, the main properties of the above limiting distribution types are shown with the attraction coefficients as they turn out during the limiting process. For the first distribution, the attraction coefficient α_n, also called the localization parameter, along with the expected value, increases proportionally with $\log_e n$, whereas the spread is unaffected. For the second distribution, the localization parameter and the expected value are unaffected by larger n but the spread, i.e. the variance, increases proportionally with $n^{1/k}$. As for the third distribution, the expected value is also unaffected, whereas the spread decreases proportionally with $n^{-1/k}$.

The three extreme value distributions that have thus been established have the following properties. For the sake of convenience, the distributions for large positive values (maxima) are presented. The minima are easily obtained by changing the variables as follows,

$$\min\{x_i\} = -\max\{-x_i\} \tag{1.148}$$

The extreme value distributions for the minima are thus completely analogous. The first distribution is either called the Gumbel distribution or Fischer–Tippet type I, after the statisticians who played a major part in the development of the theory of extreme values, and has the form

$$Q(y) = e^{-e^{-(y-\alpha)/\beta}}, \qquad -\infty < y < \infty \tag{1.149}$$

This distribution has the mean value and variance

$$E[Y] = \alpha + \beta\gamma \quad \text{and} \quad \text{Var}[Y] = \beta^2\pi^2/6 \tag{1.150}$$

where $\gamma = 0.5772\ldots$ is an irrational number called Euler's constant. The Gumbel distribution is widely applicable in problems when dealing with extreme values that can have both negative and positive values. However, it can also be applied for variables that are positive only, as its negative tail is very weak. It has been used, for example, to predict maximum wind speeds and maximum earthquake magnitudes. The second distribution is often called the Fischer–Tippet Type II or the Frêchet distribution, after the French statistician and mathematician, and has the form

$$q(y) = e^{-[(y-\alpha)/\beta]^{-k}}, \qquad \alpha < y < \infty, \qquad \alpha, \beta \geqslant 0, \quad k > 0 \tag{1.151}$$

Usually the lower limit of y is set at 0. The mean value and variance are found to be

$$\left.\begin{aligned} &E[Y] = \alpha + \beta\Gamma[1-1/k], \quad k>1 \\ \text{and} \qquad &\\ &\text{Var}[Y] = \beta^2\{\Gamma[1-2/k] - \Gamma^2[1-1/k]\}, \quad k>2 \end{aligned}\right\} \tag{1.152}$$

The Frêchet distribution is convenient when the extreme values are positive only. It has found use in wind engineering to predict maximum wind speeds to name an example, [4]. Often the coefficient α is disregarded so as to have a two-parameter distribution only. In this form the distribution is written as follows:

$$Q(y) = \mathrm{e}^{-(\beta/y)^k}, \qquad 0 \leqslant y < \infty, \quad \beta \geqslant 0, \quad k > 0 \tag{1.153}$$

The expectation and the coefficient of variation are then given as follows:

$$E[Y] = \beta\Gamma\left(1 - \frac{1}{k}\right), \quad k > 1 \qquad \text{and} \qquad V_Y = \left(\frac{\Gamma(1 - 2/k)}{\Gamma^2(1 - 1/k)} - 1\right)^{1/2}, \quad k > 2 \tag{1.154}$$

The third distribution, which is often called the Fischer–Tippet Type III or the Weibull distribution, after the Swedish engineer and statistician, has the form

$$Q(y) = \mathrm{e}^{-[(\alpha - y)/\beta]^k}, \qquad -\infty \leqslant y \leqslant \alpha, \quad \beta, k > 0, \quad \alpha > -\infty \tag{1.155}$$

but is more often used for the negative range of values in the form

$$Q(y) = 1 - \mathrm{e}^{-[(y - \alpha)/\beta]^k}, \qquad \alpha \leqslant y < \infty, \quad \beta, k > 0, \quad \alpha > -\infty \tag{1.156}$$

As in the case of the Frêchet distribution, the lower limit of y is usually set at 0. The mean value and variance of the Weibull distribution are found to be

$$E[Y] = \alpha + \beta\Gamma[1 + 1/k] \qquad \text{and} \qquad \mathrm{Var}[Y] = \beta^2\{\Gamma[1 + 2/k] - \Gamma^2[1 + 1/k]\} \tag{1.157}$$

The Weibull distribution has found use in material science dealing with material strength in its form for distributions of minima, but can also be used in applications involving maxima. It is interesting to note that type III includes the exponential distribution for $k = 1$. As above, the same argument applies for disregarding the fact that α has a two-parameter distribution.

The above presentation of the extreme value distributions has been both short and superficial. It is only included for reference, that is, for the reader to have an easy access to the main formulas. For further details, the reader should consult Gumbel, [73], or any other modern text on extreme value statistics, [87], [176]. Some interesting applications and an overview of the most recent development in the theory of extreme values are found in the proceedings of a conference commemorating the centennial anniversary of Emil Gumbel's birthday, [68].

Example 1.30 The type I distribution has been used in wind engineering to estimate maximum annual wind speeds. Based on anemometer records from England and Scotland, A. G. Davenport suggested the following distribution for maximum annual wind speeds, [5]:

$$P[V_a \leqslant v] = F_V(v) = \exp\{-\mathrm{e}^{-a(v - U)}\} \tag{1.158}$$

The parameters $1/a$ and U are examples of a kind of attraction coefficient, where a is governed by the dispersion of the data and U is the mode of the data. For extreme mean hourly wind speeds measured at the standard reference height 10 m, contours of $1/a$ and U have been determined for the British Isles. Simultaneous values for the northernmost contour (North Scotland) are

$$1/a = 9.0\,\text{m.p.h.} \qquad \text{and} \qquad U = 90\,\text{m.p.h.} \quad (40.2\,\text{m/s})$$

Determine the value of the design wind speed defined as the extreme value with a frequency of 0.02, that is, on the average, the design value will be encountered every 50 years.

Solution For large measurement series, it is often convenient to define the return period of a certain event, for instance the event that the maximum annual wind speed exceeds a certain value, say, V_c m.p.h. Let $T(c)$ be the average number of years that passes between the event of having at least V_c m.p.h. registered as the maximum annual wind speed, which is described as the result $V \geqslant V_c$. Take the measurement series

$$\ldots, V_i, \ldots, V_j, \ldots, V_k, \ldots,$$

where during the years i, j and k, the maximum wind speed was measured equal to or above V_c. The differences $k - j$ and $j - i$, which indicate the number of years between the events $V \geqslant V_c$, can be interpreted as a random variable N. The average number of years that passes between these events $T(c)$, often called the return period of the event $V \geqslant V_c$, is then defined as the expectation

$$T(c) = E[N]$$

Heuristically $1/T(c)$ is the frequency of the event $V \geqslant V_c$. Hence the probability can be evaluated as

$$P[V \leqslant V_c] = F(V_c) = c = 1 - 1/T(c) \tag{1.159}$$

Now consider the T-year maximum wind speed, that is, having a return period of T years. By Eqs. (1.158) and (1.159), the T-year maximum wind speed $V(T)$ is given by

$$V(T) = U - 1/a[\log_e(-\log_e(1 - 1/T)] = U - 1/a[\log_e(-\log_e p)] \tag{1.160}$$

where $1 - p$ is the probability or frequency $1/T(p)$. Usually, the above equation is evaluated by plotting the data on a special logarithmic paper so that Eq. (1.160) represents a straight line. The line cuts the vertical axis at the value U and has the inclination $-1/a$. If the set reference wind speed V_c is large enough, the return period $T(c)$ is comparatively large also, $(T(c) > 10)$, and the logarithm in Eq. (1.160) can be simplified applying the relation $e = \lim_{x-\infty}(1 + x)^{1/x}$, whereby

$$V(T) = U + (\log_e T)/a \tag{1.161}$$

The 50-year design wind in North Scotland is therefore given by

$$V_{50} = 90 + (\log_e 50) \times 9 = 125.2\,\text{m.p.h.} \quad (56.0\,\text{m/s})$$

Example 1.31 [240] In the North Iceland seismic zone, the following known historic earthquakes have taken place, believed to have been of magnitude 6 or above, Table 1.3. In old Icelandic annals and chronicles, the damage caused by these earthquakes

Table 1.3 *An earthquake catalogue for North Iceland*

Year	Epicentre	Magnitude	Distance to Akureyri (km)
1755	66.5°N 18.0°W	7.3	88.9
1838	66.5°N 19.0°W	6.5	97.6
1872	66.1°N 17.4°W	6.5	54.6
1885	66.2°N 17.0°W	6.3	74.5
1910	66.5°N 17.0°W	7.1	101.6
1913	66.5°N 19.5°W	6.0	108.9
1921	67.0°N 18.0°W	6.3	144.4
1934	66.0°N 18.5°W	6.3	37.9
1963	66.3°N 19.6°W	7.0	95.0
1976	66.3°N 16.6°W	6.5	95.0

is described and based on this information, the origin and magnitude of the earthquakes has been estimated, [211]. The three last earthquakes in the table have all been subjected to a scientific study based on instrumental records.

Akureyri is the main commercial centre in the zone with the geographical coordinates (65.7°N, 18.1°W). The expected maximum horizontal surface acceleration in Akureyri due to each of the above earthquakes can be calculated using the following relation:

$$A = 0.0955\exp(0.57M)\exp(-0.00587R)/R$$

in which A is the maximum acceleration as a fraction of gravity g, M is the Richter magnitude and the distance R in kilometres between the hypocentre of each earthquake to Akureyri is given by

$$R = \sqrt{D^2 + 7.3^2}$$

where D is the surface distance to the epicentre in kilometres. The surface distance between two geographical locations with the longitudes and latitudes (L_1, Φ_1) and $L_2, \Phi_2)$ is given by

$$D = 111\sqrt{(L_2 - L_1)^2 + \left((\Phi_2 - \Phi_1)\cos\left(\frac{L_2 - L_1}{2}\right)\right)^2}$$

Assuming that the maximum acceleration experienced in Akureyri for the above series of earthquakes is a Gumbel distribution random variable, find the main parameters of the distribution and the acceleration values with a return period of 50, 100 and 500 years.

Solution Using the above equation for the maximum surface acceleration, the acceleration values can be treated as the extreme distribution variable X. If the sample values are ordered in a series of decreasing values with the largest value assigned the number n, say, 250, the second largest value assigned the number 249 etc., the probability can be estimated as follows:

$$F_X(x_n) = P[X \leqslant x_n] \approx (n - \tfrac{1}{2})/N, \text{ where } N = 250$$

Table 1.4 *The maximum acceleration in Akureyri and estimated probabilities*

Year	Magnitude M	Accel. (% g)	$pn = (n - \frac{1}{2})/250$	$-\log_e(-\log_e p_n)$
1934	6.3	7.148	0.998	6.214
1872	6.5	5.104	0.994	5.113
1755	7.3	4.067	0.990	4.600
1963	7.0	3.098	0.986	4.262
1885	6.3	2.980	0.982	4.008
1910	7.1	2.948	0.978	3.806
1976	6.5	2.324	0.974	3.637
1838	6.5	2.233	0.970	3.491
1913	6.0	1.410	0.970	3.364
1921	6.3	1.026	0.966	3.065

Using the relation, Eq. (1.158),

$$A(T) = M_A - 1/a[\log_e(-\log_e(1 - 1/T)] = M_A - 1/a[\log_e(-\log_e p)]$$

From Table 1.4, the acceleration values (column 3) and the estimated log log probabilities (column 5) can be used to obtain the coefficients M_A and $1/a$ by linear regression analysis, the result being

$$M_A = -4.896 \quad \text{and} \quad 1/a = -1.946$$

The 50, 100 and 500 year acceleration values for Akureyri are then easily computed, that is,

$$A(500) = 0.07196g, A(100) = 0.04056g, A(50) = 0.02697g$$

Example 1.32 [240] At a proposed building site, the following 20 years ordered series of maximum annual wind speeds is available, which have been measured at the standard reference height 10 metres, and representing the 10 minute average value in m/s:

25.1	26.0	26.2	28.0	28.5	28.8	29.1	29.3	30.1	30.4
30.5	30.8	30.9	31.4	31.6	33.2	36.5	37.4	38.5	39.7

Find the 50, 100 and 500 year extreme values for the 10 minute wind speeds using both the Gumbel, type I and Frêchet, type II distributions.

Solution In this case, the mean value and the variance or the standard deviation of the sample values can be computed and taken as an estimate of the true values. By comparison with the same parameters of the two extreme value distributions, the attraction coefficients α and β of both distributions can then be obtained. First calculate the estimated mean value and the variance:

$$E[X] = \sum x_n/n = 31.1 \text{ m/s}$$

$$\text{Var}[X] = \sum (x_n - 31.1)^2/(n-1) = (4.099)^2 \quad (\text{m/s})^2$$

Secondly, a comparison with the Gumbel distribution parameters yields:

$$E[V] = \alpha + \beta \times 0.5772 = 31.1$$
$$\mathrm{Var}[V] = \beta^2\pi^2/6 = (4.099)^2$$

whereby

$$\alpha = 25.255 \qquad \text{and} \qquad \beta = 3.196$$

The 50, 100 and 500 year wind speeds are then given by Eq. (1.160), that is,

$$V(T) = \beta - \alpha(\log_e(-\log_e(1 - 1/T))$$

or

$$V(50) = 41.73\,\text{m/s}, \quad V(100) = 43.96\,\text{m/s}, \quad V(500) = 49.11\,\text{m/s}$$

Through a comparison with the Frêchet and type II distribution parameters,

$$F(V(T)) = \exp(-(\beta/V(T))^k = 1 - 1/T = p$$

or

$$V(T) = \beta/(-\log_e p)^{1/k}$$

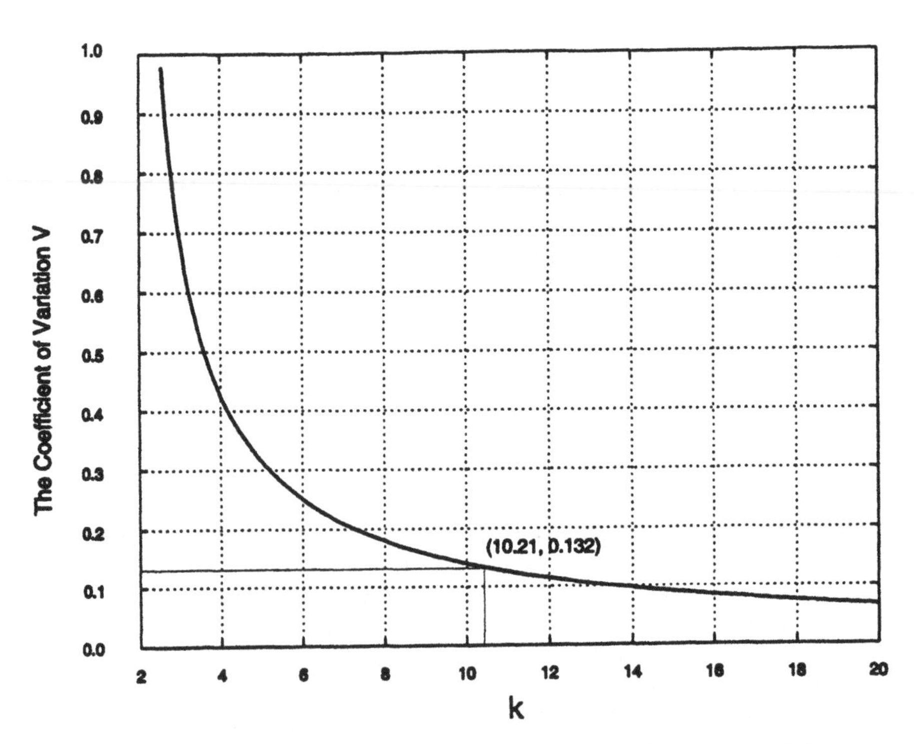

Fig. 1.17 *The coefficient of variation:* $(\Gamma(1-2/k)/\Gamma^2(1-1/k)-1)^{(1/2)}$

with the mean and coefficient of variation given by Eq. (1.154):

$$E[V] = \beta\Gamma(1 - 1/k) = 31.1\ \text{m/s}$$

and

$$V_Y = \left(\frac{\Gamma(1 - 2/k)}{\Gamma^2(1 - 1/k)} - 1\right)^{1/2} = \frac{\sigma_Y}{\mu_Y} = \frac{4.099}{31.1} = 0.132$$

In order to make the determination of the parameter k easy, a graph of the coefficient of variation and the Gamma function is shown in Figs. 1.17 and 1.18. From Fig. 1.17, k is found to be close to 10.2, whereby $\Gamma(1 - 1/k) = 1.065$ and $\beta = 29.2$. Therefore,

$$V(50) = 41.91\ \text{m/s}, \quad V(100) = 44.71\ \text{m/s}, \quad V(500) = 51.91\ \text{m/s}$$

The first value is very close to the value given before by the Gumbel distribution, whereas the 100 and 500 year values are slightly higher. Generally, the Frêchet distribution can be expected to give a more accurate description of the extreme values of maximum annual wind speeds, especially since it only accepts positive values as should be.

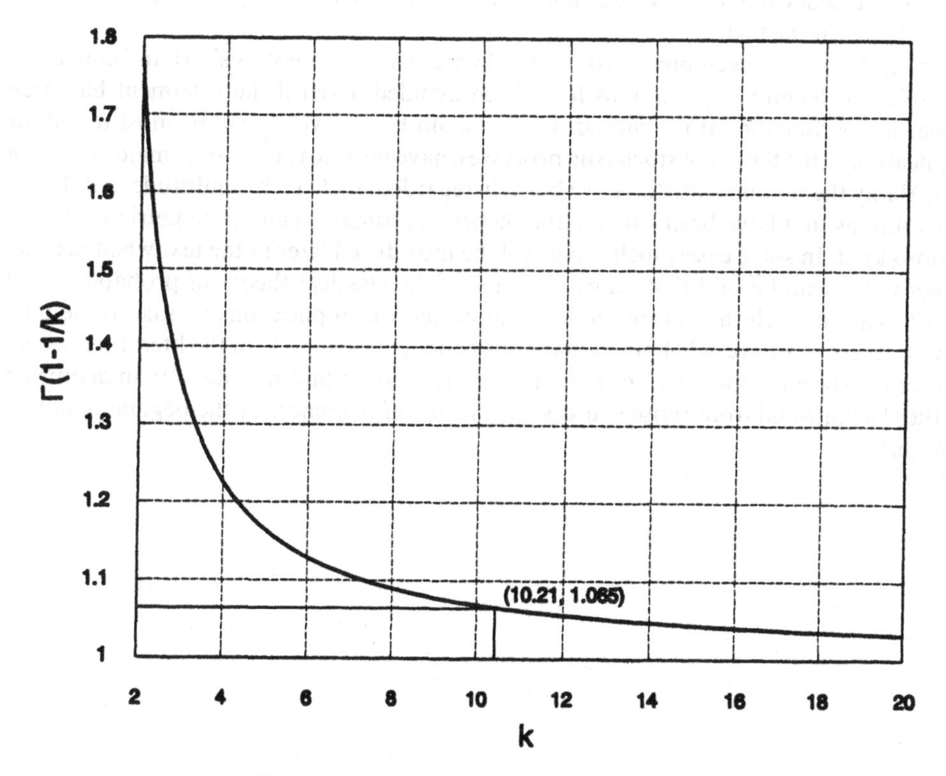

Fig. 1.18 *The gamma function,* $\Gamma(1 - 1/k)$

1.2.13 Concluding Remarks

In this chapter, the basics of probability theory have been covered to set a proper background for the material to be treated in the following chapters. The fundamental probability concept was introduced on the basis of Laplace's favourable fraction followed by the axioms of Kolmogorov. Random events and the difference between mutually exhaustive events and interrelated events were briefly explained. Conditional probabilities and Bayesian procedures were introduced, leading to the definition of random variables and their distributions. After covering the probability distributions and their parameters in some detail, marginal and conditional distributions were introduced to be able to define the concept of independence and interrelationship of random variables. The various approximations and inequalities that are particularly useful for the material to be treated in later chapters were introduced with a few practical examples, showing how approximate probabilities can be evaluated and their limited range determined. Multidimensional variables and their distributions were sketchily treated, and approximate distributions for the complicated functional relationships between random variables presented. The characteristic functions of random variables were introduced and several examples given because of the usefulness of this probabilistic concept for the treatment of stochastic processes and their parameters. The law of large numbers and the central limit theorem were subjected to a full discussion as they play a dominant role in the treatment of certain stochastic processes. Finally, the theory of extreme values and their probability distributions was discussed and a few examples given that have a bearing on the applications treated later in the text.

The text, as it has been presented in this chapter has been kept as short and concise as possible and lengthy explanations have been avoided. Even if the treatment has been somewhat perfunctory, it is believed that the main topics, which will be most useful for application in the theory of stochastic processes, have been covered. Some major topics of probability theory and statistics have been disregarded, such as the multitude of different distributions that have been studied, the theory of estimation and parametric studies of various kind. In some cases, such topics will be introduced later in the text when needed. Otherwise, it can be said that many elements of the classical theory of probability and statistics are not relevant to the subject matter and the applications treated in the text. Finally, there is an overwhelming number of good texts available in the literature, where more such specific topics are covered. The reader should find no difficulty in acquiring further background from those sources (see the list of references at the beginning of this chapter).

2 The Basic Theory of Stochastic Processes

2.1 INTRODUCTION

The mathematical concept of a stochastic or random process has proved to be a very successful mathematical tool for description of various physical and natural phenomena. Brain waves recorded on an electroscope, radio transmission noise and market fluctuations on the stock exchange, are but few examples of stochastic processes. In Fig. 2.1, three stochastic processes of interest, pertaining to engineering and geophysical applications, are shown.

The mathematical formulation of stochastic processes was already well established in the years between the two world wars. In the field of engineering, the electrical engineers quickly found wide application for this new mathematical and statistical subject in control theory and electronics. The work by S.O. Rice in the early forties is especially noteworthy, [10]. Studies over rocket hull vibrations and other random mechanical phenomena followed in the 1950s through the pioneer work of S.H. Crandall, W. Mark, D.C. Karnopp, T. Caughey and many others, [32], [43], [44], [103]. By the early sixties, random vibration in mechanical systems was a quite well covered field. In civil engineering, which by tradition is the most conservative of all engineering branches, people like A.V. Davenport, G.W. Housner and others showed in the early sixties that loads such as wind and earthquakes could only be successfully treated as random or stochastic processes, [51], [52], [70], [85].

What is then a stochastic process? A mathematical formulation may be given as follows. Consider a family of random variables $X(z)$, indexed by a parameter z varying in an index set $\{z \in Z\}$. For any value of the deterministic parameter z, the value of $X(z)$ is only described by the probability laws that govern the process. There are several ways to classify a stochastic process according to the nature of the parameter set and the random variables involved, [94]. The state space of the process is the set of possible values of an individual $X(z)$. The stochastic process has a *discrete state space* if it is countable, that is, contains a finite or an enumerable infinity of points, in other words, the variables $X(z)$ are discrete. Otherwise, the process is said to have a *continuous state space*, that is, the variables $X(z)$ are continuous random variables. If the index set Z is countable, z assumes the values of a discrete variable, and the process is said to be a *discrete parameter process.* Each change of the variable may be called an event and describes what is happening in the process at that time. Otherwise, the stochastic process is said to be a *continuous parameter process*, that is, z is a continuous variable. If the state space is discrete, then the random variables X_z are discrete and the process is often called a random parameter chain. It can then either form a discrete parameter chain or a continuous parameter chain. Of all these

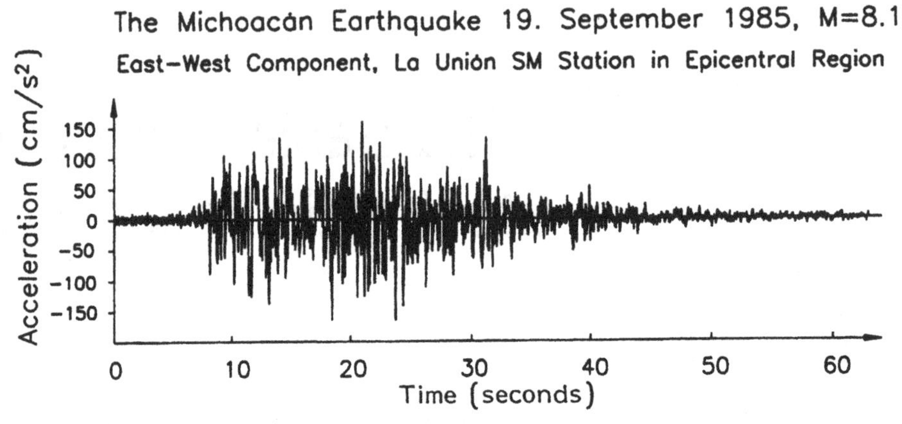

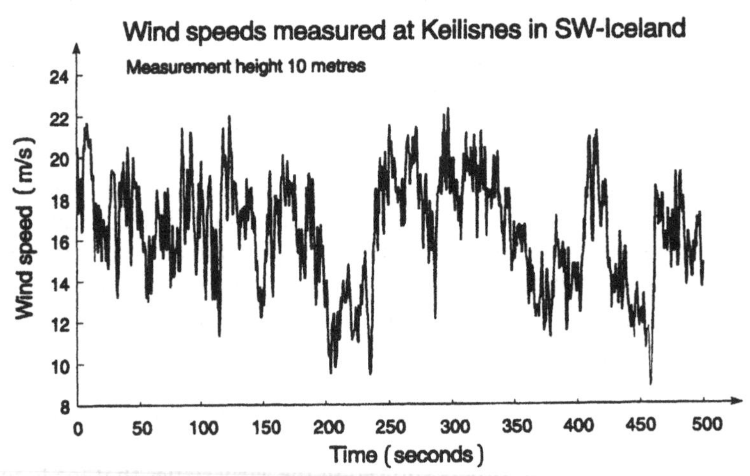

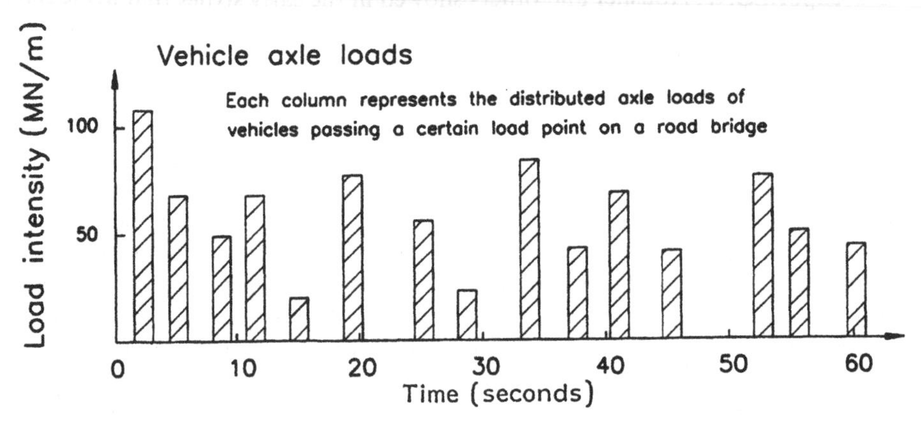

Fig. 2.1 *Examples of stochastic processes in geophysics and engineering*

Table 2.1 *Classification of stochastic processes, [94], reproduced by permission of Academic Press*

State space	Continuous	Random sequence Time series	Stochastic process Random function
	Discrete	Discrete parameter chain	Continuous parameter chain
		Discrete	Continuous
		Parameter set	

combinations, mostly continuous state space processes will be considered in the following. In Table 2.1, the above classification of stochastic processes is shown schematically following Jazwinski, [94].

Between authors, the classification of stochastic processes may vary considerably and many examples of the different interpretations of stochastic processes may be gaven. The following examples of stochastic processes may clarify the situation. For instance, noise registered by a radio antenna is an example of a stochastic process that varies continuously with time as the indexing parameter. It is therefore a random function of time. The discharge or the water flow in a river, measured at a certain point, is another example of a continuous process in time. Discrete processes can consist of series of so-called point events. They can be described by either looking at the random number of such point events, which have occurred up to an arbitrary time instant t, or by studying the random arrival time t_i of each event. For instance, the random number of child births in a population at any given date is a point event, where the number of individuals born alive rather than their arrival times are of interest. Such processes, where the number of random events that have occurred at each time rather than their arrival times is being studied, are therefore often called birth processes. On the other hand, stochastic processes such as earthquakes, heavy storms, gold strikes etc. are examples of arrival time processes. In general, discrete processes of both types, which describe the positions of series of random events in time, are usually called point processes. Finally, the basic process describing Random Walk (cf. Exercise 2.1) is an example of a discrete chain process.

Another clear classification of stochastic processes relates to the indexing parameter representing either a variation in space, that is, describing the position vector in a defined space, or a variation in time as has already been discussed. Processes of the former type are often called space series and are represented by regionalized variables that occur in space. For instance, the spatial distribution of gold strikes, where the geographical coordinates (L, Φ) of each strike are treated as random variables, is an example of a space series type of process that is sometimes called the Klondike process, [129]. When the indexing parameter relates to the time t rather than a position in space, the stochastic process is often characterized as a random time function $X(t)$ or as time series, $X(t_i), t = t_1, t_2, t_3, \ldots,$ as already mentioned. Random dynamic variables, which are time dependent, constitute a major part of the theory and treatment of stochastic processes. An example of such variables is the position vector $\{X(t), Y(t), Z(t)\}$ of a particle in Brownian motion. The spatial velocity of a turbulent gusty wind, $\{V(t)\} = \{V_x(t), V_y(t), V_z(t)\}$, is an example of a random time function. To sum up, any process, which is developing in time or space in a manner controlled by a probabilistic law, is called a stochastic process. In the following, various fundamental concepts in the mathematical theory of stochastic processes are treated somewhat sketchily and the most important definitions given in order to lay a basis for the following treatment of systems responding to random excitations. The introduction is based on the time series or random function representation of stochastic processes. However, all basic definitions and concepts can easily be adapted to space series as well.

Example 2.1 Random walk Assumes that a person is walking along a straight line, taking a step of size $\Delta > 0$, either forward or backward according to the toss of a coin. At each toss, a step forward is taken if a head (h) comes up but backwards if a tail comes up (t). If the random walk begins at position 0, what will be the position of the person after n tosses, just before taking the nth step. First define the probabilities of heads or tails coming up. Clearly, for a fair coin they are

$$P[Z=h]=p, \qquad P[Z=t]=q, \qquad p=1-q=\tfrac{1}{2}$$

where Z is a discrete random variable that can assume only these two values. Now define the random variable $W_n(Z)$ which is the random step taken after each toss of the coin, that is,

$$W_n(Z)=\begin{cases} +\Delta, & Z=h, \\ -\Delta, & Z=t, \end{cases} \quad n=1,2,\ldots \quad \text{and} \quad W_0(z)\equiv 0$$

so that

$$P[W_n(Z)=+\Delta]=p, \quad P[W_n(Z)=-\Delta]=q, \quad n=1,2,3,\ldots \quad \text{and} \quad P[W_0(z)=0]=1$$

Let the random variable X_n denote the position of the person just before taking the nth step. Then,

$$X_n=\sum_{i=0}^{n-1} W_i, \qquad n=1,2,3,\ldots, \quad X_0\equiv 0$$

Obviously $\{X_n, n=0,1,2,3,\ldots\}$ is a *discrete parameter chain*, since for each n, the position vector $X_n(\cdot)$ is a discrete random variable given by the sum of the random variables $W_n(Z)$. A possible sequence of the random steps taken up to $n=18$ is shown in Fig. 2.2, in which the variables are denoted by lower case letters to indicate that they represent a certain outcome or a realization of the process. The position function is displayed as a staircase-type of a function. It is interesting to note that the probability law governing the process, given by the discrete random variable Z, does not enter directly into the sum defining X_n.

Now assume that the probability law of the random variables W_n is different. For instance, the variables W_n ($n>0$) could be continuous IID random variables with a common probability density $p(w)$. In that case, the process generated by the sum,

$$X_n=\sum_{i=0}^{n-1} W_i, \qquad n=1,2,3,\ldots$$

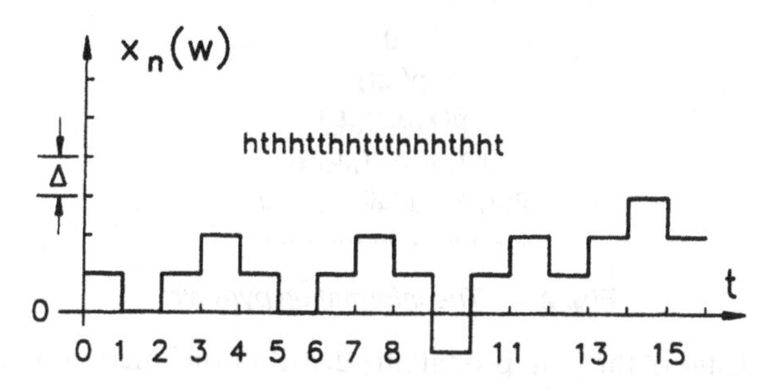

Fig. 2.2 *A one-dimensional random walk*

is a random sequence or a discrete time series if n marks each time instant t_n. For each time instant t_n, the variable X_n is a continuous random variable.

2.2 STATISTICS OF STOCHASTIC PROCESSES

2.2.1 Probability Distributions

Consider a stochastic process $\{X(t), t \in T\}$, which actually describes a family of random variables indexed by a parameter t varying in an index set T. The function $X(t)$ can for instance be the amplitude of a random load acting on a mechanical system. At any particular time instant $t = t_1$, the amplitude of the load takes on a random value which has a certain probability of occurrence, namely (Fig. 2.3),

$$P[X(t_1) \leqslant x] = F_x(x; t_1)$$

or

$$P[x < X(t_1) \leqslant x + \Delta x] = p_x(x; t_1)\Delta x \tag{2.1}$$

If information is sought as to the behaviour of $X(t)$ at two particular time instants t_1

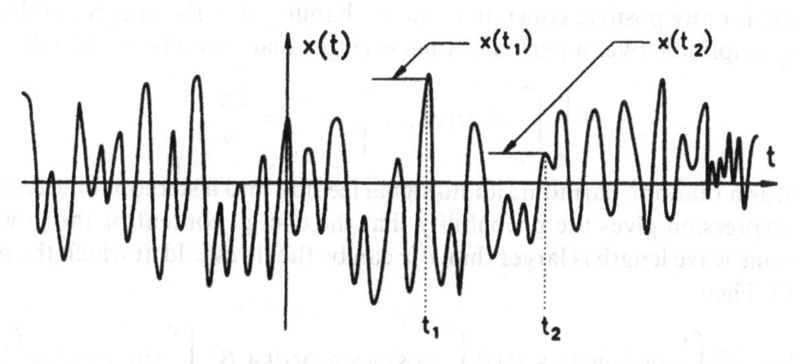

Fig. 2.3 *A physical realization of X (t)*

$$
\begin{gathered}
0 \\
p(x;t) \\
p(x_1,x_2;t_1,t_2) \\
p(x_1,x_2,x_3;t_1,t_2,t_3) \\
p(x_1,x_2,x_3,x_4;t_1,t_2,t_3,t_4) \\
\cdots\cdots\cdots\cdots\cdots\cdots
\end{gathered}
$$

Fig. 2.4 *The information pyramid*

and t_2, some knowledge of the joint probability distribution function (JPDF) is necessary, i.e. of

$$P[(X(t_1) \leqslant x_1) \cap (X(t_2) \leqslant x_2])) = F_{x_1 x_2}(x_1, x_2; t_1, t_2)$$

or

$$P[(x_1 < X(t_1) \leqslant x_1 + \Delta x_1) \cap (x_2 < X(t_2) \leqslant x_2 + \Delta x_2)] = p_{x_1 x_2}(x_1, x_2; t_1, t_2)\, \Delta x_1 \Delta x_2 \tag{2.2}$$

It should be noted that the probabilities, Eqs. (2.1) and (2.2), are in general time dependent. To obtain a full and complete description of the process, a knowledge of at least all joint distributions corresponding to any finite subset of a denumerably infinite set of time instants is necessary, as portrayed by the infinite information pyramid, Fig. 2.4. For practical purposes however, it seems plausible that the process $\{X(t),\ t \in T\}$ can adequately be described by some finite number of parameters $\{t_i\}$ representing the set T. Thus the joint probability function

$$F_{x_1 x_2 \cdot x_n}(x_1, x_2, \cdot x_n; t_{1,2}, \cdot t_n) = P[(X(t_1) \leqslant x_1) \cap (X(t_2) \leqslant x_2) \cap \cdots \cap (X(t_n) \leqslant x_n)] \tag{2.3}$$

would completely characterize the behaviour of the process. In practice, however, it is rarely possible to obtain more than the first few functions.

Example 2.2 Consider the stochastic process $\{X(t),\ t \geqslant 0\}$ defined by

$$X(t) = A \cos \omega t + B \sin \omega t$$

where the frequency ω is a positive constant, and A and B are two independent normally distributed random variables with mean 0 and variance σ^2.

In this case it is possible to find the probability of any assertion concerning the process. For instance, for any positive constant c, the probability that the integral of the squared value of the amplitude over a period of time is larger than c can be evaluated.

$$P\left[\int_0^T X^2(t)\,\mathrm{d}t > c\right], \qquad T = \frac{2\pi}{\omega}$$

Considered as a random, harmonic audio signal feeding into a stereophonic receiver, say, the above expression gives the probability that the energy content or the power in the signal over one wave length is larger than c (c can be the threshold at which the receiver is overloaded). Then,

$$\int_0^T X^2(t)\mathrm{d}t = A^2 \int_0^T \cos^2 \omega t\, \mathrm{d}t + 2AB \int_0^T \cos \omega t \sin \omega t\, \mathrm{d}t + B^2 \int_0^T \sin^2 \omega t\, \mathrm{d}t = \frac{T}{2}(A^2 + B^2)$$

and

$$P\left[\int_0^T X^2(t)\,dt > c\right] = P\left[(A^2+B^2) > \frac{c\omega}{\pi}\right]$$

Recalling, ([34], pp. 140–141), that the sum of r squares of uncorrelated standardized normal variables, i.e. $\Phi(0,1)$, has a chi-square distribution $\chi^2(r)$ (r is the number of degrees of freedom of the distribution), the probability density function of $\chi^2(r)$ is

$$p_{\chi^2(r)}(x) = \begin{cases} \dfrac{1}{2^{r/2}\Gamma(r/2)} x^{(r/2)-1} e^{-x/2} & 0 \leqslant x < \infty \\ 0 & -\infty < x \leqslant 0 \end{cases}$$

where $\Gamma(\alpha)$ is the Gamma function, ($\Gamma(\alpha+1) = \alpha\Gamma(\alpha)$, $\Gamma(1)=1$, $\Gamma(1/2)=\sqrt{\pi}$). Hence for $r=2$,

$$P\left[\int_0^T X^2(t)\,dt > c\right] = \exp\left[-\frac{c\omega}{2\pi\sigma^2}\right] = \exp\left[-\frac{c}{T\sigma^2}\right]$$

The probability of overloading the receiver is thus large for large values of T and σ^2.

Example 2.3 In Fig. 2.5, a simply supported weightless elastic beam with a rectangular cross section bh is loaded with a concentrated force at the mid span,

$$K(t) = A\cos\omega t + B\sin\omega t$$

where the amplitudes A and B are two independent and normally distributed random variables with a zero mean and standard deviation $\sigma = 20$ KN. The circular frequency ω is a real constant and t is the time. Thus the force can be interpreted as the stochastic process $\{K(t),\ t \in T\}$, which is a random function of time. The beam is made of elasto-plastic material, for instance steel, which has a yield limit X, considered to be a random variable following the Rayleigh distribution. The Rayleigh distribution has the following density:

$$p_x(x) = \begin{cases} \dfrac{x}{\tau^2}\exp\left(-\dfrac{x^2}{2\tau^2}\right), & x \geqslant 0 \\ 0 & \text{otherwise} \end{cases}$$

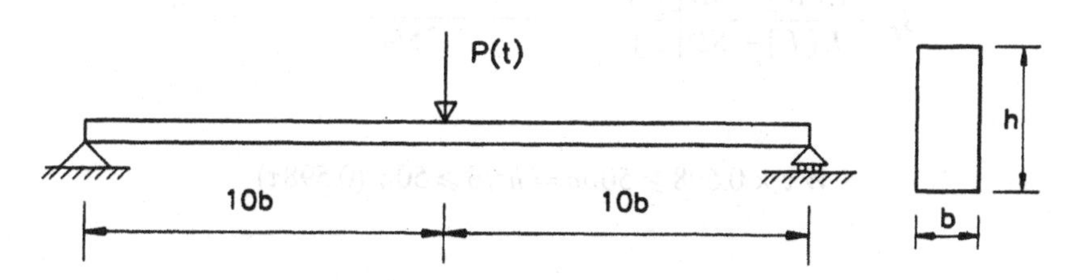

Fig. 2.5 *A simply supported beam under random load*

where $\tau = 200$ MPa is the mode of the distribution, M_3 (see Example 1.13). The Rayleigh distribution is defined the positive (or negative) values only and has the mean value and variance

$$E[X] = \tau \frac{\sqrt{2\pi}}{2} \quad \text{and} \quad \mathrm{Var}[X] = \tau^2 \frac{4-\pi}{2}$$

The beam is to be designed such that the safety factor S_F against yielding of the beam material, defined as the ratio between the mean value of the beam strength less one standard deviation and the corresponding load plus one standard deviation, is not less than two. Determine the necessary height h of the cross section of the beam.

Solution The random strength of the beam is defined as the resisting moment R of the mid-span cross section. Clearly,

$$R = XW$$

where W is the cross section modulus, $W = bh^2/6$. The mean value and the variance of the resisting moment are given through the corresponding values of the Rayleigh distribution, that is,

$$E[R] = W\tau \frac{\sqrt{2\pi}}{2} \quad \text{and} \quad \mathrm{Var}[R] = W^2\tau^2 \frac{4-\pi}{2}$$

The corresponding random load is the moment caused by the external force at the mid-span, which is given by

$$L(t) = K(t) \times 5b$$

The mean value and the variance of the load have to be obtained through the corresponding values of the stochastic process $K(t)$, that is,

$$E[L(t)] = 5b(E[A]\cos\omega t + E[B]\sin\omega t) = 0$$

$$\mathrm{Var}[L(t)] = 25b^2(\mathrm{Var}[A]\cos^2\omega t + \mathrm{Var}[B]\sin^2\omega t) = 25b^2\sigma^2$$

since at each time instant t, the cosine and sine function assume constant values. Now the factor of safety can be formed:

$$S_F = \frac{E[R] - SD[R]}{E[L] + SD[L]} = \frac{W\tau \frac{\sqrt{2\pi}}{2} - W\tau\sqrt{(4-\pi)/2}}{25b\sigma} \geqslant 2$$

and

$$W\tau \times 0.598 \geqslant 50b\sigma \Rightarrow h^2/6 \geqslant 50\sigma/(0.598\tau)$$

or

$$h \geqslant 100\,\mathrm{mm}$$

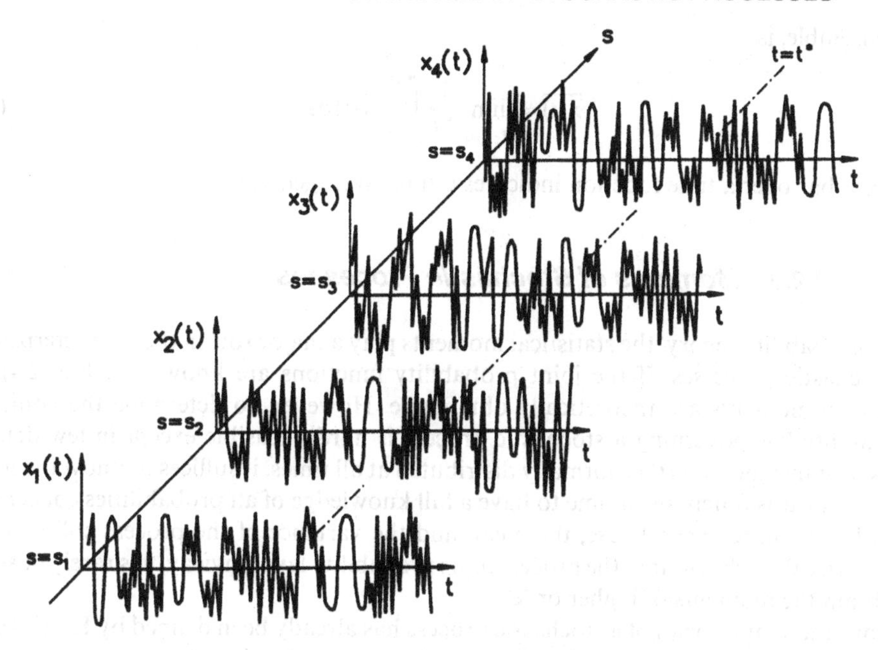

Fig. 2.6 *Sample functions $x_i(t)$*

2.2.2 Ensemble of Sample Functions

It should be noted that the stochastic process $\{X(t), t\in T\}$ can be interpreted in a way as two-dimensional (Fig. 2.6), that is, $\{X(s,t), s\in S, t\in T\}$, [170]. For a fixed value of t, $t=t^*$, $X(s,t^*)$ is a function of the probability space S (i.e. a function on the set S), that is, $X(s)$ is simply a one-dimensional random variable with sample values shown by dots along the broken line $t=t^*$. On the other hand, for a fixed $s=s^*$ in S, $X(s^*,t)$ is simply a function of t, that is, $x(t)$. In the latter case, the now deterministic function $x(t)$ is called a sample function of the process or a physical realization since sample functions can be generated and plotted to form a picture of what the process can look like in reality. Whereas the sample functions $x(t)$ are deterministic functions of time and can be treated as such, $X(s,t^*)$ for a fixed time instant $t=t^*$ is a random variable which in a probabilistic sense describes the value of all such sample functions at that particular time instant.

The sample functions are often referred to as members of an ensemble of functions. The above variable $X(s,t^*)$ is therefore the value of the process taken across the ensemble at $t=t^*$. For instance, the average value of the random variable $X=X(s,t^*)$ is called the ensemble average and is given by

$$E[X(t^*)]=\int_{-\infty}^{\infty} xp(x,t^*)\,dx \tag{2.4}$$

On the other hand, the time average of a single sample function $x(t)$, i.e. any one member of the ensemble, is

$$\overline{x(t)} = \lim_{T \to \infty} \frac{1}{2T} \int_{-T}^{T} x(t)\,\mathrm{d}t \tag{2.5}$$

where a bar over a time function indicates a time-averaged value.

2.2.3 Moments of Stochastic Processes

As in probability theory, the statistical moments play a major role in the characterization of stochastic processes. If the joint probability functions are known (cf. Fig. 2.4), all statistical moments are theoretically obtainable. However, to determine the complete probability law governing a stochastic process is hardly feasible except in few definite cases. For instance if $X(t)$ is normally distributed at all times, it suffices to know the mean and variance as functions of time to have a full knowledge of all probabilities concerning $X(t)$. In the more general case, the mean and the variance of the process still serve to summarize the behaviour of the process or its probability law. Anyway, it is rarely possible to obtain the moments of higher order.

Now, the simple mean of a stochastic process has already been defined by Eq. (2.4), the ensemble average. Simple functional expectations are defined in the same manner. For instance, let $Y(t) = g(X(t))$ be a new stochastic process. By taking notice of Eq. (1.30),

$$E[Y(t)] = \int_{-\infty}^{\infty} g(x)\,p(x;t)\,\mathrm{d}x \tag{2.6}$$

and moments of higher order are defined as

$$E[(X(t))^n] = \int_{-\infty}^{\infty} x^n p(x;t)\,\mathrm{d}x \tag{2.7}$$

The central moments are then defined in the usual manner:

$$E[(X(t) - E[X(t)])^n] = \int_{-\infty}^{\infty} (x - E[X(t)])^n p(x;t)\,\mathrm{d}x \tag{2.8}$$

The mixed second-order moments are of fundamental importance for the description of stochastic processes. Stochastic processes, which possess finite second-order moments, have found wide application in various fields since they are quite amenable for analysis. Therefore, in the following, they will be centre stage in the presentation. Let $\{X(t), t \in T\}$ be a stochastic process with finite second-order moments. Its mean value function $\mu(t)$ is defined for all t and T by

$$\mu(t) = E[X(t)] = \int_{-\infty}^{\infty} x p(x;t)\,\mathrm{d}x \tag{2.9}$$

Now take two time instances t_1 and t_2 in T. The covariance is defined as

$$K(t_1, t_2) = \mathrm{Cov}[X(t_1), X(t_2)] = E[(X(t_1) - \mu(t_1))(X(t_2) - \mu(t_2))] \tag{2.10}$$

Obviously, $K(t,t) = \mathrm{Var}[X(t)]$, the variance of $X(t)$. Another important function in the study of stochastic processes is the correlation function, which is given directly by the mixed second-order moment or

$$R(t_1, t_2) = E[X(t_1)X(t_2)] \tag{2.11}$$

Clearly the covariance function and the correlation function are related (cf. Chapter 1, Eq. [1.74])

$$K(t_1, t_2) = R(t_1, t_2) - \mu(t_1)\mu(t_2) \tag{2.12}$$

When two processes $X(t)$ and $Y(t)$ are studied simultaneously, it is possible to define the cross-correlation functions in the following way:

$$R_{XY}(t_1, t_2) = E[X(t_1)Y(t_2)]$$

and

$$R_{YX}(t_1, t_2) = E[X(t_2)Y(t_1)] \tag{2.13}$$

Thus the cross-correlation functions may give information as to how closely the two processes are correlated at two particular time instants.

2.2.4 Stationary and Evolutionary Processes

If the probability law as defined by Eqs. (2.1) or (2.3) is invariant to a shift of the time axis, the stochastic process is said to be stationary in the strict sense. Intuitively, this means that the random mechanism which generates the process does not change as time progresses. An evolutionary process is the opposite, i.e. a process which is non-stationary. A radio antenna noise is a typical example of a stationary process. It is there at all times, that is, it is has no beginning and no end, even if it may be possible to reduce its effects by proper filtering. An earthquake acceleration on the other hand is a typical example of a non-stationary process, since the earthquake starts at a certain time instant and stops at another. Physically realizable processes are not really stationary, since they have to start and stop at certain times. However, if the operation time of a physical system, which is either generating or responding to the process, is sufficiently long, the process may sometimes be treated as stationary. An example of such an interpretation is the analysis of a turbulent gusty wind. The three-dimensional wind velocity vector may be defined as $\{V_X(t) + X(t),\ Y(t),\ Z(t)\}$ where the x-axis has been chosen as the dominant or the main direction of the blowing wind. The wind velocities are thus interpreted as a unidirectional wind with the velocity $\{V_X(t), 0, 0\}$ superimposed by a turbulent gust vector $\{X(t),\ Y(t),\ Z(t)\}$. Usually it is assumed that the gust vector has zero mean, that is,

$$E[X(t)] = E[Y(t)] = E[Z(t)] = 0$$

Thus the mean velocity of the turbulent gusty wind is the vector $\{V(t), 0, 0\}$, where $V(t)$ is a deterministic function of time with a comparatively slow variation in time ('quasi-static'). For a long period of time, the gust vector $\{X(t),\ Y(t),\ Z(t)\}$ is obviously non-stationary. During extreme conditions, say, during a 10 minutes interval of the annual maximum storm, the gust vector may be interpreted as stationary. At least if the structure failed during those 10 minutes, it will never be known whether it was due to a stationary or a non-stationary disturbance.

For a more precise mathematical formulation of stationarity, first introduce processes with independent increments. Consider the stochastic process $\{X(t),\ t>0\}$ at time instances $0<t_0<t_2\cdots<t_n$. If for $i=0,1,\ldots,n$ and any t_i, the random variables $(X(t_1)-X(t_0)), (X(t_2)-X(t_1)),\ldots, (X(t_n)-X(t_{n-1}))$ are mutually independent, then the process is said to have independent increments. A stochastic process with independent increments is said to have stationary, independent increments, if for all choices of indices and any $t_0<t_1<\cdots<t_n$, the random variables $\{X(t_i)-X(t_{i-1})\}$ and $X(t_i+h)-X(t_{i-1}+h)$ have the same probability distribution irrespective of the value of h, which denotes a shift or translation of the time axis. Consequently, from a knowledge of the probability law governing $X(t)$ and $\{X(t)-X(s)\}$ for any t and s, the joint probability distribution of any n random variables, $X(t_1),\ldots,X(t_n)$ can be obtained.

Example 2.4 A stochastic process $\{X(t), t\in T\}$ is said to be a Markov process, if for $n=1,2,3,\ldots$ and any $t_m\in T$ $(m=0,1,\ldots n;\ t_0<t_1<\cdots<t_n)$, the probability law governing the process can be expressed as follows:

$$P[X_n\leqslant x_n|X_{n-1}=x_{n-1},X_{n-2}=x_{n-2},X_{n-3}=x_{n-3},\ldots,X_0=x_0]$$
$$=P[X_n\leqslant x_n|X_{n-1}=x_{n-1}] \tag{2.14}$$

where $x_0,x_1,\ldots,x_{n-1}$ are any real numbers. Accordingly, the conditional distribution Eq. (2.14) of the random variable $X_n=X(t_n)$ depends only on the last value before t_n, that is, $X_{n-1}=x_{n-1}$ but does not depend on whatever were the values of the previous amplitudes $X(t_{n-2}),\ldots,X(t_0)$. Such a process is sometimes said to be memory-less, since it does not remember its previous values, i.e. they have no influence on the conditional probability distribution Eq. (2.14). Show that stochastic processes, which have independent increment, are indeed Markov processes.

Solution Consider a stochastic process $\{X(t),\ t\geqslant 0\}$ with independent increments. Then

$$X(t_n)=X(t_0)+\sum_{i=1}^{n}(X(t_i)-X(t_{i-1}))$$

or $X(t_n)$ can be written as a sum of $(n+1)$ independent random variables. Further assume that the initial value $X(t_0)=b$ or $P[X(t_0)=b]=1$, where b is a real constant.

The conditional probability Eq. (2.14) is then given by (cf. Eq. (1.15))

$$P[(X_0=b,\ X_1=x_1,\ldots,X_{n-2}=x_{n-2},X_{n-1}=x_{n-1})\cap(X_n\leqslant x_n)] \tag{a}$$
$$=P[X_0=b,X_2-X_1=x_2-x_1,\ldots,X_n-X_{n-1}\leqslant x_n-x_{n-1}] \tag{b}$$
$$=P[X_n\leqslant x_n|X_{n-1}=x_{n-1},X_{n-2}=x_{n-2},\ldots,X_0=b]$$
$$\times P[X_0=b,X_1=x_1,\ldots,X_{n-1}=x_{n-1}] \tag{c}$$
$$=P[X_0=b]P[X_2-X_1)=(x_2-x_1)]x\cdots x\,P[(X_n-X_{n-1})$$
$$=(x_n-x_{n-1})] \tag{b1}$$
$$=P[X_0=b]P[(X_2-X_1)=(x_2-x_1)]x\cdots xP[X_n\leqslant x_n|$$
$$\times(X_{n-1}=x_{n-1},X_{n-2}=x_{n-2},\ldots,X_0=b)] \tag{c1}$$

or through equations (b1) and (c1)

$$= P[X_n \leqslant x_n | X_{n-1} = x_{n-1}, X_{n-2} = x_{n-2}, \ldots, X_0 = b] = P[(X_n - X_{n-1}) \leqslant (x_n - x_{n-1})]$$
$$= P[X_n \leqslant x_n | X_{n-1} = x_{n-1}]$$

which shows that the process is the Markov type. The condition $P[X_0 = b] = 1$ or $X_0 = b$ is necessary in order to keep X independent from $X_i - X_{i-1}$. A word of caution. A Markov process need not have independent increments. Therefore it is a sufficient but not a necessary condition.

Example 2.5 Norbert Wiener was one of the first to formulate precisely the mathematical theory of Brownian motion. Let the stochastic process $\{X(t), t > 0\}$ be the abscissa of a particle in Brownian motion at time t. The particle will be almost continuously in collision with the molecules of the fluid in which the Brownian motion takes place. Following Einstein, the abscissa increment $X(t) - X(s)$ in a time interval (s, t), which is large in comparison with the time between two succeeding molecular collisions, can be taken as the sum of the results from a large number of such collisions and to be only dependent on the interval length $(t - s)$. The increments in two non-overlapping time intervals will be independent, and finally the motion is assumed to be symmetric, that is, $E[X(t) - X(s)] = 0$.

The Wiener process is now defined as follows:

1. $\{X(t), t > 0\}$ has stationary independent increments;
2. $P[X(0) = 0] = 1$;
3. $E[X(t) - X(s)] = E[X(u)] = 0$;
4. For every $t > 0$, $X(t)$ is normally distributed.

To obtain the main parameters of the process, that is, the two first moments, it is convenient to revert to the simple process of random walk, which was discussed in Example 2.1. First let the step taken at each toss of the coin Δx as well as the time interval between tosses Δt tend to zero, that is, $\Delta x \to 0$ and $\Delta t \to 0$, whereby the random walk process approaches a continuous state space, i.e. the position variable becomes a continuous random variable. The step variable W_i has the probability density converted into a continuous form by the delta function, Eq. (1.33), which is

$$p_W(w) = p\delta(w - \Delta x) + q\delta(w + \Delta x)$$

The characteristic function of W is then obtained by Eq. (1.94) as

$$\varphi_W(\omega) = \int_{-\infty}^{\infty} e^{i\omega w} p_W \, dw = pe^{i\omega\Delta x} + qe^{-i\omega\Delta x}$$

and the characteristic function of the position variable of the random walk can be obtained directly as

$$\varphi_X(\omega) = E[e^{i\omega X}] = E[e^{i\omega(W_0 + W_1 + \cdots + W_n)}] = [pe^{i\omega\Delta x} + qe^{-i\omega\Delta x}]^n$$

since the steps are independent. Now let $p = q = \frac{1}{2}$ and note that up to the time t, $n = t/\Delta t$

steps have been taken. Therefore,

$$\varphi_X(\omega) = [(e^{i\omega\Delta x} + e^{-i\omega\Delta x})/2]^{(t/\Delta t)} = \cos(\omega\Delta x)^{(t/\Delta t)}$$

By differentiation, Eq. (1.97), the first two moments of the probability distribution of $X(t)$ are obtained:

$$E[X(t)] = 0 \quad \text{and} \quad \text{Var}[X(t)] = \varphi''_X(\omega) = t\cdot(\Delta x)^2/\Delta t$$

In order to get a sensible result when passing to the limit, $\Delta x \to 0$ and $\Delta t \to 0$, that is, a finite variance $(\Delta x)^2/\Delta t$ must have a limit, say σ^2, whereby the process $X(t)$ has a finite variance in the limit. It is interesting to note that this presumes that the step Δx must be orders of magnitude larger than the time interval Δt in which it occurs, i.e. $\Delta x = O(\Delta t^{1/2})$. With this restriction, the characteristic function becomes

$$\varphi_X(\omega) = (\cos \omega\sigma\sqrt{\Delta t})^{(t/\Delta t)}$$

Expanding the cosine term into Taylor series noting that Δt is a very small quantity,

$$\varphi_X(\omega) = \left(1 - \frac{\omega^2\sigma^2\Delta t}{2!} + O(\Delta t^4)\right)^{(t/\Delta t)} = 1 - \frac{\omega^2\sigma^2 t}{2} + o(\Delta t)$$

Now taking this to be the first terms of the Taylor expansion for the exponential function

$$e^{\Delta t} = 1 + \frac{\Delta t}{1!} + O(\Delta t^2)$$

for very small quantities Δt, the characteristic function becomes in the limit

$$\varphi_X(\omega) = \exp[-\tfrac{1}{2}\omega^2\sigma^2 t]$$

This, however, is the characteristic function of a normal variable with zero mean and variance equal to $\sigma^2 t$, Example 1.22. Thus it has been shown that the random walk process in the limit becomes a Brownian motion as it fulfils all the postulates of the Wiener process.

Brownian motion or the Wiener process is an example of a so-called Gaussian process, (cf. Chapter 6), since it is governed by the normal probability law. Gaussian processes have the important feature that the two first moments, that is, the mean and the covariance, completely define the probability law of the process. In other words, all the lines in the information pyramid, Fig. 2.4, can be obtained from the two first lines. As any linear combination of normally distributed variables yields a new normally distributed variable, the increments $X(t) - X(s)$ of a stationary Gaussian process are also normally distributed with the characteristic function

$$\varphi_{X(t)-X(s)}(\omega) = \exp[-\tfrac{1}{2}\omega^2\sigma^2|t-s|]$$

hence $\text{Var}[X(t) - X(s)] = E[(X(t) - X(s))^2] = \sigma^2|t-s|$. The probability law of the Wiener process is thus completely determined up to a parameter σ^2. This parameter is an empirical characteristic of the process and must be determined from observations. In Brownian motion, σ^2 is the mean square displacement of the particle per unit time. It was shown by Einstein that

$$\sigma^2 = \frac{4RT}{Nf}$$

where R is the gas constant, N the Avogadro's number, T the absolute temperature (^0K), and f the friction coefficient of the fluid.

As shown in the above example, processes with stationary increments will have probability functions that are only dependent upon the time shift or the translation of the time axis. A formal definition of a stationary random process can be given as follows [170]:

A stochastic process $\{X(t), t \in T\}$ is said to be

(a) *stationary of order n*, where n is a given positive integer, if for any n time instances t, $t_1, t_2, \ldots, t_n$ in T, and any time shift or translation $h \in T$ along the time axis, the n-dimensional random vectors amplitudes

$$\{X(t_1), X(t_2), \ldots, X(t_n)\} \quad \text{and} \quad \{X(t_1 + h), X(t_2 + h), \ldots X(t_n + h)\}$$

are identically distributed. The joint probability density $p(x(t), x(t + \tau_1), x(t + \tau_2), \ldots, x(t + \tau_n))$, if it exists, is then independent of t but a function of $\tau_1, \tau_2, \ldots, \tau_n$, where $\tau_i = t_i - t_{i-1}$ is the time difference between two subsequent amplitude values.

(b) *strictly stationary*, if for any integer n it is stationary of order n.

(c) *covariance stationary*, if the process is stationary of order 2 and possesses finite moments of the second order. Then its covariance kernel $K(s, t)$ is a function of the time difference only, that is,

$$K(s, t) = \Gamma(t - s)$$

or

$$\mathrm{Cov}[X(t), X(t + \tau)] = \Gamma(\tau) \tag{2.15}$$

The function $\Gamma(\tau)$ is called the autocovariance function of the process.

Recalling the information pyramid, (Fig. 2.4), the information contained in any one line in the pyramid contains all information above it since any lower-order probability density can be obtained by integration (cf. Eq. (1.46)). Therefore, stationarity of order n implies stationarity of all lower orders. For instance, a covariance stationary process has a stationary first-order probability density, which means that its mean value function must be constant since

$$E[X(t)] = \int_{-\infty}^{\infty} xp(x)\,\mathrm{d}x = \mu_X$$

as $p(x, t)$ is not a function of t. The autocorrelation function of a covariance stationary process, Eq. (2.12), can now be written as

$$R_X(\tau) = E[X(t)X(t + \tau)] = \Gamma_X(\tau) - \mu_X^2 \tag{2.16}$$

and is thus a function of the time difference only as is the autocovariance function.

In practical applications, it is seldom possible to test for more than second-order or covariance stationarity. Covariance stationary processes are comparatively easy to manage and have therefore found wide application in practical problems. If nothing else is stated, in the following, only covariance stationarity will be assumed for a stationary stochastic process.

Example 2.6 Consider the stochastic process $\{X(t), t > 0\}$, where

$$X(t) = \sum_{j=1}^{n} (A_j \cos \omega_j t + B_j \sin \omega_j t)$$

in which n is a positive integer, $\omega_1, \omega_2, \ldots, \omega_n$ are positive constants, and A_j and B_j independent random variables with zero means and equal variances, $\sigma_j^2 = E[A_j^2] = E[B_j^2]$. Find the mean value function, the covariance kernel and the autocorrelation function. Prove that the process is stationary.

Solution

(1) $\mu(t) = \sum_{j=1}^{n} E[A_j] \cos \omega_j t + E[B_j] \sin \omega_j t = 0$

(2) $K(s,t) = \mathrm{Cov}[X(s), X(t)] = E[X(t)X(s)]$

$$= E\left[\left(\sum_{i=1}^{n} A_i \cos \omega_i s + B_i \sin \omega_i s\right)\left(\sum_{j=1}^{n} A_i \cos \omega_i t + B_1 \sin \omega_i t\right)\right]$$

$$= E\left[\sum_i^n \sum_j^n \{A_i A_j \cos \omega_i s \cos \omega_j t + B_i B_j \sin \omega_i s \sin \omega_j t\right.$$

$$\left. + A_i B_j \cos \omega_i s \sin \omega_j t + A_j B_i \cos \omega_j t \sin \omega_i s\}\right]$$

Now $E[A_i A_j] = E[B_i B_j] = \sigma_j^2 \delta_{ij}$ and $E[A_i B_j] = E[A_j B_i] = 0$ for all i and j, so

$$K(s,t) = \sum_{i=1}^{n} E[A_i^2] \cos \omega_i t \cos \omega_i s + E[B_i^2] \sin \omega_i t \sin \omega_i s$$

$$= \sum_i^n \sigma_i^2 (\cos \omega_i t \cos \omega_i s + \sin \omega_i t \sin \omega_i s)$$

$$= \sum_{i=1} \sigma_i^2 \cos \omega_i (t - s) = R(t - s)$$

and the process is at least covariance stationary.

2.2.5 *Stochastic Calculus*

In the following, it will be necessary to deal with derivatives and integrals of stochastic processes. Therefore, the barest essentials of mean square calculus and stochastic differentiation and integration will be presented to prepare the background for the treatment of random excitation and response of physical systems. A discussion of the properties of the correlation functions also requires that the derivative processes, that is, the velocity process and the acceleration process can be defined for any given stochastic process $\{X(t), t\in T\}$. For a more stringent and sophisticated discussion of the mathematical aspects, the reader should for instance consult Jaswinzki, [94], Loeve, [128], Lumley, [135], Sveshnikov, [217], Yaglom, [238]. The argument to follow is mostly based on Jaswinzki's and Lumley's text.

In Section 1.2.10, two types of convergence of random sequences were introduced, that is, convergence in probability, Eq. (1.119), and convergence with probability one, Eq. (1.120). The third possibility, convergence in the mean square sense, which is often easier to handle, may be introduced as follows. Let X_n, $n=1,2,\ldots$, be a sequence of random variables. It is said to *converge in mean square* if for finite second-order moments, $E[|X_n|^2]<\infty$ for all n and $E[|X|^2]<\infty$, the following mean square limit is zero,

$$\lim_{n\to\infty} E[|X-X_n|^2]=0 \tag{2.17}$$

that is, if for any $\varepsilon>0$, there exists a number $N=N(\varepsilon)$ such that

$$E[|X_n-X|^2]<\varepsilon \qquad \text{if } n>N$$

whereby X is the limit in the mean square sense (the m.s. limit). For the sake of simplicity, the operation Eq. (2.17) can also be written using the short hand form, $\text{LIM}\{X_n\}=X$. It is of interest to rank m.s. convergence in comparison with the other two forms, Eqs. (1.119) and (1.120), where as previously discussed, convergence with probability one is stronger than convergence in probability. Using the Chebyshev inequality, Eq. (1.81),

$$P[|X-X_n|\geqslant\varepsilon]\leqslant E[|X-X_n|^2]/\varepsilon^2 \tag{2.18}$$

which shows that for every $\varepsilon>0$, m.s. convergence implies convergence in probability whereas the opposite statement is not necessarily true in certain special cases. Equation (2.18) means that for any $\varepsilon>0$ and $\delta>0$ there exists a number $N_0=N(\varepsilon,\delta)$ such that

$$P[|X-X_n|<\varepsilon]>1-\delta \qquad \text{for } n>N_0$$

which makes it particularly clear why X is the limit of the sequence X_n.

A necessary and sufficient condition for a mean square convergence is also given by the so-called Cauchy criterion

$$\lim_{n,m\to\infty} E[|X_n-X_m|^2]=0 \tag{2.19}$$

that is, the random scalar sequence $\{X_n, n=1,2,\ldots\}$ has a mean square limit if and only if the above Eq. (2.19) is true. Now, from the second-order triangular inequality, Eq. (1.78),

$$E[|X_n-X_m|^2]=E[|(X_n-X)+(X-X_m)|^2]\leqslant 2E[|X_n-X|^2]+2E[|X_m-X|^2]$$

whereby Cauchy's criterion is obviously a necessary condition since the right-hand side tends to zero as $(n,m)\to\infty$ for a random sequence that has an m.s. limit X. It can be shown that the Cauchy's criterion is also sufficient, see [94].

The following assertions about the mean square convergence can now be made. Let $\{X_n\}$, $\{Y_n\}$, $\{V_n\}$ be random sequences, and Z be a random variable, all with finite second-order moments, $\{c_n\}$ be a sequence of deterministic constants, and a, b two deterministic constants. Further let X, Y, V be the m.s. limits of the three random sequences and c be the normal limit of the sequence $\{c_n\}$, i.e. $\lim_{n\to\infty} c_n = c$. Then,

(i) $\mathrm{LIM}\{c_n\} = \lim_{n\to\infty} c_n = c$

(ii) $\mathrm{LIM}\{Z\} = Z$

(iii) $\mathrm{LIM}\{Zc_n\} = Zc$

(iv) $\mathrm{LIM}\{aX_n + bY_n\} = aX + bY$ (the LIM{*} operation is linear)

(v) $E[X] = \lim_{n\to\infty} E[X_n]$, (the two operations $E[*]$ and LIM{*} are interchangeable)

(vi) $E[XY] = \lim_{n\to\infty} E[X_n Y_n]$, (as a special case, $E[X^2] = \lim_{n\to\infty} E[X_n^2]$)

(vii) if $E[X_n Y_n] = E[V_n]$, then $E[XY] = E[V]$ (2.20)

The first three statements are all obvious. Assertion (iv) is proved by using the inequality Eq. (1.79). After arranging the terms,

$$E[|aX_n + bY_n - aX - bY|^2] = E[|a(X_n - X) + b(Y_n - Y)|^2]$$
$$\leqslant 2a^2 E[|X_n - X|^2] + 2b^2 E[|Y_n - Y|^2]$$

The last expression tends to zero as $n \to \infty$, which proves the assertion. To prove (v), the Schwartz inequality Eq. (1.76) is applied, together with Eq. (2.17), that is,

$$|E[X_n] - E[X]|^2 = |E[X_n - X]|^2 \leqslant E[|X_n - X|^2] \to 0 \qquad \text{for } n \to \infty$$

The assertion (vi) is proved by writing

$$|E[X_n Y_n] - E[XY]| = |E[(X - X_n)Y + (Y - Y_n)X - (X - X_n)(Y - Y_n)]|$$
$$\leqslant |E[(X - X_n)Y]| + |E(Y - Y_n)X]| + |E[(X - X_n)(Y - Y_n)]|$$
$$\leqslant (E[|(X - X_n)|^2]E[|Y|^2])^{(1/2)} + (E[|(Y - Y_n)|^2]E[|X|^2])^{(1/2)}$$
$$+ (E[|(X - X_n)|^2]E[|(Y - Y_n)|^2)^{(1/2)} \to 0 \qquad \text{for } n \to \infty$$

again using the Schwartz inequality. Finally, the assertion (vii) is directly inferred from (v) and (vi), that is,

$$E[V] = \lim_{n\to\infty} E[V_n] = \lim_{n\to\infty} E[X_n Y_n] = E[XY]$$

Thus the above simple rules Eqs. (2.20) that hold for the two operations, taking the expectation and seeking the m.s. convergence limit, have been grouped together for convenience.

Now, let $\{X(t),\ t\in T\}$ be a stochastic process with finite first- and second-order moments, that is, with the covariance kernel, Eq. (2.12),

$$K_X(t_1,t_2)=R_X(t_1,t_2)-\mu_X(t_1)\mu_Y(t_2) \tag{2.12}$$

The random function $X(t)$ is said to be *continuous in mean square* (m.s. continuous), at $t\in T$ if

$$\lim_{h\to 0} E[|X(t+h)-X(t)|^2]=0 \tag{2.21}$$

for $(t+h)\in T$. Now, since

$$E[|X(t+h)-X(t)|^2]=R_X(t+h,t+h)-2R_X(t+h,t)+R_X(t,t) \tag{2.22}$$

$X(t)$ is m.s. continuous at t if $R_X(t,t)$ is continuous in the ordinary sense at (t,t) as the right-hand side of Eq. (2.21) then becomes zero. This condition is also necessary. Let $X(t)$ be m.s. continuous at $t\in T$, then, since $E[*]$ is a linear operator

$$E[\lim_{h\to 0,\,k\to 0}(X(t+h)X(t+k))]=\lim_{h\to 0,\,k\to 0}E[X(t+h)X(t+k)]=E[X(t)X(t)]=R_x(t,t) \tag{2.23}$$

the autocorrelation function is continuous at (t,t). Since in the above equation, the term $t+k$ can be replaced by any $(\tau+k)\in T$, the autocorrelation function is also continuous at (t,τ). From Eq. (2.12) it is further obvious that both the covariance kernel $K(t,s)$ and the mean value function $\mu(t)$ must be continuous if the autocorrelation function is continuous.

The random function $\{X(t),\ t\in T\}$ is said to be *mean square differentiable*, m.s. differentiable, at $t\in T$ if the following limit exists:

$$\operatorname*{LIM}_{h\to 0}\left(\frac{X(t+h)-X(t)}{h}\right)=\frac{dX(t)}{dt}=\dot{X}(t) \tag{2.24}$$

From Eq. (2.24), it is obvious that

$$X(t+h)-X(t)=X(t)h+U$$

where the remainder U must satisfy the condition

$$E[U]=o(h) \qquad \text{and} \qquad E[u^2]=o(h^2)$$

so

$$\lim_{h\to 0} E[|X(t+h)-X(t)|^2]=o(h^2)$$

and $X(t)$ is m.s. continuous at $t\in T$ if it is m.s. differentiable at the same point.

As can be seen by Eq. (2.22), the autocorrelation function plays the main role in proving that the above limit Eq. (2.24) exists. Using the Cauchy criterion, Eq. (2.19), it suffices to show that the following expression has the limit zero for all $t\in T$, $(t+h)\in T$ and $(t+k)\in T$, that is,

$$\lim_{h,k\to 0} E\left[\left|\frac{X(t+h)-X(t)}{h}-\frac{X(t+k)-X(t)}{k}\right|^2\right]=0$$

Raising the expression within the brackets to the second power and taking the expectation results in

$$\lim_{h\to 0}\left[\frac{R_X(t+h,t+h)-R_X(t+h,t)-R_X(t,t+h)+R_X(t,t)}{h^2}\right]$$
$$+\lim_{k\to 0}\left[\frac{R_X(t+k,t+k)-R_X(t+k,t)-R_X(t,t+k)+R_X(t,t)}{k^2}\right]$$
$$-2\lim_{h,k\to 0}\left[\frac{R_X(t+h,t+k)-R_X(t,t+k)-(R_X(t+h,t)-R_X(t,t)}{hk}\right]$$
$$=\frac{\partial^2 R_X(t,t)}{\partial t^2}+\frac{\partial^2 R_X(t,t)}{\partial t^2}-2\frac{\partial^2 R_X(t,t)}{\partial t^2}=0 \tag{2.25}$$

which shows that if the second partial derivative of the autocorrelation function at the point (t,t) exists, it is a sufficient condition for the above stochastic derivative limit to exist as well. That the existence of the second partial derivative of the autocorrelation function is a necessary condition may be proved by the following statement, using the ground rules Eqs. (2.20). Since the process is m.s. differentiable,

$$E[\dot{X}(t)\dot{X}(s)]=\lim_{h\to 0,k\to 0}\left[\frac{E[X(t+h)-X(t)}{h}\frac{X(s+k)-X(s)}{k}\right]$$
$$=\lim_{h\to 0,k\to 0}\left[\frac{R_X(t+h,s+k)-R_X(t+h,s)-R_X(t,s+k)+R_X(t,s)}{hk}\right]$$
$$=\frac{\partial^2 R_X(t,s)}{\partial t\,\partial s} \tag{2.26}$$

which proves the statement and shows that the second partial derivatives of the autocorrelation function exist not only at every diagonal point (t,t) but also at all off-diagonal points (t,s).

It has been shown that the differentiation of stochastic process treated as random functions can be carried out in much the same manner as for usual deterministic functions. In view of the assertions, Eqs. (2.20) and the foregoing discussion, all the simple rules pertaining to the interchangeability of algebraic operations with that of differentiation are valid. Now, the introduction of the m.s. Riemann integrability of a stochastic process follows in much the same manner. Again, let $\{X(t),\ t\in T\}$ be a stochastic process with finite first- and second-order moments, that is, with the covariance kernel Eq. (2.12). Consider the time interval $[a,b]$ where $a,b\in T$, $a<b$ and the partition:

$$a=t_0<t_1<\cdots<t_{n-1}<t_n<=b$$

Further let $h = \max(t_{i+1} - t_i)$, $i = 0, 1, \ldots, n$. The stochastic process $X(t)$ is said to be m.s. Riemann integrable over the interval $[a, b]$ if the following mean square limit exists:

$$\lim_{h \to 0} E\left[\left|\sum_{i=0}^{n-1} X(t_i)(t_{i+1} - t_i) - \int_a^b X(t)\,dt\right|^2\right] = 0 \tag{2.27}$$

Consequently Eq. [2.27] defines the integral. To show under what conditions the above integral limit exists, consider another partition of the interval $[a, b]$, that is,

$$a = s_0 < s_1 < \cdots < s_{n-1} < s_n < = b$$

where $k = \max(s_{i+1} - s_i)$, $i = 0, 1, \ldots, n$. Using the Cauchy criterion Eq. (2.19), it is now sufficient to show that

$$\lim_{h,k \to 0} E\left[\left|\sum_{i=0}^{n-1} X(t_i)(t_{i+1} - t_i) - \sum_{i=0}^{m-1} X(s_i)(s_{i+1} - s_i)\right|^2\right] = 0 \tag{2.28}$$

Lifting the expression within the brackets to the second power and taking the expectation before the summation gives the following terms:

$$\sum_{i=0}^{n-1}\sum_{j=0}^{n-1} E[X(t_i)X(t_j)](t_{i+1} - t_i)(t_{j+1} - t_j) = \sum_{i=0}^{n-1}\sum_{j=0}^{n-1} R_X(t_i, t_j)(t_{i+1} - t_i)(t_{j+1} - t_j) \quad \text{(a)}$$

$$\sum_{i=0}^{m-1}\sum_{j=0}^{m-1} E[X(s_i)X(s_j)](s_{i+1} - s_i)(s_{j+1} - s_j) = \sum_{i=0}^{m-1}\sum_{j=0}^{m-1} R_X(s_i, s_j)(t_{i+1} - t_i)(t_{j+1} - t_j) \quad \text{(b)}$$

$$-2\sum_{i=0}^{n-1}\sum_{j=0}^{m-1} E[X(t_i)X(s_j)](t_{i+1} - t_i)(s_{j+1} - s_j) = -2\sum_{i=0}^{n-1}\sum_{j=0}^{m-1} R_X(t_i, s_j)(t_{i+1} - t_i)(s_{j+1} - s_j) \quad \text{(c)}$$

Adding (a), (b) and (c) gives 0 as expected. The right-hand terms are all equivalent to the Riemann integral of the autocorrelation function $R_X(t, s)$, that is,

$$\int_a^b \int_a^b R_X(t, s)\,dt\,ds$$

whereby the limit Eq. (2.28) tends to zero, which proves that the stochastic integral limit Eq. (2.27) exists if the autocorrelation function is integrable in the Riemann sense. That the autocorrelation function is Riemann integrable is also a necessary condition. If the stochastic process $X(t)$ is indeed m.s. Riemann integrable, then

$$E\left[\int_a^b X(t)\,dt \int_c^d X(s)\,ds\right] = \int_a^b \int_c^d E[X(t)X(s)]\,dt\,ds = \int_a^b \int_c^d R_X(t, s)\,dt\,ds$$

so $R_X(t, s)$ must be Riemann integrable as well. Also since

$$E\left[\int_a^b X(t)\,dt\right] = \int_a^b E[X(t)]\,dt = \int_a^b \mu_X(t)\,dt$$

both the mean value function $\mu_X(t)$ and the covariance kernel $K_X(t, s)$ by Eq. (2.12) are Riemann integrable. A further implication of this is had by noting that if the stochastic process $X(t)$ is m.s. continuous at every $t \in (a, b)$ then the mean value function and the covariance kernel as well as the autocorrelation function are continuous at every $(t, s) \in (a, b)$ (see Eqs. (2.22) and (2.23)).

Obviously, these three deterministic functions are all Riemann integrable over (a, b) since they are everywhere continuous in this interval. Therefore, if the stochastic process $X(t)$ is m.s. continuous at every $t \in (a, b)$, then in view of the above proof, it is also m.s. Riemann integrable over the same interval.

Consider now the stochastic process $Y(t)$, which is obtained from the m.s. Riemann integrable process $X(t)$ through integration, that is,

$$Y(t) = \int_a^t X(u)\,du \tag{2.29}$$

If $X(t)$ is m.s. Riemann integrable over (a, t) for every $t \in (a, b)$, then the process $Y(t)$ is m.s. continuous at every $t \in (a, b)$. The proof follows immediately from Eq. (2.21):

$$E\left[\left|\int_a^{t+h} X(u)du - \int_a^t X(u)\,du\right|^2\right] = E\left[\left|\int_t^{t+h} X(u)\,du\right|^2\right]$$

$$= \int_t^{t+h}\int_t^{t+h} R_X(u, v)\,du\,dv \to 0 \qquad \text{if } h \to 0$$

since the last integral is a continuous function of its upper limit. Furthermore, if $X(t)$ is m.s. continuous at every $t \in (a, b)$, then $Y(t)$ is m.s. differentiable at every $t \in (a, b)$ with $\dot{Y}(t) = X(t)$, since

$$E\left[\left|\frac{1}{h}\left(\int_a^{t+h} X(u)du - \int_a^t X(u)du\right) - X(t)\right|^2\right] = E\left[\left|\frac{1}{h}\int_t^{t+h} X(u)du - X(t)\right|^2\right]$$

$$= \frac{1}{h^2}\int_t^{t+h}\int_t^{t+h} R_X(u, v)du\,dv + 2E\left[X(t)\frac{1}{h}\int_t^{t+h} X(u)\,du\right] - E[X^2(t)]$$

$$\to R_X(t, t) - 2E(X^2(t)] + E[X^2(t)] = 0 \quad \text{for } h \to 0$$

which shows that the derivative of the integral function is just the integrand function as in the case of the usual deterministic functions. The integral function of an m.s. Riemann integrable stochastic process can also be obtained in the usual manner. For instance, consider the velocity process $\dot{X}(t)$, which is m.s. Riemann integrable over (a, b), then

$$X(t) - X(a) = \int_a^t \dot{X}(u)du \tag{2.30}$$

The m.s. convergence of the m.s. function to the integral limit is proved as follows:

$$E\left[\left|\left(X(t)-X(a)-\int_a^t X(u)\mathrm{d}u\right)\right|^2\right]$$

$$=E[(X(t)-X(a))^2]-2\int_a^t E[(X(t)-X(a))\dot{X}(u)]\,\mathrm{d}u+\int_a^t\int_a^t E[\dot{X}(u)\dot{X}(v)]\,\mathrm{d}u\,\mathrm{d}v$$

$$=R_X(t,t)-2R_X(t,a)+R_X(a,a)-2\int_a^t\left(\frac{\partial R_X(t,u)}{\partial u}-\frac{\partial R_X(a,u)}{\partial u}\right)\mathrm{d}u+\int_a^t\int_a^t\frac{\partial^2 R_X(u,v)}{\partial u\,\partial v}\mathrm{d}u\,\mathrm{d}v$$

$$=R_X(t,t)-2R_X(t,a)+R_X(a,a)-2(R_X(t,t)-R_X(t,a)-R_X(a,t)+R_X(a,a))$$
$$+R_X(t,t)-R_X(a,t)-R_X(t,a)+R_X(a,a)=0$$

since $R_X(t,s)$ is symmetric by Eq. (2.11). In the above evaluation of the m.s. limit of the difference between the r.h.s. and the l.h.s. of Eq. (2.30), it does not tend to zero rather it is equal to zero at all times $t\in(a,b)$. Therefore, from the Chebyshev inequality Eq. (1.81), it follows that the l.h.s. of Eq. (2.30) converges to the r.h.s. with probability one.

It has been shown that stochastic processes can be differentiated and integrated in the usual manner under the fundamental rules of ordinary calculus. The main requirement is that the correlation functions are continuous and integrable. However, the implication of these requirements is quite serious. For instance, consider the interrelationship with the joint characteristic functions (see Eq. (1.111)). The limit in the mean square sense for the first derivative of the stochastic process $X(t)$ was defined in Eq. (2.24). It clearly involves the second-order statistic of the process as the two amplitudes $X(t)$ and $X(t+h)$ have to be evaluated jointly. By Eq. (1.111), the joint characteristic function of the two amplitudes, also called a two-point characteristic function of the stochastic process $X(t)$, is given by

$$\varphi_X(\omega_1,\omega_2;t,t+h)=E[\exp\{\mathrm{i}\omega_1 X(t)+\mathrm{i}\omega_2 X(t+h)\}] \tag{2.31}$$

Therefore, the characteristic function of the derivative can be obtained as the limit of the modified characteristic function of the two amplitudes, that is,

$$\varphi_{\dot{X}}(\omega;t)=E[\exp\{\mathrm{i}\omega\dot{X}(t)\}]$$
$$=\lim_{h\to 0}E\left[\exp\left\{\mathrm{i}\omega\frac{X(t+h)-X(t)}{h}\right\}\right]=\lim_{h\to 0}\varphi_X\left(-\frac{\omega}{h},\frac{\omega}{h};t,t+h\right) \tag{2.32}$$

which clearly indicates that for the first derivative, the two-point characteristic function is needed. For higher-order derivatives, the characteristic function can be obtained in the same manner from the limit of the multi-point characteristic function of the process itself (Eq. (2.32)). It is necessary to have the full statistics of the process of the same order, since it requires the characteristic function of that order.

From the definition of an m.s. Riemann integrable process Eq. (2.29), the integral can be written as the limit of the approximating sum

$$\int_a^b X(t)\,\mathrm{d}t=\underset{h\to 0}{\mathrm{LIM}}\sum_{i=0}^{n-1}X(t_i)(t_{i+1}-t_i) \tag{2.33}$$

where $h = \max(t_{i+1} - t_i)$, $i = 0, 1, \ldots, n-1$. The characteristic function of the approximating sum in Eq. (2.33) is then given as

$$E\left[\exp\left\{i\omega \sum_{i=0}^{n-1} X(u_i)(t_{i+1} - t_1)\right\}\right] = \varphi(\omega(t_1 - t_0), \ldots, \omega(t_n - t_{n-1}); u_1, \ldots, u_n) \quad (2.34)$$

where the times $u_i \in (t_{i+1} - t_i)$. Obviously, the approximating sum in Eq. (2.33) requires the knowledge of the n-point characteristic function and letting $h \to 0$, whereby $n \to \infty$, one finally needs the characteristic function for indefinitely many points simultaneously, that is, the full statistics of the process (cf. the information pyramid, Fig. 2.4). In many practical applications, the stochastic process of interest is the response of a system subjected to a random excitation. That is, the output process is the solution of a differential equation governing the system response (cf. Chapter 3). In this case, the stochastic response process is obtained through a complicated integral of the 'forcing function' $X(t)$, Eq. (2.29). To obtain the simplest statistical information about the response process therefore requires complete statistical information about the forcing function, that is, an indefinitely-many-point characteristic function for the process $X(t)$. Usually, due to lack of such information, it is replaced by the simplest statistical information about the forcing functions under the assumption of some stochastic model.

Even though most often only scarce information is available about the forcing function, usually based on a few measurements of a sample function of the process, it is convenient to define the so-called *characteristic functional* as a natural extension of the multi-point characteristic function. From Eqs. (2.33) and (2.34), the following generalization of the exponent of the multi-point function can be made, this is,

$$\theta_0 X(t_0) + \theta_1 X(t_1) + \cdots \to \int \theta(t) X(t)\, dt$$

where the function $\theta = \theta(t)$ with the arguments $\theta_i = \theta(t_{i+1} - t_i)$ represents the values $\omega(t_{i+1} - t_i)$. The indefinitely-many-point characteristic function has been converted to the characteristic functional given by

$$\varphi_X(\theta(t)) = E\left[\exp\left\{i \int_a^b \theta(t) X(t)\, dt\right\}\right] \quad (2.35)$$

The characteristic functional will for instance play a dominant role in the treatment of compound Poisson processes, cf. Chapter 6.

2.3 ERGODIC PROCESSES

2.3.1 Definition of Ergodicity

Physical systems, which are governed by a random mechanism, will generate sample functions of a stochastic process that can be subjected to measurements and deterministic analysis. One such sample function is just a single member, a realization, of a family or an ensemble of such functions. The question therefore arises, is it possible to derive such ensemble quantities as the mean value, the covariance kernel or the autocorrelation function from a measurement and the analysis of one single observation of the stochastic process. In general, the concept of ergodicity (from the Greek, *erg* = work and *hodos* = way), is defined such that a process, which has the temporal average of one sample function that can be shown to be equal to the ensemble or the statistical averages, is said to be ergodic. Since the analysis of sample functions is often an easy and straightforward task compared with correctly assessing the probability law of a stochastic process, ergodicity is a useful and important feature of a stochastic process.

The precise mathematical formulation of ergodicity is beyond the scope of this book and will only be treated in a superficial manner. (The reader should consult for instance the work of Yaglom, [238], and that of Birkhoff and Khinchine, see Gnedenko, [71]). It suffices to state that a stochastic process is said to be ergodic if the time averages, formed from a sample function $x(t)$ of a stochastic process $\{X(t), t \in T\}$, observed over a sufficiently large interval of time $(-T, T)$, can be used as an approximation for the corresponding ensemble averages. Actually what this implies is that for sufficiently wide separations, the amplitudes of the process lose their correlation. Therefore, a sufficiently long sample function or realization of the process carries all the information pertaining to its behaviour. An infinitely long record of the process is thus divided into parts that have no correlation, so each part is representative of the whole process. For instance, if the temporal average, Eq. (2.5),

$$\overline{x(t)} = \frac{1}{2T}\int_{T}^{T} x(t)\,\mathrm{d}t = \mu, \qquad \text{a constant} \tag{2.5}$$

and the ensemble average should converge with probability one, that is,

$$P[E[X(t)] = \lim_{T \to \infty} \overline{x(t)}] = 1 \tag{2.36}$$

the process is said to be ergodic in the mean. In the same manner, the temporal autocorrelation function is defined as

$$\overline{x(t)x(t+\tau)} = \frac{1}{2T}\int_{-T}^{T} x(t)x(t+\tau)\,\mathrm{d}t = R_x(T, \tau) \tag{2.37}$$

and if it converges to the autocorrelation function with probability one, that is,

$$P_{T \to \infty}[E[R_X(T, t) = R_X(\tau)] = 1 \tag{2.38}$$

the process is ergodic with respect to the second moment. More generally, a process is said to be ergodic without any qualifying phrase if it is ergodic with respect to all its stationary functionals (Lumley, [135]). In that case, by time averaging, the expectation of any quantity, obtainable from the original process, which has an expectation, can be calculated. It should be noted that the concept of ergodicity is only meaningful for stationary processes. For instance, the temporal average, Eq. (2.5), can only be constant.

Example 2.7 Consider the stochastic process, $X(t) = A \sin(\omega t + \Phi)$, $t \geqslant 0$, where A and Φ are two independent random variables. A has a mean value μ_a and a standard deviation σ_a but Φ, the phase, is uniformly distributed between $(0, 2\pi)$. ω is a positive constant. Find the ensemble mean value, the autocorrelation function and whether the process is ergodic.

Solution

$$E[X(t)] = \int_0^{2\pi}\int_{-\infty}^{\infty} p(a)p(\varphi)\sin(\omega t+\varphi)\,\mathrm{d}a\,\mathrm{d}\varphi$$

$$= \int_{-\infty}^{\infty} p(a)a\,\mathrm{d}a \int_0^{2\pi}\frac{1}{2\pi}\sin(\omega t+\varphi)\mathrm{d}\varphi = \mu_a \times 0 = 0$$

$$R_X(t,s) = E[X(t)X(s)] = E[A^2\sin(\omega t+\Phi)\sin(\omega s+\Phi)]$$

$$= E[A^2]\frac{1}{2\pi}\int_0^{2\pi}\sin(\omega t+\varphi)\sin(\omega s+\varphi)\,\mathrm{d}\varphi$$

$$= \frac{(\mu_a^2+\sigma_a^2)}{2\pi}\int_0^{2\pi}\tfrac{1}{2}(\cos\omega(t-s) - \cos(\omega(t+s)+2\varphi))\,\mathrm{d}\varphi$$

$$= \tfrac{1}{2}(\mu_a^2+\sigma_a^2)\cos\omega(t-s)$$

or

$$R_X(\tau) = \tfrac{1}{2}(\mu_a^2+\sigma_a^2)\cos\omega t$$

$$\sigma_X^2 = E[X^2(t)] = R_X(0) = \tfrac{1}{2}(\mu_a^2+\sigma_a^2)$$

The process is covariance stationary. Now the temporal averages are formed.

$$\overline{x(t)} = \frac{1}{T}\int_0^T a\sin(\omega t+\varphi)\mathrm{d}t = 0 \qquad \text{if } T = \frac{2\pi}{\omega}n$$

$$\overline{x^2(t)} = \frac{1}{T}\int_0^T a^2\sin^2(\omega t+\varphi)\mathrm{d}t = \frac{a^2}{2} \qquad \text{if } T = \frac{2\pi}{\omega}n$$

$$\overline{x(t)x(t+\tau)} = \frac{1}{T}\int_0^T a^2\sin(\omega t+\varphi)\sin(\omega(t+\tau)+\varphi)\mathrm{d}t = \frac{a^2}{2}\cos\omega\tau$$

Since a is a sample value, both the m.s. temporal value and the temporal correlation are different for each sample function and do not converge to the corresponding ensemble averages. The mean temporal value and the ensemble mean, however, converge as $T \to \infty$. The process is ergodic with respect to the mean only.

2.3.2 *Proof of Ergodicity*

Under certain general conditions, it is possible to prove that a stochastic process is ergodic with respect to several of its stationary functionals as has already been discussed, (see for instance, Papoulis, [166] and Jazwinski, [94]). Ergodicity in a process, if it can be proven, signifies that the ensemble statistics, this is, the probability law of the process, can be inferred from one single sample function of the process. This can be a great advantage, since for a physically realizable process, it is most often easier to obtain a sample function or one single observation of the process through measurement rather than infer its probability law by any other means. In many cases, it is known from the physical aspect of the environment, which generates the process, that it is likely to be ergodic. However, it is often more an act of faith to assume a certain process to be ergodic rather than to be the result of a studied argument.

A proof of ergodicity with respect to the mean can be carried out as follows (time averaging on the process itself will yield a new random variable):

$$Z = \overline{X(t)} = \frac{1}{T}\int_0^T X(t)\,dt \tag{2.39}$$

The expectation or mean value of this variable is

$$E[Z] = \frac{1}{T}\int_0^T E[X(t)]\,dt = \frac{1}{T}\int_0^T \mu_X(t)\,dt = \mu_X$$

which for a stationary process is the same as the ensemble average. The variance of Z is obtained by evaluating the following expression:

$$\begin{aligned}\mathrm{Var}[Z] &= E\left[\frac{1}{T^2}\left(\int_0^T (X(t)-\mu_X(t))\,dt\right)\left(\int_0^T (X(s)-\mu_X(s))\,ds\right)\right] \\ &= \frac{1}{T^2}\int_0^T\int_0^T E[(X(t)-\mu_X(t))(X(s)-\mu_X(s))]\,dt\,ds\end{aligned}$$

therefore, by Eq. (2.10),

$$\mathrm{Var}[Z] = \frac{1}{T^2}\int_0^T\int_0^T K(t,s)\,dt\,ds = \frac{2}{T^2}\int_0^T\int_0^t K(t,s)\,dt\,ds \tag{2.40}$$

since $K(t,s) = K(s,t)$.

For covariance stationary processes introduce the autocovariance $\Gamma(t-s) = K(t,s) = K(s,t) = \Gamma(s-t)$ and

$$\mathrm{Var}[Z] = \frac{2}{T^2}\int_0^T\int_0^t \Gamma(t-s)\,ds\,dt \tag{2.41}$$

Now change variables to $\tau = t - s$ and $s = s$. The domain of integration will be changed as shown in Fig. 2.7. Therefore,

$$\mathrm{Var}[Z] = \frac{2}{T^2}\int_0^T\int_0^{T-t} \Gamma(\tau)\,ds\,d\tau = \frac{2}{T}\int_0^T \Gamma(\tau)\left(1-\frac{\tau}{T}\right)d\tau$$

If the autocovariance function is integrable, that is,

$$\int_{-\infty}^{\infty} \Gamma(\tau)\,d\tau = C < \infty \tag{2.42}$$

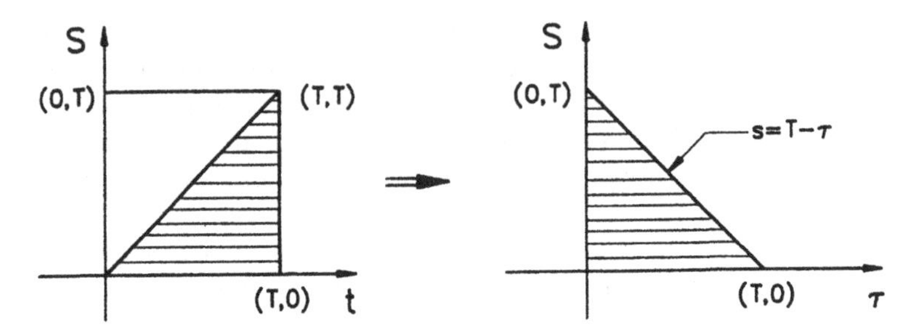

Fig. 2.7 *Domains of integration*

then

$$\lim_{T\to\infty} \text{Var}[Z] \to 0 \tag{2.43}$$

and it has been shown that the random variable Z in the limit becomes deterministic ($\sigma_Z = 0$), i.e. assumes its mean value μ_X. Hence

$$Z = \overline{X(t)} = \frac{1}{T}\int_0^T X(t)\mathrm{d}t = \mu_X$$

so time averaging on the process itself yields the ensemble average and the process is ergodic with respect to the mean. Also, since by Eq. (2.12), $\Gamma(\tau) = R(\tau) - \mu_X^2$

$$\text{Var}[Z] = \frac{2}{T}\int_0^T (R(\tau) - \mu_X^2)\left(1 - \frac{\tau}{T}\right)\mathrm{d}\tau = \frac{2}{T}\int_0^T R(\tau)\left(1 - \frac{\tau}{T}\right)\mathrm{d}\tau - \mu_X^2 \tag{2.44}$$

When $\Gamma_X(\tau)$ is integrable, lim Var$[Z] = 0$ for $T \to \infty$, which in Eq. (2.43) requires that

$$\lim_{\tau\to\infty} R_X(\tau) \to \mu_X^2 \tag{2.45}$$

This is an interesting result and shows show the autocorrelation function approaches the mean value square for very large time intervals between the two amplitudes. A physical explanation is straightforward since $R(t_1, t_2) = E[X(t_1)X(t_2)]$. For a very large time difference $t_2 - t_1$, the correlation between the two amplitudes must be very small and in the limit they will be statistically independent or

$$R_X(t_1, t_2) = E[X(t_1)X(t_2)] = E[X(t_1)]E[X(t_2)] = \mu_X^2 \tag{2.46}$$

Another interesting result can be inferred from Eqs. (2.42) and (2.43). If the autocorrelation function $R(\tau)$ is integrable, then by Eq. (2.44),

$$\lim_{T\to\infty} \text{Var}[Z] \to -\mu_x^2 \tag{2.47}$$

Since the variance always is positive or zero, an integrable autocorrelation function implies that the mean value must be zero.

Example 2.8 Consider a sample function or a realization of the process studied in Example 2.6.

$$x(t) = \sum_{j=1}^{n} (a_j \cos \omega_j t + b_j \sin \omega_j t)$$

where a_j and b_j are samples from the distribution of A_j and B_j. The time average is

$$\overline{x(t)} = \frac{1}{T}\int_0^T \left(\sum_{j=1}^{n} a_j \cos \omega_j t + b_j \sin \omega_j t\right)\mathrm{d}t = \frac{1}{T}\sum_{j=1}^{n}\left(\frac{a_j}{\omega_j}\sin \omega_j T + \frac{b_j}{\omega_j}(1-\cos \omega_j T)\right)$$

obviously $\overline{x_{T\to\infty}(t)} = 0 = E[X(t)]$ and the process appears to be mean-value-ergodic. In fact, the autocorrelation function was shown to be

$$R_X(\tau) = \sum_{j=1}^{n} \sigma_j^2 \cos \omega_j \tau$$

and

$$\int_{-\infty}^{\infty} R_X(\tau) = \lim_{T\to\infty} \int_T^T \left(\sum_{j=1}^{n} \sigma_j^2 \cos \omega_j \tau\right) \mathrm{d}\tau$$

$$= \lim_{T\to\infty} \sum_{j=1}^{n} \frac{\sigma_j^2}{\omega_j} \sin \omega_j T \leqslant \sum_{j=1}^{n} \frac{\sigma_j}{\omega_j} = \text{constant} < \infty$$

so the correlation function is integrable and hence $\mu_X = 0$ and the process is ergodic in the mean.

Example 2.9 A stationary random process $\{x(t),\ -\infty < t < \infty\}$ has the autocorrelation function shown in Fig. 2.8. If the process is ergodic with respect to the mean, determine $E[X]$ and Var $[X]$.

Solution Since the process is ergodic with respect to the mean,

$$\lim_{\tau\to\infty} R_X(\tau) = \mu_X^2 = \frac{S_0}{T}, \qquad \mu_X = \pm\sqrt{\frac{S_0}{T}}$$

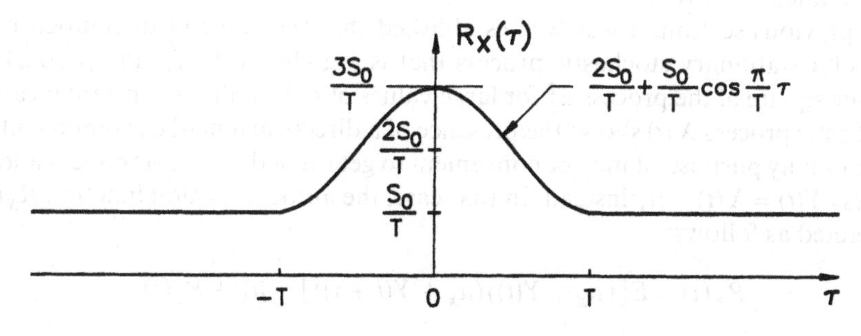

Fig. 2.8 *The auto-correlation function $R_X(\tau)$*

otherwise

$$\lim_{T\to\infty} \mathrm{Var}[Z] = \lim_{T\to\infty} \frac{2}{T}\int_0^T (R_X(\tau)-\mu_X^2)\left(1-\frac{\tau}{T}\right)\mathrm{d}\tau$$

can not tend to zero. So finally,

$$E[X^2(t)] = R_X(0) = \frac{3S_0}{T} = \sigma_X^2 + \mu_X^2$$

$$\mathrm{Var}[X(t)] = \sigma_X^2 = \frac{3S_0}{T} - \frac{S_0}{T} = \frac{2S_0}{T}$$

2.4 SOME PROPERTIES OF THE CORRELATION FUNCTIONS

A thorough study of the correlation functions of stochastic processes can be very useful, since important information about the behaviour of the processes can be derived from the properties of the correlation functions. For instance, for covariance stationary processes, the correlation functions carry all the information that is obtainable about the process. Therefore, a great part of the analysis of stochastic processes is centred on the correlation functions and their spectral representations in the frequency domain. For ergodic processes, the correlation functions can be obtained by a measurement of a single sample function of the process, when such is obtainable, which makes the correlation functions an obvious target for study. In this section, some of the main features that characterize the correlation functions will be discussed.

Many of the physical processes encountered in practice have autocorrelation functions of the form

$$R(\tau) = \mathrm{e}^{-\alpha|\tau|}\{f(\tau)\} \tag{2.48}$$

in which α is any positive constant and $f(\tau)$ is any well-behaved function of τ. For small values of t, the correlation between the two amplitudes $x(t)$ and $x(t+\tau)$ is considerable, whereas the correlation rapidly decreases with increasing separation in time between the amplitudes. When α grows large, $R(\tau)\approx 0$ and the process completely loses correlation between the amplitudes. This is characterized by very rapid and violent changes in the process as time progresses.

In the previous section, it was well established that the value of the autocorrelation function of a stationary stochastic process that is ergodic in the mean approaches the mean value square of the process μ_X^2 for large values of τ. Actually, a constant mean value of a stochastic process $X(t)$ shows the presence of a direct current (d.c.) component in the signal. For many purposes it may be convenient to get rid of the d.c. component and study the process $Y(t) = X(t) - \mu_X$ instead. In that case, the autocorrelation function $R_X(\tau)$ can be interpreted as follows:

$$R_X(\tau) = E[(\mu_X + Y(t))(\mu_X + Y(t+\tau))] = \mu_X^2 + R_Y(\tau) \tag{2.49}$$

On the other hand, an autocorrelation function that is of the type $c + f(\tau)$, where c is a positive constant, shows that there must be a d.c. component in the signal such that it has a mean value equal to $\pm\sqrt{c}$.

Now take a stationary stochastic process $X(t)$ with the autocorrelation function $R(\tau)$. By letting $\tau = -\tau$ in Eq. (2.16) and then shifting the time by $+\tau$, which does not affect the second-order statistics, the following relation is obtained:

$$R_X(-\tau) = E[X(t-\tau)X(t)] = E[X(t)X(t+\tau)] = R_X(\tau) \tag{2.50}$$

The autocorrelation function and also the autocovariance functions are even functions with respect to τ. Also, the autocorrelation function and the autocovariance have their maximum value at $\tau = 0$. In fact, by forming the expectation,

$$E[(X(t) \pm X(t+\tau))^2] \geqslant 0$$

or

$$E[X^2(t) \pm 2X(t)X(t+\tau) + X^2(t+\tau)] = R_X(0) \pm 2R_X(\tau) + R_X(0) = 2(R_X(0) \pm R_X(\tau)) \geqslant 0$$

Hence

$$R(0) \geqslant |R(\tau)| \qquad \text{for all } \tau$$

and

$$\Gamma(0) \geqslant |\Gamma(\tau)| \qquad \text{for all } \tau \tag{2.51}$$

On the foundation of stochastic calculus laid down in Section 2.2.5, the two mathematical operations, integration and differentiation, can be carried out for stochastic processes more or less in the same manner as for ordinary functions. With the same restrictions as presented in Section 2.2.5, the two operations of taking the expectation and that of differentiation can be interchanged, whereby

$$E[\dot{X}(t)] = E\left[\frac{dX(t)}{dt}\right] = \mu'_X(t) = 0 \tag{2.52}$$

The expectation of a stationary process is a constant, and hence the expectation of the derivative process (the velocity process) $\dot{X}(t)$ is zero. Now,

$$\mathrm{Cov}[\dot{X}(s)X(t)] = \frac{\partial K(s,t)}{\partial s}$$

$$\mathrm{Cov}[\dot{X}(s)\dot{X}(t)] = \frac{\partial^2 K(s,t)}{\partial s\,\partial t}$$

In the same manner, the correlation function may be differentiated to give

$$\frac{dR(\tau)}{d\tau} = E[X(t)\dot{X}(t+\tau)] = E[X(t-\tau)\dot{X}(t)] \tag{2.53}$$

Further differentiation yields

$$\frac{d^2R(\tau)}{d\tau^2} = -E[\dot{X}(t-\tau)\dot{X}(t)] = -E[\dot{X}(t)\dot{X}(t+\tau)] \tag{2.54}$$

since translation of the time axis does not change the expectation. By Eq. (2.54)

$$R''_X(\tau) = -R_{\dot{X}}(\tau) \tag{2.55}$$

that is, the second derivative of the autocorrelation function is, but for the negative sign, equal to the autocorrelation function of the velocity process.

By Eq. (2.55), it follows that if the derivative process exists, the second derivative of the autocorrelation function has a negative value at $t = 0$ and hence a negative (downward pointing) radius of curvature, $(R''_X(0) \leqslant 0)$. Differentiating the mean square value, $s_X^2 = \sigma_X^2 + \mu_X^2$, which is a constant, gives

$$\frac{d}{dt} E[X^2(t)] = 2E[\dot{X}(t)X(t)] = 2R'_X(0) = 0 \tag{2.56}$$

which shows that if the derivative process exists, the autocorrelation function must have a zero tangent at $\tau = 0$. In Fig. 2.9, the above main characteristics of the autocorrelation function are depicted.

Example 2.10 A stationary random process $X(t)$ has the autocorrelation function

$$R_X(\tau) = (1 - \tau^2)e^{-\tau^2}$$

Assuming that both the velocity process and the acceleration process exist, find the corresponding autocorrelation functions of these processes. Find the RMS (root mean square) values of all three processes.

Solution

$$\begin{aligned}
R_X(\tau) &= e^{-\tau^2}(1 - \tau^2), \qquad -\infty < \tau < \infty \\
R'_X(\tau) &= e^{-\tau^2}(2\tau^3 - 4\tau) \\
R''_X(\tau) &= e^{-\tau^2}(-4\tau^4 + 14\tau^2 - 4) \\
R'''_X(\tau) &= e^{-\tau^2}(8\tau^5 - 44\tau^3 + u + 36\tau) \\
R^{(iv)}_X(\tau) &= e^{-\tau^2}(-16\tau^6 + 128\tau^4 - 204\tau^2 + 36)
\end{aligned}$$

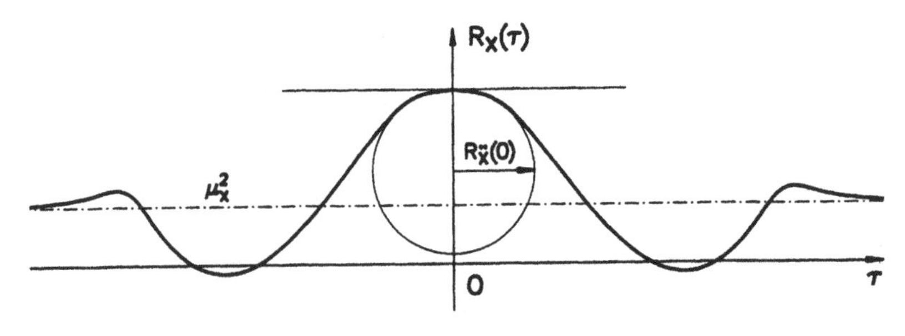

Fig. 2.9 *Main characteristics of the autocorrelation function*

so for $\dot{X}(t)$ and $\ddot{X}(t)$

$$R_{\dot{X}}(\tau) = -R''_X(\tau) = e^{-\tau^2}(-4\tau^4 + 14\tau^2 - 4)$$
$$R_{\ddot{X}}(\tau) = -R''_{\dot{X}}(\tau) = R^{(iv)}_X(\tau) = e^{-\tau^2}(-16\tau^6 + 128\tau^4 - 204\tau^2 + 36)$$

and

$$s_X = \text{RMS}[X] = \sqrt{E[X^2]} = \sqrt{R_X(0)}$$
$$s_{\dot{X}} = \text{RMS}[\dot{X}] = \sqrt{E[\dot{X}]} = \sqrt{R_{\dot{X}}(0)}$$
$$s_{\ddot{X}} = \text{RMS}[\ddot{X}] = \sqrt{E[\ddot{X}^2]} = \sqrt{R_{\ddot{X}}(0)}$$

As for the cross-correlation functions of two stochastic and stationary processes $X(t)$ and $Y(t)$, similar relation can be derived. By Eq. (2.13),

$$\begin{aligned} R_{XY}(\tau) &= E[X(t)\,Y(t+\tau)] \\ R_{YX}(\tau) &= E[Y(t)\,X(t+\tau)] \end{aligned} \tag{2.57}$$

By shifting the time by $-\tau$, it immediately follows that

$$R_{XY}(\tau) = R_{YX}(-\tau)$$
$$R_{YX}(\tau) = R_{XY}(-\tau)$$

since $R_{XY}(\tau)$ and $R_{YX}(\tau)$ in general are not the same, they are, unlike the autocorrelation function, not even in τ. Now form the correlation coefficient between $X(t)$ and $Y(t+\tau)$, that is,

$$\varrho_{XY}(\tau) = \frac{E[(X(t) - \mu_X)(Y(t+\tau) - \mu_Y)]}{\sigma_X \sigma_Y} \tag{2.58}$$

or

$$\sigma_X \sigma_Y \varrho_{XY}(\tau) = E[(X(t)\,Y(t+\tau) - \mu_X\,Y(t+\tau) - \mu_Y\,X(t) + \mu_X \mu_Y]$$
$$\sigma_X \sigma_Y \varrho_{XY}(\tau) = R_{XY}(\tau) - \mu_X \mu_Y - \mu_Y \mu_X + \mu_X \mu_Y$$

so

$$R_{XY}(\tau) = \sigma_X \sigma_Y \varrho_{XY}(\tau) + \mu_X \mu_Y$$

and

$$R_{YX}(\tau) = \sigma_Y \sigma_X \varrho_{YX}(\tau) + \mu_Y \mu_X \tag{2.59}$$

is obtained in the same manner. Since $-1 \leqslant \varrho_{XY} \leqslant 1$, where the limiting values describe perfect antiphase (-1) and perfect in-phase $(+1)$ correlation, the limiting values of the cross-correlation function must be

$$-\sigma_X \sigma_Y + \mu_X \mu_Y \leqslant R_{XY}(\tau) \leqslant \sigma_X \sigma_Y + \mu_X \mu_Y \tag{2.60}$$

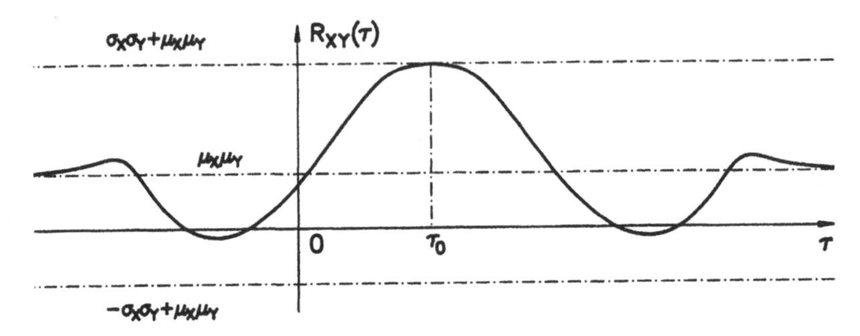

Fig. 2.10 *Properties of the cross-correlation function*

Finally, for very large separation times t, the correlation between the amplitudes of the two different processes becomes zero, that is, (see Eq. (2.44)),

$$\lim_{\tau\to\infty} R_{XY}(\tau) = \lim_{\tau\to\infty} R_{YX}(\tau) = \mu_X\mu_Y \tag{2.61}$$

The above properties are illustrated in Fig. 2.10. As shown, the cross-correlation attains a maximum value, that is, a perfect in-phase correlation, at some separation time $\tau = \tau_0$. For large τ, the oscillations fade away and the cross-correlation function approaches $\mu_X\mu_Y$. The reader may ponder over the fact that positive τ_0 (the maximum correlation) indicates that $Y(t)$ is lagging behind in phase.

2.5 SPECTRAL ANALYSIS

In connection with the response analysis of various physical systems to stochastic inputs, the methodology of harmonic analysis plays an important role. A fundamental question at once arises: is it possible to define the Fourier transform for the stochastic process $X(t)$? That is,

$$F_X(\omega) = \int_{-\infty}^{\infty} X(t)e^{-i\omega t}\,dt? \tag{2.62}$$

Unfortunately, little meaning can be attached to the integral Eq. (2.62). Neither the random function $X(t)$ nor its sample functions $x(t)$ belong to the class of functions usually treated in Fourier analysis. However, as was shown by Kolmogorov in 1941, it is possible to extend the notion of harmonic analysis to stochastic processes by assiging to each frequency ω a contribution to the power content or the energy of the process, [112]. Following Svesnikov, [217], this can be illustrated for a stationary random function $X(t)$, which can be assumed to have a mean value equal to zero without loss of generality. At least formally,* the random function can be written as a Fourier series of harmonic

*For an overview of harmonic analysis and Fourier transformation, the reader is referred to Chapter 7, where a more formal introduction to Fourier transforms and signal analysis will be presented.

exponential terms with different frequencies ω_j, that is,

$$X(t) = \sum_{j=-n}^{n} \Phi_j e^{i\omega_j t} \tag{2.63}$$

where Φ_j are zero mean random variables and can be interpreted as the complex amplitudes of the different frequency components in a random time signal. The cross-correlation function, Eq. (2.13), between the complex conjugate of $X(t_1)$ and $X(t_2)$ can be expressed in terms of a double sum of the harmonics, i.e.

$$E[X^*(t_1)X(t_2)] = E\left[\sum_{j=-n}^{n} \Phi_j^* e^{-i\omega_j t_1} \sum_{k=-n}^{n} \Phi_k e^{i\omega_k t_1}\right] = \sum_{j=-n}^{n}\sum_{k=-n}^{n} e^{i(\omega_k t_2 - \omega_j t_1)} E[\Phi_j^* \Phi_k] \tag{2.64}$$

For real functions, the above cross-correlation function is identical to the correlation function $R_X(t_1, t_2)$, Eq. (2.11), whereas for $t_1 = t_2$, it is equal to $R_X(0) = E[X^2(t)]$ as the expectation only involves the modulus of a complex function. In order to prove stationarity of the second order, it suffices to show that the correlation function is a function of the time difference only. This means that in the above double sum, only terms with identical subscripts, $j = k$, can be retained since then the exponent becomes $\omega_j(t_2 - t_1)$, and the off-diagonal terms must be equal to zero. Accordingly, the random variables Φ_j must satisfy the condition

$$E[\Phi_j^* \Phi_k] = S_j \delta_{jk} \tag{2.65}$$

where S_j are positive real coefficients and δ_{jk} is Kronecker's delta, ($\delta_{jk} = 1$ for $j = k$, $\delta_{jk} = 0$ for $j \neq k$). From electrical signal analogy, which is often used to clarify the physical meaning of the mathematical quantities associated with random functions, the set of coefficients S_j forms the energy spectrum of the signal. Clearly, S_j can be associated with the mean energy associated with the harmonic component of frequency ω_j as it is the mean value or expectation of the modulus square of the amplitude of that component. By retaining only the non-zero terms, the cross-correlation function, Eq. (2.64), reduces to

$$R_{X^*X}(t_1, t_2) = R_X(t_2 - t_1) = \sum_{j=-n}^{n} S_j e^{i\omega_j (t_2 - t_1)} \tag{2.66}$$

and the variance of the process or the average total energy is given by

$$\sigma_X^2 = R_X(0) = \sum_{j=-n}^{n} S_j \tag{2.67}$$

Therefore, for the random process $X(t)$, defined by the series Eq. (2.63), to be covariance stationary, the coefficients of the series must satisfy the condition Eq. (2.65).

The next step is to generalize the above procedure by increasing indefinitely the number of frequencies or components in the series as would be done in general Fourier analysis. Instead of a finite number of harmonic components, consider the case where the random function $X(t)$ can be represented by an infinite sum of such components with an

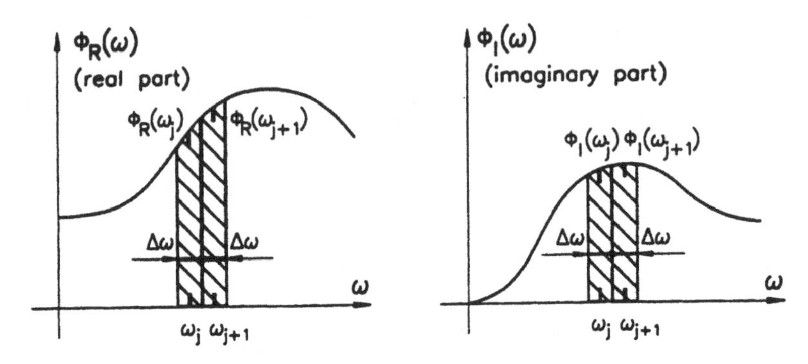

Fig. 2.11 *The random amplitude function $\Phi(\omega)$*

infinitesimal variation of the frequencies, that is, $\Delta\omega = \omega_{j+1} - \omega_j$. The amplitude of a frequency component in the frequency interval $\Delta\omega$ will then be represented by $\Delta\Phi(\omega_j) = (\Phi(\omega_{j+1}) - \Phi(\omega_j))\Delta\omega$, where a random function $\Phi(\omega)$ has been introduced instead of the discrete amplitudes Φ_j, Fig. 2.11. From Eq. (2.67), it is obvious that each harmonic component yields a positive contribution to the total energy of the process at any particular time instant. Therefore, by increasing the number of frequencies representing the process, it must be ensured that the variance stays limited or the variance of each component is correspondingly decreased. Further, the covariance stationarity of the process requires that the amplitudes of two different components from non-overlapping frequency intervals continues to be uncorrelated. In this case, the sum Eq. (2.63) will be written as an infinite series:

$$X(t) = \sum_j e^{i\omega_j t} \Delta\Phi(\omega_j) \tag{2.68}$$

Going to the limit, $\Delta\omega \to 0$, the infinite series turns into the Stieltjes integral

$$X(t) = \int_{-\infty}^{\infty} e^{i\omega t} \, d\Phi(\omega) \tag{2.69}$$

From the discussion of the differentiability of random functions, Eq. (2.24), it is obvious that the random function $\Phi(\omega)$ is not m.s. differentiable as the amplitudes are uncorrelated. Therefore, the above Stieltjes integral cannot be turned into an ordinary Riemann integral and the differential $d\Phi(\omega)$ must be retained. The properties of the infinitesimal increment $\Delta\Phi(\omega)$ can be obtained in an analogous manner as for the discrete case, Eq. (2.63), that is, the amplitudes in non-overlapping intervals must be uncorrelated so

$$E[\Delta\Phi^*(\omega_1)\Delta\Phi(\omega_2)] = S(\omega_1)\delta(\omega_1 - \omega_2)\Delta\omega_1 \Delta\omega_2 \tag{2.70}$$

where $S(\omega)$ is a positive function of its argument and corresponds to the non-negative coefficients S_j in the discrete case, Eq. (2.65), and $\delta(\omega)$ is Dirac's delta function.

Again, form the cross-correlation function, Eq. (2.64):

$$E[X^*(t_1)X(t_2)] = E\left[\int_{-\infty}^{\infty} e^{-i\omega_1 t_1}\, d\Phi^*(\omega_1) \int_{-\infty}^{\infty} e^{i\omega_2 t_2}\, d\Phi(\omega_2)\right]$$

$$= \int_{-\infty}^{\infty}\int_{-\infty}^{\infty} e^{i(\omega_2 t_2 - \omega_1 t_1)}\, E[d\Phi^*(\omega_1)d\Phi(\omega_2)]$$

$$= \int_{-\infty}^{\infty}\int_{-\infty}^{\infty} e^{i(\omega_2 t_2 - \omega_1 t_1)}\, S(\omega_1)\delta(\omega_1 - \omega_2)d\omega_1\, d\omega_2$$

$$= \int_{-\infty}^{\infty} e^{i\omega(t_2 - t_1)}\, S(\omega)d\omega = R_X(t_2 - t_1) \tag{2.71}$$

using the shifting property of the delta function. As the process is covariance stationary, it has been shown that there exists a Fourier transform relationship between the autocorrelation function $R_X(\tau)$ and a spectral non-negative function $S(\omega)$. This fundamental connection between the autocorrelation function and the spectral function $S(\omega)$ that can be regarded as a mean energy density of the random process was obtained separately by Norbert Wiener and A.J. Khinchine even before the possibility of the spectral representation of the process itself, Eq. (2.69), was indicated, [234], [106].

Considering the fundamental properties of the autocorrelation function of a stationary stochastic process with a mean value function equal to zero, Eqs. (2.48), (2.49), (2.51), it is clear that the autocorrelation function satisfies the Dirichlet condition, Eq. (7.11), or

$$\int_{-\infty}^{\infty} |R_X(\tau)|\, d\tau = c < \infty \tag{2.72}$$

In this case, the autocorrelation belongs to the class of functions which are suitable for Fourier analysis in the extended form and can be represented by the Fourier integral

$$R_X(\tau) = \int_{-\infty}^{\infty} S_X(\omega)e^{i\omega\tau}\, d\omega \tag{2.73}$$

Comparing with the result Eq. (2.71), it is now clear that the orthogonality relation Eq. (2.65) is valid when and only when a spectral function $S_X(\omega)$ exists that is the Fourier transform of the autocorrelation function, Eq. (2.73), and has the properties of an energy density spectrum, i.e. is a non-negative function of ω. Consequently, a knowledge of the autocorrelation function of the process is equivalent to knowledge of its mean energy density spectrum. In this case, every covariance stationary random process with a zero mathematical expectation has the spectral representation

$$X(t) = \int_{-\infty}^{\infty} e^{i\omega t}\, d\Phi(\omega) \tag{2.69}$$

where $\Phi(\omega)$ is a non-differentiable random function the differentials of which satisfy the orthogonality condition

$$E[d\Phi^*(\omega_1)d\Phi(\omega_2)] = S(\omega_1)\delta(\omega_1 - \omega_2)d\omega_1\, d\omega_2 \tag{2.74}$$

Consider the stationary stochastic process $\{X(t), t \in T\}$ with an autocorrelation function $R_X(\tau)$, which satisfies the Dirichlet condition. Its Fourier transform can then be written as

$$S_X(\omega) = \frac{1}{2\pi} \int_{-\infty}^{\infty} R_X(\tau) e^{-i\omega\tau} d\tau \tag{2.75}$$

which is the inverse transform of the autocorrelation function given by Eq. (2.73). In ordinary Fourier analysis, the factor $1/2\pi$ is more often kept as a part of the first Fourier integral, Eq. (2.73). However, when dealing with natural frequencies $f = \omega/2\pi$, it is more appropriate to use the above definition, cf. Eqs. (2.80) and (2.81). The thus defined spectral function $S_X(\omega)$, which has been shown to represent the energy density spectrum of the process, is called the mean square spectral density and is also referred to as the power spectral density. As has already been discussed, it is a measure of the power content of the process, $S_X(\omega)\Delta\omega$, contained in a infinitesimal frequency band of a width $\Delta\omega$ at the frequency ω. By putting $\tau = 0$ in Eq. (2.73), one obtains the mean square value of the process, that is,

$$R_X(0) = E[X^2(t)] = \int_{-\infty}^{\infty} S_X(\omega) d\omega \tag{2.76}$$

$R_X(0)$ can be interpreted as the power content of the process at any time instant t and is equal to the entire area under the power spectral density curve, Fig. 2.12. The value of the power spectral density at zero frequency on the other hand, is equal to entire area under the autocorrelation function, i.e.

$$S_X(0) = \frac{1}{2\pi} \int_{-\infty}^{\infty} R_X(\tau) d\tau \tag{2.77}$$

This implies that for finite values of the power spectral density at zero frequency, the autocorrelation function is integrable, which is only possible for processes with a zero mean value, the fundamental condition for the spectral representation Eq. (2.69) to exist. The presence of a constant mean value or d.c. component in a stochastic process obviously violates the Dirichlet condition for the autocorrelation function. However, such irregularities can be dealt with using the delta function. Thus a spike or delta function in the power spectral density at zero frequency indicates the presence of a d.c. component, i.e. a non-zero mean value.

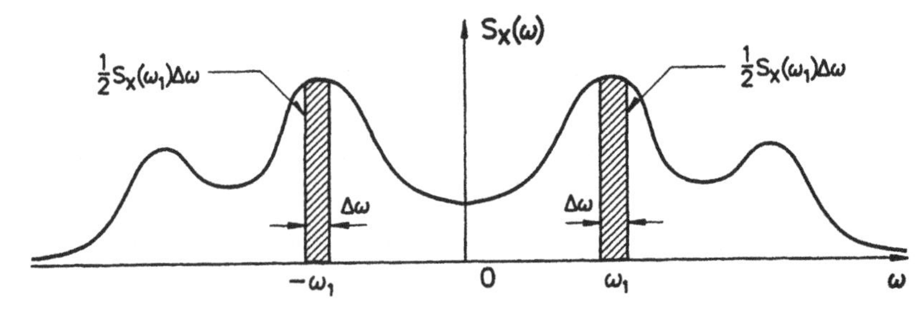

Fig. 2.12 *The power content of a random signal at any given time instant*

If the process $X(t)$ contains periodic components, the autocorrelation function also contains the same periodic components (cf. Example 2.7). Strictly speaking the power spectral density does not exist at these frequencies. However, as shown in Chapter 7, Example 7.7, the introduction of Dirac's delta function allows the periodic components to be treated separately and $S_X(\omega)$ will consist of a continuous part and a series of spikes, (impulses), at the controversial frequencies.

Example 2.11 A stationary stochastic displacement process $X(t)$ has the autocorrelation function (Example 2.10),

$$R_X(\tau) = e^{-\tau^2}(1-\tau^2)$$

Find the power spectral density of the displacement $X(t)$, the velocity $\dot{X}(t)$ and the acceleration $\ddot{X}(t)$.

Solution By Eq. (2.75)

$$S_X(\omega) = \frac{1}{2\pi}\int_{-\infty}^{\infty}(1-\tau^2)e^{-\tau^2}e^{-i\omega\tau}\,d\tau = \frac{1}{2\pi}\int_{-\infty}^{\infty}e^{-\omega^2/4}(1-\tau^2)e^{-(\tau^2+i\omega\tau-\omega^2/4)}\,d\tau$$

$$= \frac{1}{2\pi}e^{-\omega^2/4}\left[\int_{-\infty}^{\infty}\exp[-(\tau+i\omega/2)^2]\,d\tau - \int_{-\infty}^{\infty}\tau^2\exp[-(\tau+i\omega/2)^2/(2(\sqrt{\tfrac{1}{2}})^2)]\,d\tau\right]$$

$$= \frac{1}{2\pi}e^{-\omega^2/4}\left[1-\sqrt{\tfrac{1}{2}}\sqrt{2\pi}\left(\tfrac{1}{2}-\frac{\omega^2}{4}\right)\right] = \frac{1}{2\pi}e^{-\omega^2/4}\left[1-\frac{\sqrt{\pi}}{2}+\sqrt{\pi}\frac{\omega^2}{4}\right]$$

As shown in Chapter 7, Eq. (7.17), the easiest way to obtain the power spectral densities of the derivative processes is to differentiate Eq. (2.73), that is,

$$R_X(\tau) = \int_{-\infty}^{\infty} S_X(\omega)e^{i\omega\tau}\,d\omega$$

$$R_X'(\tau) = \int_{-\infty}^{\infty}(i\omega)S_X(\omega)e^{i\omega\tau}\,d\tau$$

$$-R_X''(\tau) = -\int_{-\infty}^{\infty}(-\omega^2)S_X(\omega)e^{i\omega\tau}\,d\tau = R_{\dot{X}}(\tau)$$

so

$$S_{\dot{X}}(\omega) = \omega^2 S_X(\omega) = \frac{1}{2\pi}e^{-\omega^2/4}\left[\omega^2\left(1-\frac{\sqrt{\pi}}{2}\right)+\sqrt{\pi}\frac{\omega^4}{4}\right]$$

$$R_X'''(\tau) = \int_{-\infty}^{\infty}(i\omega)^3 S_X(\omega)e^{i\omega\tau}\,d\tau$$

$$R_X^{(iv)}(\tau) = \int_{-\infty}^{\infty}\omega^4 S_X(\omega)e^{i\omega\tau}\,d\tau = -R_{\dot{X}}(\tau) = R_{\ddot{X}}(\tau)$$

$$S_{\ddot{X}}(\omega) = \omega^4 S_X(\omega) = \frac{1}{2\pi}e^{-\omega^2/4}\left[\omega^4\left(1-\frac{\sqrt{\pi}}{2}\right)+\frac{\sqrt{\pi}}{4}\omega^6\right]$$

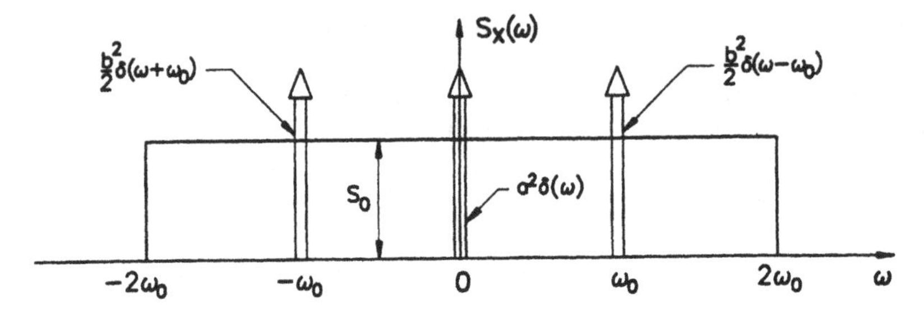

Fig. 2.13 *A power spectral density function*

Example 2.12 A stationary random process $X(t)$ is ergodic in the mean and has the power spectral density $S_X(\omega)$ shown in Fig. 2.13. Find the mean value $E[X]$, the mean square value $E[X^2]$ and determine the shape of the autocorrelation function $R_X(\tau)$.

Solution First find the autocorrelation function $R_X(\tau)$.

$$R_X(\tau) = \int_{-\infty}^{\infty} (a^2\delta(\omega) + \frac{b^2}{2}\delta(\omega \pm \omega_0) + S_0)\mathrm{e}^{\mathrm{i}\omega\tau}\mathrm{d}\omega$$

$$= a^2 + \frac{b^2}{2}\mathrm{e}^{\pm\mathrm{i}\omega_0\tau} + 4S_0\omega_0\frac{\mathrm{e}^{2\mathrm{i}\omega_0\tau} - \mathrm{e}^{-2\mathrm{i}\omega_0\tau}}{2\mathrm{i}(2\omega_0\tau)}$$

$$= a^2 + b^2\cos\omega_0\tau + 4S_0\omega_0\left(\frac{\sin(2\omega_0\tau)}{(2\omega_0\tau)}\right)$$

Therefore, the autocorrelation function consists of three parts: (1) the mean value $E[X] = \pm a$, (cf. Eq. (2.49), (2) the periodic component with period $2\pi/\omega_0$ and (3) a function of the type $(\sin x)/x$ as shown in Fig. 2.14. Also,

$$R_X(0) = a^2 + b^2\cos\omega_0\tau + 4S_0\omega_0 = E[X^2]$$

or

$$E[X^2] = \int_{-\infty}^{\infty} S_X(\omega)\mathrm{d}\omega = a^2 + b^2\cos\omega_0\tau + 4S_0\omega_0$$

Note that the mean value can immediately be picked out as the square root of the area of the spike at zero frequency, and the mean square value is the area under the density plus the area of the spikes.

It should be noted that since the autocorrelation function is both real and an even function of τ, Eqs. (2.73) and (2.75) can be rewritten as cosine transforms:

$$S_X(\omega) = \frac{1}{\pi}\int_0^{\infty} R_X(\tau)\cos(\omega\tau)\mathrm{d}\tau \qquad (2.78)$$

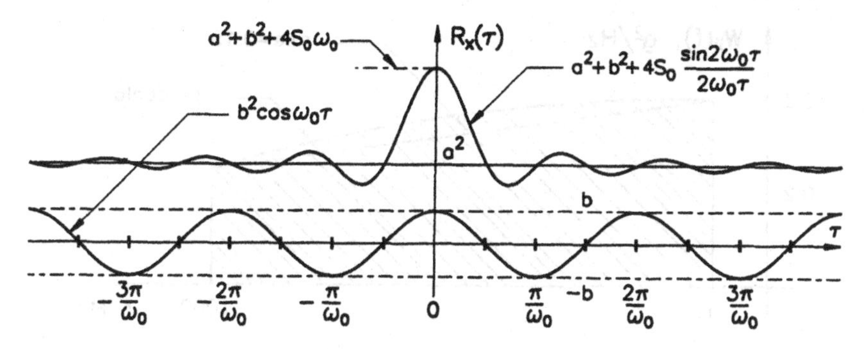

Fig. 2.14 *The autocorrelation function*

and

$$R_X(\tau) = 2\int_0^\infty S_X(\omega)\cos(\omega\tau)\mathrm{d}\omega \tag{2.79}$$

since by Eq. (2.77), $S_X(\omega)$ is also even and real. In this present form, the above equations are called the Wiener–Khinchine relations and it is noteworthy that the physically impossible negative frequencies have been avoided.

In data processing and measurement analysis, it is often more convenient to work with the natural frequency $f = \omega/2\pi$ instead of the circular frequency ω. The Wiener–Khinchine relations will then be rewritten as

$$W_X(f) = 4\int_0^\infty R_X(\tau)\cos(2\pi f\tau)\mathrm{d}\tau \tag{2.80}$$

$$R_X(\tau) = \int_0^\infty W_X(f)\cos(2\pi f\tau)\mathrm{d}f \tag{2.81}$$

where $W_X(f) = 4\pi S_X(\omega)$, the 'experimental' spectral density, is only defined for positive frequencies. The reason for keeping the factor $1/2\pi$ by first of Eqs. (2.73) and (2.75) is now clear.

Example 2.13 A stationary acceleration process has the experimental spectral density shown in Fig. 2.15. Find the mean value and the r.m.s. g level of the process.

The mean value must be zero since there is no concentration of power at zero frequency (no spike or delta function). The average power or mean square value is

$$E[\ddot{X}^2] = \int_{-\infty}^{\infty} S_{\ddot{X}}(\omega)\mathrm{d}\omega = \int_0^\infty W_{\ddot{X}}(f)\mathrm{d}f$$

$$= 0.15 \times 990 \times 2/3 + 0.5 \times 990 \times 1/2 = 346\ \mathrm{g}^2$$

$$\mathrm{RMS}[\ddot{X}(t)] = \sqrt{(346\,g^2)} = 18.6g = 182\,\mathrm{cm/s^2}$$

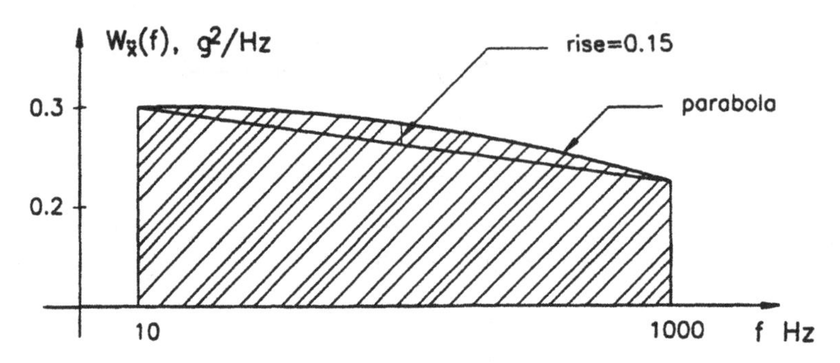

Fig. 2.15 *An experimental spectral density $W_X(f)$*

Example 2.14 White noise Consider a stationary random process, $\{X(t), -\infty < t < \infty\}$, with a power spectral density constant over the entire frequency range, Fig. 2.16. Since the spectral density is constant, the inverse Fourier transform is an impulse at the origin, that is,

$$R_X(\tau) = 2\pi S_0 \delta(\tau) \Leftrightarrow S_X(\omega) = S_0$$

This peculiar process has a constant power at all frequencies which corresponds to the energy distribution in a white light from an incandescent body, which has a spectrum that is approximately constant over the range of all visible frequencies. Therefore, any stationary random process with the above properties is given the name white noise. Obviously,

$$E[X^2(t)] = \int_{-\infty}^{\infty} S_X(\omega)\,\mathrm{d}\omega = R_X(0) = \infty$$

Therefore, the process has an infinite power and as such is nothing but a mathematical fiction. It is also completely memory-less since $R_X(\tau) = E[X(t)X(t+\tau)] \sim \delta(\tau)$.

Sample functions of white noise are characterized by a very violent oscillatory behaviour which is vividly displayed by the high-frequency components. The white noise process, whether fictitious or not, is an extremely useful tool in time series analysis. It is easily simulated or generated by a computer and through proper filtering, stationary processes with almost any prescribed spectral density functions can be constructed.

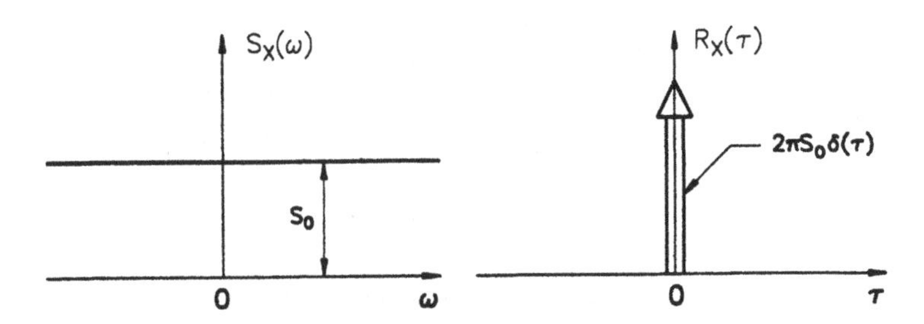

Fig. 2.16 *Power spectral density and autocorrelation function of white noise*

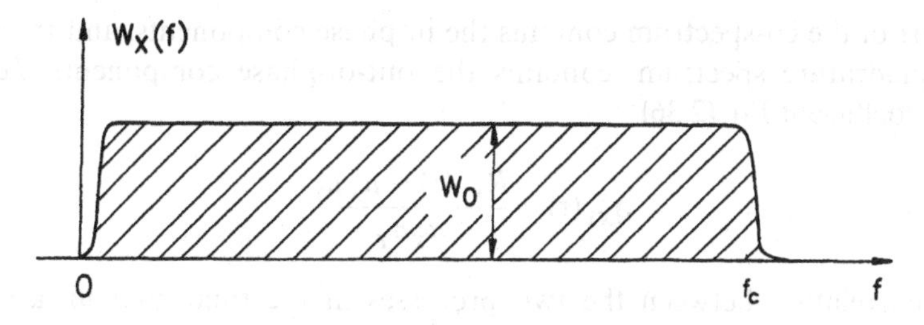

Fig. 2.17 *The power spectral density of band-limited white noise*

Example 2.15 Band-limited white noise A stationary, stochastic process $X(t)$ has a constant spectral density over a wide range of frequencies with a frequency cut-off at $f = f_c$, Fig. 2.17, that is,

$$W_X(f) = W_0 \qquad \text{for } 0 \leqslant f \leqslant f_c$$

The mean power is simply $E[X^2] = \int_0^\infty W_X(f)\mathrm{d}f = W_0 f_c$ and the autocorrelation is, Eq. (2.46),

$$R_X(\tau) = \int_0^{f_c} W_X(f)\cos(2\pi f\tau)\mathrm{d}f = \frac{W_0}{2\pi\tau}\sin(2\pi f_c\tau)$$

or since $4\pi S_0 = W_0$ and $\omega = 2\pi f$,

$$R_X(\tau) = 2S_0\omega_c \frac{\sin(\omega_c\tau)}{\omega_c\tau}$$

If the cut-off frequency ω_c grows very large, the autocorrelation function approaches the delta function and the conditions of pure white noise.

A spectrum for the cross-correlation functions of two stationary process, $X(t)$ and $Y(t)$, can defined in a similar manner as the mean spectral density, Eq. (2.75).

$$S_{XY}(\omega) = \frac{1}{2\pi}\int_{-\infty}^{\infty} R_{XY}(\tau)\mathrm{e}^{-\mathrm{i}\omega\tau}\mathrm{d}\tau \tag{2.82}$$

and

$$R_{XY}(\tau) = R_{YX}(-\tau) = \int_{-\infty}^{\infty} S_{XY}(\omega)\mathrm{e}^{\mathrm{i}\omega\tau}\mathrm{d}\omega \tag{2.83}$$

In general, $S_{YX}(\omega)$ is a complex function of ω since $R_{XY}(\tau)$ is no longer an even function of τ. It is therefore customary to break the cross-spectrum up into its real and imaginary parts, that is,

$$S_{XY}(\omega) = \mathrm{Co}_{XY}(\omega) - \mathrm{i}\,\mathrm{Qu}_{XY}(\omega)$$

and

$$S_{YX}(\omega) = S^*_{XY}(\omega) = \mathrm{Co}_{XY}(\omega) + \mathrm{i}\,\mathrm{Qu}_{XY}(\omega) \tag{2.84}$$

The real part of the co-spectrum contains the in-phase components, and the imaginary part, the quadrature spectrum, contains the out-of-phase components. Just as the correlation coefficient Eq. (2.36)

$$\varrho_{XY}(\tau)=\frac{R_{XY}(\tau)-\mu_X\mu_Y}{\sigma_X\sigma_Y}$$

shows the correlation between the two processes in the time domain, a correlation function which describes the correlation in the frequency domain can be defined. The coherence function

$$\mathrm{Coh}_{XY}(\omega)=\frac{|S_{XY}(\omega)|^2}{S_X(\omega)S_Y(\omega)} \tag{2.85}$$

provides such information. Clearly

$$-1\leqslant \mathrm{Coh}_{XY}(\omega)\leqslant 1 \tag{2.86}$$

where (-1) corresponds to perfect antiphase coherence, $(Y(t)=-\alpha X(t))$, and $(+1)$ corresponds to perfect in-phase coherence, $(Y(t)=\beta X(t))$.

Now introducing the phase function, defined as

$$\Phi_{XY}(\omega)=\mathrm{Arctan}(\mathrm{Qu}_{XY}(\omega)/\mathrm{Co}_{XY}(\omega)) \tag{2.87}$$

The co-spectrum can then be written as

$$\mathrm{Co}_{XY}(\omega)=Z_{XY}(\omega)\sqrt{S_X(\omega)S_Y(\omega)}\cos\Phi_{XY}(\omega) \tag{2.88}$$

and the quadrature spectrum as

$$\mathrm{Qu}_{XY}(\omega)=Z_{XY}(\omega)\sqrt{S_X(\omega)S_Y(\omega)}\sin\Phi_{XY}(\omega) \tag{2.89}$$

in which $Z_{XY}(\omega)=\sqrt{(\mathrm{Coh}_{XY}(\omega))}$

Example 2.16 Consider two stationary random processes

$$X(t)=A\cos(\omega_0 t+\theta)$$
$$Y(t)=B\sin(\omega_0 t+\theta-\varphi)$$

where A and B are two random amplitudes with zero mean, equal variances and correlation coefficient $\varrho=1/2$. θ is a random phase, uniformly distributed between 0 and 2π, independent from A and B, and φ is a fixed value phase lag. Find the cross-correlation function and the cross-spectrum. What is the coherence and phase between the two processes?

Solution

$$R_{XY}(\tau)=E[X(t)\,Y(t+\tau)]=E[AB\cos(\omega_0 t+\theta)\sin(\omega_0+\theta-\varphi)]$$
$$=\tfrac{1}{2}E[AB]E[\sin(2\omega_0 t+\omega_0\tau+2\theta-\varphi)+\sin(\omega_0\tau-\varphi)]=\tfrac{1}{2}\varrho\sigma^2\sin(\omega_0\tau-\varphi)$$

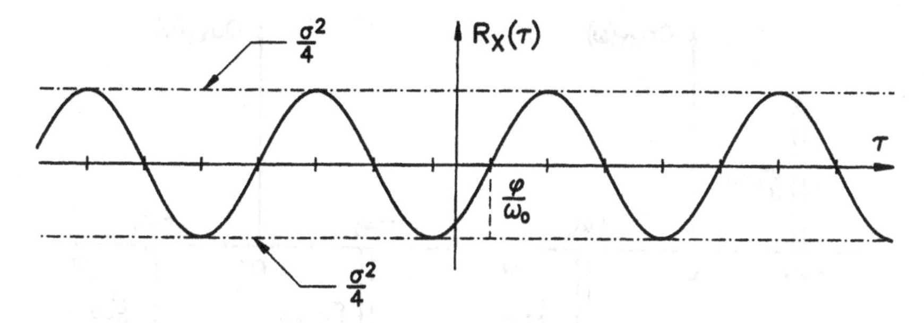

Fig. 2.18 *The cross correlation function $R_{XY}(\tau)$*

since

$$\left(+\frac{\varrho\sigma^2}{2}\int_0^{2\pi}\sin(2\omega_0 t+\omega_0\tau+2\theta-\varphi)\frac{1}{2\pi}\mathrm{d}\theta=0\right)$$

Therefore

$$R_{XY}(\tau)=\tfrac{1}{4}\sigma^2\sin(\omega_0\tau-\varphi)$$

The cross-correlation function $R_{XY}(\tau)$ is depicted in Fig. 2.18.

Note that since $Y(t)$ is trailing $X(t)$ by the phase lag φ, the positive maximum closest to the origin lies at the right-hand side of the origin.

The cross spectrum is obtained as follows:

$$S_{XY}(\omega)=\frac{1}{2\pi}\int_{-\infty}^{\infty}R_{XY}(\tau)\mathrm{e}^{-\mathrm{i}\omega\tau}\mathrm{d}\tau=\frac{\sigma^2}{8\pi}\int_{-\infty}^{\infty}\frac{\mathrm{e}^{+\mathrm{i}(\omega_0\tau-\varphi}-\mathrm{e}^{-\mathrm{i}(\omega_0\tau-\varphi)}}{2\mathrm{i}}\mathrm{e}^{-\mathrm{i}\omega\tau}\mathrm{d}\tau$$

$$=\frac{\sigma^2}{16\mathrm{i}\pi}\int_{-\infty}^{\infty}(\mathrm{e}^{-\mathrm{i}\varphi}\mathrm{e}^{\mathrm{i}(\omega_0-\omega)\tau}-\mathrm{e}^{\mathrm{i}\varphi}\mathrm{e}^{-\mathrm{i}(\omega_0+\omega)\tau})\mathrm{d}\tau$$

$$=\frac{\sigma^2}{8\mathrm{i}}[\mathrm{e}^{-\mathrm{i}\varphi}\delta(\omega_0-\omega)-\mathrm{e}^{\mathrm{i}\varphi}\delta(\omega_0+\omega)]$$

$$=\frac{\sigma^2}{8\mathrm{i}}[\mathrm{e}^{-\mathrm{i}\varphi}\delta(\omega_0-\omega)-\mathrm{e}^{\mathrm{i}\varphi}\delta(\omega_0+\omega)]$$

$$=\frac{\sigma^2}{8\mathrm{i}}[(\mathrm{i}\cos\varphi-\sin\varphi)\delta(\omega_0-\omega)+(\mathrm{i}\cos\varphi-\sin\varphi)\delta(\omega_0+\omega)]$$

so

$$\mathrm{Co}_{XY}(\omega)=-\frac{\sigma^2}{8}\sin\varphi[\delta(\omega_0-\omega)+\delta(\omega_0+\omega)]$$

and

$$\mathrm{Qu}_{XY}(\omega)=-\frac{\sigma^2}{8}\cos\varphi[\delta(\omega_0-\omega)+\delta(\omega_0+\omega)]$$

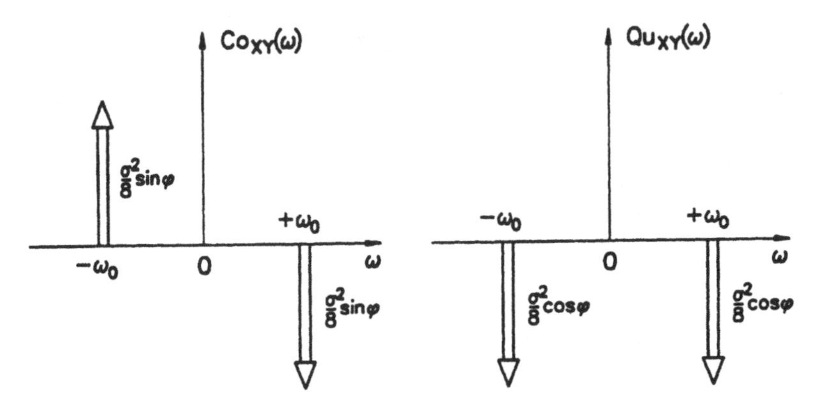

Fig. 2.19 *The co-spectrum and quadrature spectrum*

The trailing process $Y(t)$ can be written as

$$Y(t) = B\cos\varphi \sin(\omega_0 t + \theta) - B\sin\varphi \cos(\omega_0 t + \theta)$$

that is, the first part is the in-phase component and is contained in the co-spectrum, and the second part is the out-of-phase component and is contained in the quadrature spectrum as shown in Fig. 2.19.

The phase and coherence respectively are

$$\Phi_{XY}(\omega) = \text{Arctan}\left(\frac{\text{Qu}_{XY}(\omega)}{\text{Co}_{XY}(\omega)}\right) = \tan^{-1}\left(\tan\left(\frac{\pi}{2} - \varphi\right)\right) = \frac{\pi}{2} - \varphi$$

and by Eq. (2.73)

$$-\frac{\sigma^2}{8}\sin\varphi\,[\delta(\omega_0 - \omega) + \delta(\omega_0 + \omega)] = \sqrt{\text{Coh}_{XY}(\omega)}\cdot S_X(\omega)\cos\left(\frac{\pi}{2} - \varphi\right)$$

Obviously, $R_X(\tau) = R_Y(\tau) = \sigma^2 \cos\omega\tau/2$, (see Example 2.7), and

$$S_X(\omega) = S_Y(\omega) = \frac{\sigma^2}{2}[\delta(\omega_0 - \omega) + \delta(\omega_0 + \omega)]$$

so

$$\text{Coh}_{XY}(\omega) = \tfrac{1}{16}$$

Example 2.17 The wind flow around the face of a large building will be highly dependent on the spatial correlation of the gusts buffeting the face, [194], [208]. The spatial dimensions of the gusts are frequency dependent and only the low frequency gusts will have sufficiently large wavelengths to be able to engulf the face of the building. The dynamical excitation of the building, however, is mainly due to the high frequency gusts with smaller wavelengths and consequently only a portion of the face of the building is affected during a short wind gust. This lack of coherence can considerably reduce the

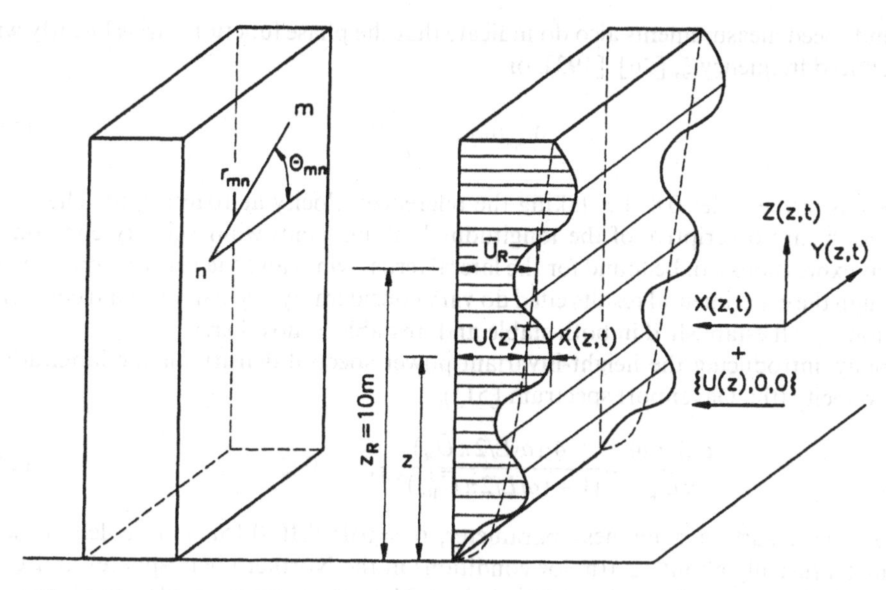

Fig. 2.20 *Wind velocity profiles and along-wind loading of a large bluff building*

pressure on the building due to the gusts. Experimental evaluation of the coherence function, [36], [193], shows that the loss of coherence is exponential with an argument proportional to the reduced frequency

$$\xi = \frac{\omega r_{mn}}{\bar{U}_R} \tag{2.90}$$

where r_{mn} is the distance between two points (m, n) on the face and $\bar{U}_R$ is the mean reference wind speed, Fig. 2.20. The coherence between the immediate wind velocities at the two points (m, n) is therefore

$$\text{Coh}_{mn}(\xi) = \exp[-2b\xi] \tag{2.91}$$

The value of the decay constant b is of the order 1–2 for the vertical separation and appears to decrease with increasing height and decreasing eddies, but to increase with increasing mean wind speed. However, these variations are rather small. More or less the same holds for the horizontal separation in which case the decay constant, now termed a, is found to be of the order 1–4. In the wind climate of the northern hemisphere, anisotropy with $a \approx 3b$ is indicated and typical values may be assumed to be $a \approx 3.8$ and $b \approx 1.3$. This may be interpreted in the manner that the decay constants in the coherence function describe an ellipse with its major and minor axis equal to $2a$ and $2b$ respectively, [208]. Introducing the angle θ_{mn} between the distance r_{mn} and the horizontal, (Fig. 2.20), the coherence function may be extended to cover all directions, that is,

$$\text{Coh}_{mn}(\omega) = \exp\left[-\frac{2ab}{\bar{U}_R}\frac{\omega r_{mn}}{\sqrt{a^2 \sin^2\theta_{mn} + b^2 \cos^2\theta_{mn}}}\right] \tag{2.92}$$

Wind speed measurements also do indicate that the phase function varies linearly with the reduced frequency ξ, [36], [193], or

$$\Phi_{mn}(\omega) = c\frac{\omega r_{mn}}{\bar{U}_R} \tag{2.93}$$

where c is of the order 0.8–1.3, taking the reference velocity at 10 m height. The above expressions are descriptive of the longitudinal (along wind) wind velocity component. Similar expressions can be found for the lateral (cross wind) and the vertical components. Although the experimental results cited do vary considerably, the narrow band coherence function is well established in both stable and unstable atmospheres.

Finally, introducing the height-invariant power spectral density for the longitudinal wind velocity (the Davenport spectrum [51]),

$$\frac{\omega S_X(\omega)}{\kappa \bar{U}_R^2} = \frac{4\cdot(\omega L/2\pi \bar{U}_R)}{(1+(\omega L/2\pi \bar{U}_R^2))^{4/3}}, \qquad 0 \leqslant \omega \leqslant \infty \tag{2.94}$$

where κ is a surface roughness parameter, ($\kappa = 0.05, 0.10, 0.15$). L is a length scale parameter (usually about 1200 m for conditions in the Northern hemisphere), and U_R is the mean reference wind velocity at 10 m height. The cross spectrum for longitudinal wind velocities (gusts) can now be constructed from the above relations:

$$\frac{\omega \mathrm{Co}_{mn}(\omega)}{\kappa \bar{U}_R^2} = \frac{4(\omega L/2\pi \bar{U}_R)^2}{(1+\omega L/2\pi \bar{U}_R^2)^{4/3}}\cdot \exp\left[-\frac{ab\omega}{\bar{U}_R}\cdot\frac{1}{\sqrt{a^2\sin^2\theta_{mn}+b^2\cos^2\theta_{mn}}}\right] \tag{2.95}$$

and the expression for the quadrature spectrum is the same except the cosine is replaced by a sine term.

Consider next a truncated sample $\{x_T(t), -T \leqslant t \leqslant T\}$ of the stochastic process $X(t)$. If x_T is considered to be zero outside the interval $[-T, T]$, the Fourier transform of the truncated sample can be defined and

$$A_X(\omega) = \int_{-\infty}^{\infty} x_T(t)\,\mathrm{e}^{-\mathrm{i}\omega t}\,\mathrm{d}t = \int_{-T}^{T} x(t)\,\mathrm{e}^{-\mathrm{i}\omega t}\,\mathrm{d}t \tag{2.96}$$

The truncated sample is an integrable function and hence the Fourier transform and the inverse transform

$$x_T(t) = \frac{1}{2\pi}\int_{-\infty}^{\infty} A_X(\omega)\,\mathrm{e}^{\mathrm{i}\omega\tau}\,\mathrm{d}\omega \tag{2.97}$$

both exist. The total energy carried by the truncated sample (the signal power) is

$$\int_{-\infty}^{\infty} x_T^2(t)\,\mathrm{d}t = \frac{1}{2\pi}\int_{-\infty}^{\infty} |A_X(\omega)|^2\,\mathrm{d}\omega \tag{2.98}$$

according to Parseval's theorem, Eq. (7.23). Therefore an energy density spectrum for the aperiodic sample function is defined as

$$\Phi_X(\omega, T) = \frac{1}{2\pi}|A_X(\omega)|^2 \tag{2.99}$$

and a power spectrum or the mean average spectrum is

$$S_X(\omega, T) = \frac{1}{2T}\Phi_X(\omega, T) = \frac{1}{4\pi T}|A_X(\omega)|^2 \qquad (2.100)$$

$S_X(\omega, T)$ is the power spectral density of the sample $x_T(t)$, and is by definition a random-variable dependent on the sample duration $2T$.

For a truncated sample $x_T(t)$, the temporal autocorrelation function $R_X(T, \tau)$, Eq. (2.37), was already introduced. It will now be shown that the temporal autocorrelation function and the mean average spectrum form a Fourier transform pair. Starting with Eqs. (2.99) and (2.97),

$$S_X(\omega, T) = \frac{A_X(\omega)A_X^*(\omega)}{4\pi T} = \frac{1}{4\pi T}\int_{-T}^{T} x(t)\,\mathrm{e}^{\mathrm{i}\omega t}\,\mathrm{d}t \int_{-T}^{T} x(s)\,\mathrm{e}^{-\mathrm{i}\omega s}\mathrm{d}s$$

or

$$S_X(\omega, T) = \frac{1}{2\pi}\frac{1}{2T}\int_{-T}^{T}\int_{-T}^{T} x(t)\,x(s)\,\mathrm{e}^{\mathrm{i}\omega(t-s)}\,\mathrm{d}t\,\mathrm{d}s$$

The following change of variables $t = t$ and $\tau = s - t$ will transform the above double integral as follows (cf. Fig. 2.22)

$$S_X(\omega, T) = \frac{1}{2\pi}\int_{-T-t}^{T-t}\left[\frac{1}{2T}\int_{-T}^{T} x(t)\,x(t+\tau)\,\mathrm{d}t\right]\mathrm{e}^{-\mathrm{i}\omega\tau}\mathrm{d}\tau \qquad (2.101)$$

and by Eq. (2.37)

$$S_X(\omega, T) = \frac{1}{2\pi}\int_{-T-t}^{T-t} R_X(\tau, T)\mathrm{e}^{-\mathrm{i}\omega\tau}\,\mathrm{d}\tau = \frac{1}{2\pi}\int_{-\infty}^{\infty} R_X(\tau, T)\mathrm{e}^{-\mathrm{i}\omega\tau}\,\mathrm{d}\tau \qquad (2.102)$$

since $R_X(\tau, T) = 0$ for $(\tau < (-T - t)) \cap (\tau > (T - t))$, so $S_X(\omega, T)$ and $R_X(\tau, T)$ form a Fourier transform pair. If in Eq. (2.100), the sample function $x(t)$ is replaced by the process itself, i.e. $X(t)$, both sides display random variables. Taking the expectations,

$$E[S_X(\omega, T)] = \frac{1}{4\pi}\int_{-T-t}^{T-t}\int_{-T}^{T} R_X(\tau)\mathrm{e}^{-\mathrm{i}\omega\tau}\,\mathrm{d}\tau \qquad (2.103)$$

where $R_X(\tau)$ is now the ensemble autocorrelation function.

The double integral Eq. (2.103) is integrated over the two regions (1) and (2) separately, Fig. 2.21. Then

$$E[S_X(\omega, T)] = \frac{1}{4\pi T}\left[\int_{-2T}^{0} R_X(\tau)\mathrm{e}^{-\mathrm{i}\omega\tau}\,\mathrm{d}\tau\int_{-T-t}^{T}\mathrm{d}t + \int_{0}^{2T} R_X(\tau)\mathrm{e}^{-\mathrm{i}\omega\tau}\,\mathrm{d}\tau\int_{-T}^{T-t}\mathrm{d}t\right]$$

$$= \frac{1}{2\pi}\int_{-2T}^{2T} R_X(\tau)\mathrm{e}^{-\mathrm{i}\omega\tau}\left(1 - \frac{|\tau|}{2T}\right)\mathrm{d}\tau$$

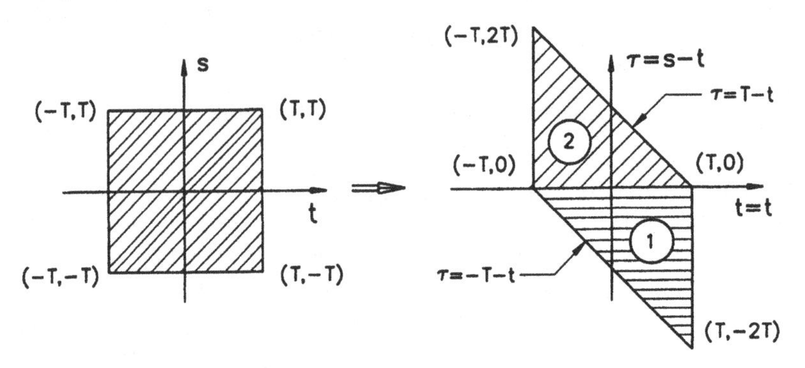

Fig. 2.21 *Domains of integration*

By letting $T \to \infty$,

$$\underset{T\to\infty}{E}\;[S_X(\omega, T)] = \frac{1}{2\pi}\int_{-\infty}^{\infty} R_X(\tau)\,\mathrm{e}^{-i\omega\tau}\,\mathrm{d}\tau = S_X(\omega) \tag{2.104}$$

So in the limit, the mean average spectrum becomes equal to the power spectral density. Thus at least formally, both the autocorrelation function and the power spectral density have been obtained from one single observation or realization of the process. However, a word of caution is in order. The manner in which the so-called periodogram function of Eq. (2.86) approaches the limit, that is,

$$\lim_{T\to\infty} \frac{1}{4\pi T}|A_X(\omega)|^2 = S_X(\omega) \tag{2.105}$$

needs to be investigated in each case. By Eq. (2.43), it was possible to prove that the variance of the estimated value became zero in the limit, whereby the expected value and the limit are one and the same with probability one. In some cases, the variance of the estimated mean average spectrum $S_X(\omega, T)$, does not approach zero as $T \to \infty$. The Gaussian process is an example of this behaviour. Therefore, measurements of the spectral density can provide questionable estimates, [44].

Finally, the concept of windowing and envelope functions should be briefly explained. Instead of working with truncated signals, it is possible to apply deterministic envelope functions that modify the process in a desired manner. Consider the stochastic process $\{X(t), -\infty < t < \infty\}$. The sample functions of this process $x(t)$ can be modified by applying an envelope function $\psi(t)$, which may have the appearance shown in Fig. 2.23, that is, it is zero for negative values of t and for values above a certain time limit T or it can approach zero for large values of t, and has a constant value in a certain time range $t_1 < t < t_2$. A new modified stochastic process $Y(t)$ is therefore defined as

$$Y(t) = \psi(t)\,X(t) \tag{2.106}$$

and the sample functions are given in the same manner. Actually it can be said that the process is viewed through a data window defined by the envelope function $\psi(t)$. In fact, the

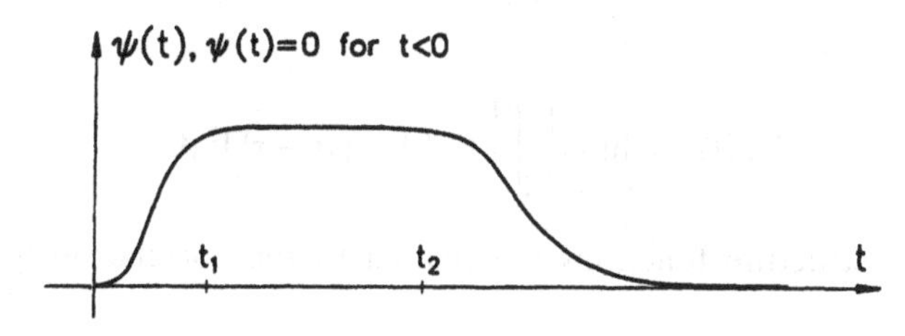

Fig. 2.22 *A deterministic envelope function*

previously studied truncated sample function $\{x_T(t),\ -T \leqslant t \leqslant T\}$ can be obtain by applying an envelope function, which is constant in the time range $-T \leqslant t \leqslant T$ and zero elsewhere. In signal analysis terminology, which will be covered in more detail in Chapter 7, it is said that the process is viewed through a box car window. The autocorrelation function of the process $Y(t)$ can be formed in the usual manner, that is,

$$R_Y(\tau) = [\psi(t)\,X(t)\,\psi(t+\tau)\,X(t+\tau)] = \varphi(t,\tau)\,R_X(\tau) \tag{2.107}$$

whereby the autocorrelation is modulated by the function $\varphi(t,\tau) = \psi(t)\,\psi(t+\tau)$.

As all random signals, which are modelled as stochastic processes, have to be studied in a truncated or a modulated form, the application of envelope functions or windowing forms an important aspect of signal treatment and evaluation. Therefore, the above ideas will be discussed further in the following chapters in relation to stochastic input–output relations of dynamic systems, filter theory and signal treatment.

2.6 STRUCTURE FUNCTIONS

In many cases, the number of data points needed to obtain reasonably stable values of the autocorrelation and cross-correlation functions with the main characteristic features standing out, such as maximum value at zero separation time with a horizontal tangent in the case of the autocorrelation function, becomes excessively large, that is, very long observation interval T has to be used. There are examples where as many as 20 000 to 50 000 data points are necessary to begin to see the characteristic features of the autocorrelation (see Schulz-DuBois and Rehberg, [186]). The concept of structure functions may become useful in data manipulations to circumvent this problem.

Structure functions of a random process were introduced by Kolmogorov in his studies of locally isotropic and homogeneous, stochastic turbulence, where he derived relations between certain longitudinal and transverse moments of second order, [112]. The autostructure function of a stochastic process, $\{X(t), t\in\mathscr{T}\}$, is defined by

$$Z_X(\tau) = \lim_{T\to\infty} \frac{1}{T}\int_0^T [x(t) - x(t+\tau)]^2\,\mathrm{d}t \tag{2.108}$$

where $x(t)$ is a sample function of the process and ergodicity is tacitly implied. Similarly the cross-structure function of two stochastic processes, $\{X(t), t\in\mathscr{T}_1\}$ and $\{Y(t), t\in\mathscr{T}_2\}$, can be defined:

$$Z_{XY}(\tau) = \lim_{T \to \infty} \frac{1}{T} \int_0^T [x(t) - y(t+\tau)]^2 \, dt \tag{2.109}$$

The two above structure functions are related to the correlation functions in the following manner:

$$Z_X(\tau) = \lim_{T \to \infty} \frac{1}{T} \int_0^T [x^2(t) + x^2(t+\tau) - 2x(t)x(t+\tau)] \, dt = 2R_X(0) - 2R_X(\tau) \tag{2.110}$$

and

$$Z_{XY}(\tau) = R_X(0) + R_Y(0) - 2R_{XY}(\tau) \tag{2.111}$$

Since the value of the correlation functions for zero lag times are constants, the structure functions are equivalent to the negative correlation functions plus a constant. Thus the information obtained by the structure functions is essentially the same. For processes that show slow fluctuations or have a time-varying trend, that is, show marked deviation from a stationary behaviour, the method of structure functions may become advantageous.

Consider a simple harmonic function, $x(t) = \sin \omega t$, Example 2.5. Without the random amplitude and phase, the process can still be shown to have the autocorrelation function, $R_X(\tau) = \frac{1}{2}\cos \omega\tau$, by taking care to integrate over an integral number of half cycles, $\omega T = n\pi$. Since this condition is difficult to realize in practice, systematic error will be introduced in the autocorrelation function for $\omega T \neq n\pi$. Near the origin, $\omega\tau < 1$, the leading term of this error can be shown to be

$$\Delta R_X(\tau) \approx -\frac{1}{2} \frac{\sin 2\omega T}{2\omega T} \tag{2.112}$$

and for the autostructure function

$$\Delta Z_X(\tau) \approx -\tfrac{1}{2}\omega^2\tau^2 \frac{\sin 2\omega T}{2\omega T} \tag{2.113}$$

which shows that the autostructure function is less distorted for $\omega\tau < 1$ and thus is less susceptible to low-frequency noise, [186].

Next consider the spectral density functions. Using the cosine transform for the autocorrelation function,

$$R_X(\tau) = 2 \int_{-\infty}^{\infty} S_X(\omega) \cos(\omega\tau) \, d\omega \tag{2.79}$$

the corresponding frequency domain representation of the autostructure function is

$$Z_X(\tau) = 4 \int_{-\infty}^{\infty} S_X(\omega)(1 - \cos(\omega\tau)) \, d\omega \tag{2.114}$$

Considering the behaviour of the above integrals for $\omega \to 0$ and $\omega \to \infty$, it is obvious that the correlation function only exists if

$$S_X(\omega) \propto \omega^{-1-\varepsilon} \qquad \text{for } \omega \to \infty$$

and

$$S_X(\omega) \propto \omega^{-1+\varepsilon} \qquad \text{for } \omega \to 0 \tag{2.115}$$

where ε is a small positive number. For the autostructure function Eq. (2.114), the condition becomes

$$S_X(\omega) \propto \omega^{-1-\varepsilon} \qquad \text{for } \omega \to \infty$$

and

$$S_X(\omega) \propto \omega^{-3+\varepsilon} \qquad \text{for } \omega \to 0 \tag{2.116}$$

which shows, that the autostructure function can tolerate more low frequency noise than the autocorrelation function. Actually, for spectral density functions that have the general behaviour

$$S(\omega) = A|\omega|^{-p} \qquad \text{where } 1+\varepsilon < p < 3-\varepsilon, \quad A \text{ is constant} \tag{2.117}$$

the autostructure function will have the form

$$Z_X(\tau) = \left[\frac{2A\pi}{\sin(\pi(p-1)/2)\cdot\Gamma(p)}\right]\cdot\tau^{p-1} \tag{2.118}$$

where $\Gamma(p)$ is the Gamma function.

Schulz-DuBois and Rehberg, [186], have compared the autocorrelation functions and the autostructure functions obtained from wall pressure data from a transonic wind tunnel using piezoelectric transducers. About 20 000 data points or terms in the sum for the autocorrelation function were needed to obtain a satisfactory result. However, even with 20 000 data points, the autocorrelation function is still not an even function of the time lag as it should be. For 2000 or 200 data points, the baseline is severely distorted, the zero lag maximum is obscured and it is difficult to discern the dominant frequency. By comparison, the autostructure function shows the dominant frequency for as little as 200 data points and has a clear minimum close to the zero lag. For 2000 data points, the autostructure function assumes a fairly regular and symmetric shape, which envelope contains information on the low-frequency portion of the process.

3 Random Excitation and Response of Simple Linear Systems

In this chapter, the basic analysis of simple vibratory systems will be covered with emphasis on the frequency and time domain representation of the system response using the Fourier transform. In the previous chapters, the Fourier transform was applied in certain instances without any further ado. As the analysis of the behaviour and response of vibratory systems is mostly based on Fourier transform techniques, which play a fundamental role in this respect, knowledge of Fourier analysis is considered to be a prerequisite. A refresher text introducing the Fourier transform is presented in Chapter 7, Section 7.1, which is more or less independent of the main text of this book and can be used to obtain sufficient knowledge of the Fourier transform. However, a solid knowledge of Fourier analysis acquired in an advanced course in mathematical analysis would be preferable. The material is organised in such a way that the two first sections are dedicated to classical dynamical analysis of simple linear mechanical systems. The third section covers stochastic input and output relations along with related topics.

Dynamical analysis of mechanical and structural systems is a well established field of science, which has its roots in mechanical physics based on Newton's law of motion. In the following, the basic equations of motions of simple linear systems with a single or at most two degrees of freedom, that is, system motion which can be described by at most two variables, will be introduced. The treatment of oscillatory motion of mechanical systems is usually divided into two parts, i.e. the *free motion or free vibrations*, which takes place when the system is moving or oscillating under the action of forces inherent in the system itself, that is, when external forces are absent. Secondly, the *forced motion or forced vibrations* when the system is responding to external forces or external excitation of any kind. In mathematical terms, this can be described as solving the homogeneous differential equation of motion without any right hand side function, and thereafter finding a particular solution to the general equation involving right hand side forcing functions. In modern computational analysis of mechanical systems undergoing vibratory motion, numerical analysis has mostly replaced the former classical methods. However, a thorough understanding of the equations and their analytical solutions is necessary to successfully set up numerical solution schemes. Moreover, applying Fourier transformation techniques offers a direct link to numerical analysis involving the Fast Fourier Transform and digital data processing, which will be covered in more detail in Chapter 7. There is a wealth of literature available on the theory of vibrations of mechanical systems in which the treatment of mechanical systems undergoing forced linear and non-linear vibration of any kind is presented in great detail, (see for instance [65], [90], [119], [157], [181] and [220].

In this chapter, the emphasis is therefore more on linking the simple classical equations of motion of simple mechanical systems to stochastical excitations.

3.1 GENERAL OUTLINE

The mathematical description of excitation–response characteristics of structural systems subjected to stochastic inputs, follows closely the standard theory of mechanical vibrations with some shift in emphasis. Due to the fact that the theory was first utilized by electrical engineers, who applied it to electronic signals passing through electronic devices or circuitry, the terminology used is heavily influenced by electrical engineering jargon. It is of course immaterial what kind of notations are used to describe the behaviour of any physical system, subjected to an excitation of one kind or another. The basic fact that is common for all such systems is their mathematical description based on the theory of differential equations.

Any physical system whose behaviour, characterized by an arbitrary response quantity that can be described by a single variable can in general terms be represented by the state equation

$$L[X(t)] = F(t) \tag{3.1}$$

where the differential operator $L[*]$ describes the main properties of the system, that is, acts as a mathematical model for the system. $X(t)$ is the response or the system output when it is subjected to an excitation or input $F(t)$, stochastic or deterministic, which in turn may be derived by a similar governing equation or generating mechanism. This behaviour is illustrated in Fig. 3.1, which shows a block diagram representation of Eq. (3.1). The operation performed on the input signal $F(t)$ is the inverse operation $L^{-1}[*]$, that is, the output signal $X(t)$ is obtained by the inverse equation

$$X(t) = L^{-1}[F(t)] \tag{3.2}$$

The inverse operation Eq. (3.2) is sometimes referred to as filtering of the signal input $F(t)$, whereby the 'black box' of Fig. 3.1 is called a filter.

Consider a filter which in response to an input, $\{f(t), -\infty < t < \infty\}$, produces an output $\{x(t), -\infty < t < \infty\}$. If the filter output $x(t)$ is invariant to a time shift τ of the filter input $f(t)$, that is, if for $\{f(t+\tau), -\infty < t < \infty\}$ the output is $\{x(t+\tau), -\infty < t < \infty\}$, the filter is said to be time invariant or the physical system represented by the filter is stable. The filter (and/or the physical system) is said to be linear if its response to an input signal, which is a linear combination of many signals $y_r(t)$, that is,

$$f(t) = \sum_{r=1}^{n} a_r y_r(t) \tag{3.3}$$

is given by the same linear combination. Therefore the output signal $x(t)$ is given by

$$x(t) = \sum_{r=1}^{n} a_r z_r(t) \tag{3.4}$$

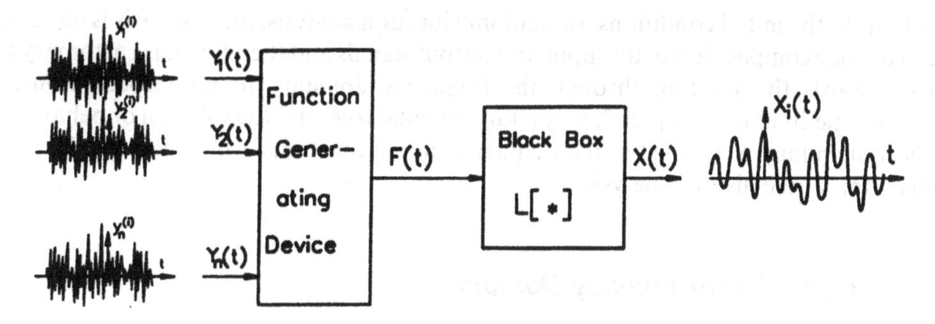

Fig. 3.1 *Input-output processes of a simple linear system*

where each $z_r(t)$ is the response to each $y_r(t)$ acting alone, i.e.

$$z_r(t) = L^{-1}[y_r(t)] \tag{3.5}$$

For complex functions as a special case, the real and imaginary parts can be separated, i.e. for a complex input $F(t) = F_R(t) + iF_I(t)$, the output is

$$\begin{aligned} X_R(t) &= L^{-1}[F_R(t)] \\ X_I(t) &= L^{-1}[F_I(t)] \end{aligned} \tag{3.6}$$

where $X_R(t)$ and $X_I(t)$ are the real and imaginary parts of the output signal.

It should be noted that a linear, time invariant filter may very well be represented by a non-linear operator $L[*]$. The linear operator $L[*]$ on the other hand will be time invariant and act as a linear filter. In the case of a linear operator, Eq. (3.1) assumes the form of a simple differential equation with constant coefficients, that is,

$$a_n \frac{d^n x}{dt^n} + \cdots + a_2 \frac{d^2 x}{dt^2} + a_1 \frac{dx}{dt} + a_0 x = f(t) \tag{3.7}$$

In the following the discussion will mainly pertain to systems that can be described by this basic equation, that is, linear systems with one state variable (linear systems with a single degree of freedom).

3.2 VIBRATION TRANSMISSION OF LINEAR SYSTEMS

In the classical analysis of mechanical or structural vibration, the treatment is centred about how to set up the equations of motion for vibratory systems and find the time domain solution, that is, the system response to any excitation or forcing function. This usually involves the analysis or solution of equations of the same type as Eq. (3.7), which can be performed in many different ways. The classical approach, however, is to solve the homogeneous equation first, that is, for $f(t) \equiv 0$, and then find one particular solution to the inhomogeneous equation and add up the two solutions after adjusting any integration constants to the initial conditions. In random vibration analysis, the emphasis is more on the frequency composition of the input and output signals and therefore it is customary to work towards the solution through the frequency domain. In this respect, Fourier transform methods (cf. Chapter 7) play a fundamental role, and the solution or behaviour of the system may be studied in the frequency domain or transformed back to the time domain for a time history analysis.

3.2.1 *The Frequency Domain*

Considering equations of the same type as Eq. (3.7), the methods indicated in Chapter 7 show that the frequency domain representation of Eq. (3.7) is equivalent to

$$X(\omega)\{a_n(\mathrm{i}\omega)^n + \cdots + a_2(\mathrm{i}\omega)^2 + a_1(\mathrm{i}\omega) + a_0\} = F(\omega) \tag{3.8}$$

where $X(\omega)$ and $F(\omega)$ are the Fourier transforms of the response and the excitation respectively, that is,

$$X(\omega) = \int_{-\infty}^{\infty} x(t)\,\mathrm{e}^{-\mathrm{i}\omega t}\,\mathrm{d}t, \qquad F(\omega) = \int_{-\infty}^{\infty} f(t)\,\mathrm{e}^{-\mathrm{i}\omega t}\,\mathrm{d}t$$

also,

$$X(\omega) = H(\omega)F(\omega) \tag{3.9}$$

where

$$H(\omega) = \frac{1}{a_n(\mathrm{i}\omega)^n + \cdots + a_2(\mathrm{i}\omega)^2 + a_1(\mathrm{i}\omega) + a_0} \tag{3.10}$$

is a complex function of the frequency ω which depends only on the system constants, $\{a_n, \ldots, a_2, a_1, a_0\}$. It is therefore a function, characteristic of the system, but independent of the excitation.

The function $H(\omega)$ has been given many different names such as the complex frequency response, the mechanical admittance, the transmissibility function, the reactance etc. In the following, it will mostly be called by the first name, the (complex) frequency response, which is better explained in Example 3.1.

Example 3.1 A vibratory system is governed by the equation of motion Eq. (3.7). Given that the excitation is the harmonic force

$$f(t) = f_0 \cos(\omega_{\mathrm{d}} t + \theta) \tag{3.11}$$

find the particular solution or the system response.

It is most often more convenient to treat harmonic components as complex harmonics, that is, writing

$$f(t) = A\,\mathrm{e}^{\mathrm{i}\omega_{\mathrm{d}} t} = (A_{\mathrm{R}} + \mathrm{i}A_{\mathrm{I}})(\cos\omega_{\mathrm{d}} t + \mathrm{i}\sin\omega_{\mathrm{d}} t) = F_{\mathrm{R}}(t) + \mathrm{i}F_{\mathrm{I}}(t) \tag{3.12}$$

whereby the real part of the complex harmonic is the actual forcing function or

$$f(t) = F_R(t) = A_R \cos\omega_d t - A_I \sin\omega_d t = f_0 \cos(\omega_d t + \theta)$$
$$f_0 = \sqrt{(A_R^2 + A_I^2)} \quad \text{and} \quad \theta = \text{Arctan}(A_I/A_R) \tag{3.13}$$

Therefore, study instead the equation

$$L[x] = A\exp(i\omega_d t) \tag{3.14}$$

Clearly, a particular solution has the form

$$x = B\exp(i\omega_d t) \tag{3.15}$$

which, inserted in Eqs. (3.14) or (3.7), gives

$$B\{a_n(i\omega_d)^n + \cdots + a_2(i\omega_1)^2 + a_1(i\omega_d) + a_0\}\exp(i\omega_{d1} t) = A\exp(i\omega_d t)$$

or, by Eq. (3.10),

$$B = H(\omega_d)A = (H_R(\omega_d) + iH_I(\omega_d))(A_R + iA_I) \tag{3.16}$$

That is, the frequency response shows the magnification of a complex harmonic passed through the system, given by

$$\left|\frac{B}{A}\right| = |H(\omega_d)| = \sqrt{H_R^2 + H_I^2} \tag{3.17}$$

In power amplifiers, it is customary to measure the magnification of the power in the input signal at a particular frequency, when passed through the amplifier (sinus watts). In this context, Eq. (3.17) is written as

$$\left|\frac{B}{A}\right|^2 = |H(\omega_1)|^2 = G(\omega_1) \tag{3.18}$$

The modulus squared of the frequency response, called $G(\omega_d)$, is often referred to as the amplifier gain. The frequency response $H(\omega_d)$ also carries information on the phase of the response. Equations (3.15) and (3.16) together give

$$x(t) = H(\omega_d)A\,e^{i\omega dt} = H(\omega_d)f_0\,e^{i\omega dt + i\theta}$$

using the relations Eq. (3.13). Also, writing $H(\omega_d) = |H(\omega_d)|\,e^{i\varphi}$

$$x(t) = f_0|H(\omega_d)|\,e^{i(\omega dt + \theta + \varphi)} \tag{3.19}$$

where

$$\varphi = \text{Arctan}\left[\frac{H_I(\omega_d)}{H_R(\omega_d)}\right] \tag{3.20}$$

is the phase shift angle, this is the angle by which the response is leading or lagging behind the excitation.

Finally, by Eq. (3.6), the real part of the response, Eq. (3.19), is the response to the real part of the excitation, that is the harmonic force Eq. (3.11) and

$$x_R(t) = f_0 |H(\omega_d)| \cos(\omega_d t + \theta + \varphi) \tag{3.21}$$

It has been shown that the frequency domain solution Eq. (3.9) gives information about the intensity of the response at each particular frequency. Moreover, a time domain solution can be constructed without further analysing the system equation by discretizing the continuous amplitude spectrum. In fact, following the procedures outlined in Chapter 7 (see Eq. (7.12)),

$$x(t) \simeq \frac{1}{\pi} \sum_{k=1}^{n} (\sqrt{X_R^2(\omega_K) + X_I^2(\omega_k)})\, \Delta\omega \cos(\omega_k t + \varphi_k) \tag{3.22}$$

where

$$\begin{aligned} X_R(\omega_k) &= H_R(\omega_k) F_R(\omega_k) - H_I(\omega_k) F_I(\omega_k) \\ X_I(\omega_k) &= H_R(\omega_k) F_I(\omega_k) + H_I(\omega_k) F_R(\omega_k) \end{aligned} \tag{3.23}$$

whereby the phase angle φ_k is the sum of the phase angles of the excitation and the frequency response, that is,

$$\varphi_k = \theta_F + \theta_H \tag{3.24}$$

The approximate response Eq. (3.22) is easily generated numerically. However, probably a more simple approach is to convert to digital analysis and use the Fast Fourier transform (see Section 7.4).

Example 3.2 In the mechanical system shown, Fig. 3.2, the excitation is the displacement $x(t)$ of the massless driving cart and the response is the displacement $y(t)$ of the mass m. Find the differential equation of motion relating $y(t)$ to $x(t)$, and for:

$$c_1 = c_2 = c/2, \quad k_1 = k_2 = k/3, \quad m = 3c^2/4k$$

find the frequency response function $H(\omega)$.

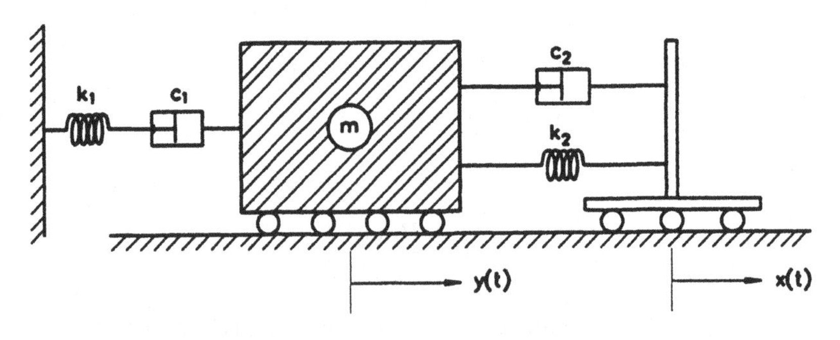

Fig. 3.2 *A mechnical vibratory system*

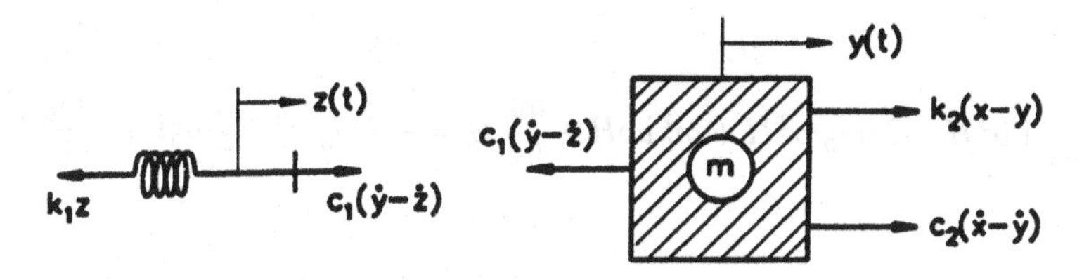

Fig. 3.3 *The internal forces and displacements*

Solution Introduce an auxiliary displacement variable $z(t)$ in between the spring element k_1, and the dashpot c_1. The forces and displacements acting on the system are sketched in Fig. 3.3.

By Newton's second law of motion,

$$m\ddot{y} = k_2(x - y) + c_2(\dot{x} - \dot{y}) - c_1(\dot{y} - \dot{z}) \tag{a}$$

and

$$k_1 z = c_1(\dot{y} - \dot{z}) \tag{b}$$

Eqs. (a) and (b) combined give

$$m\ddot{y} = k_2(x - y) + c_2(\dot{x} - \dot{y}) - k_1 z \tag{c}$$

Differentiation of Eq. (c),

$$m\dddot{y} = k_2(\dot{x} - \dot{y}) + c_2(\ddot{x} - \ddot{y}) - k_1 \dot{z} \tag{d}$$

Multiplying Eq. (a) by k_1/c_1 yields

$$(k_1/c_1)m\ddot{y} = k_1 k_2/c_1(x - y) + k_1 c_2/c_1(\dot{x} - \dot{y}) - k_1(\dot{y} - \dot{z}) \tag{e}$$

Adding Eqs. (d) and (e), and arranging terms,

$$\begin{aligned} &m\dddot{y} + (c_2 + k_1/c_1 m)\ddot{y} + (k_1 + k_2 + k_1 c_2/c_1)\dot{y} + k_1 k_2/c_1 y \\ &= c_2\ddot{x} + (k_2 + k_1 c_2/c_1)\dot{x} + k_1 k_2/c_1 x \end{aligned} \tag{f}$$

To evaluate the frequency response $H(\omega)$, let $x = e^{i\omega t}$ and obtain $Y(t) = H(\omega)e^{i\omega t}$. Further introduce $c_1 = c_2 = c/2$, $k_1 = k_2 = k/3$, $m = 3c^2/4k$, $\lambda = c/c_{cr}$, the ratio of critical damping, $c_{cr} = 2\sqrt{(km)}$, $\omega_0^2 = k/m$, $2\lambda\omega_0 = c/m$. Then Eq. (f) reduces to

$$m\dddot{y} + c\ddot{y} + k\dot{y} + \frac{2k^2}{9c}y = \frac{c}{2}\ddot{x} + \tfrac{2}{3}k\dot{x} + \frac{2k^2}{9c}x$$

or

$$\dddot{y} + 2\lambda\omega_0\ddot{y} + \omega_0^2 + \dot{y} + \frac{\omega_0^3}{9\lambda}y = \lambda\omega_0\ddot{x} + \tfrac{2}{3}\omega_0^2\dot{x} + \frac{\omega_0^3}{9\lambda}x$$

and

$$-i\omega^3 H - 2\lambda\omega_0\omega^2 H + \omega_0^2 i\omega H + \frac{\omega_0^3}{9\lambda}H = -\lambda\omega_0\omega^2 + \tfrac{2}{3}\omega_0^2 i\omega + \frac{\omega_0^3}{9\lambda}$$

whereby,

$$H(\omega) = \frac{\left(\lambda\omega_0\omega^2 - \dfrac{\omega_0^2}{9\lambda}\right) - \tfrac{2}{3}i\omega_0^2\omega}{\left(2\lambda\omega_0\omega^2 - \dfrac{\omega_0^3}{9\lambda}\right) + i(\omega^3 - \omega\omega_0^2)}$$

3.2.2 The Time Domain

To find the solution of equations of the same type as Eq. (3.7) in the time domain, there are, as already mentioned, several possible avenues available. The classical approach is to seek the homogeneous solution $x_h(t)$ and then find and add to it any particular solution $x_p(t)$, that is (cf. Eq. (3.2)),

$$x(t) = x_h(t) + x_p(t) = L^{-1}[f(t)] \tag{3.25}$$

Example 3.3 In Fig. 3.4, the construction frame of a turbine pedestal is shown, which is being excited by the eccentric rotating mass μ at a natural frequency $f = 50$ Hz and eccentricity e. Ignoring the vertical excitation due to the very large vertical stiffness of the two columns and making the usual assumptions regarding a vibration model for the horizontal motion $x(t)$ of the frame, the equation governing the horizontal motion can be established as follows:

$$m\ddot{x}(t) + c\dot{x}(t) + kx(t) = f_H(t) \tag{3.26}$$

where m is the vibrating mass (turbine and pedestal), k is the equivalent horizontal stiffness of the structural frame and c is an internal viscous damping coefficient. If the horizontal

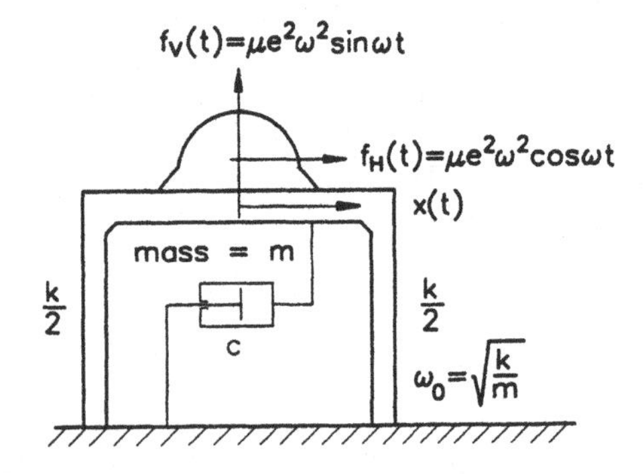

Fig. 3.4 *A turbine pedestal under working loads*

displacement at time $t = t_0$ (when final running speed of turbine is obtained) is x_0 and the horizontal velocity $\dot{x}_0$, both referred to the centre of the vibrating mass, find the complete solution to the above equation, (Eq. (3.26)). This equation is the classical equation for almost any vibratory system with one degree of freedom and will be used to demonstrate the solution techniques for both the time and frequency domains.

First solve the homogeneous equation

$$m\ddot{x}(t) + c\dot{x}(t) + kx(t) = 0$$

or

$$\ddot{x} + 2\lambda\omega_0\dot{x} + \omega_0^2 x = 0 \tag{3.27}$$

by introducing the ratio of critical damping $\lambda = c/c_{\mathrm{cr}}$, where $c_{\mathrm{cr}} = 2\sqrt{(km)}$. Assume that the solution is of the form

$$x_{\mathrm{h}}(t) = \mathrm{e}^{-\lambda\omega_0 t}(A\cos\overline{\omega_0}t + B\sin\overline{\omega_0}t) \tag{3.28}$$

in which the damped frequency

$$\overline{\omega_0} = \omega_0\sqrt{(1-\lambda^2)}.$$

To obtain the two integration constants A and B, the initial conditions corresponding to the two relations,

$$x_0 = x_h(t_0) = \mathrm{e}^{-\lambda\omega_0 t_0}(A\cos\overline{\omega_0}t_0 + B\sin\overline{\omega_0}t_0)$$

$$\dot{x}_0 = \dot{x}_h(t_0) = \mathrm{e}^{-\lambda\omega_0 t_0}[\lambda\omega_0(A\cos\overline{\omega_0}t_0 + B\sin\overline{\omega_0}t_0) + \overline{\omega_0}(B\cos\overline{\omega_0}t - A\sin\overline{\omega_0}t)]$$

are utilized and solving the above equations for A and B yields

$$A = \mathrm{e}^{-\lambda\omega_0 t_0}\left(x_0\sin\overline{\omega_0}t + \frac{\dot{x}_0}{\overline{\omega_0}}\cos\omega_0 to + \frac{\lambda\omega_0}{\overline{\omega_0}}x_0\cos\overline{\omega_0}t_0\right) \tag{3.29a}$$

and

$$B = \mathrm{e}^{-\lambda\omega_0 t_0}\left(x_0\cos\overline{\omega_0}t - \frac{\dot{x}_0}{\overline{\omega_0}}\sin\overline{\omega_0}t_0 + \frac{\lambda\omega_0}{\overline{\omega_0}}x_0\sin\overline{\omega_0}t_0\right) \tag{3.29b}$$

Combining Eqs. (3.29) and the homogeneous solution Eq. (3.28), it is written as

$$x_{\mathrm{h}}(t) = \mathrm{e}^{-\lambda\omega_0(t-t_0)}\left\{x_0\left(\cos\overline{\omega_0}(t-t_0) + \frac{\lambda\omega_0}{\overline{\omega_0}}\sin\overline{\omega_0}(t-t_0)\right) + \frac{\dot{x}_0}{\overline{\omega_0}}\sin\overline{\omega_0}(t-t_0)\right\} \tag{3.30}$$

Introducing two functions

$$h(t) = \frac{1}{\overline{\omega_0}}\mathrm{e}^{-\lambda\omega_0 t}\sin\overline{\omega_0}t \tag{3.31}$$

and

$$a(t) = \frac{1}{\omega_0^2}\left(1 - e^{-\lambda\omega_0 t}\left(\cos\overline{\omega}_0 t + \frac{\lambda\omega_0}{\overline{\omega}_0}\sin\overline{\omega}_0 t\right)\right) \tag{3.32}$$

the importance of which will be discussed in Ex. 3.4, the homogeneous solution is also written as

$$x_h(t) = x_0(1 - \omega_0^2 a(t - t_0)) + \dot{x}h(t - t_0) \tag{3.33}$$

Now, all that is needed to do is to find one single particular solution of Eq. (3.26) and add that to the homogeneous solution Eq. (3.30) or (3.33). One such particular solution has already been found for more general equations, that is, applying Eq. (3.21),

$$x_p(t) = \mu e^2 \omega |H(\omega)| \cos(\omega t + \varphi)$$

since the phase angle of the exciting force is 0. The complex frequency response of the system, Eq. (3.26), is

$$\begin{aligned} H(\omega) &= \frac{1}{m(i\omega)^2 + c(i\omega) + k} = \frac{1}{m}\frac{1}{(\omega_0^2 - \omega^2) + 2\lambda i\omega_0\omega} \\ &= \frac{1}{m}\frac{(\omega_0^2 - \omega^2) - 2\lambda i\omega_0\omega}{[(\omega_0^2 - \omega^2)^2 + 4\lambda^2\omega_0^2\omega^2]} = H_R(\omega) + iH_I(\omega) \end{aligned} \tag{3.34}$$

Therefore, the final complete solution is ($\alpha = \mu e^2/m$)

$$x(t) = x_h(t) + \frac{\alpha\omega\cos(\omega t - \varphi)}{\sqrt{((\omega_0^2 - \omega^2)^2 + 4\lambda^2\omega_0^2\omega^2)}} \tag{3.35a}$$

$$\varphi = \text{Arctan}\left[\frac{2\lambda\omega_0\omega}{(\omega_0^2 - \omega^2)}\right] \tag{3.35b}$$

Because of the damping, the homogeneous part in Eq. (3.35) will decay and finally lose importance. The remainder, the steady-state response, shows that the horizontal response is in perfect resonance with the driving force and the amplitude of the response is governed by the frequency response $H(\omega)$. The response, that is the output signal, is found to be lagging behind the driving force by the phase φ.

As the excitation or the input signal $f(t)$ becomes more complicated, a particular solution is not so easily obtained. The classical approach, which is usually treated in textbooks on vibration theory (see e.g. [37] and [109]), is based on looking at the excitation as an infinite series of force impulses and finding the solution for each particular impulse. Then, by adding up all such solutions, the final solution is obtained in the form of a Duhamel integral. However, a much more elegant method can be selected by applying the Fourier analysis techniques described in Chapter 7.

The time domain solution can be derived from the frequency domain solution in two different ways. Firstly, for equations of the same type as Eq. (3.7), the solution in the frequency domain

$$X(\omega) = H(\omega)F(\omega) \tag{3.9}$$

is of the form where the convolution theorem can be used. In fact, by Eq. (7.22), the solution in the time domain is

$$x(t) = \int_{-\infty}^{\infty} f(u)\,h(t-u)\,du \tag{3.36}$$

where the functions, $x(t) \leftrightarrow X(\omega)$, $f(t) \leftrightarrow F(\omega)$ and $h(t) \leftrightarrow H(\omega)$ are Fourier transform pairs. The time function $h(t)$ in the response, Eq. (3.36), is therefore obtainable as the inverse transform of the complex frequency response $H(\omega)$. It is a functional characteristic of the vibrating system and independent on the excitation, that is

$$h(t) = \frac{1}{2\pi}\int_{-\infty}^{\infty} H(\omega)e^{i\omega t} - d\omega \tag{3.37}$$

If the excitation $f(t)$ is taken as an impulse or equivalent to the delta function $\delta(t)$, the response is easily obtained from Eq. (3.36):

$$x(t) = \int_{-\infty}^{\infty} \delta(u)\,h(t-u)\,du = h(t) \tag{3.38}$$

so $h(t)$ is the system response to an impulse type excitation. The function $h(t)$ is also called the impulse response function and by Eq. (3.7),

$$a_n h^{(n)}(t) + \cdots + a_2 h''(t) + a_1 h'(t) + a_0 h(t) = \delta(t) \tag{3.39}$$

Rather than finding the impulse response as the inverse transform, Eq. (3.37), it is often more easy to solve the above equation, Eq. (3.39), directly. In order to do so it is, however, more convenient to work with the system admittance function $a(t)$ which is defined as follows:

$$a'(t) = h(t) \rightarrow a(t) = \int h(t)\,dt \tag{3.40}$$

or the admittance is the undefined integral of the impulse response.

In Chapter 7, the delta function is introduced as a Gaussian probability density with a zero mean and a standard deviation σ approaching zero. Taking a look at the corresponding cumulative distribution, it will in the limit ($\sigma \rightarrow 0$) have the appearance shown in Fig. 3.5. Therefore,

$$1(t) = \int_{-\infty}^{t} \delta(u)du \qquad \text{or} \qquad \delta(t) = 1'(t) \tag{3.41}$$

where the function $1(t)$ is called *Heaviside's unit function*. Therefore, introducing Eqs. (3.40) and (3.41) after integrating Eq. (3.39) once with respect to t, a new equation for the

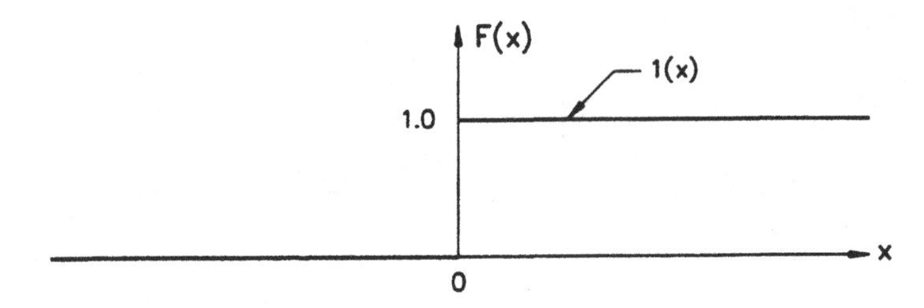

Fig. 3.5 *A Gaussian distribution with zero variance*

admittance $a(t)$ is found, that is,

$$a_n a^{(n)}(t) + \cdots + a_2 a''(t) + a_1 a'(t) + a_0 a(t) = 1(t) \tag{3.42}$$

The solution of Eq. (3.42) can now be obtained in a simple way since a particular solution is either zero for negative times or $1/a_0$ for positive times. Care should be taken, however, so set up proper initial conditions for the admittance $a(t)$.

Example 3.4 Find the admittance $a(t)$ and the impulse response $h(t)$ of the system discussed in Example 3.3.

Solution By Eq. (3.27), Eq. (3.42) takes the form

$$a''(t) + 2\lambda\omega_0 a'(t) + \omega_0^2 a(t) = 1(t)$$

For positive time, $t > 0$, the complete solution is

$$a(t) = \frac{1}{\omega_0^2} + e^{-\lambda\omega_0 t}(A\cos\overline{\omega_0}t + B\sin\overline{\omega_0}t)$$

For negative time, $t \leqslant 0$, the system is at rest, since only a particular solution of the general equation, Eq. (3.7), is being sought. Therefore

$$a(0) = a'(0) = 0 \text{ and } a(-|t|) \equiv 0$$

or

$$a(0) = 1/\omega_0^2 + A \rightarrow A = -1/\omega_0^2$$

$$a'(0) = B\overline{\omega_0} - \lambda\omega_0 A \rightarrow B = -\lambda/\omega_0\overline{\omega_0}$$

Therefore

$$a(t) = \frac{1}{\omega_0^2}\left(1 - e^{-\lambda\omega_0 t}\left(\cos\overline{\omega_0}t + \frac{\lambda\omega_0}{\overline{\omega_0}}\sin\overline{\omega_0}t\right)\right) \tag{3.32}$$

and since $h(t) = a'(t)$,

$$h(t) = \begin{cases} 1/\overline{\omega_0}\, e^{-\lambda\omega_0 t} \sin \overline{\omega_0} t & t \geqslant 0 \\ 0 & t < 0 \end{cases} \tag{3.31}$$

and the functions $h(t)$ and $a(t)$, already found in Example 3.3, are easily recognized.

A second alternative is to seek the time domain solution directly as the inverse Fourier transform of the frequency domain solution Eq. (3.9). This would especially appear to be advantageous if, in the state equation, when transformed into the frequency domain, the complex frequency response and the transform of the excitation can not be separated as in Eq. (3.9). Then,

$$x(t) = \frac{1}{2\pi} \int_{-\infty}^{\infty} \left[\int_{-\infty}^{\infty} L^{-1}[f(s)]\, e^{-i\omega s}\, ds \right] e^{i\omega t}\, d\omega \tag{3.43}$$

The solution to the inverse Fourier transform Eq. (3.43) can be sought in tables of Fourier integrals or in larger mathematical handbooks, [72], [232] or, if convenient, constructed by solving the above integral with the aid of Cauchy's integral theorem or the residue theorem for complex analytical functions (see Chapter 7, Example 7.4).

Example 3.5 Given the equation of the simple harmonic oscillator discussed in Example 3.3, i.e.

$$m\ddot{x}(t) + c\dot{x}(t) + kx(t) = f(t)$$

where the excitation $f(t)$ has the Fourier transform $F(\omega)$ without singular points, find the response $x(t)$ by unlocking the integral, Eq. (3.43).

Solution The frequency domain solution is

$$X(\omega) = H(\omega)F(\omega) = \frac{1}{m} \frac{F(\omega)}{(\omega_0^2 - \omega^2) + 2\lambda i \omega_0 \omega}$$

To find

$$x(t) = \frac{1}{m2\pi} \int_{-\infty}^{\infty} \frac{F(\omega)\, e^{i\omega t}\, d\omega}{(\omega_0^2 - \omega^2) + 2\lambda i \omega_0 \omega}$$

consider the complex contour integral, (Fig. 3.6), that is,

$$\oint g(z)dz = \frac{1}{m} \oint \frac{F(z)e^{izt}dz}{(\omega_0^2 - z^2) + 2\lambda i \omega_0 z}$$

Considering the two integration paths shown in Fig. 3.6, the poles of the integrand function are the singular points which make the integrand function infinite. Since $F(\omega)$ does not have any singular points, the poles are

$$z_{1,2} = \begin{cases} i\lambda\omega_0 + \omega_0\sqrt{1-\lambda^2} = i\lambda\omega_0 + \overline{\omega_0} \\ i\lambda\omega_0 - \omega_0\sqrt{1-\lambda^2} = i\lambda\omega_0 - \overline{\omega_0} \end{cases}$$

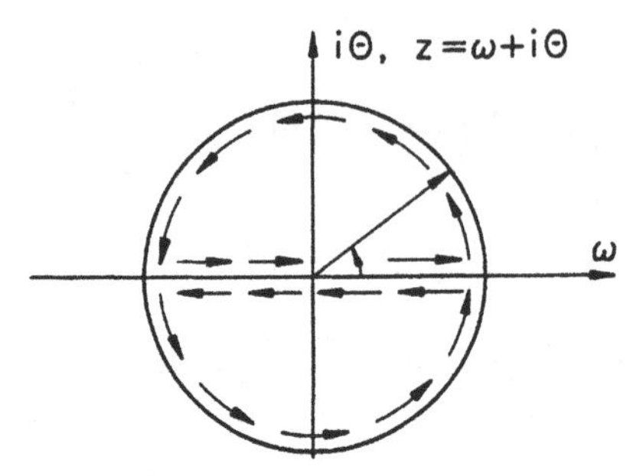

Fig. 3.6 *Contour integration paths*

Both poles are thus found to be within the integration path (1). The residues at the two poles are

$$\operatorname{Res}[z=z_1]=\lim_{z\to z_1}(z-z_1)g(z)=\frac{F(z_1)\mathrm{e}^{\mathrm{i}z_1t}}{(z_1-z_2)}=\frac{F(\mathrm{i}\lambda\omega_0+\overline{\omega_0})}{2\overline{\omega_0}}\mathrm{e}^{-\lambda\omega_0t}\,\mathrm{e}^{\mathrm{i}\overline{\omega_0}t}$$

and

$$\operatorname{Res}[z=z_2]=\frac{F(z_2)\mathrm{e}^{\mathrm{i}z_2t}}{(z_2-z_1)}=\frac{F(\mathrm{i}\lambda\omega_0-\overline{\omega_0})}{-2\overline{\omega_0}}\mathrm{e}^{-\lambda\omega_0t}\,\mathrm{e}^{-\mathrm{i}\overline{\omega_0}t}$$

The according to the residue theorem,

$$\oint g(z)\mathrm{d}z=-2\pi\mathrm{i}\sum\qquad\text{(all residues inside contour)}$$

Now first evaluate the left-hand side. For both paths, the contour integral is split into two parts. Firstly, along the ω-axis,

$$\int_{-R}^{R}\frac{F(\omega)\mathrm{e}^{\mathrm{i}\omega t}\,\mathrm{d}\omega}{(\omega_0^2-\omega^2)+2\lambda\mathrm{i}\omega_0\omega}$$

which for $R\to\infty$ becomes the integral being sought. Along the circle, the integral is

$$\int_{v_1}^{v_2}\frac{F(R\mathrm{e}^{\mathrm{i}v})\mathrm{e}^{\mathrm{i}(R\mathrm{e}^{\mathrm{i}v})t}R\mathrm{e}^{\mathrm{i}v}\mathrm{d}v}{(\omega_0^2+R^2\mathrm{e}^{\mathrm{i}2v})+2\lambda\mathrm{i}\omega_0R\mathrm{e}^{\mathrm{i}v}}$$

For path (1), this integral is evaluated as

$$\int_0^{\pi}\frac{F(R\mathrm{e}^{\mathrm{i}v})\mathrm{e}^{\mathrm{i}tR\cos v}\mathrm{e}^{-tR\sin v}R\mathrm{e}^{\mathrm{i}v}\mathrm{d}v}{(\omega_0^2+R^2\mathrm{e}^{\mathrm{i}2v})+2\lambda\mathrm{i}\omega_0R\mathrm{e}^{\mathrm{i}v}}\leqslant\int_0^{\pi}\frac{|F(R\mathrm{e}^{\mathrm{i}v})\mathrm{e}^{-tR\sin v}\mathrm{d}v}{\left|\left(\dfrac{\omega_0^2}{R+R}\mathrm{e}^{\mathrm{i}2v}\right)+2\lambda\mathrm{i}\omega_0\mathrm{e}^{\mathrm{i}v}\right|}$$

Since $\sin v\geqslant 0$ in the range $[0,\pi]$, the integral will tend to zero as $R\to\infty$ only if $t\geqslant 0$. Thus for positive time, path (1) must be chosen whereby in the limit, the contour integral becomes the sought after integral. For path (2), the above integral can be evaluated as

$$\int_{\pi}^{2\pi} \frac{|F(Re^{iv})| e^{-tR\sin v} dv}{\left| \left(\frac{\omega_0^2}{R} + Re^{i2v} \right) + 2\lambda i \omega_0 e^{iv} \right|}$$

Since $\sin v$ is negative in the range $[\pi, 2\pi]$, the integral can only be brought to zero as $R \to \infty$ if $t < 0$, that is for negative time, path (2) must be chosen to retain only the sought after integral.

Now turn to the right-hand side. First look at the functional values $F(i\lambda\omega_0 \pm \overline{\omega_0})$ which enter into the residual values. By the definition of the Fourier transform,

$$F(i\lambda\omega_0 \pm \overline{\omega_0}) = \int_{-\infty}^{\infty} f(u) e^{-i(i\lambda\omega_0 \pm \overline{\omega_0})u} du = \int_{-\infty}^{\infty} f(u) e^{\lambda\omega_0 u} e^{\mp \overline{\omega_0} u} du$$

$$= \int_{-\infty}^{\infty} f(u) e^{\lambda\omega_0 u} (\cos \overline{\omega_0} u \mp \sin \overline{\omega_0} u) du = F_R \mp i F_I$$

For path (1) (when $t \geqslant 0$), the sum of the residues is

$$\frac{e^{-\lambda\omega_0 t}}{2\overline{\omega_0} m} [(F_R - iF_I)(\cos \overline{\omega_0} t + i \sin \overline{\omega_0} t) - (F_R + iF_I)(\cos \overline{\omega_0} t - i \sin \overline{\omega_0} t)]$$

$$= i \frac{e^{-\lambda\omega_0 t}}{2\overline{\omega_0} m} [F_R \sin \overline{\omega_0} t - F_I \cos \overline{\omega_0} t]$$

with the above expressions for F_R and F_I; note that in order to maintain generality in the residual sum, the time argument in F_R and F_I must not be called t. Therefore the residual sum for path (1), $t \geqslant 0$, is

$$i \frac{e^{-\lambda\omega_0 t}}{m\overline{\omega_0}} \left[\left(\int_{-\infty}^{\infty} f(u) e^{\lambda\omega_0 u} \cos \overline{\omega_0} u \, du \right) \sin \overline{\omega_0} t - \left(\int_{-\infty}^{\infty} f(u) e^{\lambda\omega_0 u} \sin \overline{\omega_0} u \, du \right) \cos \overline{\omega_0} t \right]$$

which can be contracted as

$$\frac{i}{m\overline{\omega_0}} \int_{-\infty}^{\infty} f(u) e^{-\lambda\omega_0 (t-u)} \sin \overline{\omega_0} (t-u) du$$

For path (2) and $t < 0$, the sum of residues is zero since there are no poles in the lower half-plane. The final result is therefore as follows:

$$\frac{1}{2\pi m} \int_{-\infty}^{\infty} \frac{F(\omega) e^{i\omega t} d\omega}{(\omega_0^2 - \omega^2) + 2\lambda i \omega_0 \omega} = \frac{1}{m\overline{\omega_0}} \int_{-\infty}^{\infty} f(u) e - \lambda\omega_0 (t-u) \sin \overline{\omega_0} (t-u) du \quad \text{for } t \geqslant 0$$

$$= 0 \qquad \text{for } t < 0$$

whereby the time domain solution $x(t)$ is obtained. It should be hastily mentioned that the above most complicated procedure could have been considerably simplified by first applying the convolution theorem and then only inverting the Fourier transform $H(\omega)$. In fact, letting the excitation $f(t)$ be the delta function or the impulse $m\delta(t)$, $F(\omega)=m$. The above result then immediately gives.

$$h(t)=\frac{1}{2\pi}\int_{-\infty}^{\infty}\frac{e^{i\omega t}\,d\omega}{(\omega_0^2-\omega^2)+2\lambda i\omega_0\omega}=\frac{1}{\overline{\omega}_0}e^{-\lambda\omega_0 t}\sin\overline{\omega}_0 t \qquad \text{for } t\geqslant 0$$

$$=0 \qquad \text{for } t<0$$

which is the previously obtained result Eq. (3.31).

3.3 STOCHASTIC INPUT AND OUTPUT RELATIONS

A proper background has now been laid for the treatment of vibratory systems subjected to stochastic excitation. Surely, when any physical system, electrical, mechanical or structural, is subjected to loads or an excitation of stochastic character, the system response or output will also be stochastic. In this case, the analysis of any response characteristic of the system, for instance deflection, stress etc., can only be treated statistically through knowledge of the probability laws governing the input process. Two fundamentally different approaches are possible. Firstly, for a given stochastic input or excitation process, for which either the probability law is known or a mathematical, statistical model for the process is available, the system response or output process may be analysed directly through the system response functions. A second alternative is to convert to purely numerical analysis. A mathematical model describing the input process is sought even if its probability law has not been completely determined. The process may be simulated and generated numerically whereby an ensemble of sample functions, or a subset thereof, is obtained. The system response to all such sample functions is obtained by numerical integration and the probabilistic law and hence any statistical quantity concerning the response are inferred numerically from the response data material. This approach, the so-called Monte Carlo technique, has been applied by many authors. Goldberg *et al.* [70], Housner and Jennings [85], to name some earlier attempts, and Penzien's work, [8], is especially noteworthy in connection with artificial earthquake generation (see also [125], [192], [209], [210] and [239]). Then Chiu and other authors, [35], [118], have studied the simulation and generation of wind speed record, to name but a few.

Various textbooks are available which offer a clear and lucid treatment of random vibration transmission through mechanical systems [9], [43], [120], [122], [152], [154] and [180]. Based on these and numerous other sources, the various mathematical relations, which form the basis for treatment of the response of mechanical and structural systems when subjected to random excitations, are presented in this chapter.

3.3.1 The Basic Input–Output Relations

Consider a simple physical vibratory system of one degree of freedom subjected to a random excitation given as the stochastic process $\{X(t),\ -\infty<t<\infty\}$. The system response is the output process $\{Y(t), t\in T\}$. Now, for ordinary functions of time, the system response would be given by the differential equation Eq. (3.7), which has the formal time domain solution Eq. (3.36). This implies that the excitation process must be both differentiable up to the second order and integrable as well. In Section 2.2.5 it was shown that a stochastic process is mean square differentiable if its auto-correlation function has a valid second partial derivative at all points (t, s). (The velocity process $\dot{X}(t)$ has to have a valid second partial derivative as well.) Furthermore, the process is mean square Riemann integrable if its autocorrelation function is Riemann integrable. In that case, the process $X(t)$ can be treated as an ordinary function of time, allowing the usual manipulation through integration and differentiation. In the following it will be assumed that the stochastic processes being treated have autocorrelation functions which satisfy the above conditions. This clearly allows an easy treatment of the input–output relations in the time domain along the lines shown in Section 3.1. On the other hand, going to the frequency domain, relations of the type Eqs. (3.9) and (3.41) break down by write of the fact that stochastic functions are not in general Fourier transformable.

Accepting for a fact that the response process $Y(t)$ is indeed given by

$$Y(t) = \int_{-\infty}^{\infty} X(u)\, h(t-u)\, du \tag{3.44}$$

where $h(t)$ is the impulse response function of the system considered, a few words concerning the integration bounds are in order. From the discussion of the admittance and the impulse response, it has already been noticed that $h(t) \equiv 0$ for $t < 0$ or in Eq. (3.44), the dummy time u can not exceed the real time t, so

$$Y(t) = \int_{-\infty}^{t} X(u)\, h(t-u)\, du \tag{3.45}$$

Thus the infinite upper limit has been replaced by the operation time t. In order words, a physically realizable system must be causal, i.e. it can only have backwards-oriented memory and can not predict the future. All man-made systems have finite operation time. The system has to be manufactured and put into operation at, say, time $t = 0$, that is,

$$Y(t) = \int_{0}^{t} X(u)\, h(t-u)\, du \tag{3.46}$$

It should be noted, however, that the stochastic process $X(u)$ can be operative at all times, that is for $(-\infty < u < \infty)$ in all three versions of the above integral. However, generally the universal form, Eq. (3.44), is to be preferred for any mathematical treatment.

Taking the expectation of both sides of Eq. (3.44), the mean value of the output process $\mu_Y(t)$ is obtained, i.e.

$$\mu_Y(t) = E[Y(t)] = E\left[\int_{-\infty}^{\infty} X(u)\, h(t-u)\, du\right]$$

$$= \int_{-\infty}^{\infty} E[X(u)]\, h(t-u)\, du = \int_{-\infty}^{\infty} \mu_X(u)\, h(t-u)\, du \tag{3.47}$$

Example 3.6 If the input process $X(t)$ is stationary, its mean value μ_X is a constant. Then by Eq. (3.47),

$$E[Y(t)] = \mu_Y(t) = \int_{-\infty}^{\infty} \mu_X h(t-u) du = \mu_X \int_{-\infty}^{\infty} h(t-u)\, du$$

Now $h(t-u)$ is the inverse Fourier transform of $H(\omega)$ with the time $(t-u)$, Eq. (3.37). Therefore, replacing t by $(t-u)$ in the corresponding Fourier transform,

$$H(\omega)=\int_{-\infty}^{\infty} h(t-u)\,\mathrm{e}^{-\mathrm{i}\omega(t-u)}\,\mathrm{d}(t-u)$$

whereby

$$\mu_Y(t)=\mu_Y=\mu_X\int_{-\infty}^{\infty} h(t-u)\,\mathrm{d}u=\mu_X H(0)$$

Consider for instance the simple oscillator of Example 3.3. By Eq. (3.34), $H(0)=1/m\omega_0^2$ and

$$\mu_Y(t)=\mu_X/m\omega_0^2=\mu_X/k$$

which shows that the mean level of the input process is transmitted as any static load acting on the system. The random excursions of the input process have no influence on the mean value of the output process.

To evaluate the correlation functions and the spectral density of the output process, consider once more the input–output relation Eq. (3.44), subtract the mean value by Eq. (3.47) and form the double integral

$$\begin{aligned}(Y(t)-\mu_Y(t))(Y(s)-\mu_Y(s)) &= \left[\int_{-\infty}^{\infty}(X(u)-\mu_X(u))h(t-u)\mathrm{d}u\right]\left(\int_{-\infty}^{\infty}(X(v)-\mu_X(v))h(s-v)\mathrm{d}v\right)\\ &= \int_{-\infty}^{\infty}\int_{-\infty}^{\infty}(X(u)-\mu_X(u))(X(v)-\mu_X(v))\,h(t-u)\,h(s-v)\,\mathrm{d}u\,\mathrm{d}v\end{aligned} \tag{3.48}$$

Taking the expectation of both sides,

$$E[(Y(t)-\mu_Y(t))(Y(s)-\mu_Y(s))]=K_Y(t,s)=\int_{-\infty}^{\infty}\int_{-\infty}^{\infty} K_X(u,v)\,h(t-u)\,h(s-v)\,\mathrm{d}v\,\mathrm{d}v \tag{3.49}$$

which gives the relation between the covariance kernels of the input and output processes. For covariance stationary input processes, the covariance kernel can be replaced by the autocovariance function $\Gamma_X(u-v)$. By introducing the following change of variables, $\theta=t-u$, $\varphi=s-v$, $\tau=t-s$, the double integral Eq. (3.49) will be changed to

$$K_Y(t,s)=\int_{-\infty}^{\infty}\int_{-\infty}^{\infty} K_X(t-\theta,s-\varphi)\,h(\theta)\,h(\varphi)\mathrm{d}\varphi\,\mathrm{d}\varphi$$

or (3.50)

$$\Gamma_Y(\tau)=\int_{-\infty}^{\infty}\int_{-\infty}^{\infty}\Gamma_X(\tau+\varphi-\theta)\,h(\theta)\,h(\varphi)\mathrm{d}\theta\,\mathrm{d}\varphi$$

Since $\Gamma_X(s-\varphi-t+\theta)=\Gamma_X(-(s-\varphi-t+\theta))=K_X(*,*)$, and through the right-hand side of Eq. (3.50), $K_Y(t,s)$ is a function of the time difference $\tau=t-s$ only. Hence $Y(t)$ is also covariance stationary.

Now introduce the autocorrelation function $R_X(t)$. Eq. (3.50) is rewritten as

$$R_Y(\tau)+\mu_Y^2=\int_{-\infty}^{\infty}\int_{-\infty}^{\infty}R_X(t+\varphi-\theta)\,h(\theta)\,h(\varphi)\mathrm{d}\theta\,\mathrm{d}\varphi+\mu_X^2\int_{-\infty}^{\infty}\int_{-\infty}^{\infty}h(\theta)\,h(\varphi)\mathrm{d}\theta\,\mathrm{d}\varphi$$

since both the mean values are now constant due to the stationarity. Then by Eq. (3.48),

$$R_Y(\tau)=\int_{-\infty}^{\infty}\int_{-\infty}^{\infty}R_X(t+\varphi-\theta)\,h(\theta)\,h(\varphi)\mathrm{d}\theta\,\mathrm{d}\varphi \tag{3.51}$$

By Fourier transforming both sides of Eq. (3.51), an equivalent relation is obtained in the frequency domain:

$$\begin{aligned}\frac{1}{2\pi}\int_{-\infty}^{\infty}R_Y(\tau)\mathrm{e}^{-i\omega\tau}\mathrm{d}\tau&=\frac{1}{2\pi}\int_{-\infty}^{\infty}\mathrm{d}\tau\int_{-\infty}^{\infty}h(\theta)\mathrm{d}\theta\int_{-\infty}^{\infty}[R_X(t+\varphi-\theta)\,h(\varphi)\\&\quad\times\mathrm{e}^{-i\omega(\tau+\varphi-\theta)}\mathrm{e}^{-i\omega\theta}\mathrm{e}^{i\omega\varphi}]\mathrm{d}\varphi\\&=\int_{-\infty}^{\infty}h(\theta)\mathrm{e}^{-i\omega\theta}\mathrm{d}\theta\int_{-\infty}^{\infty}h(\varphi)\mathrm{e}^{-i\omega\varphi}\mathrm{d}\varphi\\&\quad\times\left[\frac{1}{2\pi}\int_{-\infty}^{\infty}R_X(t+\varphi-\theta)\mathrm{e}^{-i\omega(\tau+\varphi-\theta)}\mathrm{d}(\tau+\varphi-\theta)\right]\end{aligned}$$

or

$$S_Y(\omega)=H(\omega)H^*(\omega)S_X(\omega)=|H(\omega)|^2\,S_X(\omega) \tag{3.52}$$

by introducing the power spectral density, Eq. (2.41), and noting that $H(-\omega)=H^*(\omega)$ is the complex conjugate of $H(\omega)$.

Equations (3.51) and (3.52) are among the most important and famous results in the theory of random vibrations. Through a knowledge of the system functions $h(t)$ and $H(\omega)$, the characteristics of the response for any given stationary random excitation are determined.

Example 3.7 Again, consider the oscillator of Example 3.3, but with a unit mass $m=1$. Let the oscillator be subjected to a random excitation which has the characteristics of white noise with a power spectral density $S_X(\omega)=S_0$, a constant. Find the autocorrelation function and the spectral density of the response. What is the mean value and the root mean square value of the response?

Solution The autocorrelation of the excitation, a white noise process with $S_X(\omega)=S_0$, is $R_X(t)=2\pi S_0\delta(t)$, that is, the delta function. By Eq. (3.51)

$$R_Y(\tau)=\int_{-\infty}^{\infty}\int_{-\infty}^{\infty}2\pi S_0\delta(t+\varphi-\theta)\,h(\theta)\,h(\varphi)\mathrm{d}\theta\,\mathrm{d}\varphi=2\pi S_0\int_{-\infty}^{\infty}h(\varphi)\,h(\tau+\varphi)\mathrm{d}\varphi$$

or by Eq. (3.31)

$$R_Y(\tau) = \frac{2\pi S_0}{\overline{\omega_0^2}} \int_0^\infty e^{-\lambda\omega_0\varphi} e^{-\lambda\omega_0(\varphi+\tau)} \sin\overline{\omega_0}\varphi \sin\overline{\omega_0}(\tau+\varphi) d\varphi$$

$$= \frac{2\pi S_0}{\overline{\omega_0^2}} e^{-\lambda\omega_0|\tau|} \int_0^\infty e^{-2\omega_0\varphi} \tfrac{1}{2}[\cos\overline{\omega_0}\tau - \cos\overline{\omega_0}(\tau+2\varphi)] d\varphi$$

$$= \frac{\pi S_0 e^{-\lambda\omega_0|\tau|}}{2\lambda\omega_0^3}\left[\cos\overline{\omega_0}\tau + \frac{\lambda\omega_0}{\overline{\omega_0}} \sin\overline{\omega_0}|\tau|\right]$$

where use has been made of the standard integrals

$$\int_0^\infty e^{-ax}\cos bx\, dx = \frac{a}{a^2+b^2}, \quad \int_0^\infty e^{-ax}\sin bx\, dx = \frac{b}{a^2+b^2}$$

The spectral density on the other hand is simply

$$S_Y(\omega) = |H(\omega)|^2 S_0 = \frac{S_0}{(\omega_0^2-\omega^2)^2 + 4\lambda^2\omega_0^2\omega^2}$$

In passing it may be noted that passing a scaled white noise signal with a spectral density $S_0 = 1$ through a filter with a gain function $G(\omega)$, the spectral density of the output signal is exactly the gain function, i.e. $S_Y(\omega) = G(\omega)$. As for the mean value, $\mu_Y = 0$, since $S_Y(\omega)$ has no power concentration at $\omega = 0$. Also, $\mu_X = 0$ for the same reason.

The mean square value can be obtained in two ways:

$$E[Y^2] = \begin{cases} R_Y(0) = \dfrac{\pi S_0}{2\lambda\omega_0^3} \\ \displaystyle\int_{-\infty}^\infty S_Y(\omega) d\omega = \int_{-\infty}^\infty \frac{S_0\, d\omega}{(\omega_0^2-\omega^2)^2 + 4\lambda^2\omega_0^2\omega^2} \end{cases}$$

The integral can be evaluated by applying Cauchy's integral theorem (see Example 3.5), and the same result is found. Thus the root mean square response is

$$\text{RMS}[Y] = \sqrt{E[Y^2]} = \frac{1}{\omega_0}\sqrt{\frac{\pi S_0}{2\lambda\omega_0}}$$

In general the frequency domain information carried by the power spectral density is often more easily obtained than the time domain information supplied by the autocorrelation function. The opposite, of course, is the case when only measured sample functions are available. This means that integrals of the type $\int_{-\infty}^{\infty} |H(\omega)|^2 d\omega$ frequently occur in problems of the above nature. Such integrals are best handled by using Cauchy's integral theorem or by getting hold of some elaborate integral tables, [72]. For integrand functions of the type

$$H(\omega) = \frac{b_0 + (i\omega)b_1 + \cdots + (i\omega)^{n-1}b_{n-1}}{a_0 + (i\omega)a_1 + \cdots + (i\omega)^{n-1}a_{n-1} + (i\omega)^n a_n} \tag{3.53}$$

in which $\{a_i\}$ and $\{b_i\}$ are real constants, the integral $I_n = \int_{-\infty}^{\infty} |H(\omega)|^2 \mathrm{d}\omega$ has been evaluated and tabulated results are given in references [43], [91] and [152].

For stable vibratory systems, (negative damping), the poles of the integrand function, that is, the roots of the characteristic equation

$$a_0 + (\mathrm{i}\omega)a_1 + \cdots + a_n(\mathrm{i}\omega)^n = 0$$

must all lie in the upper half of the ω-plane (see Example 3.5). Provided this condition is satisfied, the following table of results is obtained (from Newland, [152]).

For $n = 1$

$$H(\omega) = \frac{b_0}{a_0 + (\mathrm{i}\omega)a_1}, \qquad I_1 = \frac{\pi b_0^2}{a_0 a_1}$$

For $n = 2$

$$H(\omega) = \frac{b_0 + (\mathrm{i}\omega)b_1}{a_0 + (\mathrm{i}\omega)a_1 - \omega^2 a_2}, \qquad I_2 = \frac{\pi(a_0 b_1^2 + a_2 b_0^2)}{a_0 a_1 a_2}$$

For $n = 3$

$$H(\omega) = \frac{b_0 + (\mathrm{i}\omega)b_1 - \omega^2 b_2}{a_0 + (\mathrm{i}\omega)a_1 - \omega^2 a_2 - \mathrm{i}\omega^3 a_3}, \qquad I_3 = \frac{\pi(a_0 a_3(2b_0 b_2 - b_1^2) - a_0 a_1 b_2^2 - a_2 a_3 b_0^2)}{a_0 a_3(a_0 a_3 - a_1 a_2)}$$

For $n = 4$

$$H(\omega) = \frac{b_0 + (\mathrm{i}\omega)b_1 - \omega^2 b_2 - \mathrm{i}\omega^3 b_3}{a_0 + (\mathrm{i}\omega)a_1 - \omega^2 a_2 - \mathrm{i}\omega^3 a_3 + \omega^4 a_4}$$

$$I_3 = \frac{\pi[(a_0 b_3^2(a_0 a_3 - a_1 a_2) + a_0 a_1 a_4(2b_1 b_3 - b_2^2) - a_0 a_3 a_4(b_1^2 - 2b_0 b_2) + a_4 b_0^2(a_1 a_4 - a_2 a_3))]}{a_0 a_4(a_0 a_3^2 + a_1^2 a_4 - a_1 a_2 a_3)}$$

For $n = 5$, 6, 7, see reference [91].

Example 3.8 In the mechanical vibratory system of Example 3.2, let the displacement of the driving cart $X(t)$ be a white noise process with a power spectral density $S_X(\omega) = S_0$, a constant. Find the mean square response of the mass m.

Solution The complex frequency response of the mass m was found to be

$$H(\omega) = \frac{(\omega_0^3/9\lambda) + (\mathrm{i}\omega)\frac{2}{3}\omega_0^2 - \omega^2(\lambda\omega_0)}{(\omega_0^3/9\lambda) + (\mathrm{i}\omega)\omega_0^2 - \omega^2(2\lambda\omega_0) - \mathrm{i}\omega^3}$$

The mean square response $E[Y^2] = \sigma_Y^2$ is given by

$$E[Y^2] = \int_{-\infty}^{\infty} S_0 |H(\omega)|^2 \mathrm{d}\omega = S_0 I_3$$

so

$$E[Y^2] = \pi S_0 \frac{a_0 a_3(2b_0 b_2 - b_1^2) - a_0 a_1 b_2^2 - a_2 a_3 b_0^2}{a_0 a_3(a_0 a_3 - a_1 a_2)}$$

in which $b_0 = \omega_0^3/9\lambda$, $b_1 = \frac{2}{3}\omega_0^2$, $b_2 = \lambda\omega_0$, $a_0 = \omega_0^3/9\lambda$, $a_1 = \omega_0^2$, $a_2 = 2\lambda\omega_0$, $a_3 = 1$

$$E[Y^2] = \pi S_0 \frac{[2(\omega_0^3/9\lambda)\lambda\omega_0 - \frac{4}{9}\omega_0^4] - \omega_0^2\lambda^2\omega_0^2 - (2\lambda\omega_0/\omega_0^3)9\lambda(\omega_0^6/(9\lambda)^2)}{(\omega_0^3/9\lambda) - \omega_0^2 2\lambda\omega_0}$$

and

$$E[Y^2] = \pi S_0 \frac{\frac{4}{9}\omega_0^4 + \lambda^2\omega_0^4}{2\lambda\omega_0^3 - (\omega_0^3/9\lambda)} = \pi S_0 \omega_0 \frac{4\lambda + 9\lambda^3}{18\lambda^2 - 1}$$

As before, the mean square value is mostly dependent on the frequency and the ratio of critical damping in the system.

Example 3.9 A simply supported uniform beam with mass per unit length equal to m is subjected to a stochastic load $P(t)$ acting at the centre of the beam (Fig. 3.7). The beam is completely elastic throughout the motion with rigidity or bending stiffeness EI. For sufficiently accurate results regarding the mid-span section, the vibrating beam may be treated as system with a single degree of freedom with the mass $mL/2 = ml$ concentrated at the midpoint and the equivalent stiffness $k = (\pi/2l)^4\ EIl$. The shape of the deflection is sinusoidal with the beam-length as one half period. A 10% critical inner damping should also be taken into account. The stochastic load $P(t)$ consists of a d.c. component of magnitude P_0 and a fluctuating part which has the characteristic of a stationary random process $X(t)$ with the power spectral density

$$S_X(\omega) = \frac{\omega_0^2 S_0}{\omega^2 + \omega_0^2}$$

where ω_0 is characteristic frequency and S_0 a scaling factor.

Find the mean value and the root mean square value of the mid-span deflection, when the force is suddenly switched on at $t = 0$.

Solution Firstly it should be realized that the beam will undergo deflection due to its own weight, where the mid-span deflection is

$$v(l, *) = \frac{5}{384}\frac{mg\,L^4}{EI} = \frac{5}{24}\frac{mg\,l^4}{EI}$$

Over this static deflection is superimposed, firstly the contribution due to the d.c. component and secondly the purely random contribution.

The equation of motion for the mid-span point is

$$m\ddot{Y}(t) + c\,\dot{Y}(t) + k\,Y(t) = P_0 + X(t)$$

and the beam deflection is given by

$$V(z, t) = Y(t)\sin(\pi z/2l) \Rightarrow V(L/2, t) = Y(t)$$

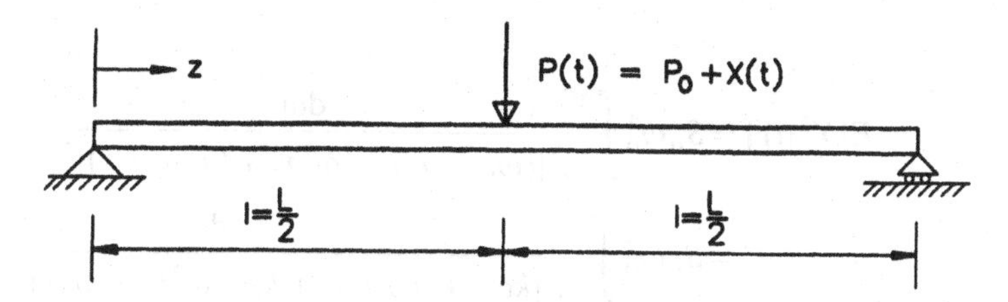

Fig. 3.7 *A simply supported beam loaded at its centre*

in which $\overline{m} = ml$, $c = (\lambda = 0\,005)2\sqrt{(km)}$ and $k = (\pi/2l)^4\,EIl$. Now, the deflection $Y_1\,(t)$, due to the step-function loading $P_0 l(t)$, is obtained using Eqs. (3.42) and (3.32).

$$Y_1\,(t) = \frac{P_0}{\overline{m}}\,a(t) = \frac{2P_0(2l)^3}{\pi^4 EI}\left(1 - e^{-\lambda\omega_1 t}\left(\cos\overline{\omega_1}\,t + \frac{\lambda\omega_1}{\overline{\omega}^1}\sin\overline{\omega_1}\,t\right)\right)$$

and

$$\omega_1 = \left(\frac{\pi}{2l}\right)^2\sqrt{\frac{EI}{\overline{m}}}$$

Since $S_X(0) = S_0$, there is no power concentration at zero frequency and $E[X)] = 0$, so there are no further d.c. components present. Moreover, the mean value of the deflection $Y_2\,(t)$ due to the random part is also zero. The total mean value of the mid-span deflection is therefore given by $Y_1\,(t)$ which is time dependent! The output is no longer stationary. However, the free vibrations in $Y_1\,(t)$ will quickly decay and vanish. After only four cycles, $(4 \times 2\pi/\omega_1)$, the exponential decay factor is of the value 0.08, and the total stationary mean value is

$$E[Y] = \frac{2P_0(2l)^3}{\pi^4 EI} \simeq \frac{P_0(2l)^3}{48\,EI}$$

The mean square value $E[Y^2]$, after the free vibrations in the beam have died out, is only dependent on the random part $X(t)$. In fact

$$S_{Y_2}(\omega) = |H(\omega)|^2\,S_X(\omega)$$

and

$$E[Y^2(t)] = E[Y_2^2(t)] = \int_{-\infty}^{\infty} S_{Y_2}(\omega)\,d\omega$$

Now

$$S_X(\omega) = \frac{\omega_0^2 S_0}{\omega^2 + \omega_0^2} = S_0\omega_0^2\frac{1}{|\omega_0 + i\omega|^2}$$

and

$$|H(\omega)|^2 = \frac{1}{|-\omega^2\overline{m} + (i\omega)c + k|^2}$$

so

$$E[Y^2(t)] = S_0\omega_0^2 \int_{-\infty}^{\infty} \frac{d\omega}{|(\omega_0 + i\omega)(-\omega^2\overline{m} + (i\omega)c + k)|^2}$$

$$= S_0\omega_0^2 \int_{-\infty}^{\infty} \frac{d\omega}{|k\omega_0 + (i\omega)(\omega_0 + k) - \omega^2(c + \omega_0\overline{m}) - i\omega^3\overline{m}|^2}$$

$$= S_0\omega_0^2 I_3$$

where $b_0 = 1$, $b_1 = b_2 = 0$, $a_0 = k\omega_0$, $a_1 = c\omega_0 + k$, $a_2 = c + \omega_0\overline{m}$ and $a_3 = \overline{m}$. Therefore,

$$E[Y^2(t)] = S_0\omega_0^2 \frac{\pi a_2 b_0^2}{a_0(a_1 a_2 - a_0 a_3)} = \pi S_0\omega_0^2 \frac{c + \omega_0\overline{m}}{ck(c\omega_0 + k + \omega_0^2\overline{m})}$$

Finally the root mean square response is

$$\text{RMS}[Y^2(t)] = \sqrt{\frac{\pi S_0}{kc}} \left(\frac{1 + 2\lambda\left(\dfrac{\omega_1}{\omega_0}\right)}{1 + 2\lambda\left(\dfrac{\omega_1}{\omega_2}\right) + \left(\dfrac{\omega_1}{\omega_0}\right)^2} \right)^{1/2}$$

3.3.2 *Linear Pass-band Filters*

In most problems, the input spectral density does not have such convenient form to allow the output spectral density to be treated as a single function of the type $|H(\omega)|^2$. For lightly damped systems, the frequency response function squared, that is, the gain function $G(\omega) = |H(\omega)|^2$ has a strong and narrow peak about the fundamental frequency ω_0. If the input power spectral density is a reasonable well behaved and smooth function of ω, the major contribution to the mean square value of the output comes from the value of $S_X(\omega)$, that is, the input spectral density evaluated at $\omega = \omega_0$, the fundamental frequency. Or (see Fig. 3.8)

$$E[Y^2(t)] = \int_{-\infty}^{\infty} |H(\omega)|^2 S_X(\omega) d\omega \simeq S_X(\omega_0) \int_{-\infty}^{\infty} |H(\omega)|^2 d\omega \tag{3.54}$$

Consider a linear, time-invariant filter with the gain function $G(\omega)$ and a centre frequency ω_0. If a random stationary input signal $X(t)$ with a power spectral density $S_X(\omega)$ is passed through the filter, the output signal $Y(t)$ has the spectral density, (cf. Eq. (3.52)),

$$S_Y(\omega) = G(\omega) S_X(\omega) \cong G(\omega) S_X(\omega_0) \tag{3.55}$$

This relationship is illustrated in Fig. 3.8. The implication of Eq. (3.55) is that for sufficiently sharp resonance peaks, (the shaded area), the area under the curve $S_Y(\omega) = G(\omega) S_X(\omega)$ is approximately the same as the area under the gain function times the constant spectral density $S_X(\omega_0)$.

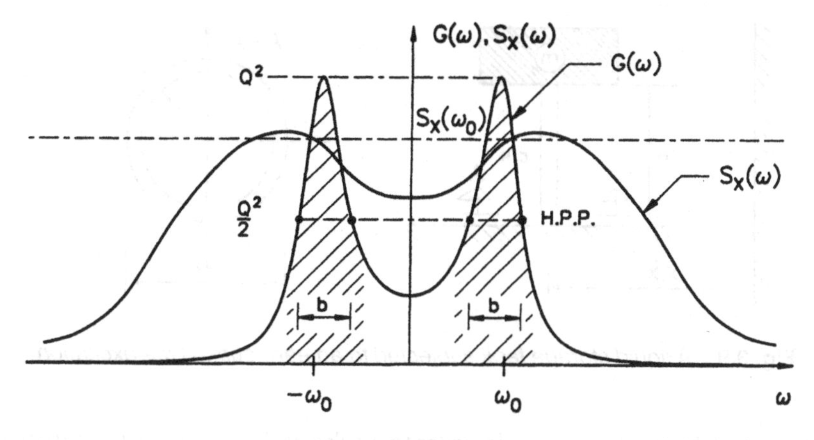

Fig. 3.8 *Gain and spectral density functions*

To study the effect of the damping factor of the filter, $\lambda = c/c_{cr}$, consider a filter with the well-known gain function,

$$\omega_0^4 G(\omega) = \frac{1}{(1-x^2)^2 + 4\lambda^2 x^2}, \qquad x = \frac{\omega}{\omega_0} \tag{3.56}$$

i.e. that of the simple oscillator. For the small value of damping to be of any interest, the resonance peaks are of the magnitude

$$Q^2 = \tfrac{1}{4}\lambda^2 \qquad \text{for } \omega \simeq \omega_0 \tag{3.57}$$

The factor $Q = \frac{1}{2}\lambda$, the quality factor of the filter, is a measure the damping of the filter. The frequency band width of the filter is usually defined as the width of the resonance peak at the half power points (HPP) where $\omega_0^4 \cdot G(\omega_{1/2}) = Q^2/2$. The half power frequencies are then given by

$$\frac{1}{8\lambda^2} = \frac{1}{(1-x^2)^2 + 4\lambda^2 x^2} \Rightarrow x^2 = 1 - 2\lambda^2 \pm \sqrt{1 + \mathrm{j}}$$

or for small values of λ, $x \simeq \sqrt{(1 \pm 2\lambda)}$, and

$$\omega_{1/2} = \omega_0(1 \pm \lambda) = \omega_0(1 \pm Q/2) \tag{3.58}$$

Hence the filter bandwidth is

$$b = 2\lambda\omega_0 = \omega_0/Q \tag{3.59}$$

For large values of b, that is the resonance peak is flat and wide, the filter is said to be a broad band filter. If b is small, the resonance peak is sharp and narrow, and the filter is said to be a narrow band filter. Systems, for which $Q > 10$, such as many mechanical vibration systems and most civil engineering structures, are definitely of the narrow pass-band type. Thus the output signal or the response will be a narrow band process, with the whole power confined in a narrow frequency band about the system natural frequency ω_0. The approximation leading to Eq. (3.55) is valid and it is sufficient to know the value of

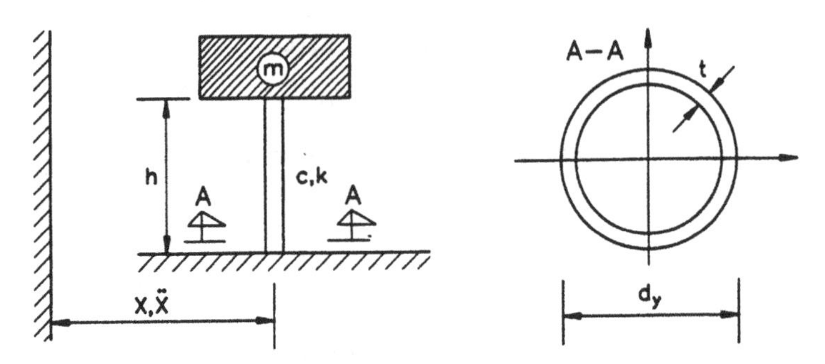

Fig. 3.9 *A liquid storage tank subjected to an earthquake type excitation*

the input spectral density for $\omega = \omega_0$ in order to work out the second order statistics of the system response.

For the simple linear oscillator, the mean square response $E[Y^2(t)]$ to a stationary stochastic input process with a power spectral density $S_X(\omega)$ is sufficiently accurately obtained for $Q \geqslant 10$ by (see Ex. 3.7)

$$E[Y^2(t)] = S_X(\omega_0) \int_{-\infty}^{\infty} |H(\omega)|^2 \, d\omega = \frac{\pi S_X(\omega_0)}{2\lambda\omega_0^3} \tag{3.60}$$

Example 3.10 An elevated tank-like structure, Fig. 3.9, is subjected to a base acceleration which is a band-limited white noise process between 0.1 Hz and 100 Hz (see Ex. 212), with an r.m.s. acceleration level of $\frac{1}{3}g$. The following data pertaining to the tank are given:

mass $m = 2.55 \times 10^6$ kg, damping coefficient $c = 6.4 \times 10^6$ kg/s, height $h = 30$ m

The shaft supporting the tank is to be treated as a weightless linear-elastic cantilever beam with an elastic modulus, $E = 30$ kN/mm^2, a circular cross section of diameter $d_y = 4.0$ m and wall thickness $t = 0.3$ m. Find the r.m.s. stress level in the extreme fibres of the shaft at the base.

Solution The equation of motion governing the horizontal deflection of the tank $Y(t)$ is essentially the well-known equation

$$m\ddot{Y}(t) + c\dot{Y}(t) + kY(t) = -m\ddot{X}(t)$$

The spring constant of the shaft $k = 3EI/h^3$, that is,

$$k = \frac{3E}{h^3}\left(\frac{\pi}{8} d_y^3 t\right) = \frac{3 \times 3 \times 10^4 \times 7.54}{30^4} = 25.15 \text{ MN/m}$$

The fundamental frequency is

$$\omega_0 = \sqrt{\frac{k}{m}} = \sqrt{\frac{25.15}{2.55}} = \pi \text{ r.p.s.}, \; T_0 = 2.0 \text{ s}$$

and the ratio of critical damping, $\lambda = c/(2\omega_0 m) = 0.04$. The excitation has a power spectral density which is constant in the range $\omega_1 = 0.1 \times 2\pi$ to $\omega_2 = 100 \times 2\pi$ with a mean square

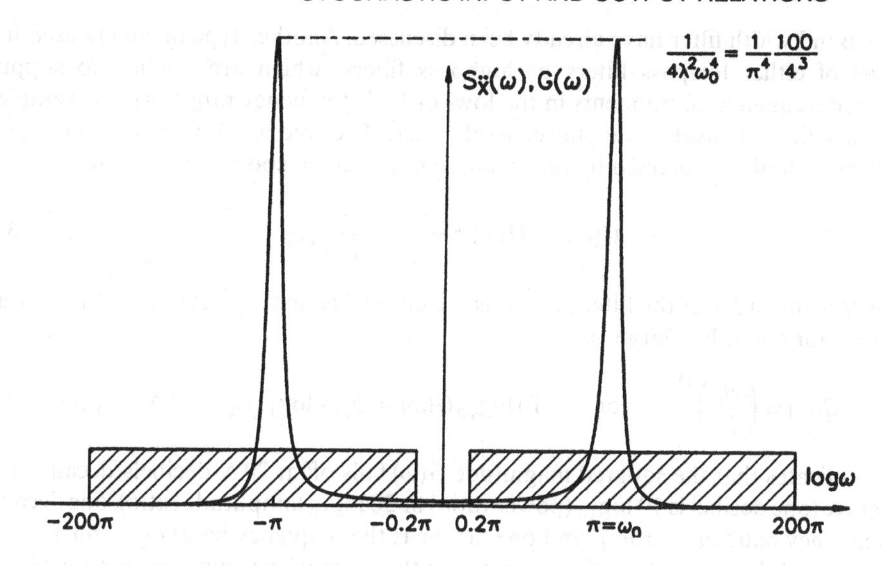

Fig. 3.10 *Tank frequency response and excitation spectral density*

level, $E[\ddot{X}^2] = \frac{1}{9}g^2$, Fig. 3.10. Therefore

$$E[\ddot{X}] = 0 \text{ (no concentration at } \omega = 0), \Rightarrow E[Y] = 0$$

$$E[\ddot{X}^2] = \tfrac{1}{9}g^2 = S_0 \times 2(100 - 0.1)2\pi \simeq 400\pi S_0 \Rightarrow S_0 = g^2/(3600\pi)$$

Since $\lambda = 0.04$ or $Q = 1(2\lambda) = 12.5$, the approximation Eq. (3.35) is valid and sufficiently accurate.

$$E[Y^2(t)] = S_0 \int_{-\infty}^{\infty} |H(\omega)|^2 \,\mathrm{d}\omega = \frac{\pi S_0}{2\lambda\omega_0^3} = \frac{g^2}{7200\,\lambda\omega_0^3}$$

$$\mathrm{RMS}[Y] = \frac{\sqrt{g^2}}{7200\,\lambda\omega_0^3} = \frac{9.81}{\sqrt{7200 \times 0.04\pi^3}} = 0.10\,\mathrm{m}$$

Now due to the weight alone, the normal stress in the base section is

$$\sigma_N = \frac{N}{A} = \frac{2.55 \times 9.81}{\pi \times 4 \times 0.3} = 6.63\,\mathrm{MN/m^2}$$

over this stress is superimposed the r.m.s. stress due to the base acceleration, that is

$$\sigma_{\ddot{X}} = \frac{M}{W} = \frac{k \times \mathrm{RMS}[Y] \times h}{(\pi/4)d_y^2 t} = \frac{25.15 \times 0.10 \times 30}{3.77} = 20\,\mathrm{MN/m^2} \qquad \text{(root mean square)}$$

So finally, the r.m.s. stress level in the extreme fibres of the shaft is 26.63 $\mathrm{MN/m^2}$ in compression and 13.37 $\mathrm{MN/m^2}$ in tension.

Linear continuous filters, also called analogue filters to distinguish them from discrete or digital filters, which will be discussed in Chapter 7, serve as an important tool for shaping and modifying the response of a system, for instance a record from a measurement device. Band width filter have already been discussed. Another type of widely used filter consists of either lowpass filters or highpass filters, which are applied to suppress unwanted frequency components in the low- or high-frequency range. As an example of a lowpass filter, consider the Butterworth filter. The frequency response function of a Butterworth filter is defined by the squared magnitude response or the gain, that is,

$$G(\omega) = |H(\omega)|^2 = \frac{1}{1 + (\omega/\omega_c)^{2N}} \tag{3.61}$$

where N is the order of the filter and ω_c is the cutoff frequency. Obviously, $G(0) = 1$ and $G(\omega_c) = \frac{1}{2}$ for all N. For large ω,

$$G(\omega) \approx \left(\frac{\omega_c}{\omega}\right)^{2N} \quad \text{or} \quad 10\log_{10} G(\omega) = 20\,N\log_{10}\omega_c - 20\,N\log_{10}\omega \tag{3.62}$$

which indicates that the squared magnitude $G(\omega)$ falls off at $20\,N$ decibels/decade or $6N$ dB/octave (one decibel is $10\log_{10}$(power ratio) or $20\log_{10}$(amplitude ratio), one decade is the frequency band $\omega_2 = 10\omega_1$, and one octave is the frequency band $\omega_2 = 2\omega_1$).

Location of the roots of the denominator or the poles of the gain function Eq. (3.61) is fundamental for the stability of the filter. The poles of the Butterworth filter are best determined by analysing the corresponding gain function in the s-plane ($s = \sigma + \mathrm{i}\omega$ for Laplace transforms, see Section 7.2.6), i.e.

$$H(s)\,H(-s) = \frac{1}{1 + (-s^2/\omega_c^2)^N} \tag{3.63}$$

By Eq. (3.63), they are located on a circle of radius ω_c at equally spaced points, which are given by

$$-\frac{s^2}{\omega_c^2} = (-1)^{1/N} = \mathrm{e}^{\mathrm{i}(2k+1)\pi/2N}, \qquad k = 0, 1, 2, \ldots, N$$

whereby

$$s_k = \omega_c\,\mathrm{e}^{\mathrm{i}\pi/2}\,\mathrm{e}^{\mathrm{i}(2k+1)\pi/2N}, \qquad k = 0, 1, 2, \ldots, N-1 \tag{3.64}$$

The frequency response of the Butterworth filter is shown in Fig. 3.11 for $N = 1$ (-20 dB/decade), $N = 3$ (-60 dB/decade), $N = 5$ (-100 dB/decade) and $N = 7$ (-140 dB/decade). The magnitude square function is monotonic in both the passband and the stopband of the filter (see Fig. 7.16). The order N of the filter needed to attenuate a specified frequency component $\mathrm{e}^{\mathrm{i}\omega_0}$ by a certain ratio δ is given by

$$\frac{1}{1 + (\omega_0/\omega_c)^{2N}} = \delta_2^2 = \frac{1}{\sqrt{1+\delta^2}}$$

or

$$N = \frac{\log_{10}[(1/\delta_2^2) - 1]}{2\log_{10}(\omega_0/\omega_c)} \tag{3.65}$$

Example 3.11 Find the order and the poles of a lowpass Butterworth filter that has a -3 dB bandwidth of 600 Hz ($10\log_{10} G(600) = -3$) and an attenuation of 60 dB at 1200 Hz.

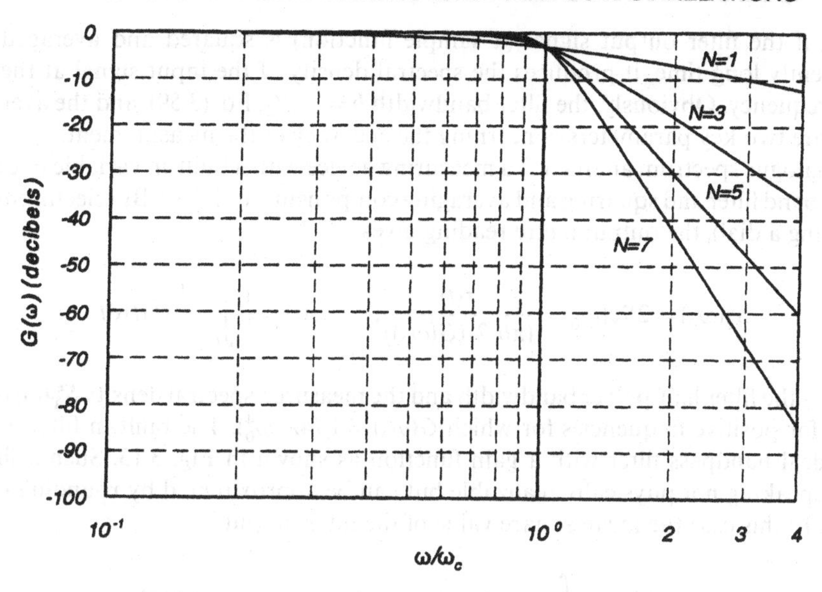

Fig. 3.11 *Frequency response of a Butterworth filter*

Solution The critical frequencies are the cutoff frequency $f_c = 600$ Hz and the stopband frequency $f_s = 1200$ Hz, that is

$$\omega_c = 2\pi f_c = 1200\pi \qquad \text{and} \qquad \omega_s = 2\pi f_s = 2400\pi$$

The attenuation δ is specified at 60 dB whereby

$$20\log_{10}\delta = 60 \qquad \text{or} \qquad \delta = 1000$$

The parameter $\delta_2 = 1/(1+\delta^2)^{1/2} = 0.001 = 10^{-3}$. Hence by Eq. (3.65),

$$N = \frac{\log_{10}[10^6 - 1]}{2\log_{10}2} = 9.96$$

Therefore to satisfy the requirements of the filter specifications, the order of the filter is $N = 10$. The pole locations are given by Eq. (3.64)

$$s_k = 1200\pi e^{i\pi/2}\, e^{i(2k+1)\pi/20}, \qquad k = 0, 1, 2, \ldots, 9$$

For ergodic processes, Eq. (3.55) offers an opportunity to measure the spectral density at discrete frequencies. In fact, using a narrow-band filter with the gain function Eq. (3.57), the spectral density of an ergodic signal passed through the filter is evaluated at the filter centre frequency ω_0 as

$$S_X(\omega_0) = \frac{2\lambda\omega_0^3}{\pi} E[Y^2(t)] \simeq \frac{2}{\pi}\lambda\omega_0^3 \frac{1}{T}\int_0^T y^2(t)\,dt \tag{3.66}$$

That is, if the filter output signal (a sample function) is squared and averaged over a sufficiently long time, it produces the spectral density of the input signal at the filter centre frequency. Obviously, the filter bandwidth $b = \omega_0/Q$, Eq. (3.59), and the averaging time T are two key parameters concerning the accuracy of the measurement.

An analogue spectrum analyser is a measuring device with a built-in variable frequency narrow-band filter and squaring and averaging components, Fig. 3.12. By selecting ω_0 and Q, (turning a dial), the output metre reading gives

$$S_X^e(\omega_0) = 2S_X(\omega_0) = \frac{z(t)}{(\pi b/2)(G(\omega_0))}, \qquad z(t) = \frac{1}{T}\int_t^{T+t} y^2(t)\,dt \tag{3.67}$$

where b is the filter half-power bandwidth and the measured spectral density $S_X^e(\omega)$ is only defined for positive frequencies for which $G(\omega_0) = 1/(4\lambda^2\omega_0^4)$. The built-in filter is a so-called ideal bandpass filter with a gain function as shown in Fig. 3.13. Such a filter is strictly speaking not physically realizable but can be approximated by rounding off the corners. In this case the mean square value of the filter output is

$$E[Y^2(t)] = \int_{-\infty}^{\infty} |H(\omega)|^2 S_X(\omega)\,d\omega = S_X(\omega_0) \times 2BH_0^2 \tag{3.68}$$

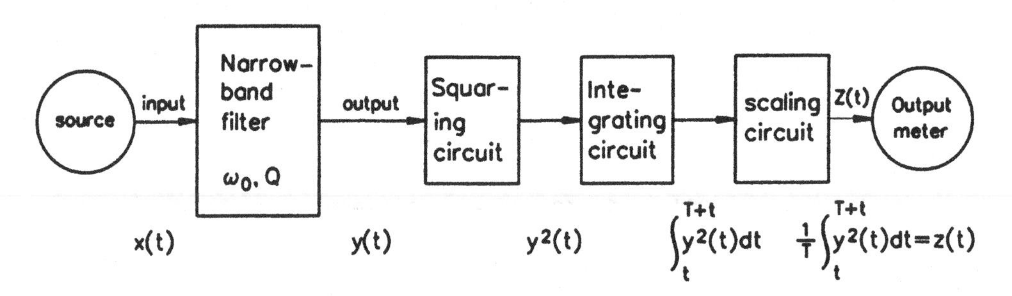

Fig. 3.12 *Schematic diagram of an analogue spectrum analyser*

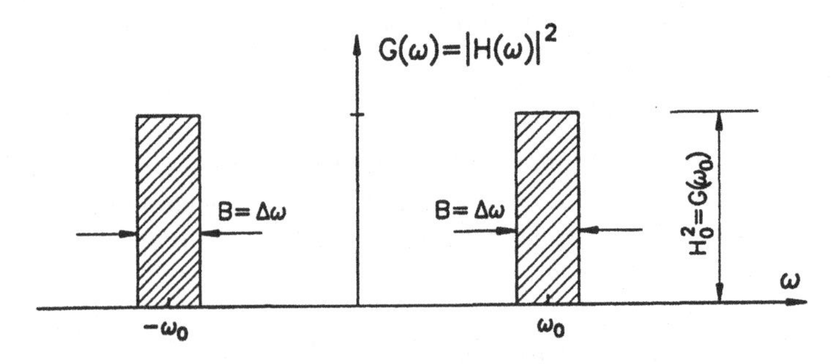

Fig. 3.13 *The gain of an ideal pass-band filter*

or by Eq. (3.67)

$$S_X^e(\omega_0) = \frac{z(t)}{BH_0^2} \tag{3.69}$$

The bandwidth of the ideal pass-band filter thus corresponds to $\pi/2$ times the half-power bandwidth of the narrow-band filter previously used. $B_{eq} = \pi/2 \times b$ is also referred to as the equivalent bandwidth or the noise bandwidth of the filter and is about 57% larger than the half-power bandwidth to compensate for the noise associated with the frequency components passed through, outside the filter bandwidth (see Section 7.2.7).

Finally, the output metre reading $z(t)$ is a function of the starting time t. To study the accuracy of the measurement Eq. (3.69), form the random process

$$Z(t) = \frac{1}{T}\int_{t-T}^{t} Y^2(t)\mathrm{d}t = BH_0^2 S_X^e(\omega_0) \tag{3.70}$$

by time-averaging on the output process $Y(t)$ itself. Taking the expectation,

$$E[Z] = \frac{1}{T}\int_{t-T}^{t} E[Y^2(t)]\mathrm{d}t = E[Y^2] = \int_{-\infty}^{\infty} |H(\omega)|^2 S_X(\omega)\mathrm{d}\omega = S_X(\omega_0) \times 2BH \tag{3.71}$$

So the mean value of this new process is the value being sought. Since a time-averaged value with a finite averaging time T to estimate the spectral density is being used, that is, Eq. (3.66) instead of Eq. (3.71), how good is the estimate, i.e. how large is the variance? Now,

$$\mathrm{Var}\,[Z] = (E[Z^2] - (E[Z]^2)) \tag{3.72}$$

Crandall, [44], has shown that for a random process governed by a Gaussian probability law (see Section 6.1), the difficult expression Eq. (3.72) can be evaluated to yield

$$\sigma_Z^2 = \mathrm{Var}[Z] \simeq 2\pi/T(1/B)(E[Y^2])^2 = (2\pi/T)BH_0^4(S_X^e(\omega_0))^2 \tag{3.73}$$

The coefficient of variation of the measurement is

$$V_Z = \frac{\sigma_Z}{\mu_Z} = \sqrt{\frac{2\pi B}{T}\frac{H_0^2 S_X^e(\omega_0)}{BH_0^2 S_X^e(\omega_0)}} = \sqrt{\frac{2\pi}{BT}} \tag{3.74}$$

Usually, in experimental analysis, the natural frequency $f = \omega/2\pi$ is preferred, whereby the filter bandwidth in natural frequency units is $B^f = B^\omega/2\pi$. Thus, Eq. (3.74) finally becomes

$$V_Z = 1/\sqrt{(BT)} \tag{3.75}$$

which is the accuracy or uncertainty of the measurement.

For a good resolution of the filter, a small bandwidth is needed, that is, $B \to 0$. Otherwise, the narrow band condition for Eq. (3.55) breaks down. This, however, makes the variation coefficient extremely large and ruins the accuracy of the measurement unless the averaging time T is taken very large, which is hardly feasible. A usual compromise is to select $BT \sim 1$.

3.3.3 Influence of Finite Operation Times

In Exercise 3.9, it was briefly touched upon that the vibrating system needed time to reach stationary response. In fact, the output or response of a physically realizable system, that is with finite operating time, is always non stationary! By Eq. (3.46), the response $Y(t)$ of a finite operation time system to a stationary random excitation $X(t)$ is

$$Y(t) = \int_0^t X(u)\,h(t-u)\,du \tag{3.46}$$

The mean value of the response is obtained by taking the expectation of both sides or

$$E[Y(t)] = \int_0^t E[X(u)]\,h(t-u)du = \mu_X \int_0^t h(t-u)du = \mu_X \int_0^t h(u)du \tag{3.76}$$

Clearly the mean value of the response is time dependent even if μ_X is a constant. By introducing the admittance function $a(t)$, the mean value $E[Y(t)]$ can also be written as (see Eq. (3.40))

$$\mu_Y(t) = \mu_X a(t) \tag{3.77}$$

To the mean value would also be added any free vibration components present for other reasons (initial conditions).

Example 3.12 A simple linear oscillator (m, c, k) is suddenly subjected to a stationary random excitation $X(t)$ at time $t = t_0$. Given that the oscillator was already in motion at that time with amplitude $y_0 = y(t_0)$ and velocity $v_0 = \dot{y}(t_0)$, find the mean value of the total response $Y(t)$ of the oscillator.

Solution The motion of the oscillator mass at time t_0 is prescribed by Eq. (3.33), and due to the random excitation as the only external force after $t = t_0$, the mean value of the response is given by Eq. (3.77). Therefore, the mean value of the total response $Y(t)$ is

$$\mu_Y(t) = y_0(1 - \omega_0^2 a(t-t_0)) + v_0 h(t-t_0) + \mu_X a(t-t_0)$$

or

$$\mu_Y(t) = y_0 + (\mu_X - \omega_0^2 y_0)\,a(t-t_0) + v_0 h(t-t_0)$$

and the mean value deviates from the stationary mean value μ_X/ω_0^2 by a harmonic term which has the constant non-decaying amplitude $((y_0 - \mu_X/\omega_0)^2 + (v_0/\omega_0)^2)^{1/2}$. Even if $y_0 = v_0 = 0$, that is, the system is at rest at the onset of the random excitation, the mean value is

$$\mu_Y(t) = \mu_X/\omega_0^2(1 - \cos\omega_0 t)$$

and never reaches a stationary value. In the case $0 < \lambda < 1$,

$$\mu_Y(t) \to \mu_X/\omega_0^2 \qquad \text{for } \lambda\omega_0 t \to \infty$$

and the rapidity with which the mean value becomes stationary depends on the magnitude of $\omega_0\lambda$.

The above example shows clearly how physically realizable systems will show non-stationary behaviour regarding the mean value, even if the excitation is a perfectly stationary stochastic process. However, due to the damping inherent in all physical systems, the steady state mean value response of the system becomes stationary after a while. Much the same applies to the higher moments. Therefore, form the covariance kernel of the response of a finite operation time system:

$$K_Y(t,s) = \mathrm{Cov}[Y(t), Y(s)] = E\left[\left(\int_0^t X(u)\,h(t-u)\,\mathrm{d}u\right)\left(\int_0^s X(v)\,h(s-v)\mathrm{d}v\right)\right]$$

$$= \int_0^t \int_0^s h(t-u)\,h(s-v)\,E[X(u)X(v)]\,\mathrm{d}u\,\mathrm{d}v$$

or

$$K_Y(t,s) = \int_0^t \int_0^s h(t-u)\,h(s-v)\,K_X(u,v)\mathrm{d}u\,\mathrm{d}v \tag{3.78}$$

which relates the covariance kernels. If and when $X(t)$ is stationary, $\Gamma_X(v-u) = K_X(u,v)$, and using the same transformation as when deriving Eq. (3.50),

$$K_Y(t,s) = \int_0^t \int_0^s \Gamma_X(\varphi - \theta + (t-s))\,h(\theta)\,h(\varphi)\mathrm{d}\theta\mathrm{d}\varphi \tag{3.79}$$

The variance of $Y(t)$, $\sigma_Y^2(t)$ is then given by

$$\sigma_Y^2(t) = \int_0^t \int_0^t (R_X(\varphi-\theta) - \mu_X^2)\,h(\theta)\,h(\varphi)\mathrm{d}\theta\,\mathrm{d}\varphi \tag{3.80}$$

Whether the covariance kernel or the variance can reach stationary values clearly depends on the behaviour of the impulse response function $h(t)$.

Example 3.13 Find the variance of the response of the linear oscillator of Example 3.12, if $\mu_X = 0$, and $t_0 = 0$.

Solution By Eq. (3.80),

$$\sigma_Y^2(t) = \int_0^t \int_0^t R_X(\varphi-\theta)\,h(\theta)h(\varphi)\mathrm{d}\theta\,\mathrm{d}\varphi$$

Introducing the inverse transform

$$R_Y(\varphi-\theta) = \int_{-\infty}^{\infty} S_X(\omega)\mathrm{e}^{\mathrm{i}\omega(\varphi-\theta)}\mathrm{d}\omega$$

and writing

$$H(\omega) = \int_0^{\infty} h(\theta)\mathrm{e}^{-\mathrm{i}\omega\theta}\mathrm{d}\theta, \qquad H(-\omega) = \int_0^{\infty} h(\varphi)\mathrm{e}^{\mathrm{i}\omega\varphi}\mathrm{d}\varphi$$

the variance is obtained through the following expression

$$\sigma_Y^2(t) = \int_{-\infty}^{\infty} |H(\omega)|^2 S_X(\omega)\left[1 - \frac{H(-\omega)}{|H|^2}\int_t^{\infty} h(\theta)e^{-i\omega\theta}\,d\theta\right.$$

$$\left. - \frac{H(\omega)}{|H|^2}\int_t^{\infty} h(\varphi)e^{i\omega\varphi}d\varphi + \frac{1}{|H|^2}\int_t^{\infty} h(\theta)e^{-i\omega\theta}d\theta \int_t^{\infty} h(\theta)e^{-i\omega\theta}d\theta \int_t^{\infty} h(\varphi)e^{i\omega\theta}d\varphi\right]d\omega$$

$$= \int_{-\infty}^{\infty} |H(\omega)|^2 S_X(\omega)\left\{1 - \frac{2(\omega_0^2-\omega^2)}{|H|^2}\int_t^{\infty} h(x)\cos\omega x\,dx + \frac{4\lambda\omega_0\omega}{|H|^2}\int_t^{\infty} h(x)\sin\omega x\,dx\right.$$

$$\left. + \frac{1}{|H|^2}\left[\left(\int_t^{\infty} h(x)\cos\omega x\,dx\right)^2 + \left(\int_t^{\infty} h(x)\sin\omega x\,dx\right)^2\right]\right\}d\omega$$

This rather intimidating looking equation has been resolved by Caughey and Stumpf, [32], and the result is

$$\sigma_Y^2(t) = \int_{-\infty}^{\infty} |H(\omega)|^2 S_X(\omega)\left\{1 + e^{-2\lambda\omega_0 t}\left[1 + \frac{2\lambda\omega_0}{\overline{\omega_0}}\sin\overline{\omega_0}t\cos\overline{\omega_0}t\right.\right.$$

$$- e^{\lambda\omega_0 t}\left(2\cos\overline{\omega_0}t + \frac{2\lambda\omega_0}{\overline{\omega_0}}\sin\overline{\omega_0}t\right)\cos\omega t$$

$$\left.\left. - e^{\lambda\omega_0 t}\frac{2\omega}{\overline{\omega_0}}\sin\overline{\omega_0}t\sin\omega t + \frac{(\lambda\omega_0)^2 - \overline{\omega_0^2} + \omega^2}{\overline{\omega_0^2}}\sin^2\overline{\omega_0}t\right]\right\}d\omega$$

The above expression shows clearly the time dependence of the variance. However, as time progresses, the variance converges asymptotically towards a stationary value just as the mean value.

To investigate the rapidity of the convergence, study the case where the input process is a white noise with constant spectral density S_0. Then $R_X(\tau) = 2\pi S_0\delta(\tau)$ and

$$\sigma_Y^2(t) = 2\pi S_0\int_0^t (h(\theta))^2\,d\theta = \frac{2\pi S_0}{\overline{\omega_0^2}}\int_0^t e^{-2\lambda\omega_0}\sin^2\overline{\omega_0}\,d\theta$$

$$= \frac{\pi S_0}{2\lambda\omega_0^3}\left[1 - e^{-2\lambda\omega_0 t}\left\{1 + \frac{\lambda\omega_0}{\overline{\omega_0}}\sin 2\overline{\omega_0}t + \frac{2\lambda^2\omega_0^2}{(\overline{\omega_0})^2}\sin^2\overline{\omega_0}t\right\}\right]$$

The same expression is also obtained if small damping is assumed, and the approximation Eq. (3.55) is applied in conjunction with the other more complicated version of the variance.

If the damping is zero ($\lambda = 0$), the variance becomes

$$\sigma_Y^2(t) = \lim_{\lambda\to 0}\left[\frac{\pi S_0}{2\lambda\omega_0^3}\left(1 - e^{-2\lambda\omega_0 t}\left\{1 + \frac{\lambda\omega_0}{\overline{\omega_0}}\sin 2\overline{\omega_0}t + \frac{2\lambda^2\omega_0^2}{(\overline{\omega_0})^2}\sin^2\overline{\omega_0}t\right\}\right)\right] = \frac{f(\lambda)}{g(\lambda)} = \frac{0}{0}$$

Therefore,

$$\sigma_Y^2(t) = \lim_{\lambda\to 0}\frac{f'(\lambda)}{g'(\lambda)} = \frac{\pi S_0}{2\omega_0^3}(2\omega_0 t - \sin 2\omega_0 t)$$

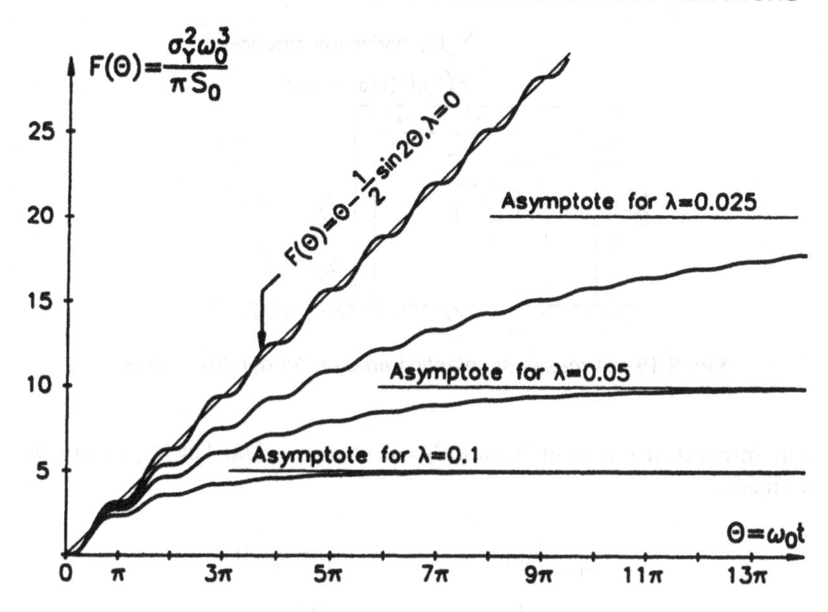

Fig. 3.14 *Stationary limits for non-stationary variance, [32], reproduced by permission of ASME*

The behaviour of the variance for damping ratios $\lambda = 0$, $\lambda = 0.025$, $\lambda = 0.05$ and $\lambda = 0.10$ is shown in Fig. 3.14. As seen in Fig. 3.14, the variance in case of $\lambda = 0$ never reaches a stationary value. For $\lambda = 0.1$ the system output becomes stationary in roughly three cycles (6π). For $\lambda = 0.05$ and 0.025 it takes a little longer or about 7 cycles (14π) and 13 cycles (26π) respectively. Since all physically realizable systems have inherent damping values, no matter how carefully designed, a non-stationary steady-state response to a stationary excitation is excluded. It is only a matter of time, when a stationary steady state is reached.

Example 3.14 Consider a massless elastic frame which at $t = 0$ is suddenly subjected to a horizontal, stationary random load $F(t)$ with an uniform spectral density S_0. Find the mean value and the mean square value of the horizontal deflection of the frame (Fig. 3.15).

Solution The dynamical aspect of this problem will be studied in more detail in Example 7.4. The equation of motion governing the horizontal deflexion $Y(t)$ is found to be

$$c\dot{Y}(t) + kY(t) = F(t)$$

The admittance $a(t)$ is obtained as the solution of the equation

$$c\dot{a}(t) + ka(t) = 1(t)$$

which has the complete solution

$$a(t) = l/k + A\,e^{-kt/c}$$

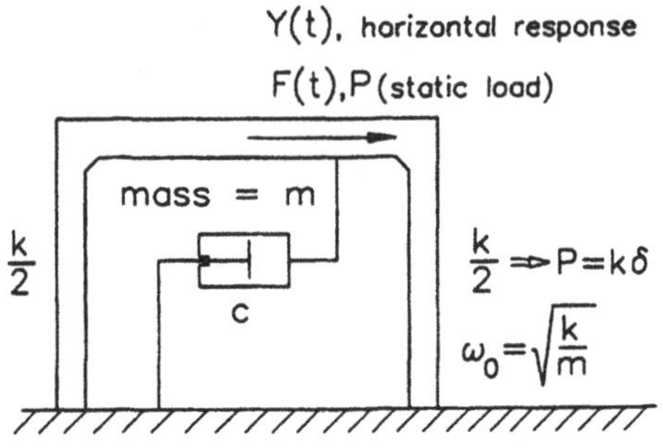

Fig. 3.15 *A massless, elastic frame under dynamic loads*

where A is an integration constant. Now $a(0)=0$, so $A=-1/k$. The impulse response $h(t)$ is then obtained as

$$h(t)=\begin{cases} a'(t)=\dfrac{1}{c}\mathrm{e}^{-kt/c} & \text{for } t\geqslant 0 \\ 0 & \text{for } t<0 \end{cases}$$

For $t\geqslant 0$, $F(t)$ is white noise with $R_{\mathrm{F}}(t)=2\pi S_0\delta(\tau)$

(i)

$$E[Y(t)]=\int_0^t E[F(u)]\,h(t-u)\,\mathrm{d}u=0$$

since $E[F(u)]=0$.

(ii) By Eq. (3.80),

$$E[Y^2(t)]=\int_0^t\int_0^t 2\pi S_0\delta(\varphi-\theta)\,h(\varphi)\mathrm{d}\theta\,\mathrm{d}\varphi$$

$$=2\Pi S_0\int_0^t h^2(\theta)\mathrm{d}\theta=2\pi S_0\int_0^t\frac{1}{c^2}\mathrm{e}^{-(2k/c)u}\,\mathrm{d}u=\frac{\pi S_0}{kc}(1-\mathrm{e}^{-(2k/c)t})$$

In Fig. 3.16 it is shown how the variance reaches a stationary limit in the course of time.

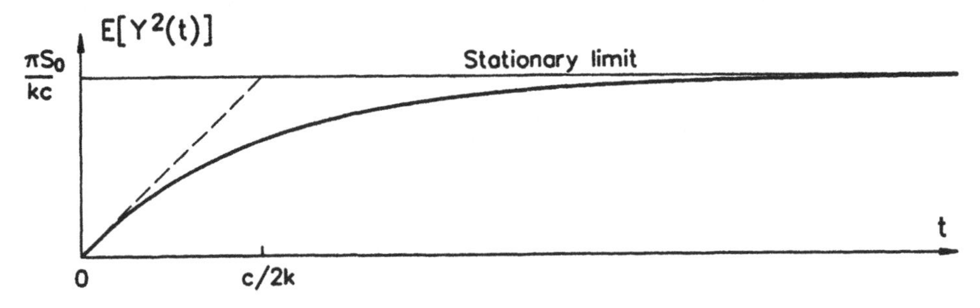

Fig. 3.16 *The time evolution of the main square*

4 Random Excursions and Failure Probabilities

4.1 INTRODUCTION

In the previous chapter, the input–output relations of linear systems were studied, that is, the methods available to find and analyse the response process of a linear physical system when subjected to a random excitation. The main parameters of interest were the mean value and the standard deviation or the mean square value of the input and output processes. However, in many engineering applications, it is necessary to evaluate the maximum response or the probability that the maximum response exceeds a certain limit that is precarious to the system. The system may break down if the maximum response (deflection, stress or any other relevant response quantity) goes beyond a certain threshold value only once. In this case, one may talk about the first exceedance or first passage probability. In other cases, the system behaviour may be such that the physical device in question will tolerate few response excursions above or beyond the threshold value before breaking down or suffering damage. Accumulation of damage may take place in the system for such cycles that produce response above the threshold. The breakdown of physical systems, due to material fatigue during such random excursions beyond the threshold value, is a well-known problem in material physics.

Obviously, these and related matters have to do with the probabilistic behaviour of the stochastic process at large amplitudes, that is, the frequency of crossing of certain amplitude limits and the distribution of peak values. Of major importance will be the frequency composition of the process, that is, whether it is of the narrow-band type with a predominant frequency or a broad-band process with a wide frequency range. The former will be typical in situations when treating the response of lightly damped systems, where the fundamental system frequency will dominate the response. A broad-band situation may arise when dealing with any kind of a broad-band filter. For instance, the shaping of earthquake records, when the incident waves are passed through thick layers of different soils, will produce broad-band seismograms. In the following, few of these problems will be addressed and examples are given for certain simple well-known cases. As only stationary stochastic processes are amenable to analysis, it is a tacit presumption that the processes being dealt with are all stationary. The development of the theory and handling of such problems that arise in connection with maximum values and peak distribution of random functions dates back to the early work of Rice [177], Cartwright and Longuet-Higgins, [30], Powell [74], Crandall, [43], and many others. The following discussion is mostly along the lines set by Crandall and others [43], [9], [45], and similar notations are used.

4.2 THRESHOLD CROSSING RATES

Consider a sample function $x(t)$ of stochastic process $\{X(t),\ t\in\mathscr{T}\}$, Fig. 4.1. If the frequency composition of the process is sufficiently narrow, that is, the sample functions have their power contained in a reasonably narrow frequency band, whereby the amplitude peaks are not too irregular, the behaviour of the function near its extreme values can be described in a meaningful way. Of immediate interest is to find out about the number of times the sample function crosses a certain amplitude level $x = a$ in a given time interval T. Now let a random variable $N_a^+(T)$ denote the number of upward crossings of the line a, that is, crossings with a positive slope. The mean value of this variable is called $n_a^+(T)$ such that

$$n_a^+(T) = E[N_a^+(T)] \tag{4.1}$$

Since the process is stationary, taking twice as long an interval will not alter the statistics, so,

$$n_a^+(2T) = 2n_a^+(T) \tag{4.2}$$

Therefore, the average number of upward crossings is proportional to the time interval T, that is,

$$n_a^+(T) = v_a^+ T \tag{4.3}$$

where v_a^+ is the average frequency of upward crossings of the threshold $x = a$. The probability law which governs the frequency of up-crossings can now be derived from the underlying probability distribution of the stochastic process itself.

Let $p_{X\dot{X}}(x, \dot{x})$ be the joint probability density of the amplitudes and the velocities of $X(t)$, and consider the situation during an actual up-crossing in the time interval $(t, t + \Delta t)$, Fig. 4.2. In order to have an up-crossing, the velocity which corresponds to the slope α of the amplitude has to be larger than the angle β, otherwise the up-crossing takes place outside the interval. Therefore,

$$\dot{x} = \tan\alpha \geqslant \tan\beta = (a - x(t))/\Delta t \tag{4.4}$$

This condition, together with the obvious requirement that $x(t) < a$, sets up the region for

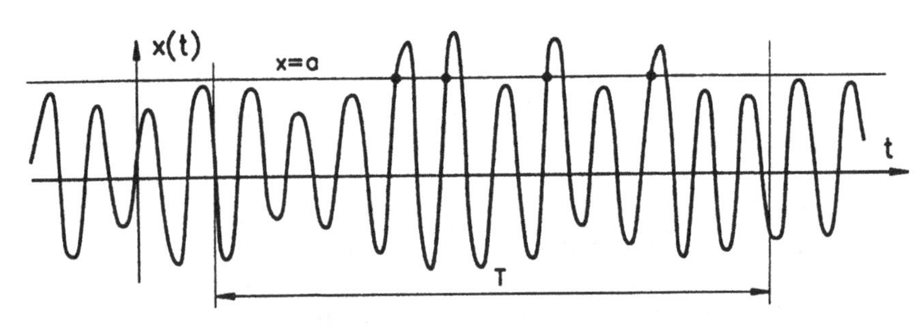

Fig. 4.1 *The crossings of a stochastic process of a preset level $x = a$*

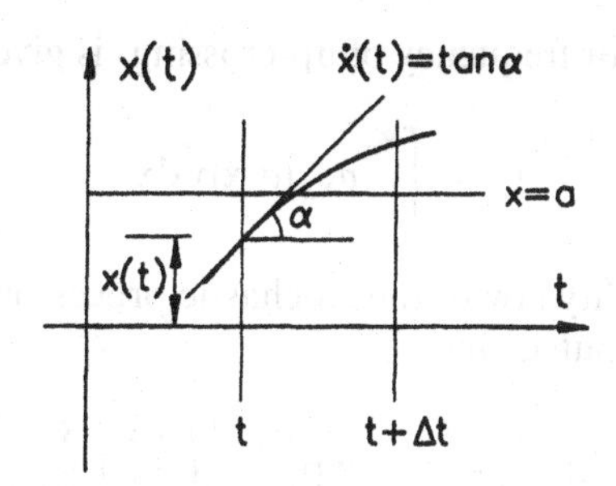

Fig. 4.2 *Up-crossing of the line x = a*

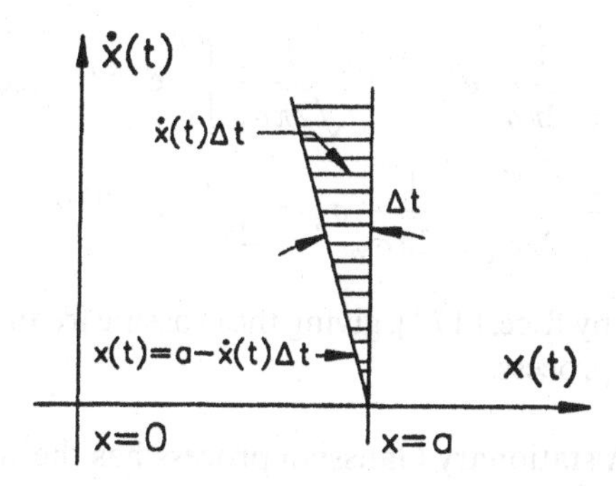

Fig. 4.3 *The favourable region for up-crossing of x = a*

the favourable fraction over which the probability density has to be integrated to obtain the probability of up-crossing in the specified interval. In fact, the favourable region in the $\dot{x}x$-plane is obviously the wedge formed by the two lines, $x = a$ and $x(t) = a - \dot{x}(t)\Delta t$, Fig. 4.3. All points within the wedge fulfil the condition Eq. (4.4) and constitute a favourable combination for up-crossing. Therefore,

$$P[\text{"up-crossing of the line } x = a \text{ at a time } t \text{ in } (t, t+\Delta t)\text{"}]$$

$$= P_{\nu_a^+\Delta t} = \nu_a^+ \Delta t = \int_0^\infty \int_{a-\dot{x}\Delta t}^a P_{X\dot{X}}(x, \dot{x}) \mathrm{d}x \, \mathrm{d}\dot{x}$$

since the probability will be approximately the same as the mean crossing rate times the interval length if Δt is very small. In that case,

$$\int_{a-\dot{x}\Delta t}^a p_{X\dot{X}}(x, \dot{x}) \mathrm{d}x \, \mathrm{d}\dot{x} = p_{X\dot{X}}(x = a, \dot{x}) \dot{x} \Delta t$$

so the double integral is evaluated as

$$\Delta t \int_0^\infty p_{X\dot{X}}(a, \dot{x}) \dot{x} \, \mathrm{d}\dot{x}$$

and the mean crossing rate or frequency of up-crossings is given by

$$\nu_a^+ = \int_0^\infty p_{X\dot{X}}(a, \dot{x})\dot{x}\,\mathrm{d}\dot{x} \tag{4.5}$$

If the underlying probability law of the stochastic process is Gaussian or normal, the above integral is easily computed since

$$p(x, \dot{x}) = \frac{1}{2\pi\sigma_X\sigma_{\dot{X}}}\exp\left(-\frac{1}{2}\left(\frac{x^2}{\sigma_X^2} + \frac{\dot{x}^2}{\sigma_{\dot{X}}^2}\right)\right) \tag{4.6}$$

whereby

$$\begin{aligned} \nu_a^+ &= \frac{1}{\sqrt{2\pi}\sigma_X}\mathrm{e}^{-a^2/2\sigma_{X^2}}\frac{1}{\sqrt{2\pi}\sigma_{\dot{X}}}\int_0^\infty \mathrm{e}^{-\dot{x}^2/2\sigma_{\dot{X}}^2}\,\dot{x}\,\mathrm{d}\dot{x} \\ &= \frac{\mathrm{e}^{-a^2/2\sigma_{X^2}}}{\sqrt{2\pi}\sigma_X}\frac{1}{\sqrt{2\pi}\sigma_{\dot{X}}}\sigma_{\dot{X}^2} = \frac{1}{2\pi}\frac{\sigma_{\dot{X}}}{\sigma_X}e^{-a^2/2\sigma_X} \end{aligned} \tag{4.7}$$

which is the famous result by Rice, [177], giving the crossing frequency of the level $x = a$ of a Gaussian narrow-band process.

Example 4.1 A stationary Gaussian process has the spectral density function sketched in Fig. 4.4. Determine the expected rate of zero up-crossings.

Solution Now by Eq. (4.5),

$$\nu_0^+ = \int_0^\infty p(0, \dot{x})\dot{x}\,\mathrm{d}\dot{x} = \frac{1}{2\pi}\frac{\sigma_{\dot{X}}}{\sigma_X}$$

$$\sigma_X^2 = E[X^2] = \int_{-\infty}^{\infty} S_X(\omega)\mathrm{d}\omega = 2\int_{\omega_0}^{2\omega_0} S_0\,\mathrm{d}\omega = 2S_0\omega_0$$

and

$$\sigma_{\dot{X}}^2 = E[\dot{X}^2] = \int_{-\infty}^{\infty} \omega^2 S_X(\omega)\mathrm{d}\omega = 2\int_{\omega_0}^{2\omega_0} \omega^2 S_0\,\mathrm{d}\omega = 2S_0\left(\frac{8\omega_0^3}{3} - \frac{\omega_0^3}{3}\right)$$

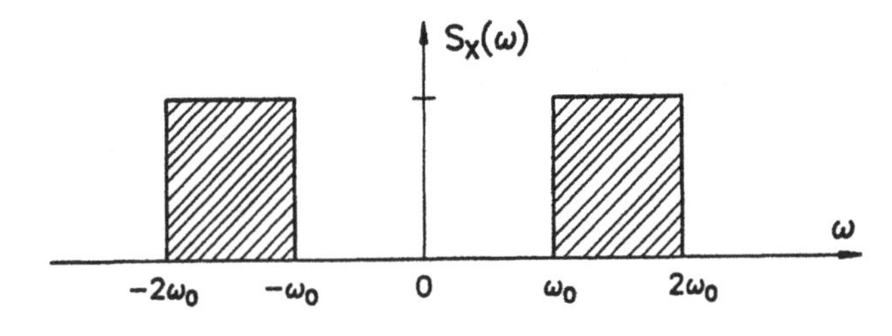

Fig. 4.4 *The spectral density of a Gaussian process*

whereby

$$\nu_0^+ = \frac{1}{2\pi}\left(\frac{14S_0\omega_0^3}{3\times 2S_0\omega_0}\right)^{1/2} = \frac{\omega_0}{2\pi}\sqrt{\frac{7}{3}}$$

so the crossing frequency dependent only on the main frequency of the signal, that is ω_0.

Example 4.2 A stationary stochastic process $X(t)$ has the following joint probability density function for x and $\dot{x}$, (evaluated at the same time),

$$p(x,\dot{x}) = \begin{cases} \dfrac{1}{4\alpha\beta} & \text{for } \begin{cases} -\alpha < x < \alpha \\ -\beta < \dot{x} < \beta \end{cases} \\ 0 & \text{elsewhere} \end{cases}$$

find the expected rate of up-crossings of the level $x = a,\ -\infty < a < \infty$.

Now,

$$\nu_a^+ = \int_0^\infty p(a,\dot{x})\dot{x}\,\mathrm{d}\dot{x}$$

so

$$\nu_a^+ = \int_0^\beta \frac{1}{4\alpha\beta}\dot{x}\,\mathrm{d}\dot{x} = \frac{\beta}{8\alpha} \qquad \text{for } -\alpha < a < \alpha \quad \text{or} \quad |a| < \alpha$$

$$\nu_a^+ = 0 \qquad \text{for } |a| > \alpha$$

4.3 THE PROBABILITY DISTRIBUTION OF MAXIMA

Again consider a sample function $x(t)$ of a stochastic process $\{X(t),\ t\in\mathscr{T}\}$. Within the interval of observation T, the function will show many peaks of positive and negative amplitudes, Fig. 4.5. It can therefore be advantageous to define the positive and negative maxima and minima of the function as shown in the figure. Now, what are the conditions that the function obtains a maximum value within a small interval $(t, t+\Delta t)$? Firstly, the velocity $\dot{x}(t)$, that is the slope, has to change sign. In other words, the velocity has to make a zero crossing in the interval $(t, t+\Delta t)$. If it is to be a positive maximum, the crossing has to be from above as shown in Fig. 4.6. Therefore, the conditions for a positive maximum can be set up as follows:

$$\tan\alpha = \dot{x} > 0$$

$$|\tan\beta| = |\ddot{x}| = (\dot{x}/\gamma\Delta t) \tag{4.8}$$

or

$$\dot{x} = \gamma\Delta t|\ddot{x}|, \qquad \gamma < 1$$

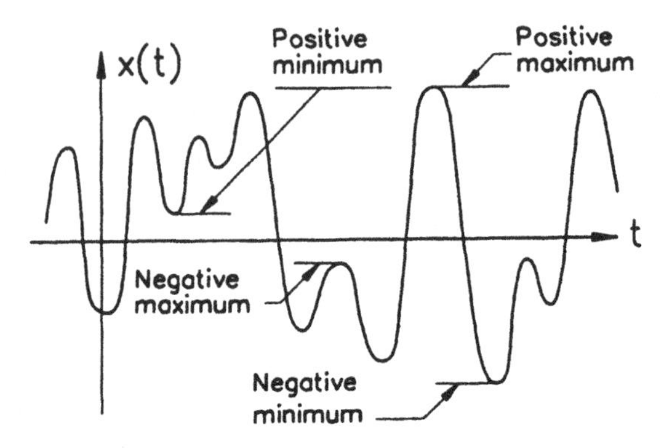

Fig. 4.5 *Maxima and minima of a sample function*

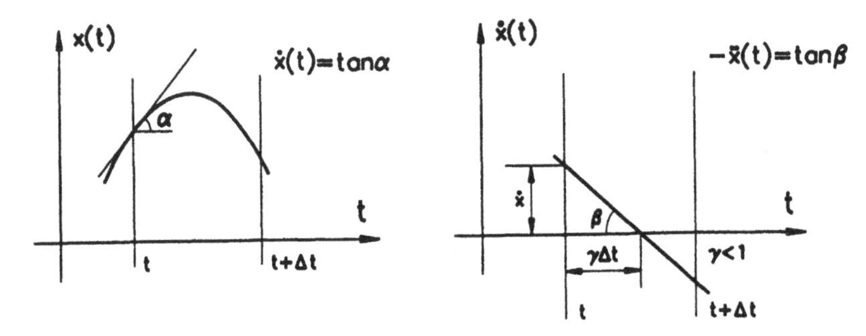

Fig. 4.6 *Conditions for a positive maximum in an interval (t, t + Δ t)*

Now form the three random variables $X = U$, $V = \dot{X}$ and $W = \ddot{X}$. Since X is a stationary process, the velocity and acceleration processes are also stationary (Example 2.8). Actually, differentiating a stochastic process can only improve stationarity, whereas integration makes matters worse. The joint probability density of these variables is

$$p_{X\dot{X}\ddot{X}}(x, \dot{x}, \ddot{x}) = p(u, v, w) \tag{4.9}$$

The probability for a maximum to occur in the small time interval $(t, t + \Delta t)$ and be within the range $(u, u + \Delta u)$ is then calculated by integrating the density Eq. (4.9) over a favourable volume in the (u, v, w) space, Fig. 4.7. From the conditions, Eq. (4.8), it can be seen that the favourable volume consists of a wedge, which is clamped between the two vertical planes $x = u$ and $x = u + \Delta u$, and v must be within the shaded area, that is, made up by the lines $v = 0$ and $v = |w|\Delta t$. Carrying out the implied integration,

$$P[\text{“maximum in } (u, u + \Delta u) \text{ at time } t \text{ in } (t, u + \Delta t)\text{”}] =$$

$$P_M = \int_{x=u}^{x=u+\Delta u} \int_{w=-\infty}^{0} \int_{v=0}^{|w|\Delta t} p(x, v, w) \mathrm{d}v \, \mathrm{d}w \, \mathrm{d}x$$

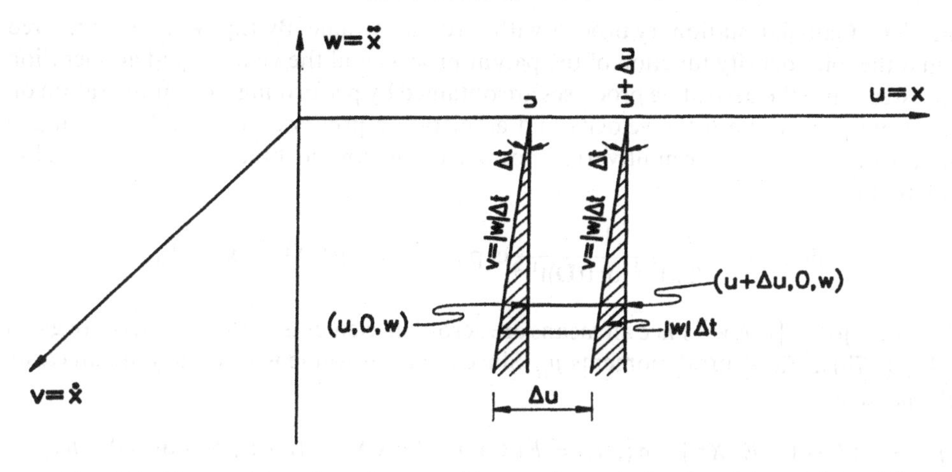

Fig. 4.7 *Favourable volume for a positive maximum in the time interval (t, t+ Δ t)*

For very small Δt,

$$\int_{v=0}^{|w|\Delta t} p(x, v, w)\mathrm{d}v \doteq p(x, v=0, w)\,|w|\,\Delta t$$

and

$$P_{\mathrm{M}} \doteq \left[\int_{w=-\infty}^{0} p(x, 0, w)|w|\,\Delta t\,\mathrm{d}w\right](u+\Delta u-u)$$
$$=\left[-\int_{-\infty}^{0} p(u, 0, w)\,w\,\mathrm{d}w\right]\Delta u\,\Delta t$$

$P_{\mathrm{M}}/\Delta t$ must be interpreted as the frequency of maxima in the interval $(u, u+\Delta u)$. The frequency of maxima in the whole range $(-\infty < u < \infty)$ can therefore be obtained as

$$\mu_0 = \int_{-\infty}^{\infty}\left[-\int_{-\infty}^{0} p(u, 0, w)w\,\mathrm{d}w\right]\mathrm{d}u \tag{4.10}$$

Finally seek the favourable fraction, that is, the frequency of maxima in $(u, u+\Delta u)$ divided by the frequency of maxima any place, which is equal to $p_{\mathrm{M}}(u)\Delta u$, and the probability density function for the maxima is found to be

$$p_{\mathrm{M}}(u) = \frac{\left[-\int_{-\infty}^{0} p(u, 0, w)w\,\mathrm{d}w\right]}{\int_{-\infty}^{\infty}\left[-\int_{-\infty}^{0} p(u, 0, w)w\,\mathrm{d}w\right]\mathrm{d}u} \tag{4.11}$$

Stationary stochastic processes that are Gaussian, that is, their probabilistic behaviour is governed by the normal distribution, are of particular interest in this context since the above density function can be evaluated to yield a comparatively simple explicit expression. (For an overview of Gaussian stochastic processes, see Section 6.1). Therefore, consider a Gaussian stationary process with zero mean value. By Eq. (4.11), it is required to find the joint density function of the parent process and the velocity and acceleration processes. Since the derivative processes are obtained by performing a linear operation on the parent process, both the velocity and acceleration processes are also Gaussian (see Section 6.1). Moreover, the multivariate normal density for the three variables is given Eq. (6.8), that is,

$$p(u, v, w) = \frac{1}{(2\pi)^{3/2}(\det(\mathbf{D}))^{1/2}} \exp\left[-\tfrac{1}{2}\{\mathbf{x}-\boldsymbol{\mu}\}^{\mathrm{T}}\mathbf{D}^{-1}\{\mathbf{x}-\boldsymbol{\mu}\}\right]$$

where $(\mathbf{x}-\boldsymbol{\mu})^{\mathrm{T}} = \{u, v, w\}$ since all means are zero. The elements of the $\mathbf{D}$ matrix are given by Eq. (6.7b) as the central moments μ_{ij}. Since the mean value is zero, they are all easily calculated as

$$\mu_{11} = E[UU] = E[X^2] = \sigma_X^2,\ \mu_{12} = E[UV] = E[X\dot{X}] = 1/2\mathrm{d}E[X^2]/\mathrm{d}t = 0 = \mu_{21}$$
$$\mu_{13} = E[UW] = E[X\ddot{X}] = -E[\dot{X}^2] = -\sigma_{\dot{X}}^2 = \mu_{31}, \mu_{22} = E[VV] = E[\dot{X}^2] = \sigma_{\dot{X}}^2$$
$$\mu_{23} = E[VW] = E[\dot{X}\ddot{X}] = 0 = \mu_{32}, \mu_{33} = E[WW] = E[\ddot{X}^2] = \sigma_{\ddot{X}}^2$$

Now introduce the so-called spectral moments

$$m_n = \int_{-\infty}^{\infty} \omega^n S_X(\omega)\mathrm{d}\omega \tag{4.12}$$

where $S_X(\omega)$ is the spectral density of the parent process. Obviously,

$$m_0 = \sigma_X^2, \qquad m_2 = \sigma_{\dot{X}}^2, \qquad m_4 = \sigma_{\ddot{X}}^2 \tag{4.13}$$

so the $\mathbf{D}$ matrix is written as follows:

$$\mathbf{D} = \begin{Bmatrix} m_0 & 0 & -m_2 \\ 0 & m_2 & 0 \\ -m_2 & 0 & m_4 \end{Bmatrix}, \qquad \det(\mathbf{D}) = m_2(m_0 m_4 - m_2^2) \tag{4.14}$$

and the inverted matrix is

$$\mathbf{D}^{-1} = \frac{1}{m_2(m_0 m_4 - m_2^2)} \begin{Bmatrix} m_2 m_4 & 0 & m_2^2 \\ 0 & m_0 m_4 - m_2^2 & 0 \\ m_2^2 & 0 & m_2^2 \end{Bmatrix} \tag{4.15}$$

whereby the density is

$$p(u, v, w) = \frac{1}{(2\pi)^{3/2}(\det(\mathbf{D}))^{1/2}} \exp\left[-\frac{1}{2}\frac{m_2 m_4 u^2 + m_0 m_4 v^2 + m_2 m_0 w^2 + 2m_2^2 uw}{m_2(m_0 m_4 - m_2^2)}\right] \tag{4.16}$$

and

$$p(u, 0, w) = \frac{1}{(2\pi)^{3/2}(\det(\mathbf{D}))^{1/2}} \exp\left[-\frac{1}{2}\frac{\dfrac{u_2}{m_0} + \dfrac{2m_2 uw}{m_0 m_4} + \dfrac{w^2}{m_4}}{(m_0 m_4 - m_2^2)/(m_0 m_4)}\right] \tag{4.17}$$

The value of the denominator expression $m_0 m_4 - m_2^2$ is always positive, which makes the density form correct. In fact,

$$m_0 m_4 - m_2^2 = \lim_{c \to \infty} \left[\int_{-c}^{c} S_X(u) du \int_{-c}^{c} v^4 S_X(v) dv - \int_{-c}^{c} u^2 S_X(u) du \int_{-c}^{c} v^2 S_X(v) dv \right]$$

$$= \lim_{c \to \infty} \left[(S_X^*)^4 \int_{-c}^{c} \int_{-c}^{c} (v^4 - u^2 v^2) du\, dv \right] = \lim_{c \to \infty} \left[(S_X^*)^4 \frac{16}{25} c^4 \right] > 0$$

where S_X^* is a maximum value for the spectral density in the entire range. Now compute the frequency of maxima anywhere, inserting Eq. (4.17) into Eq. (4.10):

$$\mu_0 = \frac{-1}{(2\pi)^{3/2} \sqrt{m_2(m_0 m_4 - m_2^2)}} \int_{\infty}^{\infty} \int_{-\infty}^{0} \exp\left[-\frac{1}{2} \frac{\frac{u^2}{m_0} + \frac{2m_2 uw}{m_0 m_4} + \frac{w^2}{m_4}}{(m_0 m_4 - m_2^2)/(m_0 m_4)} \right] w\, dw\, du \tag{4.18}$$

The integral can be evaluated by first splitting up the exponential as follows:

$$\exp\left[-\frac{1}{2} \frac{\frac{u^2}{m_0} + \frac{2m_2 uw}{m_0 m_4} + \frac{w^2}{m_4}}{(m_0 m_4 - m_2^2)/(m_0 m_4)} \right] = \exp\left[-\frac{1}{2} \frac{[u + (m_2/m_4) w]^2}{(m_0 - m_2^2/m_4)} \right] \exp\left[-\frac{1}{2} \frac{w^2}{m_4} \right]$$

By integrating first over the entire range, $(-\infty < u < \infty)$, the first exponential must yield the value

$$\sigma \sqrt{2\pi} = \sqrt{2\pi} \sqrt{\frac{m_0 m_4 - m_2^2}{m_4}}$$

The second integral therefore involves only the second exponential times w, that is,

$$\int_{-\infty}^{0} \exp\left[-\frac{1}{2} \frac{w^2}{m_4} \right] w\, dw = -m_4$$

The final result is therefore

$$\mu_0 = \frac{(-1)}{(2\pi)^{3/2} \sqrt{m_2(m_0 m_4 - m_2^2)}} \sqrt{2\pi} \sqrt{\frac{m_0 m_4 - m_2^2}{m_4}} (-m_4) = \frac{1}{2\pi} \left(\frac{m_4}{m_2} \right)^{1/2} = \frac{1}{2\pi} \frac{\sigma_{\ddot{X}}}{\sigma_{\dot{X}}} \tag{4.19}$$

giving the frequency of maxima of the Gaussian process anywhere. By comparison with Eq. (4.7), the frequency of maxima is the same as the down-crossing rate of the velocity process $\dot{X}(t)$ as it should be.

The probability density function of the maxima of a stationary Gaussian process can now be obtained by inserting Eqs. (4.17) and (4.18) into Eq. (4.11). After some lengthy unwieldy arithmetics, using similar tactics as when leading to Eq. (4.18), the result is

$$p_M(\eta) = \frac{1}{\sqrt{2\pi}} \left[\varepsilon e^{-\eta^2/2\varepsilon^2} + \sqrt{(1-\varepsilon^2)}\, \eta \exp\left[-\frac{\eta^2}{2} \right] \int_{-\infty}^{[\eta\sqrt{1-\varepsilon^2}/\varepsilon]} e^{-(x^2/2)} dx \right] \tag{4.20}$$

in which

$$\eta = \frac{u}{\sqrt{m_0}} = \frac{u}{\sigma_X} \quad \text{and} \quad \varepsilon^2 = \frac{m_0 m_4 - m_2^2}{m_0 m_4} = 1 - \left(\frac{\nu_0}{\mu_0}\right)^2 \tag{4.21}$$

The parameter ε, introduced by Cartwright and Longuet-Higgins [11], is called the bandwidth parameter. By Eq. (4.18),

$$0 < \varepsilon < 1 \tag{4.22}$$

The bandwidth parameter is directly related to the frequency bandwidth of the process. For $\varepsilon \to 0$, $\nu_0^+ = \mu_0^-$, that is, 'one maximum or peak for each up-crossing', which corresponds to a perfect narrow-band behaviour. In this case, the probability density Eq. (4.20) simplifies to the Rayleigh probability density, (cf. Ex. (2.3)),

$$p_M(\eta) = \eta \exp\left(-\frac{\eta^2}{2}\right) \tag{4.23}$$

The Rayleigh density is defined only for positive values of η (or only for negative values), and has a mean value equal to $\sqrt{2\pi/2}$ and variance equal to $(4 - \pi)/2$. The mode of the distribution is equal to 1, whereby the maximum value of the density corresponds to $u = \sigma_X$, the standard deviation of the parent process. Most of the maxima will have a value about this magnitude, Fig. 4.8. If $\varepsilon \to 1$ on the other hand, $\nu_0^+/\mu_0^- \to 0$, and the behaviour of the process corresponds to a very broad or unlimited bandwidth. What actually happens is that the frequency of maxima (or the down-crossing rate of $\dot{x}$) goes to infinity. In that case, the probability density ($\varepsilon = 1$) becomes

$$p_M(\eta) = \frac{1}{2\pi} \exp\left[-\frac{\eta^2}{2}\right] \tag{4.24}$$

which is the normal or Gaussian density. For intermediate values of ε, the densities move from left to right to the positive values of the Rayleigh density as seen in Fig. 4.8.

A simple method of evaluating ε for a given sample function of duration T can be devised as follows. Count the number N of positive and negative maxima inside the interval of length T, Fig. 4.5. The number of negative maxima is called N^- and the ratio

$$r = \frac{N^-}{N} \tag{4.25}$$

is therefore the proportion of negative maxima in the total and must be equal to

$$r = \int_{-\infty}^{0} p_M(\eta) \mathrm{d}\eta \tag{4.26}$$

By integrating the density, i.e. Eq. (4.20), over the range of negative maxima, $-\infty < \eta < 0$,

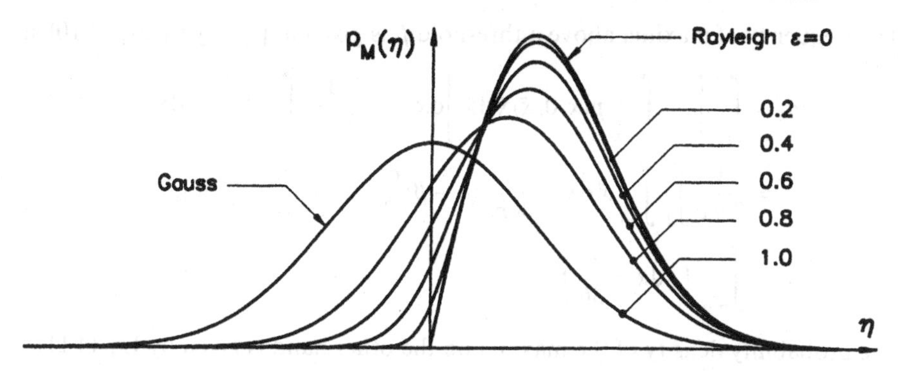

Fig. 4.8 *Distributions of maxima of processes with different bandwidths*

and neglecting higher-order terms, the result is

$$\varepsilon^2 = 4r(1 - r) \tag{4.27}$$

The ratio r as defined in Eq. (4.25), introduced into Eq. (4.27), will then give a very close approximation to the bandwidth parameter.

Example 4.3 The joint probability density of a stationary random process $X(t)$ and its derivative processes $\dot{X}(t)$ and $\ddot{X}(t)$ at the same time is given by

$$p(x, \dot{x}, \ddot{x}) = 1/(8abc), \ |x| \leqslant a, \ |\dot{x}| \leqslant b, \ |\ddot{x}| \leqslant c$$

Find

(a) the expected rate at which the process crosses a threshold β with a positive slope;
(b) the expected number of maxima of $x(t)$ per unit time above the level β;
(c) The probability density of the maxima and the probability density of the envelope function of the process.

Solution

(a) To obtain the crossing frequency Eq. (4.5), the marginal distribution or density for the amplitude and the velocity is needed. By Eq. (1.57),

$$p(x, \dot{x}) = \int_{-\infty}^{\infty} p(x, \dot{x}, \ddot{x}) \mathrm{d}\ddot{x} = \frac{1}{8abc} \int_{-c}^{c} \mathrm{d}\ddot{x} = \frac{1}{4ab}$$

so therefore by Eq. (4.5)

$$\begin{aligned} v_\beta^+ &= \int_0^\infty p(\beta, \dot{x}) \dot{x} \, \mathrm{d}\dot{x} = \frac{1}{4ab} \int_0^b \dot{x} \, \mathrm{d}\dot{x} = \frac{1}{4ab} \frac{b^2}{2} = \frac{b}{8a} \qquad && \text{if } |\beta| \leqslant a \\ (v_0^+ = v_\beta^+) & \qquad\qquad\qquad\qquad\qquad\qquad\qquad = 0 && \text{if } |\beta| > a \end{aligned}$$

(b) The frequency of maxima above a threshold β is given implicitly by Eq. (4.10) as

$$\mu_\beta = \int_\beta^\infty \left[-\int_{-\infty}^0 p(x,0,\ddot{x})\ddot{x}\,d\ddot{x} \right] dx = -\frac{1}{8abc}\int_\beta^\infty \int_{-\infty}^0 \ddot{x}\,d\ddot{x}\,dx$$

$$= \frac{1}{8abc}\int_\beta^a \int_{-c}^0 \ddot{x}\,d\ddot{x} = \frac{1}{8abc}(a-\beta)\frac{c^2}{2} = \frac{(a-\beta)c}{16ab} \qquad \text{if } |\beta| \leqslant a$$

$$\left(\mu_0 = \int_{-\infty}^\infty [\,]dx = \frac{c}{8b}\right) \qquad = 0 \qquad \text{if } |\beta| > a$$

(c) The probability density of the maxima on the other hand is given by Eq. (4.11)

$$p_M(u) = -\int_{-\infty}^0 p(u,0,w)w\,dw\frac{1}{\mu_0} = -\frac{1}{8abc}\int_{-c}^0 w\,dw\frac{1}{c/8b} = \frac{c}{16ab}\frac{8b}{c} = \frac{1}{2a}$$

for $|u| \leqslant a$.

(d) Following Crandall, [45], the envelope process $A(t)$ of a broad-band stochastic process $X(t)$ is implicitly given by

$$V(A) = \tfrac{1}{2}\dot{X}^2 + V(X)$$

that is, the energy functional $V(A)$ is the sum of the kinetic and potential energy per unit mass of the moving object. Therefore, $A(t)$ would be the amplitude or displacement if the total energy were converted entirely to potential energy. For linear behaviour of the system, the potential energy can be given as $V(X) = \frac{1}{2}kX^2$. In this case, the envelope is simply given by

$$\alpha^2 = x^2 + 1/k\dot{x}^2 = x^2 + \dot{x}^2/\omega_0^2, \qquad \text{where } \omega_0^2 = \sqrt{((m=1)/k)}$$

The probability of the envelope being less than a certain amplitude α can be written as

$$F_A(\alpha) = P[A(t) \leqslant \alpha] = 4\int_0^\alpha dx \int_0^{\sqrt{2(V(\alpha)-V(x))}} p(x,\dot{x})d\dot{x}$$

The probability density of the envelope process is therefore given by

$$p_A(\alpha) = \frac{dF_A(\alpha)}{d\alpha} = 4V'(\alpha)\int_0^\alpha 2\sqrt{V(\alpha)-V(x)}\,p(x,\sqrt{2(V(\alpha)-V(x))})dx$$

Now assuming that $V(X) = \frac{1}{2}kX^2$, whereby, $V'(\alpha) = k\alpha$, the density is given by

$$p_A(\alpha) = \frac{k\alpha}{ab}\int_0^\alpha \frac{dx}{\sqrt{k\alpha^2 - kx^2}} = \frac{\sqrt{k}}{ab}\alpha\left[\sin^{-1}\left(\frac{x}{\alpha}\right)\right]_0^\alpha = \frac{\pi\sqrt{k}}{2ab}\alpha, \qquad 0 \leqslant \alpha \leqslant a$$

4.4 PEAK DISTRIBUTIONS

In many applications, one is concerned with the probability distribution of the amplitude peaks rather than the maxima of the sample function of a stochastic process. Intuitively, one would expect the peak distribution to be closely related to the distribution of maxima as it well is. Following Powell, [174], it is easy to make use of the previously obtained crossing frequencies to find the peak distribution of a narrow-band process. Consider once more the sample function $x(t)$ observed within a time interval of length T. The probability that a peak chosen at random has a magnitude exceeding $x = a$ is

$$P[\text{peak} > a] = \int_a^\infty p_p(u)\mathrm{d}u \tag{4.28}$$

where $p_p(u)$ is the peak probability density. Now in the time interval T, the process will have $(\nu_0^+ T)$ zero crossings on the average and $(\nu_a^+ T)$ up-crossings of the line $x = a$ on the average. The favourable fraction for peak larger than $x = a$ is therefore simply

$$\frac{\nu_a^+ T}{\nu_0^+ T} = \frac{\nu_a^+}{\nu_0^+} = \int_a^\infty p_p(u)\mathrm{d}u \tag{4.29}$$

Differentiating the above equation with respect to a gives the peak density as

$$-p_p(a) = \frac{1}{\nu_0^+}\frac{\mathrm{d}\nu_a^+}{\mathrm{d}a} \tag{4.30}$$

which is the general expression for the peak probability density of a narrow-band process. The above argument presupposes a resonably smooth behaviour of the narrow-band process, such that all maxima occur above $x = 0$ and all minima below $x = 0$, that is, each cycle crosses the zero line.

Now for a Gaussian process, the crossing frequencies are given by Eq. (4.7), so the probability density is obtained from Eq. (4.30),

$$-p_p(a) = \frac{\mathrm{d}e^{-a^2/(2\sigma_{X^2})}}{\mathrm{d}a} = -\frac{a}{\sigma_{X^2}}e^{-a^2/(2\sigma_{X^2})}$$

which is the Rayleigh distribution already found for the distribution of maxima of a stationary narrow-band Gaussian process, Eq. (4.23). This result should not come as any surprise. The probability that any peak chosen at random is less than the level a is therefore given by Eqs. (4.28) and (4.31) as

$$P[X_{\text{peak}} \leqslant a] = 1 - \int_0^a \frac{u}{\sigma_X^2}e^{-u^2/(2\sigma_X^2)}\mathrm{d}u$$
$$= 1 - \exp\left[-\frac{a^2}{2\sigma_{X^2}}\right] \tag{4.32}$$

In a number of cases where the underlying probability distribution of the stochastic process is uncertain or deviates from the Gaussian distribution, the peak distribution may depart significantly from the Rayleigh distribution, [187]. In the general case, the Weibull distribution may give a better result. Actually, the Rayleigh distribution is a special case of the Weibull distribution, Eq. (1.156), with the coefficient $k = 2$ as will be shown in the following.

$$\Phi(y) = 1 - \exp\left[-\left(\frac{y}{\beta}\right)^k\right], \quad 0 \leqslant y < \infty, \quad k, \beta > 0 \tag{1.156}$$

Let a_0 be the median peak height, for which

$$P[X_{\text{peak}} \leqslant a_0] = P_{\text{p}}(a_0) = 1 - \exp\left[-\frac{a_0^2}{2\sigma_{X^2}}\right] = \tfrac{1}{2} \tag{4.33}$$

The median peak can then be expressed in terms of the standard deviation of the underlying probability distribution, that is,

$$a_0 = \sigma_X \sqrt{2 \log_e 2} \tag{4.34}$$

Introducing the ratio of the peaks to the median peak, the distribution, Eq. (4.32), is rewritten as

$$P\left[\frac{X_{\text{peak}}}{a_0} \leqslant \frac{a}{a_0}\right] = P_{\text{p}}\left(\frac{a}{a_0}\right) = 1 - \exp\left[-\log_e 2\left(\frac{a}{a_0}\right)^2\right] \tag{4.35}$$

Comparing Eqs. (4.35) and (1.135), $k = 2$ and $a_0 = \sigma_X \sqrt{2 \log_e 2} = \beta$, so the peak distribution is also a Weibull distribution with the above parameters. It is therefore natural to generalize the distribution by accepting different values of k. By replacing the number 2 by k in Eq. (4.35), the generalized distribution is

$$P\left[\frac{X_{\text{peak}}}{a_0} \leqslant \frac{a}{a_0}\right] = P_{\text{p}}\left(\frac{a}{a_0}\right) = 1 - \exp\left[-\log_e 2\left(\frac{a}{a_0}\right)^k\right] \tag{4.36}$$

For $a = a_0$, the corresponding probability according to Eq. (4.36) is

$$P\left[\frac{X_{\text{peak}}}{a_0} \leqslant 1\right] = 1 - \exp[\log_e 2] = \tfrac{1}{2}$$

so the median peak remains unchanged regardless of the value of k. The probability density of the Weibull peak distribution is given by the derivative, that is,

$$p_{\text{p}}\left(\frac{a}{a_0}\right) = k(\log_e 2)\left(\frac{a}{a_0}\right)^{k-1} \exp\left[-\log_e 2\left(\frac{a}{a_0}\right)^k\right] \tag{4.37}$$

and is plotted for different values of k in Fig. 4.9 ($k = 1, 2, 4, 10$). From the figure it is seen that for $k = 10$, most of the peaks are close to the median peak and as k grows very large, all the peaks occur at the median peak height a_0. This is the situation when the stochastic process is represented by a harmonic signal of a constant amplitude but various phases. For $k < 2$, there are more large peaks and more small peaks than in the case of the Rayleigh distribution, for which $k = 2$. For $k = 1$, the density of peaks decreases monotonically from its maximum value at zero height towards zero at very large peak heights. However, the probability of having small peaks and large peaks is greater than in the case k is larger than one.

Consider the highest peak that may occur in the time interval T. As before, the average number of cycles and hence peaks will be $\nu_0^+ T$, where ν_0^+ is the frequency of positive slope crossings of the zero line. If the level $a_{\max}$ is preset such that the average number of upward crossings of $a_{\max}$ in the time interval T is 1, $\nu_{a\max}^+ T = 1$, then there will be only one peak on

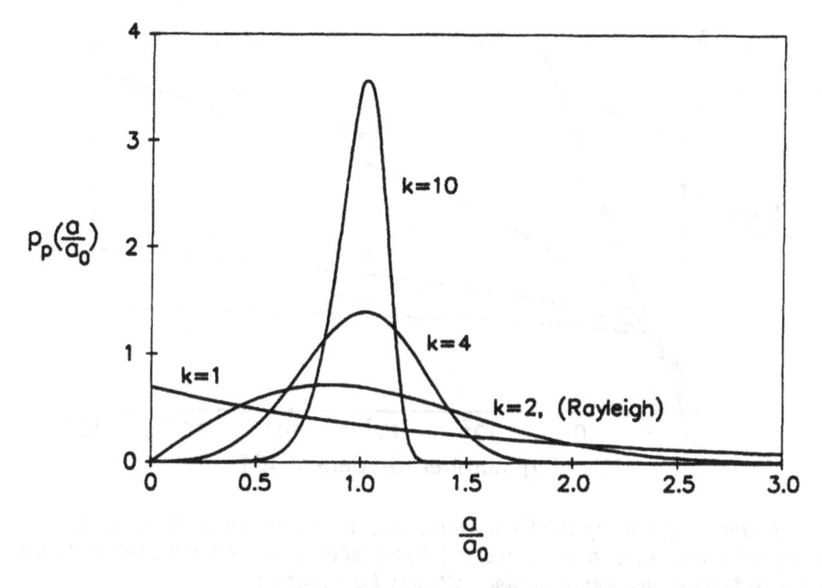

Fig. 4.9 *Weibull peak probability densities, (after Newland, [152]), reproduced by permission of Addison Wesley Longman Ltd*

the average greater than a_{max} and all other peaks will be smaller. The value of a_{max} can be computed and thus a lower bound for the extreme peaks can be obtained as follows. From Eqs. (4.29) and (4.36), the probability that the maximum peak is larger than a_{max} is

$$P[X_{peak} > a_{max}] = \frac{\nu_{a_{max}}^{+} T}{\nu_0^{+} T} = \frac{1}{\nu_0^{+} T} = e^{-(\log_e 2)(a_{max}/a_0)^k}$$

whereby

$$\frac{a_{max}}{a_0} = \left(\frac{\log_e(\nu_0^{+} T)}{\log_e 2} \right)^{1/k} \tag{4.38}$$

Equation (4.38) is the general result for any narrow-band process that has the fundamental attribute of having one full cycle for each up-crossing of the zero axis, and gives the maximum level, which only one peak on the average will reach during the time interval T, as a function of T and the zero crossing frequency. The behaviour is shown in Fig. 4.10. For $\nu_0^{+} T = 1$, the maximum level is obviously zero for all the curves. That is, if there is only going to be one peak in T, it must have a height greater than zero. For $\nu_0^{+} T = 2$, the maximum level is equal to the median peak level for all k, $a_{max} = a_0$, so all the curves pass through point (2, 1). This is because if there are only two peaks in T, on the average one will be above the median peak height and one will be below. Hence the median peak height is exceeded on the average by one peak in every two.

In a particular application, the Weibull coefficient k is the key parameter that has to be determined. One simple method for determining k, is by plotting the probability of the peaks against the level a on a suitable logarithmic paper. In fact, taking the natural logarithm twice of both sides of Eq. (4.36),

$$\log_e[-\log_e\{P[X_{peak} > a]\}] = \log_e \log_e 2 + k \log_e a - k \log_e a_0 \tag{4.39}$$

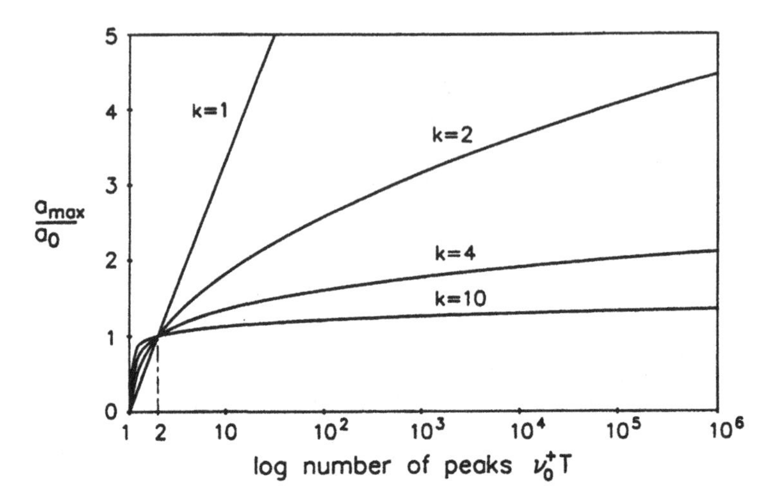

Fig. 4.10 *Relative maximum peak level values, a_{max}/a_0, exceeded on the average by only one peak in every ($\nu_0^+ T$) peaks as a function of ($\nu_0^+ T$) for different values of the Weibull coefficient k, [61], reproduced by permission of Addison Wesley Longman Ltd*

The slope of the graph of $\log_e[-\log_e\{P[X_{peak} > a]\}]$ against $\log_e a$ is the Weibull coefficient k and the zero intercept gives the median peak height, since

$$\log_e[-\log_e\{P[X_{peak} > a]\}] = 0 \qquad \text{for } a = a_1 \tag{4.40}$$

gives

$$a_0 = a_1(\log_e 2)^{1/k} \tag{4.41}$$

This procedure has for instance been applied by Melbourne for the analysis of the probability distribution of maximum wind loads, [143].

Example 4.4 A stationary narrow band process $X(t)$ has the following joint distribution of the amplitude $u = x(t)$ and the velocity $v = \dot{x}(t)$,

$$p(u,v) = \begin{cases} \dfrac{T}{4\beta^2} & \text{when} \quad \begin{cases} -\beta < u < \beta \\ -\beta < vT < \beta \end{cases} \\ 0 & \text{elsewhere} \end{cases}$$

where β is a positive constant having the dimension of $x(t)$ and T is another positive constant having the dimension of time.

Find the crossing frequencies of the amplitude levels $x = 0$ and $x = a$. Also, find the peak probability density and the envelope probability density.

Solution From Eq. (4.5),

$$\nu_a^+ = \nu_0^+ = \int_0^\infty p(a,v)v\,\mathrm{d}x\,\mathrm{d}v = \frac{T}{4\beta^2}\int_0^\infty v\,\mathrm{d}v = \begin{cases} \dfrac{1}{8T} & \text{if } |a| \leqslant \beta \\ 0 & \text{if } |a| > \beta \end{cases}$$

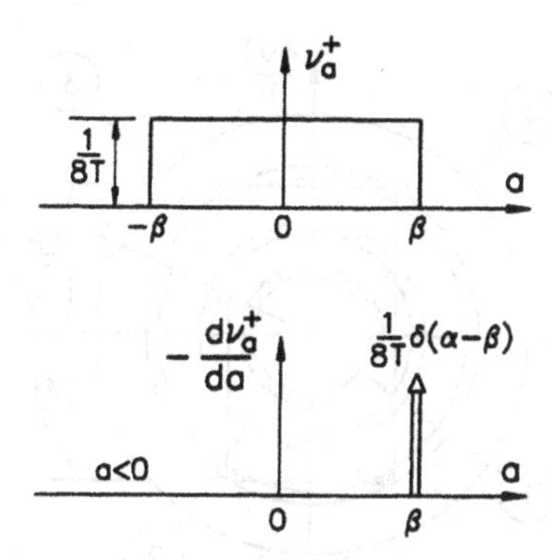

Fig. 4.11 *The crossing frequency and peak density*

Because of the form of the joint density, the two crossing frequencies are equal, so obviously the sample functions of this process show an oscillating behaviour with an angular frequency $\omega_0 = 2\pi\nu_0^+ = 2\pi/8T$, whereby the main period of the process is $8T$. The peak density is given by Eqs. (4.28) and (4.30),

$$p_p(a) = \frac{d}{da}\left(1 - \frac{\nu_a^+}{\nu_0^+}\right) = -\frac{1}{\nu_0^+}\frac{d\nu_a^+}{da} = \frac{8T}{8T}\delta(a-\beta) = \delta(a-\beta)$$

where the result of the above differentiation is made obvious through Fig. 4.11.

Following the explanations given in Example 4.2, the envelope distribution is found by taking a look at the magnitude of the radius vector in a $(x, \dot{x}/\omega_0)$ diagram. The radius vector, which forms the envelope of the sample function, is given by

$$r^2 = x^2 + \frac{v^2}{\omega_0^2} = x^2 + \frac{16v^2T^2}{\pi^2}$$

The joint probability density for the amplitude and the modified velocity are consequently given by the expression

$$p_r\left(u, \frac{4vT}{\pi}\right) = \begin{cases} \dfrac{\pi}{4T}\dfrac{T}{4\beta^2} = \dfrac{\pi}{16\beta^2} & \text{if } \begin{cases} |u| < \beta \\ \left|\dfrac{4vT}{\pi}\right| < \dfrac{4\beta}{\pi} \end{cases} \\ 0 & \text{elsewhere} \end{cases}$$

Now there are four different possibilities for evaluating the above density function for each of the different ranges of the parameters T and β, Fig. 4.12,

1) when $0 < a < \beta$, $p_e(a)da = 2\pi a\, da\, \pi/(16\beta^2) = \pi a/(4\beta^2)\pi/2$ (Fig. 4.13)
2) when $\beta < a < 4\beta/\pi$, $4\beta/\pi = \pi a/(4\beta^2)\sin^{-1}(\beta/a)$
   since $\theta = \sin^{-1}(\beta/a)$ (Fig. 4.14)
3) when $4\beta/\pi < a < \beta\sqrt{(1+16/\pi^2)}$
   $p_e(a)da = \pi/(16\beta^2)a\, da\, 4\theta = \pi a/(4\beta^2) - [\cos^{-1}(\beta/a) - \sin^{-1}(\beta/a]$
   since $\theta = \cos^{-1}(\beta/a) - \sin^{-1}(\beta/a)$
4) when $\beta\sqrt{(1+16/\pi^2)} < a$, $p_e(a) = 0$

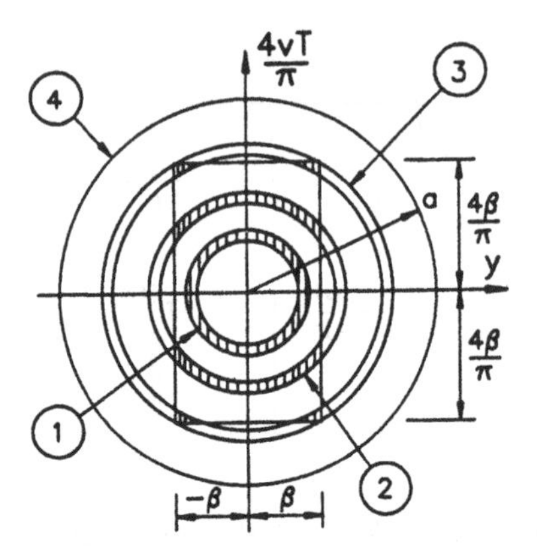

Fig. 4.12 *The envelope probability regions*

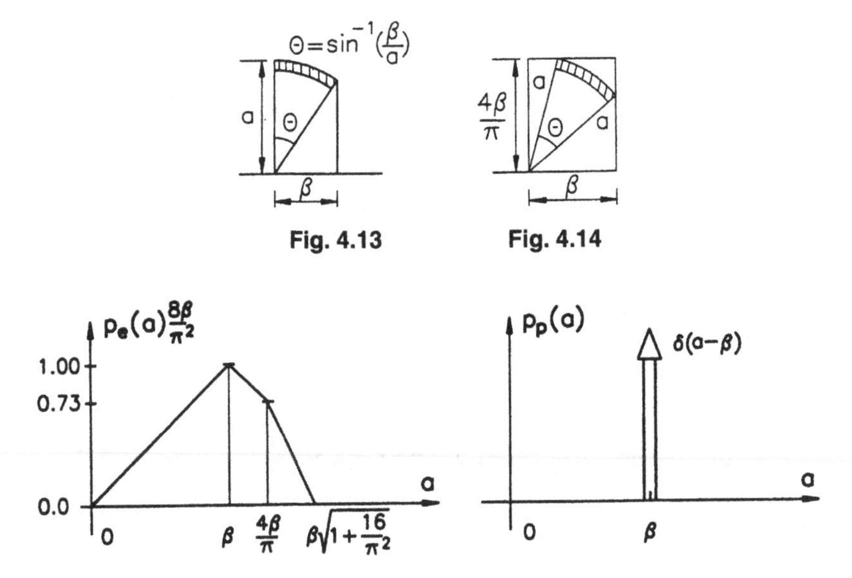

Fig. 4.15 *The probability densities for the peaks and the envelope*

The peak and envelope densities are shown in Fig. 4.15, and an imaginary sample function for $X(t)$ is sketched in Fig. 4.16.

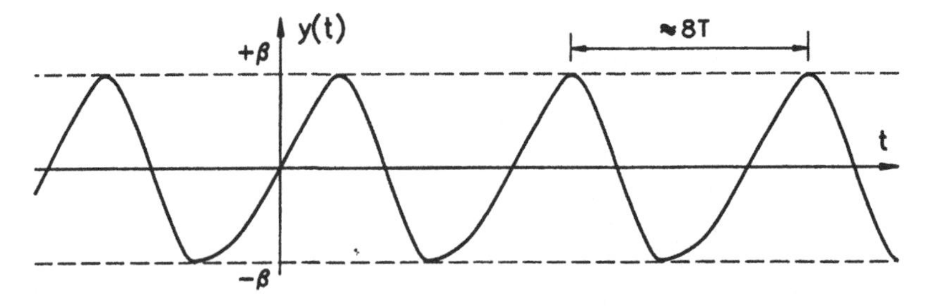

Fig. 4.16 *An imaginary sample function, x(t)*

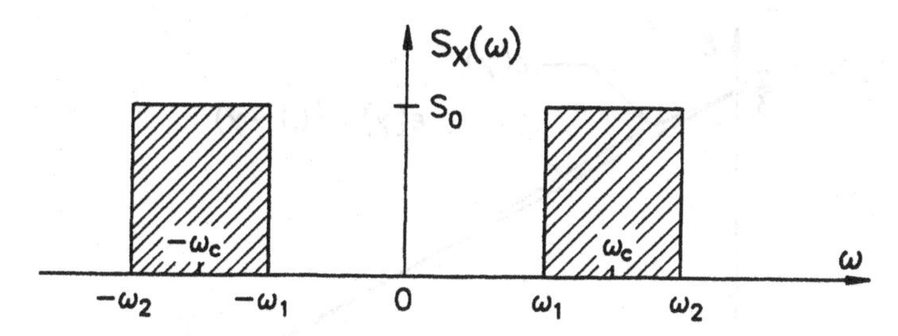

Fig. 4.17 *The spectral density of a stationary gaussian process*

Example 4.5 A Gaussian stationary process has the spectral density shown in Fig. 4.17.

(a) Find the bandwidth parameter ε as a function of the frequency ratio $\gamma = \omega_1/\omega_2$ and show by a sketch what this function looks like.
(b) Given that the function $\varepsilon(\gamma)$ found in (a) is approximated by the following expression

$$\varepsilon(\gamma) = \tfrac{2}{3}(1-\gamma)$$

find the crossing rate ν_0^+ for $\gamma = 0.5$.

Solution

(a) Calculate the spectral moments, Eq. (4.12):

$$m_0 = \int_{-\infty}^{\infty} S_X(\omega)d\omega = 2S_0(\omega_2 - \omega_1) = 2S_0\omega_2(1-\gamma) = 4S_0\omega_c\frac{1-\gamma}{1+\gamma}$$

$$m_2 = \int_{-\infty}^{\infty} \omega^2 S_X(\omega)d\omega = 2S_0(\omega_2^3 - \omega_1^3) \times \tfrac{1}{3}$$

$$m_4 = \int_{-\infty}^{\infty} \omega^4 S_X(\omega)d\omega = 2S_0(\omega_2^5 - \omega_1^5) \times \tfrac{1}{5}$$

whereby the square of the bandwidth parameter, ε, given by Eq. (4.20), is

$$\varepsilon^2 = \frac{m_0 m_4 - m_2^2}{m_0 m_4} = \frac{(\omega_2-\omega_1)(\omega_2^5-\omega_1^5)\times 9 - 5(\omega_2^3-\omega_1^3)^2}{(\omega_2-\omega_1)(\omega_2^5-\omega_1^5)\times 9}$$

or

$$\varepsilon^2 = 1 - \frac{5}{9}\frac{(\omega_2^3-\omega_1^3)}{(\omega_2-\omega_1)(\omega_2^5-\omega_1^5)} = 1 - \frac{5}{9}\frac{(1-\gamma^3)^2}{(1-\gamma)(1-\gamma^5)}$$

$$\varepsilon = \sqrt{\frac{4}{9} - \frac{5}{9}\gamma\frac{(1+\gamma-\gamma^2-\gamma^3)}{(1-\gamma^5)}}$$

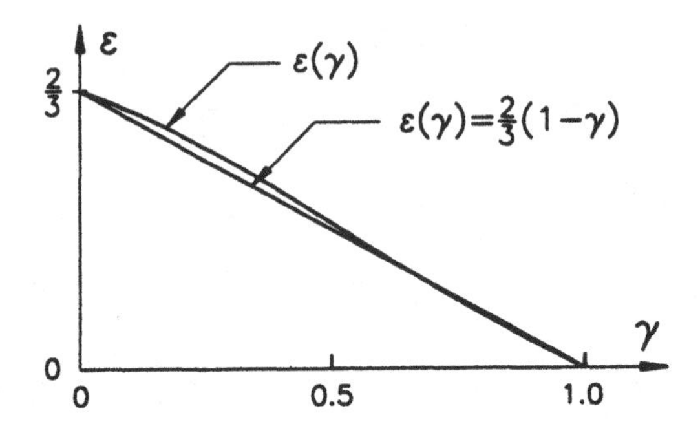

Fig. 4.18 *The bandwidth parameter ε as a function of $\gamma = \omega_2/\omega_1$*

This last function or square root giving ε is plotted in Fig. 4.18. Inside the square root the fraction involving γ becomes undefined when $\gamma \to 1$. However, L'Hôpital's rule can be applied such that

$$\lim_{\gamma \to 1} \varepsilon(\gamma) = 0$$

(b) The zero-crossing frequency is given by Eqs. (4.7) and (4.13),

$$\nu_0^+ = \frac{1}{2\pi}\sqrt{\frac{m_2}{m_0}}\frac{1}{2\pi}\sqrt{\frac{1}{3}\frac{\omega_2^3 - \omega_1^3}{\omega_2 - \omega_1}} = \frac{\omega_2}{2\pi}\sqrt{\frac{1}{3}(1 + 0.5 + 0{,}25)} = \frac{\omega_2}{2\pi} \times 0.7638$$

Another way of finding ν_0^+ is to use Eq. (4.21), noting that $\gamma = 0.5$, whereby $\varepsilon = \frac{1}{3}$

$$\varepsilon^2 = 1 - \left(\frac{\nu_0}{\mu_0}\right)^2, \qquad \nu_0^+ = \mu_0\sqrt{1 - \varepsilon^2} = \mu_0\sqrt{\frac{8}{9}}$$

also

$$\mu_0 = \frac{1}{2\pi}\sqrt{\frac{m_4}{m_2}} = \frac{1}{2\pi}\sqrt{\frac{3}{5}\frac{\omega_2^5 - \omega_1^5}{\omega_2^3 - \omega_1^3}} = \frac{\omega_2}{2\pi}\sqrt{\frac{3}{5}\frac{1 - (0.5)^5}{1 - (0.5)^3}} = \frac{\omega_2}{2\pi} \times 0.815 \text{ so } \nu_0^+ = \frac{\omega_2}{2\pi} 0.7684$$

4.4.1 Distribution of Extreme Peaks

The above discussion has made it clear how the probability distributions for the maxima and peaks of a stochastic process can be obtained. Also, a simple method to obtain the probability distribution of the largest peaks, using the Weibull distribution, was introduced. A different approach to describe the probability distribution of the extreme peaks is shown in the following, based on the statistics of maxima of a random broad-band function as was presented in Section 4.3. Thus, the probability distribution of the extreme maxima of the process is sought, which from the above discussion can be taken as a measure of the extreme peaks. Therefore, picking up the thread with Eqs. (4.20) and

(4.21), again consider a sample function $x(t)$ in the time interval of length T of the stochastic process $X(t)$. Assuming that N independent maxima, $\{H_i\}$, are observed in the time interval T, each having the same probability density $p_M(\xi)$, Eq. (4.20), the probability distribution of the extreme peak or maximum is obtained in the same manner as when dealing with extreme values, Example 1.6, that is,

$$P[\text{all } N \text{ maxima} \leqslant \xi] = P_E[\xi] = P[(H_1 \leqslant \xi) \cap (H_2 \leqslant \xi) \cdots \cap (H_N \leqslant \xi)] = [P_M(\xi)]^N \tag{4.42}$$

where

$$P_E[\xi] = [P_M(\xi)]^N = \left(\int_{-\infty}^{\xi} p_M(\eta) \mathrm{d}\eta \right)^N \tag{4.43}$$

Obviously, the probability distribution indicated by Eqs. (4.42) and (4.43) belongs to the class of extreme value distributions, [73]. Following the classical derivation by Cartwright and Longuet-Higgins [30], and Davenport's interpretation of some of their results, [52], the expression Eq. (4.43) together with the distribution of maxima Eq. (4.20), can be evaluated to yield the following simple expression for the distribution of extreme peaks:

$$P_E(\xi) = [P_M(\xi)]^N = \exp[-\nu T \exp[-\tfrac{1}{2}\xi^2]] \tag{4.44}$$

where

$$N = \nu T, \; \nu = \mu_0 \sqrt{(1-\varepsilon^2)} \qquad \text{and} \qquad \xi = u/\sigma_X \tag{4.45}$$

that is, μ_0 is the frequency of maxima anywhere, Eq. (4.19), and ν is the average frequency of maxima in the time interval of length T. Introduce the cumulative distribution function of large maxima $Q(\xi)$, that is, the probability of having a large maximum exceeding a certain level ξ:

$$Q(\xi) = \int_{\xi}^{\infty} p_M(\eta) \mathrm{d}\eta = 1 - P_M[\xi] \tag{4.46}$$

From Eq. (4.20), $Q(\xi)$ is given by

$$Q(\xi) = \int_{\xi}^{\infty} p_M(\eta) \mathrm{d}\eta = \frac{1}{\sqrt{2\pi}} \left[\int_{\xi/\varepsilon}^{\infty} \mathrm{e}^{-x^2/2} + \sqrt{(1-\varepsilon^2)} \exp\left[-\frac{\xi^2}{2} \right] \int_{-\infty}^{[\xi\sqrt{1-\varepsilon^2}/\varepsilon]} \mathrm{e}^{-(x^2/2)} \mathrm{d}x \right] \tag{4.47}$$

For large values of ξ, $\varepsilon \neq 1$, Eq. (4.47) can be simplified to yield

$$Q(\xi) = \sqrt{(1-\varepsilon^2)} \mathrm{e}^{-\xi^2/2} + O\left(\frac{1}{\xi^3} \mathrm{e}^{-\xi^2/(2\varepsilon^2)} \right) \tag{4.48}$$

as the integral

$$\int_{x}^{\infty} \mathrm{e}^{-x^2/2} \mathrm{d}x = \mathrm{e}^{-x^2/2} \left[\frac{1}{x} + O\left(\frac{1}{x^3} \right) \right]$$

has the approximation shown for large values of x. Now consider a sample of N maxima. The probability that the largest of the N maxima has a value less than or equal to the level η is given by Eqs. (4.43) and (4.46), that is,

$$P_E(\eta) = [1 - Q(\eta)]^N \tag{4.49}$$

Since $Q(\eta)$ is the probability of a large maximum exceeding the level η, it can be represented by a small number z/N in the range $0 < z/N \leqslant 1$ where $0 < z \leqslant N$ and N is large. Equation (4.49) can now be simplified using the limit value for the exponential function, that is,

$$P_E[\eta] = [1 - Q(\eta)]^N = \left[\left[1 - \frac{z}{N}\right]^{-N/z}\right]^{-z} = e^{-z} \quad \text{for } N \text{ large} \tag{4.50}$$

Therefore, from Eqs. (4.45) and (4.48),

$$z = NQ(\eta) = N\sqrt{(1-\varepsilon^2)}\exp[-\eta^2/2] = \nu T\exp[-\eta^2/2] \tag{4.51}$$

so in the end, putting this result into Eq. (4.50),

$$P_E(\eta) = [P_M(\eta)]^N = \exp[-\nu T\exp[-\eta^2/2]] \tag{4.44}$$

and thus Eq. (4.44) has been obtained.

Equation (4.44) bears a strong resemblance to the Gumbel or Type I extreme value distribution, Eq. (1.31), as was to be expected. If the process is purely narrow-band, $\varepsilon = 0$, and ν is the zero crossing frequency $\nu_0^+ = \mu_0$. If the bandwidth is unlimited, $\varepsilon = 1$, and the frequency of maxima goes to infinity as already seen by the discussion of Eq. (4.21). The probability density of the distribution Eq. (4.44) can be found by differentiation or

$$p_E(\xi) = \frac{dP_E}{d\xi} = NP_M^{N-1}\,p_M(\xi) = e^{-\psi}\frac{d\psi}{d\xi} \tag{4.52}$$

where

$$\psi = \nu T\exp[-\tfrac{1}{2}\xi^2] \quad \text{or} \quad \xi = \sqrt{(2(\log_e \nu T - \log_e \psi))} \tag{4.53}$$

The mean value and mean square value of the peak distribution can by now be obtained as follows, that is,

$$E[\Xi] = \int_{-\infty}^{\infty} \xi p_E(\xi)d\xi = \int_{-\infty}^{\infty} \xi dP_E(\xi) = \int_{-\infty}^{\infty} \xi e^{-\psi}d\psi \tag{4.54}$$

and

$$E[\Xi^2] = \int_{-\infty}^{\infty} \xi^2 e^{-\psi}d\psi \tag{4.55}$$

By expanding the latter of the expressions in Eq. (4.53) into series, it is possible to evaluate the above integrals to obtain explicit expressions for the mean value and mean square value, that is, by the expansion,

$$\xi = \sqrt{2\log_e \nu T}\left[1 - \frac{\log_e \psi}{2\log_e \nu T} + \text{higher-order terms}\right] \tag{4.56}$$

By dropping all the higher-order terms, the integrals Eqs. (4.54) and (4.55) are evaluated using the standard integrals often encountered in extreme value statistics, [41], namely

$$\int_0^\infty \log_e \psi e^{-\psi} d\psi = \gamma \quad (=0.5772) \qquad \text{and} \qquad \int_0^\infty \log_e^2 \psi e^{-\psi} d\psi = \frac{\pi^2}{6} + \gamma^2 \tag{4.57}$$

and the following result is obtained,

$$E[\Xi] \doteq \sqrt{2\log_e \nu T} + \frac{(\gamma = 0.5772)}{\sqrt{2\log_e \nu T}} \tag{4.58}$$

and

$$E[\Xi^2] = \sqrt{2\log_e \nu T} + 2\gamma + \frac{\pi^2}{6}\frac{1}{2\log_e \nu T} + \frac{\gamma^2}{\sqrt{2\log_e \nu T}} \tag{4.59}$$

where γ is Euler's constant, which is a real fraction with an infinite number of digits. The value 0.5772 gives the first four digits. The variance is then given by the simple expression

$$\text{Var}[\Xi] = \sigma_\Xi^2 = \frac{\pi^2}{6}\frac{1}{2\log_e \nu T} \tag{4.60}$$

Obviously, the spread of the distribution is very small for large values of νT, that is, the main frequency of the signal is large and/or the duration T is large.

If $\varepsilon \to 1$, that is, the bandwidth of the process is unlimited, the above approximation breaks down. However, it is still possible to obtain the mean value of the extreme peaks as shown in [30], i.e.

$$E[\Xi] = M + \frac{\gamma M}{1 + M^2} \approx \sqrt{2\log_e \frac{\nu T}{\sqrt{2\pi}}} \tag{4.61}$$

where M is the mode of the distribution for Ξ given by

$$\sqrt{(2\pi)} M \exp[M^2/2] = N \Rightarrow M^2 = \log_e(N^2/2\pi) - \log_e M^2$$

or by the approximation

$$M = [\log_e((\nu T)^2/(2\pi)) - \log_e(\log_e((\nu T)^2/(2\pi))]^{1/2} \tag{4.62}$$

and taking only the dominant term, the approximation for the mean extreme peak given by Eq. (4.61) is obtained.

It should be remembered that in the above expressions, all peak values are dimensionless and have to be multiplied by the standard deviation of the parent process to get the proper values, Eq. (4.21). Thus $X_{max} = \Xi\sigma_X$. In Fig. 4.19, various extreme peak distributions are shown for different values of νT.

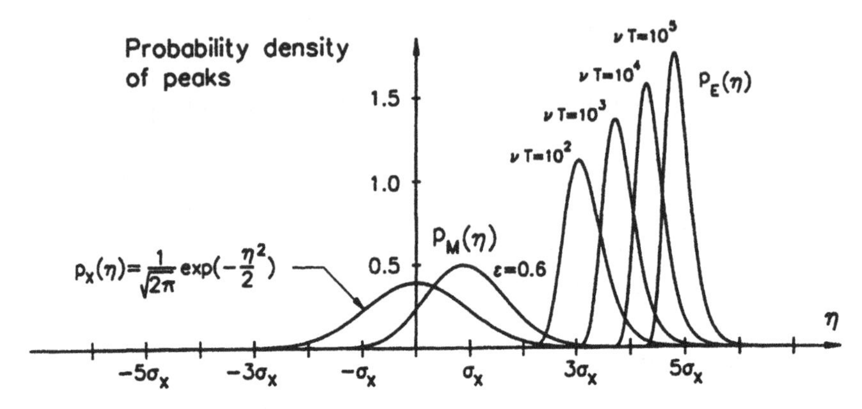

Fig. 4.19 *Extreme peak distributions*

Example 4.6 For the Gaussian stationary process given in Example 4.5, consider sample functions that are observed in a time interval of 5 minutes. Given that the centre frequency of the spectral density in Fig. 4.15 is 60 Hz with $\gamma = 0.5$, find the average value and standard deviation of the extreme values or largest peaks in the 5 minute interval.

Solution The centre frequency

$$\omega_c = 2\pi f_c = \tfrac{1}{2}(\omega_2 + \omega_1) = \omega_2 \frac{1+\gamma}{2} = \tfrac{3}{4}\omega_2$$

Also, $\varepsilon = 2/3\ (1-\gamma) = 1/3$,

$$\nu = \nu_0^+ \sqrt{1-\varepsilon^2} = \frac{\omega_2}{2\pi} 0.764\sqrt{1-(\tfrac{1}{3})^2} = \frac{4\omega_c}{3(2\pi)} 0.764\sqrt{\tfrac{8}{9}} = 57.62\ \text{Hz}, \qquad \nu T = 17\,286.87$$

Then by Eq. (4.39),

$$E[\Xi] = \sqrt{2\log_e \nu T} + \frac{0.5772}{\sqrt{2\log_e \nu T}} = 4.42 + \frac{0.5772}{4.42} = 4.55$$

$$E[X_{\text{peak}}] = 4.55\sigma_X = 4.55\sqrt{4S_0\omega_c \frac{1-\gamma}{1+\gamma}} = 102\sqrt{S_0}$$

and

$$\text{SD}[X_{\text{peak}}] = \sigma_\xi \cdot \sigma_X = \pi \frac{\sigma_X}{\sqrt{6}\sqrt{2\log_e \nu T}} = 0.29\sigma_X = 0.29\sqrt{4S_0\omega_c \frac{1-\gamma}{1+\gamma}} = 6.51\sqrt{S_0}$$

4.4.2 Design Factors

Classical structural design methods are based on simple provisions against overloading of a structural member. The external actions, whether natural loads such as wind, snow or earthquakes or anthropogenic loads such as life loads on floors etc., are interpreted as deterministic loads for which certain design values are provided by official structural design codes. The design strength of the structural members is based on simple laboratory experiments of sample specimens. With proper safety levels assigned by the various local and national codes, the design process can be described by the following simple formula:

$$\frac{R}{n_{\mathrm{R}}} = n_{\mathrm{L}} L \tag{4.63}$$

where R is the material strength or resistance of a structural member and L is the corresponding action due to both external and possible internal loads (thermal stresses, differential settlements etc.). R and L must be compatible, that is, if R is a resistance in terms of a material stress then L is a stress due to external loads at the same point in the structure. Likewise, if R is a member capacity (strength) at a certain cross section then L is the corresponding action (moment, axial force etc.) at the same place. The safety factors n_{R} and n_{L} are chosen using mostly experience and heuristic arguments ensuring that the so-called allowable stresses are well within the elastic range (actually about half the yield stress of the material), whereby the design is based on the normal daily state of the structure.

Gradually, as ultimate limit state theories became better formulated, describing the state of the structure at its breaking point, the need for probabilistic definition of both the loads and the material resistance was obvious. Nowadays, modern structural design codes are based on probabilistic assessment of both the loads and the resistance together with the structural safety requirements formulated as the reliability of the structure, i.e. the accepted probability of failure. In short, these new probabilistic methods can be described as follows. The characteristic or nominal resistance of a structural member is defined as

$$P[R \leqslant R_{\mathrm{p}}] = p \tag{4.64}$$

where p is the probability that the actual resistance R is less than the nominal resistance. Usually, the value for p is chosen around 0.005–0.01 whereby the characteristic (when actual numbers for p and q have been chosen) strength is the lower 5‰ to 1% fractile. Similarly, the characteristic or nominal load, the ultimate breaking load, is defined as

$$P[L > L_q] = q \tag{4.65}$$

where the probability of larger loads than the characteristic load is usually chosen around 0.95–0.99, that is, the characteristic load is the upper 5%–1% fractile, or a load that can be expected every 20–100 years on the average. Finally, the reliability of the structure is defined as the probability that R is greater than L or

$$P[R > L] = P[(R-L) > 0] = p_{\mathrm{s}} = 1 - p_{\mathrm{f}} \tag{4.66}$$

where p_{s} is the probability of safety and hence p_{f} is the failure probability, usually chosen around 10^{-n} with $5 < n < 8$ according to the importance and expected lifetime of the structure.

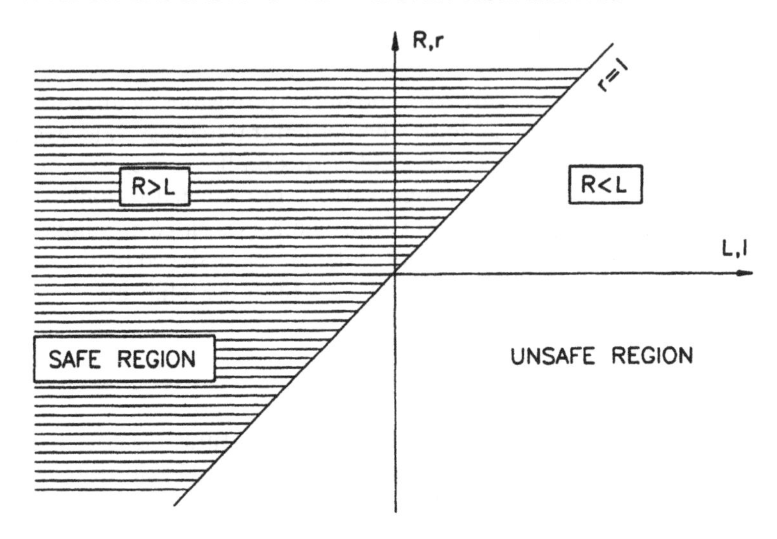

Fig. 4.20 *Safe combinations of R and L (R > L)*

The safety and failure probabilities (p_s, p_f) can be computed if the joint probability distribution

$$F_{RL}(r, l) = \int_{-\infty}^{r} \int_{-\infty}^{l} p_{RL}(x, y) \mathrm{d}x \, \mathrm{d}y \tag{4.67}$$

is known. In fact, integrating the joint probability density Eq. (4.67) over the 'safe' region, shown in Fig. 4.20,

$$\begin{aligned} p_s = 1 - p_f &= \iint_{R>L} p_{RL}(x, y) \mathrm{d}x \, \mathrm{d}y \\ &= \int_{-\infty}^{\infty} \mathrm{d}x \int_{-\infty}^{r} p_{RL}(x, y) \mathrm{d}y = \int_{-\infty}^{\infty} \mathrm{d}y \int_{l}^{\infty} p_{RL}(x, y) \mathrm{d}x \end{aligned} \tag{4.68}$$

In most cases, the two random variables R (the resistance) and L (the load) have nothing to do with another, i.e. they are independent. Then, $p_{RL}(r, l) = p_R(r) \cdot p_L(l)$, and the above integral can be evaluated as

$$p_s = \int_{-\infty}^{\infty} p_R(r) \int_{-\infty}^{r} p_L(l) \mathrm{d}l = \int_{-\infty}^{\infty} p_R(r) F_L(r) \mathrm{d}r \tag{4.69}$$

or

$$p_s = \int_{-\infty}^{\infty} p_L(l) \mathrm{d}l \int_{l}^{-\infty} p_R(r) \mathrm{d}r = 1 - \int_{-\infty}^{\infty} p_L(l) F_R(l) \mathrm{d}l \tag{4.70}$$

The above integrals can be difficult to evaluate as knowledge of the probability distributions for both the loads and the resistance is often scarce. However, many

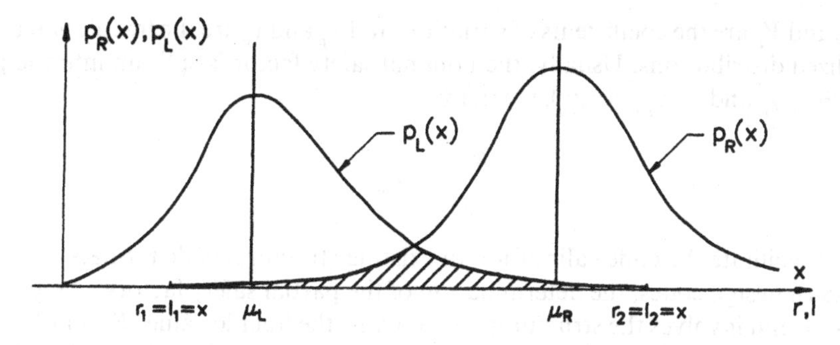

Fig. 4.21 *A Water diagram over the load and resistance probability densities*

conclusions can be drawn from the probabilistic structure of the safety probability, which is indicated by the above integrals. For instance, the so-called Warner diagram is shown in Fig. 4.21. For $x_1 < s_1 = r_1$, the probability of a lower resistance has become so small that the outcome, i.e. the probability of safety, is no longer influenced by lower values. For $x_2 > s_2 = r_2$, on the other hand, the probability of higher loads has become so small that p_s is no longer influenced by higher loads. Therefore, the following interpretation can be given to the shaded area between the two densities, (Fig. 4.21):

(a) $R > r_2$, failure is 'impossible'
(b) $r_1 < R < r_2$, failure is possible
(c) $R < r_1$, failure is 'certain'

In any kind of structural design provisions, it is ensured that the mean value $\mu_R > \mu_S$. For this reason r_1 will always be less than r_2 $(r_1 < r_2)$. It is therefore only in the range $r_1 < R < r_2$ that the structural safety is uncertain but this is also the range within which all structural systems can be expected to fall.

Since the mean values μ_R and μ_L are likely to fall within the critical range, it is necessary to set up the design criterion with the fractile values (Eqs. (4.64) and (4.65)) instead. Based on the original design criterion ('code format') Eq. (4.63), the new design criterion is defined as

$$R_p = \gamma_{pq} L_q \tag{4.71}$$

The factor of safety γ_{pq} is called the nominal safety factor. If the mean values are used instead of the fractile values,

$$\mu_R = \gamma_0 \mu_L \tag{4.72}$$

where γ_0 is the so-called central safety factor. The nominal safety factor γ_{pq} can be related to the central safety factor γ_0 through the standardized distributions for R and L. In fact, by Eqs. (1.50), (4.71) and (4.72),

$$\gamma_{pq} = \frac{R_p}{L_q} = \frac{\mu_R(1 + x_p V_R)}{\mu_L(1 + y_q V_q)} = \gamma_0 \frac{1 + x_p V_R}{1 + y_q V_q} \tag{4.73}$$

where V_R and V_L are the coefficients of variation, and x_p and y_q are the fractile values of the standardized distributions. Usually, the nominal safety factor is split up into the partial safety factors γ_p and γ_q ($\gamma_{pq} = \gamma_p \gamma_q$), whereby

$$\frac{R_p}{\gamma_p} = \gamma_q L_q \tag{4.74}$$

in order to facilitate the code calibration to existing structural code formats.

In modern design codes, the determination of the partial safety factors γ_p and γ_q is the central issue and involves the structural reliability p_s, the fractile values R_p and L_q and the so-called safety index β. For this reason it is convenient to introduce the random variable

$$Z = R - L \tag{4.75}$$

that is, the difference between the resistance and the load at any particular point in the structure. Clearly, the reliability is given by

$$p_s = P[Z > 0] = F_Z(0) \tag{4.76}$$

in which $F_Z(z)$ is the unknown distribution of Z. By standardizing the distribution of Z, the reliability Eq. (4.76) is given by

$$p_s = P\left[\frac{Z-\mu_Z}{\sigma_Z} > \frac{0-\mu_Z}{\sigma_Z}\right] = P\left[U > \frac{\mu_Z}{\sigma_Z}\right] = 1 - P\left[U \leqslant \frac{-1}{V_Z}\right] = 1 - P[U \leqslant -\beta] \tag{4.77}$$

in which $U = (Z - \mu_Z)/\sigma_Z$ and $V_Z = \sigma_Z/\mu_Z$, and a new quantity, the safety index $\beta = 1/V_Z$ has been introduced. The reliability is therefore given by

$$p_s = 1 - F_U(-\beta) \tag{4.78}$$

As $F_U(-\beta_2) < F_U(-\beta_1)$ for $\beta_2 > \beta_1$, the reliability increases for increasing values of β.

The safety index is used to set up and calibrate new code formats. For instance, if $F_U(u)$ is a normal distribution then $\beta = 4$ gives the reliability $p_s = 1 - 3 \times 10^{-5}$. From Eq. (4.78) it is obvious that the negative tail of the distribution is mostly involved in the determination of p_s. The tail is therefore often approximated with that of the exponential distribution, that is,

$$\log_{10} F_U(u) = a + bu \tag{4.79}$$

in which $b \approx 2$ for a normally distributed and a Gumbel distribution variable.

The partial safety factors can easily be related to the safety index. Since,

$$\beta = \frac{1}{V_Z} = \frac{\mu_R - \mu_L}{\sqrt{\sigma_R^2 + \sigma_L^2}} \quad \text{and} \quad \gamma_0 = \frac{\mu_R}{\mu_L}$$

therefore,

$$\mu_R^2 - 2\mu_R\mu_L + \mu_L^2 = \beta^2\sigma_R^2 + \beta^2\sigma_L^2$$

or

$$\gamma_0^2(1-\beta^2 V_R^2)-2\gamma_0+(1-\beta^2 V_L^2)=0 \tag{4.80}$$

Equation (4.80), which is a quadratic equation in γ_0, has two distinct roots, that is,

$$\gamma_0=\frac{1\pm\beta\sqrt{V_R^2+V_L^2-\beta^2 V_R^2 V_L^2}}{1-\beta^2 V_R^2} \tag{4.81}$$

Obviously, only the larger root with the positive sign is of interest since $\gamma_0>1$. The nominal safety factor is now given by Eqs. (4.73) and (4.81), that is,

$$\gamma_{pq}=\frac{1+\beta\sqrt{V_R^2+V_L^2-\beta^2 V_R^2 V_L^2}}{1-\beta^2 V_R^2}\cdot\frac{1+k_p V_R}{1+k_q V_L} \tag{4.82}$$

In Fig. 4.22, the central and nominal safety factors are depicted for different values of the coefficient of variation for the loads and the resistance respectively. The reliability index is fixed at four ($\beta=4$), and as fractile values, k_p for the resistance and k_q for the load, the 10% lower fractile and the 2% upper fractile of the standardized normal distribution are selected ($p=0.1$ and $q=0.98$). The central safety factor is more sensible to variation in the statistical data than the nominal safety factor, which stays in the range $1.0<\gamma_{pq}<2.0$ for all practical values of the coefficients of variation of both the resistance and the loads.

Finally, the partial safety factors γ_p and γ_q, which were defined in Eq. (4.74), can be interpreted in the following manner. From the definition of the safety index β,

$$\mu_R-\mu_L=\beta(\sigma_R^2+\sigma_L^2)^{1/2}=\beta\sigma_Z \tag{4.83}$$

Obviously, the standard deviations σ_R, σ_L and σ_Z form a vector diagram such that $\sigma_Z=\alpha_1\sigma_R+\alpha_2\sigma_L$ where α_1 and α_2 are the direction cosines of the vector diagram. Therefore,

$$\mu_R-\mu_L=\beta\alpha_1\sigma_R+\beta\alpha_2\sigma_L \text{ or } \mu_R(1-\beta\alpha_1 V_R)=\mu_L(1+\beta\alpha_2 V_L) \tag{4.84}$$

From the definition of fractile values, $\mu_R=R_p/(1+x_p V_R)$ and $\mu_L=L_q/(1+y_q V_L)$. Thus,

$$R_p\frac{1-\beta\alpha_1 V_R}{1+x_p V_R}=L_q\frac{1+\beta\alpha_2 V_L}{1+y_q V_L} \tag{4.85}$$

Comparing the above Eq. (4.85) with Eq. (4.73), the nominal safety factors are given by

$$\gamma_{pq}=\frac{1+x_p V_R}{1-\beta\alpha_1 V_R}\cdot\frac{1+\beta\alpha_2 V_L}{1+y_q V_L}=\gamma_p\gamma_q$$

It is therefore natural to assume that the partial safety factors can be defined by the expressions

$$\gamma_p=\frac{1+x_p V_R}{1-\beta\alpha_1 V_R} \quad\text{and}\quad \gamma_q=\frac{1+\beta\alpha_2 V_L}{1+y_q V_L} \tag{4.86}$$

For a more detailed discussion, the reader is referred to the specialist literature (e.g.

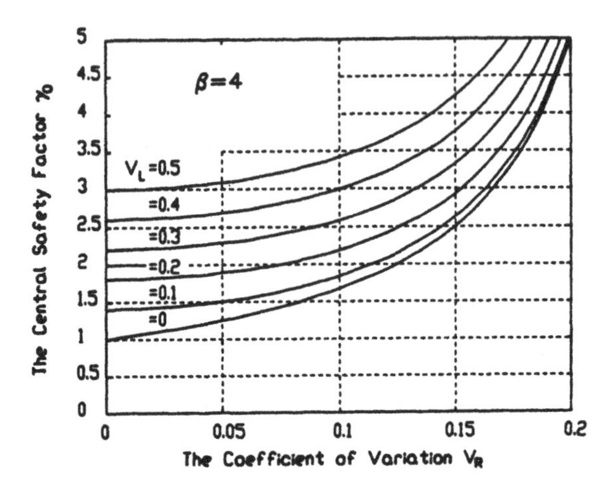

Fig. 4.22a *The central safety factor* γ_0

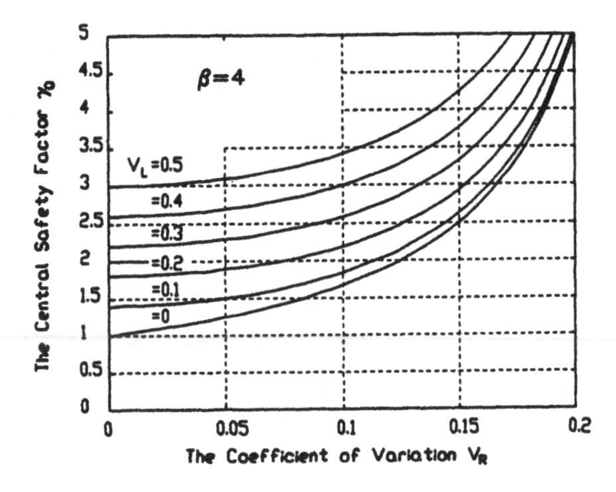

Fig. 4.22b *The nominal safety factor* γ_{pq}

Thoft-Christensen and Baker, [219]). A comprehensive account of characteristic loads and material resistance together with definition of limit states for design is found in the 1978 edition of CEB's Model Code for Concrete Structures, [120].

Example 4.7 A partial safety factor for concrete strength can be defined as follows. The coefficient of variation of the compressive strength of concrete cylinder specimens is usually found to be around 20%. The reliability or the safety index can be selected to be four ($\beta = 4$), and the two direction cosines are assumed to have equal values, i.e. 0.7. Using a characteristic concrete strength corresponding to $p = 0.1$, i.e. the 10% lower fractile, the corresponding standardized normal fractile value $k_p = -1.28$, and the partial safety factor is then given by Eq. (4.86), that is,

$$\gamma_p = \frac{1 + k_p V_R}{1 - \beta \alpha_1 V_R} = \frac{1 + (-1.28)0.20}{1 - 4 \times 0.7 \times 0.20} = 1.77 \cong 1.8$$

which is a common value for concrete material safety factors.

Example 4.8 Figure 4.23 shows a cantilevered beam (ABD), which is simply supported at A and B. The beam is subjected to a uniform load Q KN/m and two singular loads P_1 at C and P_2 at the end D. These loads are random variables with the following mean values and standard deviations:

$$E[Q] = 32/l \text{ KN/m} \quad \sigma_Q = 8/l \text{ KN/m}$$
$$E[P_1] = 32 \text{ KN} \quad \sigma_{P_1} = 4 \text{ KN}$$
$$E[P_2] = 16 \text{ KN} \quad \sigma_{P_2} = 4 \text{ KN}$$

The two forces P_1 and P_2 are statistically dependent with $\text{Cov}[P_1 P_2] = 8 \text{ KN}^2$, whereas other covariances are zero. The resistance of the beam is defined for the two cross sections at B and C. The beam is said to have reached its load bearing capacity, i.e. the beam fails if either

$$M_C \geqslant \tfrac{2}{3} M_F \quad \text{or} \quad M_B \geqslant M_F$$

where M_B and M_C are the random moments at B and C due to the external loads, whereas M_F is the plastic moment capacity of the beam, which is a random variable with the coefficient of variation $V_M = 0.15$.

If the safety index $\beta = 3$ (corresponding to short-term design), determine the necessary mean value and standard deviation of the moment capacity M_F. For $\beta = 4$ (long-term design), calculate the maximum acceptable mean values of the singular forces P_1 and P_2 if $E[M_F] = 50\, l$ KN m, and all other quantities are kept equal.

Solution The two moments M_B and M_C are easily computed. Thus,

$$M_C = P_1 l/4 - P_2 l/4 + Q l^2/16 \quad \text{and} \quad M_B = P_2 l/2 + Q l^2/8$$

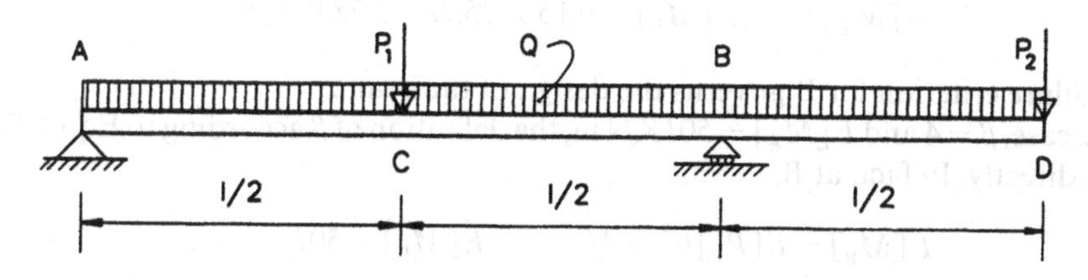

Fig. 4.23 *A cantilevered beam under external loads*

The mean values and standard deviations are given by

$$E[M_C] = E[P_1 l/4 - P_2 l/4 + Ql^2/16] = 8l - 4l + 2l = 6l$$

$$E[M_B] = E[P_2 l/2 + Ql^2/8] = 8l + 4l = 12l$$

and

$$\text{Var}[M_C] = \text{Var}[P_1]l^2/16 + \text{Var}[P_2]l^2/16 + \text{Var}[Q](l^2/16)^2 - 2\,\text{Cov}[P_1 P_2]l^2/16$$
$$= l^2 + l^2 + l^2/4 - l^2 = 5l^2/4$$

$$\text{Var}[M_B] = \text{Var}[P_2 l/2 + Ql^2/8] = 4l^2 + l^2 = 5l^2$$

The coefficients of variation are now obtained, that is,

$$V_{M\text{C}} = \sqrt{\frac{\text{Var}[M_C]}{(E[M_C])^2}} = \sqrt{\frac{5}{4 \times 36}} \cong 0.19$$

$$V_{M\text{B}} = \sqrt{\frac{\text{Var}[M_B]}{(E[M_B])^2}} = \sqrt{\frac{5}{144}} \cong 0.19$$

Since there are two failure criteria, that is, either the moment M_C is larger than two thirds of the capacity of the beam or M_B is larger than the capacity, the central safety factor γ_0 can be computed in two different ways using Eqs. (4.72) and (4.81).

For C,

$$\gamma_0 = \frac{E[R]}{E[L]} = \frac{2E[M_F]}{3 \times 6l} = \frac{1 + \beta\sqrt{V_{M\text{F}}^2 + V_{M\text{C}}^2 - \beta^2 V_{M\text{F}}^2 V_{M\text{C}}^2}}{1 - \beta^2 V_{M\text{F}}^2}$$

$$= \frac{1 + 3\sqrt{0.15^2 + 0.19^2 - 3^2 \times 0.15^2 \times 0.19^2}}{1 - 3^2 \times 0.15^2 = 2.1}$$

whereby

$$E[M_F] = 2.1 \times 9l = 18.9l$$

For B, $\gamma_0 = 2.1$ as the coefficients of variation are the same. However,

$$E[M_F] = 2.1\,E[M_B] = 2.1 \times 2l = 25.2l\,\text{KN m}$$

and

$$\sigma[M_F] = V_M E[M_B] = 0.15 \times 25.2l = 3.78\,\text{KN m}$$

so the failure criterion for B governs the design of the beam.

In the case, $\beta = 4$ and $E[M_F] = 50l$ KN m, the definition of β according to Eq. (4.77) can be used directly. In fact, at B,

$$E[M_B] = E[P_2]l/2 + 4l \qquad E[M_F] = 50l$$

$$\text{Var}[M_B] = 5l^2 \qquad \text{Var}[M_F] = (0.15 \times 50)^2 = (7.5l)^2$$

whereby,

$$4 = \frac{\mu_R - \mu_L}{\sqrt{\sigma_R^2 + \sigma_L^2}} = \frac{50l - 4l - E[P_2]\frac{l}{2}}{\sqrt{(7.5l)^2 + 5l^2}} = \frac{46l - E[P_2]\frac{l}{2}}{7.83l}$$

which gives

$$E[P_2] = 29.36\,\text{KN}$$

At C, $E[R] = 2 \times 50l/3$ and $\text{Var}[R] = (V_M E[R])^2 = 0.15 \times 2 \times 50l/3 = 25l^2$. Also,

$$E[M_C] = E[P_1]l/4 - E[P_2]l/4 + E[Q]l^2/16] = E[P_1]l/4 - 29.36l/4 + 2l = E[P_1]l/4 - 5.34l$$

and

$$\text{Var}[M_C] = 5l^2/4$$

so

$$4 = \frac{\frac{2}{3}50l - E[P_1]\frac{l}{2} + 5.34l}{\sqrt{25l^2 + \frac{5}{4}l^2}} = \frac{27.99l - E[P_1]\frac{l}{4}}{5.12l}$$

whereby,

$$E[P_1] = 29.98\,\text{KN}$$

Thus the mean values for the single forces P_1 and P_2 must not exceed 30 KN to maintain the reliability of the beam at the safety index $\beta = 4$.

4.4.3 Overload Design Factors

In the previous section, a short overview of modern design methods based on characteristic loads and resistances with suitable safety factors was presented. It was shown how the characteristic values of the static design loads and the corresponding structural resistances were defined, taken as fractile values with small probability of exceedance. Designing mechanical and structural systems to withstand expected loads during the lifetime of a system is a highly complex subject. Part of the expected loads is 'static' by nature, i.e. can be described as the d.c. component of an otherwise time-dependent load signal. The verification of sufficient system resistance to static loads is a straightforward matter and comprises the basis of all structural and mechanical systems design. However, the time-dependent loads often constitute the major threat to the system, and they are also more difficult to handle. Therefore, in the case of dynamic excitation, the ability of the system to withstand extreme signal peaks will be crucial in the system design. Thus, the determination of the extreme load peaks is most useful and more important in connection with the design of physical systems responding to random excitation. If the system has a threshold value which must not be exceeded, the extreme peak distribution can serve as a tool to both evaluate the survival possibilities when the response process has been established, and also as a means to define appropriate design factors to prevent overloading of the system. In the previous sections, proper background has been established for a simple and straightforward design method to guard against extreme load peaks. An example of such application follows.

Let $Y(t)$ be the random stationary response process of a linear physical system (stress, deflection, spring force, strain etc.) when subjected to a stationary random Gaussian excitation. Then

$$Y_{\max} = \bar{Y} + (Y_{\text{peak}} - \bar{Y})$$

where $E[Y] = \bar{Y}$ is the response to the d.c.-component or the excitation or the static part. As seen in Fig. 4.17, the probability distribution of the extreme peaks $(Y_{\text{peak}} - \bar{Y})$ is fairly narrow for the reason that the standard deviation, Eq. (4.60), is very small for long duration times. Therefore for all practical purposes, the extreme value $(Y_{\text{peak}} - \bar{Y})$ can simply be replaced by the mean value of the extreme peak distribution, Eqs. (4.58) and (4.61), so

$$Y_{\max} = \bar{Y} + \sigma_Y E[\Xi_Y]$$

or

$$Y_{\max} = \bar{Y}(1 + V_Y g(\nu T)) = G\bar{Y} \tag{4.87}$$

in which $V_Y = \sigma_Y / \bar{Y}$ is the coefficient of variation of the response process. The function $g(\nu T) = E[\Xi_Y]$ is called the peak function or peak factor, and G is then the overload design factor.

This procedure has been applied with considerable success in more recent design codes for heavy wind loads, [58]. In Fig. 4.24, various values of the peak factor in the case of

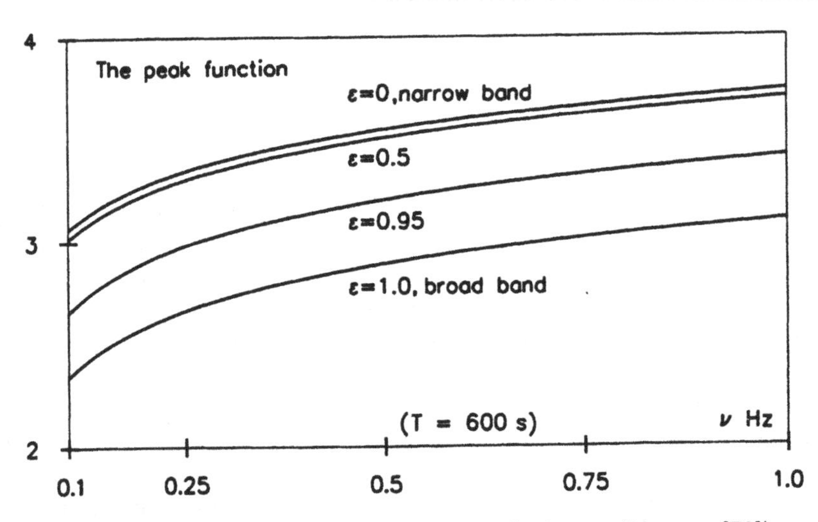

Fig. 4.24 *The peak factor for various bandwidths, (after Dyrbye and Hansen, [72]), reproduced by permission of Statens Byggeforskningsinstitut, Denmark*

heavy wind loads is shown, where the observation or duration time $T = 10$ minutes, which is standard in wind speed assessment.

4.5 FAILURE MODELS

4.5.1 Introduction

In the middle of the nineteenth century, the German railway engineer, August Wöhler, made interesting fatigue tests on railway track steel profiles and discovered the fundamental behaviour and endurance of structural steel members under repeated loading, [235]. For very high loads, producing high stress levels in the members, a single excursion above a threshold value, usually the yield level of the material, denoted by S_Y, is sufficient to destroy the load-bearing capacity of the member. For lower stress levels, caused by repeated loading of a sinusoidal character, after a certain number N of load cycles, the material would yield or break even though the stress level throughout the loading cycles is lower than the yield stress or the material strength. At a certain low stress level, the members would be able to sustain an unlimited number of load cycles without showing any effect of damage. This fundamental behaviour is usually described using the so-called Wöhler diagram, where the stress level S is plotted against the number of sinusoidal load cycles, Fig. 4.25. The fatigue strength of the material is then given as a function of the number of cycles N necessary to produce yielding in the material. Usually, the lower limit of the fatigue strength is set at 2 million cycles, $N = 2 \times 10^6$, that is, if the material can not be broken during 2 million cycles, it can sustain the stress level indefinitely. This behaviour can be approximated by the following functional relationship:

$$\frac{N}{N_1} = \left(\frac{S}{S_1}\right)^{-\alpha} \tag{4.88}$$

where S_1 and N_1 are reference levels to be set.

If a system is subjected to a random excitation, a possible fatigue failure takes place as the combination of many stress cycles of different amplitudes and frequencies so the

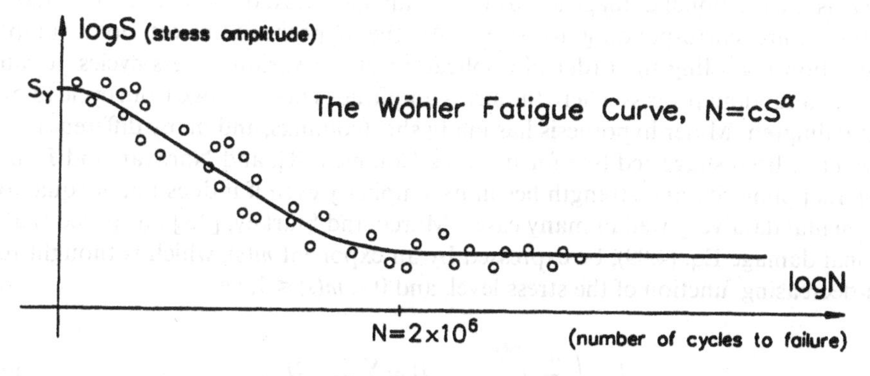

Fig. 4.25 *A logarithmic plot of the Wöhler S–N fatigue curve*

Wöhler curve is not directly applicable. Unfortunately, the basic mechanism of fatigue failure is not yet sufficiently well understood to allow a clear and precise treatment of what happens in a system when subjected to a random type of loading. Recent developments and investigations look promising, however, as more knowledge of the failure mechanism in different kind of materials accumulates. The breaking up of the material is believed to start by sliding of atomic layers, which is caused by a combination of dislocations due to local stress concentrations. The spreading of such dislocations under cyclic loading causes localized plastic deformation in the material. Microscopic cracks are formed that grow together to produce major cracks. Therefore, crack initiation and crack growth are the main reason for fatigue damage accumulation, and sophisticated fatigue models must address these two factors. For a recent and full discussion of the fatigue problem, the reader is referred to Collins, [38], Kachanov, [97] and Sobczyk and Spencer, [204], where a 'state of the art' description of the most current theories and development is presented.

Many hypotheses have been suggested to cover the fatigue failure of systems subjected to irregular stress levels. The best-known hypothesis, and most widely used, is the Palmgren–Miner hypothesis of accumulated damage, which was proposed by the German engineer A. Palmgren, [165], in the mid-1920s and independently but later by the American engineer, M.A. Miner, [149], working with aluminium alloys specimens. The Palmgren–Miner hypothesis states that a system which is subjected to n stress cycles of stress amplitude S will have suffered a fractional damage equal to n/N where N is the number of cycles required to 'break' the system at stress level S according to Wöhler's fatigue curve. It is also assumed that the fractional damage of this stress level can be added to the corresponding fractional damage of other stress levels if several separate stress cycles can be identified in the system response. If the system response can be broken down into groups of harmonic cycles with n_i cycles occurring at stress levels S_i, $i = 1, 2, 3, \ldots$, which would each need N_i constant cycles to break or incur fatigue failure in the material (see Fig. 4.25), the hypothesis states that the total damage produced is the sum of the fractional damages n_i/N_i for each group. Failure of the system is then expected when the accumulated sum of all such fractional damages reaches unity, that is,

$$d_i = \frac{n_i}{N_i}, \qquad D = \sum_i d_i = 1 \tag{4.89}$$

where d_i is the fractional damage and D is the damage index, $0 \leqslant D \leqslant 1$, a 100% damage or total failure corresponding to $D = 1$. As the hypothesis makes no assumption or restriction regarding the order of application of the various stress cycles, it can be applied to a random process where the stress amplitudes may change from cycle to cycle.

The Palmgren–Miner hypothesis has many shortcomings, and many different modifications have been suggested (see for instance Collins, [38], and Narayan and Roberts, [150]). Its popularity and strength lies in its simplicity even if it does not adequately fit experimental data very well in many cases. Marco and Starkey, [13], proposed that the fractional damage Eq. (4.89), be expressed by an exponent $m(s)$, which is thought to be a non-decreasing function of the stress level, and $0 < m(s) \leqslant 1$, i.e.

$$d_i = \left(\frac{n_i}{N_i}\right)^{m(s)}, \qquad D = \sum_i d_i = D_{\text{crit}} \tag{4.90}$$

As before, failure is defined when the accumulated damage index D reaches a certain critical value, which can be in the range $0.5 \leqslant D_{crit} \leqslant 2.0$. If the exponent $m(s)$ is independent of the stress level, the hypothesis reduces to a modified Palmgren–Miner relation with a constant exponent.

A different approach to cumulative damage was proposed by Henry, [82]. He introduces the original fatigue limit or endurance limit E_0 of the unloaded virgin material. This fatigue limit will be reduced after repeated loading because of accumulated damage to, say, E, and the cumulative damage is defined as

$$D = \frac{E_0 - E}{E_0} \tag{4.91}$$

The S–N curve is given a new form, that is,

$$N = \frac{k}{S - E} \tag{4.92}$$

in which k is a constant depending on the material, and postulated to be proportional to the new endurance limit E for any degree of damage that has accumulated in the specimen. If the stress level is increased by, say, a constant C, the endurance limit is decreased by the same constant, that is $E^* = E/C$. The number of cycles N to fatigue is also reduced by C, that is,

$$N = \frac{k}{C(S - E)} = \frac{k^*}{S - E}$$

whereby

$$\frac{k}{k^*} = \frac{E}{E^*} \tag{4.93}$$

in which k^*, E^* correspond to a new damage state, and the S–N curve be shifted accordingly. If n overstress cycles are applied at stress level S, the endurance life or number of cycles to fatigue becomes $N - n$, i.e.

$$N - n = \frac{k^*}{S - E^*} \tag{4.94}$$

Now, combining Eqs. (4.91)–(4.94), the fatigue or endurance limit E is given by

$$E = \frac{S(1 - n/N)}{\left(\frac{S - E_0}{E_0}\right) + \left(1 - \frac{n}{N}\right)} \tag{4.95}$$

and the fractional or cumulative damage is given by

$$D = \frac{(n/N)}{1 + \left(\frac{E_0}{S - E_0}\right)\left(1 - \frac{n}{N}\right)} = \frac{\beta}{1 + \frac{1 - \beta}{\gamma}} \tag{4.96}$$

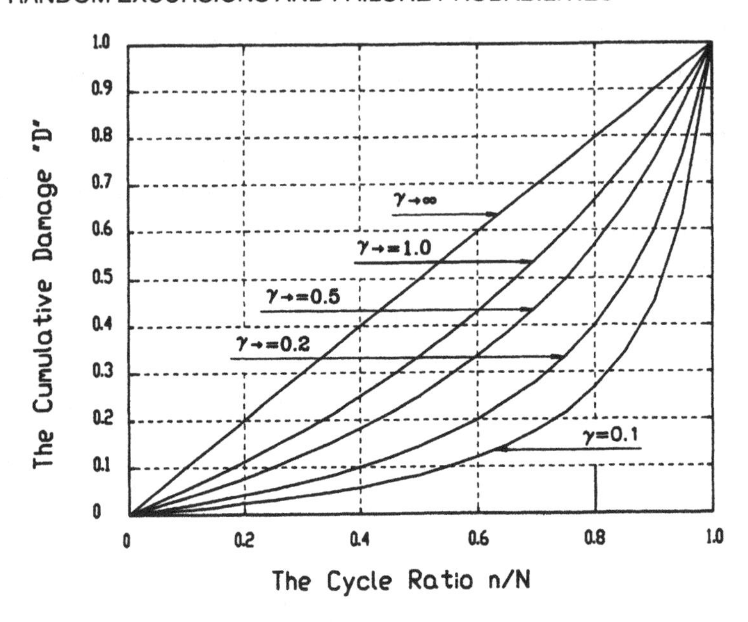

Fig. 4.26 *Cumulative damage as a function of cycle ratio n/N, (after Henry, [82]), reproduced by permission of ASME*

where, in Henry's original notation, $\beta = n/N$ and $\gamma = (S - E_0)/E_0$ is the overstress ratio. In Fig. 4.26 the cumulative damage is shown as a function of the cycle ratio, $\beta = n/N$, for overstress ratios $\gamma = 0.1, 0.2, 0.5$ and 1.0.

4.5.2 Accumulation of Damage due to Random Excitation

Assume that a lightly damped resonant system is subjected to a 'dose' of random load of duration T. Even if the stochastic load process is a broad-band process, the response processes will be narrow-band if the quality factor of the system is sufficiently large, $Q \geqslant 10$, Eq. (3.57). A narrow-band stress process that has been obtained as the response of a resonant system subjected to a broad-band excitation will have on the average $\nu_0^+ T$ stress cycles in the interval of observation or duration T. By Eq. (4.28), the probability that the peak values are in the stress range $(s, s + \Delta s)$ is $p_p(s)\Delta s$. Therefore, on the average, $(\nu_0^+ T)\,(p_p(s)\Delta s)$ peaks reach the stress level s during the time T. For each such peak, a corresponding stress cycle will cause an incremental damage, which can be denoted $\delta(s)$, according to the cumulative damage hypothesis discussed in the last section. Therefore, the average cumulative damage due to stress cycles with peaks in the range $(s, s + \Delta s)$ is $(\nu_0^+ T)\,\delta(s)\,p_p(s)\Delta s$. By summing up all stress cycles at all possible stress levels, the average or expected total damage resulting from all stress peaks during the time interval T is

$$E[D(S)] = \int_{s=0}^{\infty} \mathrm{d}D(S) = (\nu_0^+ T) \int_0^{\infty} \delta(s)\,p_p(s)\mathrm{d}s \tag{4.97}$$

Now, using the Palmgren–Miner hypothesis Eq. (4.89) and the S–N curve, Fig. 4.25, the incremental damage $\delta(s)$ can be expressed as

$$\delta(s) = \frac{1}{N(s)} = \frac{s^{\alpha}}{c} \tag{4.98}$$

whereby the expected total damage during the time T is given by

$$E[D(S)] = \frac{(\nu_0^+ T)}{c} \int_0^{\infty} s^{\alpha} p_{\mathrm{p}}(s) \mathrm{d}s \tag{4.99}$$

It is reasonable to assume that in Eq. (4.98), the average accumulated damage rate δ per unit time is

$$E[\delta] = E[D(S)]/T, \qquad D(S) = \delta T$$

and (4.100)

$$E[\delta]/\nu_0^+$$

is the accumulated average damage per unit time and per cycle.

If the random loading is sustained for a prolonged duration T, the average value of the damage, Eq. (4.90), reaches the value of unity and the system fails, (cf. Eq. (4.89)). Eq. (4.99) can therefore be used to calculate the expected lifetime of a resonant system when subjected to bursts of random excitation by setting $E[D(S)]$ equal to one, that is,

$$T = \frac{c}{\nu_0^+ \int_0^{\infty} s^{\alpha} p_{\mathrm{p}}(s) d} \tag{4.101}$$

This calculation, however, is subject to statistical errors, which have been assessed by Crandall and Mark, [43]. Also, experimental errors are involved due to ignorance of the true mechanism of fatigue. In practice, it can be assumed that provided the bandwidth of the resonant response peak is not too small and the number of cycles to failure is sufficiently large, ($\lambda N > 10^3$), the experimental error is the significant one. The lifetime can therefore be expected to lie within the range of values of the order $0.3\,T$ to $3\,T$, where T is the value given by Eq. (4.101), Crandall, [46], [47].

If the stress process is narrow-band Gaussian, the peak distribution is the Rayleigh distribution given by Eq. (4.31). In this case, an explicit expression for the average accumulated damage can be obtained. By inserting the Rayleigh density, Eq. (4.99) becomes

$$\begin{aligned} E[D(S)] &= \frac{(\nu_0^+ T)}{c} \int_0^{\infty} s^{\alpha} \frac{s}{\sigma_S^2} \mathrm{e}^{-s^2/2\sigma_S^2} \mathrm{d}s \\ &= \frac{(\nu_0^+ T)}{c} \sigma_S^{\alpha} 2^{\alpha/2} \int_0^{\infty} \left(\frac{s^2}{2\sigma_S^2}\right)^{\alpha/2} \mathrm{e}^{-S^2/2\sigma_S^2} \mathrm{d}\left(\frac{s^2}{2\sigma_S^2}\right) = \frac{(\nu_0^+ T)}{c} \sigma_S^{\alpha} 2^{\alpha/2} \Gamma\left(\frac{\alpha+2}{2}\right) \end{aligned} \tag{4.102}$$

which gives the expected total damage during the time interval T in a closed form.

Example 4.9 A linear elastic system is governed by the simple equation of a linear inverted pendulum, (cf. Examples 3.3 and 3.7). The system is lightly damped with $\lambda = 0.01$ and has a natural undamped frequency of 50 Hz. The material in the elastic spring member, resisting the motion, is known to have a Wöhler curve that can be represented by the equation $(N/N_1) = (S/S_1)^{8.0}$. This system has to be designed to be able to function without a fatigue failure of the elastic spring member when subjected to a dose of random excitation of 5 minutes duration in the form of a forced random acceleration of the base of the structure, $\ddot{x}(t)$. The acceleration process $\ddot{X}(t)$ has an experimental spectral density $W(f)$, which is constant equal to $0.3\, g^2/\text{Hz}$ throughout the frequency range 2–2400 Hz.

A prototype of the system has been built and it is desired to test it to see if it meets the specifications. If the system is tested on a shaking table, what is the r.m.s. acceleration level that has to be supplied? Alternatively, the system could be tested using a simple sinusoidal shaker, that could supply a constant acceleration amplitude up to the level of $10\, g$ to the base of the structure of the system at any chosen frequency close to the 50 Hz frequency of the system. In this case, estimate the required amplitude level of the shaker at 50 Hz to produce the same amount of damage in the system in 5 minutes as it would receive according to the Palmgren–Miner hypothesis from the specified random excitation.

Solution The system equation can be written as follows:

$$\ddot{Y} + 2\lambda\omega_0 \dot{Y} + \omega_0^2 Y = \begin{cases} -\ddot{X} & \text{case (a)} \\ g_0 \cos(2\pi f_0 t + \varphi) = G e^{i2\pi f_0 t} & \text{case (b)} \end{cases}$$

In case (a), the r.m.s. level of the shaking table is simply $(0.3(2400 - 2))^{1/2} = 26.8\, g$, which produces a mean square value of the system response, $Y(t)$, which is

$$\sigma_Y^2 = \frac{W_0}{8\lambda(2\pi f_0)^3} = \frac{0.3 \times 9810^2}{8 \times 0.01(2\pi 50)^3} = 11.6258\,\text{mm}^2 \qquad \text{or} \qquad \sigma_Y = 3.41\,\text{mm}$$

In case (b), a 'sinusoidal' input with amplitude g_0 (as a multiple of g).

$$Y(t) = H(\omega) G e^{i2\pi f t} \Rightarrow Y_G(f)\cos(2\pi f + \theta)$$

$$Y_G(f_0) = \left(\frac{g_0}{|(\omega^2 - \omega_0^2) + i2\lambda\omega\omega_0|} \right)_{\omega=\omega_0} \Rightarrow Y_G(f_0) = \frac{g_0}{2\lambda\omega_0^2}$$

whereby

$$Y_G(f_0) = \frac{g_0 g}{2 \times 0.01(100\pi)^2} = 4.97 g_0 \qquad (g = 9810\,\text{mm/s}^2)$$

In both cases (a) and (b), the stress induced in the spring-like element is proportional to the relative displacement Y or $S = kY$ where k is a constant (the same for both cases) which only depends on the system configuration. Therefore,

$$\text{case (a): } \sigma_S = k\sigma_Y \qquad \text{and} \qquad \text{case (b): } S = kY$$

The average damage produced in the system by a Gaussian narrow-band stress process,

$S(t)$, over a time T, Eqs. (4.47) and (4.49),

$$E[D(T)] = E[\delta]T = T\nu_0^+(\sqrt{2}\sigma_S)^\alpha/c\Gamma[(\alpha+2)/2]$$

where the Wöhler curve is given as $NS^\alpha = c$. Now it is proposed to produce the same amount of damage by a sinusoidal shaker, (case (b)). Therefore calculate the damage rates (damage per unit time and per cycle) for both cases and put them equal.

$$\frac{1}{N} = \frac{S^\alpha}{c} = \frac{(\sqrt{2}\sigma_S)^\alpha}{c}\Gamma\left(\frac{\alpha+2}{2}\right)$$

as $1/N$ is the damage due to one sinusoidal cycle and $E[D(T)]/(T\nu_0^+)$ the damage per unit time and cycle in the random case. Now since, $(N/N_1) = (S/S_1)^{8.0}$, $\alpha = 8$ so

$$S^8 = (\sqrt{2})^8\sigma_S^8\Gamma(5) = 2^4 \times 4!\sigma_S^8 = (2.1)^8\sigma_S^8$$

or

$$S = 2.1\sigma_S \Rightarrow Y = 2.1\sigma_Y$$

whereby

$$4.97G = 2.1 \times 3.41 \Rightarrow G = 1.44g$$

With this amplitude, the sinusoidal shaker, when tuned to resonance, $(f = f_0)$, will produce Palmgren–Miner damage in the system at the same average rate as the specified broad-band acceleration with an r.m.s. level of 26.8 g.

Example 4.10 A delicate instrument is subjected to a random load environment, which produces a random stress in a spring-like element, proportional to the relative displacement between the mass element of the instrument and its mounting points. The damping constant of the instrument is very small ($Q > 20$).

1. In a fatigue limit test of the instrument, it was found that it needed 100 000 cycles of sinusoidal stress of level 6.9 N/mm^2 to break the spring element and for stress levels less than 4.5 N/mm^2, no fatigue limit was found (there were indications however, that 2 million cycles of this last stress level could just barely break the element). Find and sketch the fatigue stress curve ("the Wöhler curve") of the spring element. What is the approximate yield stress of the material?
2. The instrument is now subjected to a stationary random excitation. It is found that the random stress produced in the spring element has the following joint density function of the stress $X(t)$ and the stress derivative $\dot{X}(t)$:

$$p(x,\dot{x}) = \frac{1}{4ab}\exp\left[-\frac{|x|}{a} - \frac{|\dot{x}|}{b}\right], \qquad -\infty < (x,\dot{x}) < \infty, \quad a,b > 0$$

   (a) Find the natural frequency of the instrument and the stress peak distribution density.
   (b) Assuming Palmgren–Miner type of damage due to the stationary random stress, find the expected fatigue life of the instrument.

(c) Determine the fatigue life time in terms of the level a, if the spring element is made of the material described in 1. If the specifications for the instruments call for a fatigue life corresponding to a 30 h dose of random load, what is the level a in N/mm^2?

Solution

1. A simplified outline of the Wöhler curve of the form $NS^\alpha = c$ is shown in Fig. 4.27, making use of the information given. The exponential power is given by

$$(10^5)(6.9)^\alpha = (4.5)^\alpha 2 \times 10^6$$

or

$$(0.652)\, \alpha = 0.05, \qquad \alpha = 7$$

The constant c can now be calculated:

$$\begin{aligned} c &= (10^5)(6.9)^7 = (4.5)^7 2 \times 10^6 \\ &= 7.46 \times 10^{10} \end{aligned}$$

so the yield level is given by

$$1 \cdot S_Y^7 = 7.46 \times 10^{10}$$

or

$$S_Y = 35.75\,\text{N/mm}^2$$

2. (a) As before, (Ex. 4.4), the stress peak distribution can be obtained from Eqs. (4.5) and (4.30). Therefore, first find the crossing frequency of a level $x = \xi$.

$$p(x, \dot{x}) = \frac{1}{4ab} \exp\left(-\frac{|x|}{a} - \frac{|\dot{x}|}{b} \right)$$

$$\nu_\xi^+ = \int_0^\infty \dot{x}\, p(\xi, \dot{x})\, \mathrm{d}\dot{x} = \frac{1}{4ab} \exp\left(-\frac{|\xi|}{a} \right) \int_0^\infty \mathrm{e}^{-\dot{x}/b}\, \dot{x}\, \mathrm{d}\dot{x}$$

$$= \frac{1}{4ab} \exp\left(-\frac{|\xi|}{a} \right) (-b^2 \mathrm{e}^{-\dot{x}/b}\, \dot{x} - b^2 \mathrm{e}^{-\dot{x}/b}) = \frac{b}{4a} \mathrm{e}^{-|\xi|/a}$$

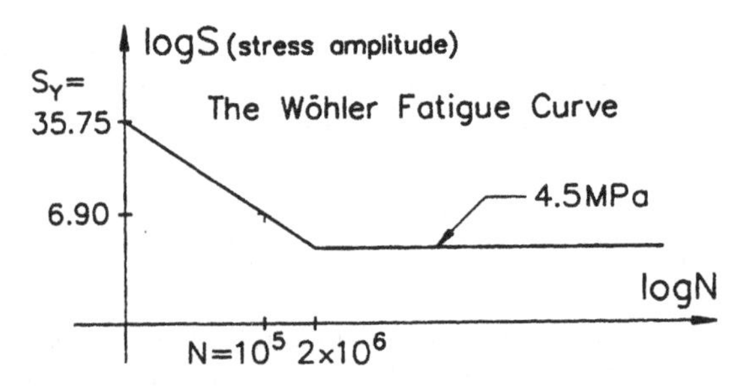

Fig. 4.27 *The Wöhler fatigue curve*

The peak density is then given by

$$p_p(\xi) = -\frac{1}{v_0^+}\frac{\mathrm{d}v_\xi^+}{\mathrm{d}\xi} = -\frac{1}{a}\mathrm{e}^{-\xi/a} \quad \text{for } \xi > 0, \qquad 0 \text{ elsewhere}$$

The main frequency of the system is obviously directly connected to the zero-crossing frequency since the response is highly narrow band, ($Q > 20$), therefore

$$v_0^+ = b/(4a), \qquad \omega_0 = 2\pi b/(4a), \qquad f_0 = b/(4a)$$

(b) From the Wöhler curve, $N\xi^\alpha = c$, where $\alpha = 7$ and $c = 7.46 \times 10^{10}$. Therefore, by Eqs. (4.46) and (4.47),

$$E[\delta] = \frac{v_0^+}{c}\frac{1}{a}\int_0^\infty \mathrm{e}^{-\xi/a}\xi^\alpha \mathrm{d}\xi = \frac{v_0^+}{c}\frac{1}{a}a^{\alpha+1}\int_0^\infty \mathrm{e}^{-z}z^\alpha \mathrm{d}z = \frac{v_0^+}{c}a^\alpha\Gamma(\alpha+1)$$

so

$$T_\mathrm{F} = \frac{1}{E[\delta]} = \frac{c}{v_0^+ a^\alpha \Gamma(\alpha+1)} = \frac{4c}{ba^{\alpha-1}\Gamma(\alpha+1)}$$

(c) Since $\alpha = 7$, $\Gamma(8) = 7 \times 6 \times 5 \times 4 \times 3 \times 2 = 5040$, $v_0^+ = f_0 = 1/T_0 = b/(4a) = 10$, the fatigue life is found as

$$T_\mathrm{F} = \frac{4 \times 10^5}{ba^{7-1}5040} = 1.984a^{-7} = \frac{4a}{b}\cdot\frac{19.84}{a^7} = 1.984a^{-7}$$

Therefore, if the instrument is to have a fatigue life corresponding to 30 h dose random excitation, the level a is found to be

$$30 \times 60 \times 60 = 1.984\,a^{-7}, \qquad a = 1.45\ \mathrm{N/mm^2}$$

4.5.3 Failure due to Random Excursions

Aside from the accumulated damage produced by adding up a random number of stress cycles of different composition, a typical cause of damage is an aberrant behaviour of the system, when it spends to much time outside a prescribed stress range or limits of acceptable system response. A special case of this kind of behaviour is a total failure or damage of the system when it crosses the limit for the first time, a so-called first passage failure problem, Fig. 4.28. This kind of situation frequently occurs in sensitive instruments and electronic equipment when subjected to a random environment. It is also possible to imagine a situation when a mechanical system will break down due to plastic deformation, having had deflections for too long beyond its elastic limits. In some cases, crossing such a level only once suffices to destroy the structure. The treatment of the above and related problems can lead to very complicated analysis, and closed form solutions are very difficult to obtain, see Ditlevsen, [54]. If the problem can be treated as a Markov diffusion process (see Section 6.3.1), closed form solutions can be obtained by interpreting the first passage time as corresponding to an absorbing boundary of the outside backward

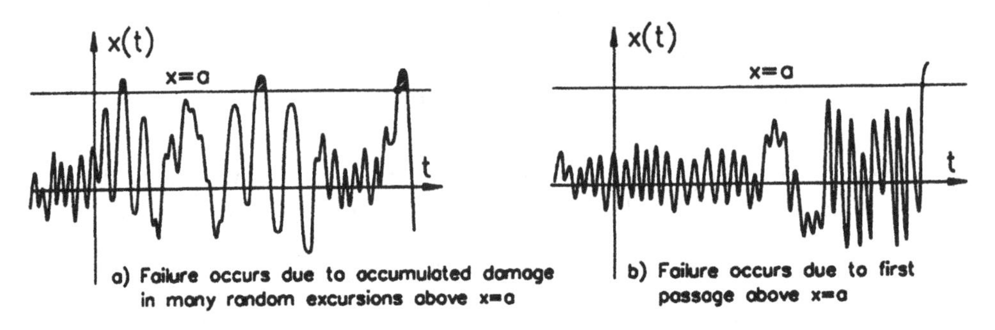

Fig. 4.28 *Failure modes of vibratory systems under random excitation, (after Newland, [61]), reproduced by permission of Addison Wesley Longman Ltd*

Kolmogorov equation, Eq. (6.103). For a full discussion, the reader is referred to Lin and Cai, [123]. It can be stated, however, that failure of vibratory systems due to fatigue and random excursions above and beyond prescribed levels, is a field that needs further investigation and research effort. In the following, a simple and limited interpretation of such random excursion failure problems is presented following Crandall and Mark, [43].

Consider a physical system, which has the response sample function $y(t)$ that is ergodic and is observed in a very long sample time interval T. Let a be a fixed positive threshold value. The system need not fail even if the response occasionally crosses the threshold a (an isolated malfunction). However, if the response variable $y(t)$ stays above the threshold a for sufficiently long time, failure may occur due to accumulated damage. Therefore, system failure is defined to occur when the fraction of time for which $y(t) > a$ is greater than a preset fraction ε. In other words, defining the fraction of elapsed time above a as FET (a), failure occurs when FET $(a) > \varepsilon$. Now, the proportion of time the response $y(t)$ spends above the limit a, that is, the fraction of elapsed time of excursions above the critical threshold divided by the total time, is simply given by the expression

$$\mathrm{FET}(a) = \int_a^\infty p_Y(y)\mathrm{d}y \tag{4.103}$$

in which $p_Y(y)$ is the probability density function of the response process. If the response process is a Gaussian process with zero mean and standard deviation σ_Y, then the fraction of elapsed time above a is

$$\begin{aligned}\mathrm{FET}(a) &= \int_a^\infty p_Y(u)\mathrm{d}u = \frac{1}{2} - \frac{1}{\sigma_Y\sqrt{2\pi}}\int_0^a \mathrm{e}^{-u^2/2\sigma_Y^2}\mathrm{d}u \\ &= \frac{1}{2}\left[1 - \mathrm{Erf}\left(\frac{a}{\sigma_Y}\right)\right]\end{aligned} \tag{4.104}$$

using Eqs. (4.103) and (1.47), and the error function defined by Eq. (1.124). For large values of the critical level a, that is, $\sigma_Y \gg a$, (for $\sigma_Y = a/3$, the error is of the order 10), the error function can be simplified in the following manner. For large values of x/σ, the integral

$$\frac{2}{\sigma\sqrt{2\pi}}\int_0^x \exp\left[-\frac{u^2}{2\sigma^2}\right]du$$

$$\frac{2}{\sigma\sqrt{2\pi}}\int_0^x \exp\left[-\frac{u^2}{2\sigma^2}\right]du \simeq 1 - \sqrt{\frac{2}{\pi}}\frac{\sigma\exp[-x^2/2\sigma^2]}{x}\left(1 - \frac{\sigma^2}{x^2} + \frac{1\times 3\sigma^4}{x^4}\right)$$

By retaining the first term only, $(\sigma^2/x^2 \approx 0)$,

$$\mathrm{Erf}\left(\frac{a}{\sigma_Y}\right) \simeq 1 - \sqrt{\frac{2}{\pi}}\frac{\sigma_Y}{a}\,\mathrm{e}^{-a^2/2\sigma_Y^2}$$

whereby the fraction of elapsed time above the critical level is given by

$$\mathrm{FET}(a) \simeq \frac{1}{\sqrt{2\pi}}\frac{\sigma_Y}{a}\,\mathrm{e}^{-a^2/2\sigma_Y^2} \qquad \text{for } a \gg \sigma_Y \tag{4.105}$$

Based on this expression, the fraction of elapsed time spent above the threshold a for different values of a/σ_Y is easily computed. For instance, if $a/\sigma_Y = \{3.09, 3.72, 4.27\}$, then $\mathrm{FET}(a) = \{0.001, 0.0001, 0.00001\}$. Thus, if the present failure fraction $\varepsilon = 0.001$, all Gaussian ergodic response processes with $\sigma_Y > 3.09$ would cause the system to fail whereas responses with smaller r.m.s. values would be safe. For this particular failure model, the response level causing failure is thus adequately described by a single parameter σ_Y in the case of Gaussian ergodic response condition. Finally, it should be mentioned that the basic relation Eq. (4.103) is only strictly correct for an infinite time interval T. For a finite time interval T, the fraction $\mathrm{FET}_T(a)$ becomes a random variable which assumes different values for different sample response functions. The ensemble average is given by Eq. (4.103), that is,

$$E[\mathrm{FET}_T(a)] = \mathrm{FET}(a) \tag{4.106}$$

To unravel the distribution of the random variable $\mathrm{FET}_T(a)$ seems to be a difficult problem with an elusive solution. Qualitative considerations indicate that the fluctuations in the time spent above the threshold grow in proportion to $\sqrt{T}$ as the interval length T increases and the fluctuations in $\mathrm{FET}(a)$ would therefore decrease in proportion to $1/\sqrt{T}$.

In the above discussion, the system failed due to accumulation of damage as the response $y(t)$ stayed too long above the threshold $y(t) = a$. Now, failure may occur the first time the response passes the threshold, which can be either negative or positive. This kind of system behaviour relates to the first exceedance problem in probability theory, which requires the knowledge of the first passage distribution or density $p(T_1)$ where T_1 is the first exceedance time from the starting of operation. The determination of the first exceedance distribution constitutes one of the more difficult problems in extreme value statistics, and a precise solution has turned out be elusive, [54]. One of the problems with first passage time failures is that the process may not have reached stationary conditions before it crosses the level. Disregarding this possibility, consider the behaviour of the system in the time interval of length T. For any preset level $x = a$, there will be an average number of up-crossings of that level equal to $\nu_a^+ T$. For a Gaussian random process, the uncrossing rate is given by Eq. (4.7); however, it is not immediately clear how to relate this information to the first up-crossing of the threshold. If the level a is set larger and larger, there will be fewer and fewer up-crossings until only one or no crossing of the level is likely to occur. Formally, the probability of first passage or up-crossing of the level $y = a$ during the time interval T, called $P_1(T)$, can be expressed as

$$P_1(T) = P[T_1 \leqslant T] = 1 - \int_T^\infty p_1(t)\mathrm{d}t \tag{4.107}$$

where T_1 is the time to the first up-crossing and $p_1(t)$ is the probability density of T_1. Take the up-crossing of a to be a rare event. In this case it may be assumed to be a Poisson event with the intensity ν_a^+ (cf. Section 6.2, Eq. (6.33)), whereby the probability density of the Poisson event is given by (Example 6.4)

$$p_1(t) = \nu_a^+ \exp[-\nu_a^+ t] \tag{4.108}$$

Therefore,

$$P_1(T) = 1 - (\nu_a^+) \int_T^\infty \mathrm{e}^{-(\nu_a^+ t)}\,\mathrm{d}t = 1 - \mathrm{e}^{-(\nu_a^+ T)} \tag{4.109}$$

which is the probability of failure corresponding to at least one up-crossing taking place in the time interval T, disregarding the possibility that for a small but finite number of sample responses, failure may occur immediately at $t = 0$. If the response process is Gaussian, then the up-crossing frequency is given by Eq. (4.7), or

$$\nu_a^+ = \nu_0^+ \exp[-a^2/(2\sigma_Y^2)] \tag{4.7}$$

and the probability of failure is given by

$$P_1(T) = 1 - \mathrm{e}^{-(\nu_a^+ T)} = 1 - \exp\left[-\nu_0^+ T \exp\left(-\frac{a^2}{2\sigma_Y^2}\right)\right] \tag{4.110}$$

The above result is strictly valid for ergodic processes only and should be treated with caution. For instance, since large amplitude peaks usually follow one another in clusters, the interval between such clusters will be longer than the average time between crossings and the probability of first crossing will be less than that given by Eqs. (4.109) and (4.110). For instance, the mean value of the time to failure is obtained from the density, Eq. (4.108),

$$E[T_1] = \nu_a^+ \int_0^\infty t\,\mathrm{e}^{-(\nu_a^+ t)}\,\mathrm{d}t = \frac{1}{\nu_a^+} \tag{4.111}$$

and the variance is given by

$$\sigma_{T_1}^2 = \nu_a^+ \int_0^\infty t^2\,\mathrm{e}^{-(\nu_a^+ t)}\mathrm{d}t - \left(\frac{1}{\nu_a^+}\right)^2 = \frac{1}{(\nu_a^+)^2} \tag{4.112}$$

so the standard deviation and the mean value of the first passage time are equal.

According to the above model, (cf. Eq. (4.109)), failure always occurs if the time interval T is sufficiently long. By restricting the operation time of the system, the failure probability can be made as small as required. For instance, the probability of failure in the time interval $0 < t < T_0$ is given by Eq. (4.109) as

$$P_1(T_0) = 1 - \exp(-\nu_a^+ T_0) = 10^{-n}$$

in which the order of n relates to the acceptable system reliability. Therefore, by restricting the operation time such that

$$T_0 \leqslant \log_e(1 - 10^{-n})$$

the risk of first passage failure can be reduced to acceptable limits.

The assumption of the first exceedance times to be Poisson events requires that the different time instances t_j of up-crossings of the threshold $y = a$ are statistically independent. For a narrow-band process this is unlikely as there is a tendency for clusters of peaks of similar magnitudes, i.e. the envelope is slowly varying. The number of cycles per cluster will be larger as the frequency bandwidth of the process becomes narrower. For a fixed bandwidth, however, the level a can be set high enough so that on the average there will be only one up-crossing per cluster. In this case, the Poisson assumption may be a good assumption. On the whole, the Poisson assumption is on the conservative side as it predicts failure sooner than a model which takes into account the clumping of the up-crossings.

Example 4.11 A schematic diagram of an electrical relay is shown in Fig. 4.29. The moving contact of the relay acts as a cantilever beam when it is in the open position. By testing it has been found out that the natural frequency of this element is 100 Hz and the resonance peak has a quality factor $Q = 10$.

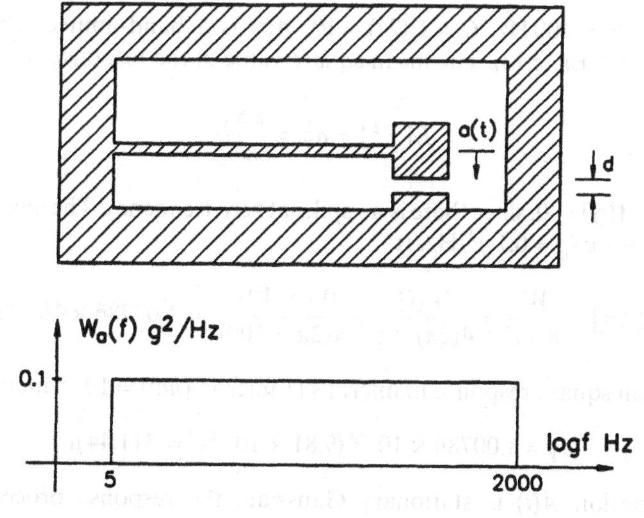

Fig. 4.29 *An electrical relay under random loading*

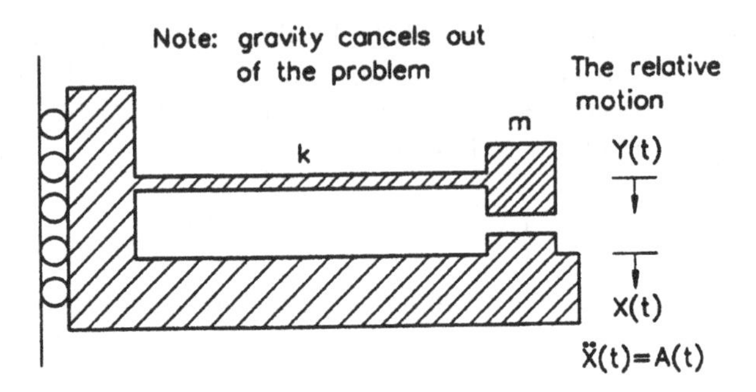

Fig. 4.30 *The relative motion of the relay*

Estimate the minimum clearance between the 'hammer' and the 'anvil' if the relay is to have a reliability $R = 0.999$ against accidental contact during two minutes of exposure to a random excitation, which produces an acceleration of the hammer in the direction shown that is a stationary Gaussian stochastic process with an experimental spectral density shown in Fig. 4.29.

Solution The system is essentially the same as an inverted elastic pendulum with a forced motion of the base of the pendulum, Fig. 4.30. Assuming that the damping is velocity/viscous with the damping constant c, the equation of motion of the hammer is

$$m(\ddot{Y} - \ddot{X}) + c\dot{Y} + k\dot{Y} = 0$$

or

$$\ddot{Y} + 2\lambda\omega_0\dot{Y} + \omega_0^2 Y = \ddot{X} = A(t)$$

Given: $f_0 = \omega_0/2\pi = 100\,\text{Hz}$, $Q = 1/(2\lambda) = 10$, $A(t)$ is a band-limited white noise with $W_a(f)$ as shown (cf. Ex. 2.13). The mean square value of the response is

$$E[Y^2] = \sigma_Y^2 \cong \frac{\pi S_0}{2\lambda\omega_0^3}$$

$E[Y] = 0$ as $E[A(t)] = 0$, since there is no spike at zero frequency. The mean square value, noting that $W_0 = 4\pi S_0 = 0.1\,g^2/\text{Hz}$, is

$$E[Y^2] = \frac{W_0}{8\lambda\omega_0^3} = \frac{W_0 Q}{4(2\pi)^3 f_0^3} = \frac{0.1 \times 10\,g^2}{4(2\pi \times 100)^3} = 1.00786 \times 10^{-9}\,g^2$$

so the root mean square response in microns (1 micron (μm) $= 10^{-6}$ metres),

$$\sigma_Y = 1.00786 \times 10^{-9}\,(9.81 \times 10^{-6})^2 = 311.44\,\mu$$

Since the excitation $A(t)$ is stationary Gaussian, the response process $Y(t)$ is also stationary Gaussian (cf. Section 6.1).

The operation time T_s of the system is 2 minutes. The reliability $R = 0.999$ is the probability of not having a contact during the 2 minutes, called $P_s(T_s)$, where the probability of first passage or down-crossing of level $y = -d$ during the operation time T_s, Eq. (4.96), is

$$P[T_1 \leqslant T_s] = 1 - P[T_1 > T_s] = 1 - \int_{T_s}^{\infty} p_1(t)\,dt = 1 - P_s(T_s)$$

and the crossing frequency, Eq. (4.7), is

$$\nu_d^- = \nu_0^- \exp[-d^2/(2\sigma_Y^2)]$$

Therefore,

$$P_s(T_s) = (\nu_d^-)\int_{T_s}^{\infty} e^{-(\nu_d^- t)}\,dt = e^{-(\nu_d^- T_s)} = \exp\left[-\nu_0^- T \exp\left(-\frac{d^2}{2\sigma_Y^2}\right)\right] = R$$

or

$$\log_e R = -\nu_0^- T_s \exp[-d^2/(2\sigma_Y^2)\,d^2/(2\sigma_Y^2)] = -2\log_e R[-\log_e R/(\nu_0^- T_s)]$$

Now, with $R = 0.999$, $T_s = 120\text{s}$ and $\nu_0^- = 100\,\text{Hz}$ ('narrow-band condition')

$$d/(2\sigma_Y) = 5.71$$

which gives $d = 1.78\,\text{mm}$

5 Random Excitation and Response of Multiple and Continuous Systems

Dynamical analysis of complex structural systems has developed rapidly during the last fifteen years with ever increasing computer power and extensive software packages for structural analysis, which have become freely available. Before the computer era, dynamics and vibration of structures was a highly specialized field, which was only being studied by a select few engineers, whereas most structures were being analysed using statically equivalent methods. Such so-called static equivalent design methods often involved the determination of the fundamental frequency of the structural system together with its first vibration form, which was obtained by using rough approximative methods. As for dynamical analysis of structural systems, the equations of motions were rarely solved directly through numerical analysis because of lack of computing power. The standard procedure was to obtain an analytic solution of the equations of motion, which required that many restrictions had to be imposed on the mathematical model for the dynamical behaviour of the structure. In most cases, linear behaviour would be presupposed in order to have a system of linear differential equations governing the response of the system to external dynamical excitation. Linear analysis would encompass the determination of the classical normal vibration modes of the system and its corresponding eigenvalues or frequencies. If the system could be described as a multiple discrete mass system, that is, a system having a finite number of moving mass points with concentrated mass, the normal modes-equal in number to the number of discrete mass points times the degrees of freedom of motion at each-point are easily computed using matrix methods. The solution or calculation of any response quantity of the system (deflection, rotation, stress etc.) is then given as a linear combination or a weighted sum of all the normal modes of the system. For linear continuous mass system, that is, the mass distribution is continuous rather than discrete, a normal mode approach is still viable resulting in an infinite number of modes and frequencies, which combined together, forming an infinite sum, yield the "solution", that is any response quantity required. A number of excellent textbooks is available describing the above classical methods of which a few are listed in the bibliography and reference list, that is, Clough and Penzien [37], Fertis [65], Klotter [109], Jacobsen and Ayre [90], Norris *et al.*, [157], Rogers [181], Langen and Sigbjörnsson [119] and Thomson [220].

In the following a short overview of the classical methods of dynamical analysis will be presented as they still form the basis for the understanding of the dynamical behaviour of structural systems. Moreover, some mention will be made of more modern methods involving direct numerical integration of the equations of motions utilizing large software packages such as ALGOR, [6].

5.1 DISCRETE LINEAR SYSTEMS. THE NORMAL MODE APPROACH

Consider a linear n-degree of freedom vibratory system with a consistent mass matrix $\mathbf{m}$, damping matric $\mathbf{c}$ and stiffness matrix $\mathbf{k}$. When the system is subjected to random force $\mathbf{F}(t)$, the equations of motion describing any system response characteristic, for instance the random displacement, can be written in the following matrix form:

$$\mathbf{m}\ddot{\mathbf{Y}} + \mathbf{c}\dot{\mathbf{Y}} + \mathbf{k}\mathbf{Y} = \mathbf{F}(t) \tag{5.1}$$

where $\mathbf{Y}(t)$ is the random displacement response vector of the system to the random excitation vector $\mathbf{F}(t)$ which in the following is assumed to have stationary elements only.

Equation (5.1) can be solved using Lagrange's generalized coordinates or by breaking the system up into its normal modes. Consider damping matrices only that can be written as:

$$\mathbf{m}^{-1}\mathbf{c} = \sum_i a_i(\mathbf{m}^{-1}\mathbf{k})^i \tag{5.2}$$

where the a_i's are any set of real constants. Since the matrix $\mathbf{m}^{-1}\mathbf{c}$ is formed as a polynomial of the matrix $\mathbf{m}^{-1}\mathbf{k}$ they posses the same set of eigenvectors, and their eigenvalues are related by the same polynomial. Calling the ith eigenvector $\mathbf{V}_i$ and its transpose $\mathbf{V}_i^{\mathrm{T}}$, the following relations are all easily established,

$$(\mathbf{k} - \mathbf{m}\omega_1^2)\mathbf{V}_i = \mathbf{0} \tag{5.3}$$

$$\mathbf{V}_i^{\mathrm{T}}\mathbf{m}\mathbf{V}_j = \begin{cases} M_i & i=j \\ 0 & i \neq j \end{cases}, \quad \mathbf{V}_i^{\mathrm{T}}\mathbf{c}\mathbf{V}_j = \begin{cases} C_i & i=j \\ 0 & i \neq j \end{cases}, \quad \mathbf{V}_i^{\mathrm{T}}\mathbf{k}\mathbf{V}_j = \begin{cases} K_i & i=j \\ 0 & i \neq j \end{cases} \tag{5.4}$$

Hence, introducing the transformation

$$\mathbf{Y}(t) = \mathbf{V}\mathbf{Q}(t) = \sum_{i=1}^{n} \mathbf{V}_i\mathbf{Q}_i(t) \tag{5.5}$$

into Eq. (5.1), where $\mathbf{V}$ is the matrix of the eigenvectors, column-wise, and $\mathbf{Q}(t)$ are the generalized coordinates, and premultiplying the equation by $\mathbf{V}^{\mathrm{T}}$, that is, the transposed eigenvector matrix (the eigenvectors are now row-wise), the following equation emerges:

$$\mathbf{V}_i^{\mathrm{T}}\mathbf{m}\mathbf{V}\ddot{\mathbf{Q}}(t) + \mathbf{V}^{\mathrm{T}}\mathbf{c}\mathbf{V}\dot{\mathbf{Q}}(t) + \mathbf{V}^{\mathrm{T}}\mathbf{k}\mathbf{V}\mathbf{Q}(t) = \mathbf{V}^{\mathrm{T}}\mathbf{F}(t) \tag{5.6}$$

Making use of the orthogonality conditions, Eq. (5.4), Eq. (5.6) collapses to a more simple form; it is said to be uncoupled in n-space:

$$\lceil M_i \rfloor\ddot{\mathbf{Q}}(t) + \lceil C_i \rfloor\dot{\mathbf{Q}}(t) + \lceil K_i \rfloor\mathbf{Q}(t) = \mathbf{V}^{\mathrm{T}}\mathbf{F}(t) \tag{5.7}$$

where $\lceil A_i \rfloor$ designates a diagonal matrix with zero off-diagonal elements. The ith normal coordinate $Q_i(t)$ therefore satisfies the equation

$$M_i\ddot{Q}(t) + C_i\dot{Q}(t) + K_iQ(t) = \mathbf{V}_i^{\mathrm{T}}\mathbf{F}(t) = P_i(t) \tag{5.8}$$

where

$$P_i(t) = \mathbf{V}_i^T\mathbf{F}(t) = (V_{i1}F_1(t) + V_{i2}F_2(t) + \cdots + V_{in}F_n(t)) \tag{5.9}$$

is the generalized force.

The generalized or normal coordinate $Q_i(t)$ is obtained by Eq. (5.8),

$$Q_i = \frac{1}{M_i}\int_{-\infty}^{\infty} P_i(u)h_i(t-u)\mathrm{d}u \tag{5.10}$$

where

$$h_i(t) = \begin{cases} \dfrac{1}{\overline{\omega}_i}\exp(-\lambda\omega_i)\sin\overline{\omega}_i t & t \geqslant 0 \\ = 0 & t > 0 \end{cases} \tag{5.11}$$

is the ith mode impulse response function in which

$$\omega_i^2 = K_i/M_i, \qquad 2\lambda_i\omega_i = C_i/M_i, \qquad \overline{\omega}_i = \omega_i\sqrt{1-\lambda^2} \tag{5.12}$$

Neglecting any transient terms, the solution of the system of Eqs. (5.1) is completely given by Eqs. (5.5) and (5.9).

Now, any response quantity $\mathbf{Z}(t)$ of the system (stress, displacement etc.) can be linearly related to the normal coordinates $Q_i(t)$. Thus

$$Z(t) = \sum_{i=1}^{m} B_i Q_i(t) = \sum_{i=1}^{n} Z(t) \tag{5.13}$$

where the B_i are constants and $Z_i(t)$ are called the modal responses (ith normal mode contribution to the total response). Usually the series involved in Eqs. (5.5) and (5.13) are rapidly convergent and only the few first normal mode contributions are needed.

5.1.1 Second-order Statistics of Systems with Many Degrees of Freedom

The autocorrelation function of any response quantity Eq. (5.13) can now be obtained as follows:

$$R_Z(\tau) = E[Z(t)Z(t+\tau)] = E\left[\sum_i^n\sum_j^n B_iB_jQ_i(t)Q_j(t+\tau)\right]$$

$$= \sum_i^n\sum_j^n E[B_iB_jQ_i(t)Q_j(t+\tau)] = \sum_i^n\sum_j^n R_{Z_i}R_{Z_j} \tag{5.14}$$

where the cross-correlation functions are given by the following double integral:

$$R_{Z_iZ_j} = E[Z_i(t)Z_j(t+\tau)] = \frac{B_i}{M_i}\frac{B_j}{M_j}\int_{-\infty}^{\infty}\int_{-\infty}^{\infty} R_{P_iP_j}(\tau-(v-u))h_i(u)h_j(v)\mathrm{d}u\,\mathrm{d}v \tag{5.15}$$

For a system having well-separated frequencies and light damping of all modes, the response process $Z_i(t)$ of the ith mode will be close to being statistically independent from the response process $Z_j(t)$ of the jth mode. Therefore, the autocorrelation function of the total response can usually be approximated by the relation

$$R_Z(t) \simeq \sum_{i}^{n} R_{Z_i Z_j}(\tau) \tag{5.16}$$

involving the autocorrelation functions of the modal responses only. The variance σ_Z^2 is obtained for $\tau = 0$ in Eq. (5.16), that is,

$$\sigma_Z^2 = (\sigma_1^2 + \sigma_2^2 + \cdots + \sigma_n^2)$$

or

$$\text{RMS}[Z] = ((\text{RMS}[Z_1])^2 + (\text{RMS}[Z_2]^2 + \cdots + (\text{RMS}[Z_n]^2)^{1/2} \tag{5.17}$$

which is called the root mean square evaluation or prediction of the maximum response Z_{max}. Finally by Eq. (5.9), the autocorrelation function

$$R_{P_i P_j}(\tau) = \sum_{r=1}^{n} \sum_{s=1}^{n} V_{ir} V_{js} R_{F_r F_s}(\tau) = \mathbf{V}_i^{\mathrm{T}} R_F \mathbf{V}_j \tag{5.18}$$

where

$$\mathbf{R}_{\mathbf{F}} = \{R_{F_r F_s}(\tau)\} \tag{5.19}$$

is the correlation matrix of the random force vector $\mathbf{F}(t)$. Thus, Eqs. (5.18) and (5.19) with Eqs. (5.14) and (5.15) give the desired autocorrelation function of the response quantity $\mathbf{Z}(t)$.

The mean spectral density of the response $\mathbf{Z}(t)$ is obtained through the relation

$$S_Z(\omega) = \frac{1}{2\pi} \int_{-\infty}^{\infty} R_Z(\tau) e^{-i\omega\tau} d\tau$$

or by Eq. (5.14)

$$S_Z(\omega) = \sum_{i=1}^{n} \sum_{j=1}^{n} S_{Z_i Z_j}(\omega) \tag{5.20}$$

where by Eq. (5.15), the cross-spectral density

$$S_{Z_i Z_j}(\omega) = \frac{B_i}{M_i} H_i^*(\omega) S_{P_i P_j}(\omega) H_j(\omega) \frac{B_j}{M_j} \tag{5.21}$$

in which $S_{P_i P_j}(\omega)$ is the cross-spectral density of the generalized forces and

$$H_i(\omega) = \frac{1}{\omega_i^2 - \omega^2 + i2\lambda_i \omega_i \omega} \tag{5.22}$$

are the complex frequency responses of each of the normal modes, ($H_i^*(\omega) = H_i(-\omega)$, the complex conjugate).

For lightly damped systems with well-separated frequencies, the product $H_i(\omega)H_j^*(\omega)$ is very small except when $i=j$. Therefore, approximately

$$S_Z(\omega) = \sum_{i=1}^{n} S_{Z_iZ_i}(\omega) \tag{5.23}$$

involving only the mean square spectral densities

$$S_{Z_iZ_i}(\omega) = \frac{B_i^2}{M_i^2}|\mathrm{H}_i(\omega)|^2 S_{P_iP_i}(\omega) \tag{5.24}$$

Thus by Eq. (5.9)

$$S_{P_iP_j}(\omega) = \sum_{r=1}^{n}\sum_{s=1}^{n} V_{ir}V_{js}S_{F_rF_s}(\omega) = V_i^{\mathrm{T}} S_F V_j \tag{5.25}$$

where

$$\mathbf{S}_{\mathbf{F}} = \{S_{F_rF_s}(\omega)\} \tag{5.26}$$

is the spectral density matrix of the random excitation vector $\mathbf{F}(t)$. It should be noted that when the matrices $\mathbf{R}_F(\tau)$ and $\mathbf{S}_F(\omega)$ are being established, the force matrix

$$\mathbf{F}(t)\mathbf{F}^{\mathrm{T}}(t) = \begin{bmatrix} F_1(t) \\ \vdots \\ F_1(t) \end{bmatrix} \{F_1(t) \cdots F_1(t)\}$$

has to be formed.

Example 5.1 A uniform inverted L-shaped member of mass μ per unit length and flexural stiffness EI in its plane is discretised as shown in Fig. 5.1:

$$\mathbf{P} = \mathbf{K}\boldsymbol{\Delta}, \qquad \boldsymbol{\Delta} = \mathbf{C}\mathbf{P}$$

in which $\mathbf{K}$ is the stiffness matrix and $\mathbf{C}$ is the damping matrix. $\mathbf{P}$ is a force vector and $\boldsymbol{\Delta}$ is a displacement vector, both with components in the directions 1, 2, 3 as shown.

The system is subjected to simultaneous stationary random base accelerations $a_x(t)$ and $a_y(t)$ having mean square spectral densities and cross spectral densities given by

$$S_{a_xa_x}(\omega) = S_0, S_{a_ya_y}(\omega) = 0.5S_0$$

$$S_{a_xa_y}(\omega) = S_{a_ya_x}(\omega) = C_0 S_0$$

where C_0 is a real constant.

Assuming modal damping of the uncoupled form where the ratio of critical damping in each normal mode equals a constant λ, determine the variance of the base moment $M_{\mathrm{B}}(t)$ expressed in terms of μ, L, EI, λ, S_0 and C_0. What is the range of possible numerical values for C_0?

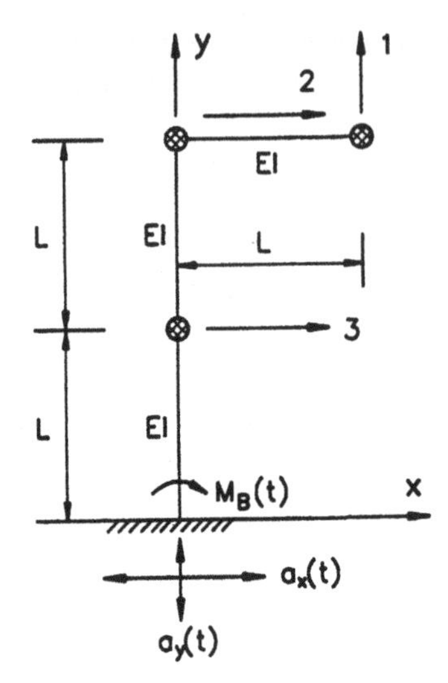

Fig. 5.1 *A mass-less uniform L-shaped beam column*

Solution

1. Compute the flexibility and mass matrices.

$$\mathbf{m} = \frac{\mu L}{2}\begin{Bmatrix} 1 & 0 & 0 \\ 0 & 3 & 0 \\ 0 & 0 & 2 \end{Bmatrix}$$

$$\mathbf{c} = \frac{L^3}{6EI}\begin{Bmatrix} 14 & 12 & 3 \\ 12 & 16 & 5 \\ 3 & 5 & 2 \end{Bmatrix}$$

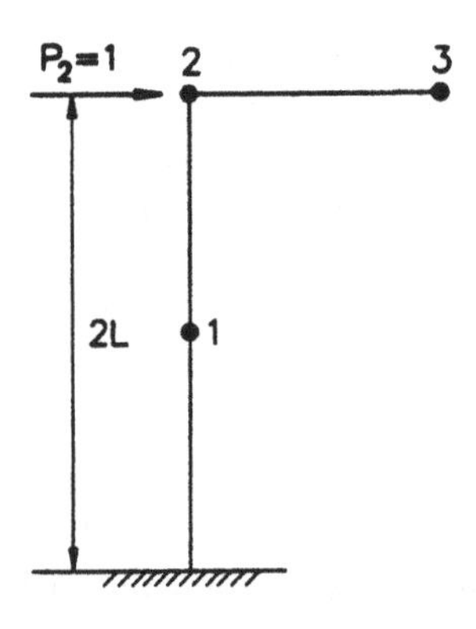

Fig. 5.2 *Determination of element C_{22}*

The mass matrix is already given. For example, to find the element C_{22} in the flexibility matrix look at

$$\Delta = \mathbf{CP} \text{ where } \mathbf{P} = \{0, 1, 0\} \Rightarrow \Delta_2 = C_{22}$$

$$\Delta_2 = \frac{1}{3}\frac{P_2(2L)^3}{EI} = 16\frac{L^3}{6EI}$$

and so forth. Therefore,

$$\mathbf{k} = \mathbf{c}^{-1} = \frac{3EI}{13L^3}\begin{Bmatrix} 7 & -9 & 12 \\ -9 & 19 & -34 \\ 12 & -34 & 80 \end{Bmatrix}$$

Solving the system of equations given by

$$(\mathbf{k} - \omega^2\mathbf{m})\mathbf{V} = \mathbf{0}$$

the following mode shapes and frequencies are obtained:

$$\mathbf{V}_1 = \begin{Bmatrix} -0.807 \\ 1.000 \\ 0.307 \end{Bmatrix}, \quad \mathbf{V}_2 = \begin{Bmatrix} -1.000 \\ -0.213 \\ -0.280 \end{Bmatrix}, \quad \mathbf{V}_3 = \begin{Bmatrix} -0.306 \\ -0.304 \\ 1.000 \end{Bmatrix}$$

$$\omega_1^2 = 0.197\frac{EI}{\mu L^4}, \qquad \omega_2^2 = 2.566\frac{EI}{\mu L^4}, \qquad \omega_3^2 = 21858\frac{EI}{\mu L^4}$$

The generalized masses, $\{\mathbf{M}_i\} = \{\mathbf{V}_i^{\mathrm{T}}\mathbf{m}\mathbf{V}_i\}$,

$$M_1 = 3.84(\mu L/2), \qquad M_2 = 1.29(\mu L/2), \qquad M_3 = 2.41(\mu L/2)$$

The uncoupled equations of motions are

$$M_i\ddot{Q}_i + C_i\dot{Q} + K_iQ_i = -\mathbf{V}_i^{\mathrm{T}}\mathbf{ma}(t) = P_i(t)$$

or

$$\ddot{Q}_i + 2\lambda\omega_i\dot{Q}_i + \omega_1^2Q_i = -\frac{1}{M_i}(\mathbf{V}_i^{\mathrm{T}}\mathbf{ma}(t)) = \frac{P_i(t)}{M_i}$$

where

$$\mathbf{a}^{\mathrm{T}}(t) = \{a_y(t), a_x(t), a_x(t)\}$$

Now the response quantity is the base moment $M_{\mathrm{B}}(t)$. To find $M_{\mathrm{B}}(t)$, first compute the elastic spring forces induced in the system, resisting the motion $\mathbf{Y}(t)$.

$$\mathbf{F}^{\mathrm{s}}(t) = \mathbf{kY}(t) = \mathbf{kVQ}(t)$$

Obviously,

$$M_{\mathrm{B}}(t) = \{-L, 2L, L\}\,\mathbf{F}^{\mathrm{s}}$$

$$\mathbf{F}^{\mathrm{s}}(t) = \mathbf{kV}_1\mathbf{Q}_1(t) + \mathbf{kV}_2\mathbf{Q}_2(t) + \mathbf{kV}_3\mathbf{Q}_3(t)$$

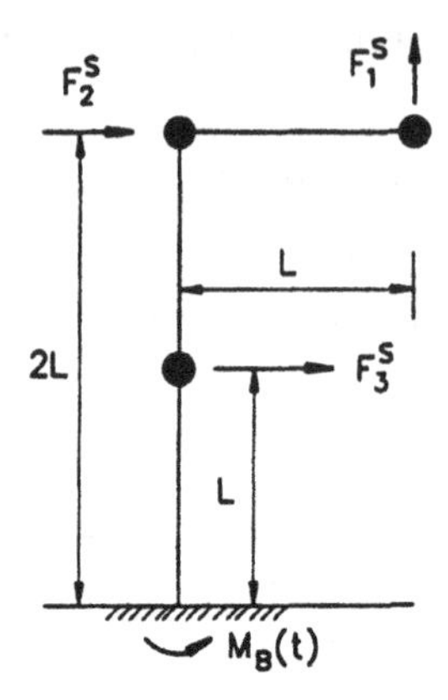

Fig. 5.3 *The base moment $M_B(t)$ and elastic spring forces*

also since $\mathbf{kV} = \omega^2\mathbf{mV}$

$$\mathbf{F}^s(t) = \omega_1^2\mathbf{mV}_1Q_1 + \omega_2^2\mathbf{mV}_2Q_2 + \omega_3^2\mathbf{mV}_2Q_3$$

so

$$\mathbf{F}^s(t) = \mathbf{mV}[\omega_i^2Q_i(t)]$$

Therefore,

$$M_B(t) = \sum M_{nB}(t) = \sum B_nQ_n(t) = \{-L, 2L, L\}mV[\omega_n^2Q_n(t)]$$

so

$$B_n = \{-L, 2L, L]\mathbf{mV}_n\omega_n^2 = \omega_n^2(\mu L^2/2)\{-3{,}84, 2{,}58, 2{,}41\}\mathbf{V}_n, n = 1, 2, 3$$

and

$$\{B_n\} = (EI)/(2L^2)\{0.1458, 6.7118, 6.1858\}$$

The mean value of $M_B(t)$ is zero since the acceleration processes are signals of white noise type without concentration at zero frequency. Then,

$$\sigma_{M_B(t)} = \int_{-\infty}^{\infty} S_{M_B(t)}(\omega)\mathrm{d}\omega$$

and

$$S_{M_B}(\omega) = \sum_{m=1}^{3}\sum_{n=1}^{3} B_mH_m^*(\omega)S_{P_mP_n}(\omega)H_n(\omega)B_n$$

where

$$S_{P_mP_n}(\omega) = \mathbf{V}_m^T\{S_{F_mF_n}(\omega)\}\mathbf{V}_n = \mathbf{V}_m^T\mathbf{m}\{S_{a_ma_n}(\omega)\}\mathbf{mV}_n$$

or

$$S_{P_mP_n}(\omega) = \mathbf{V}_m^T\mathbf{m}\begin{Bmatrix} \frac{S_0}{2} & C_0S_0 & C_0S_0 \\ C_0S_0 & S_0 & S_0 \\ C_0S_0 & S_0 & S_0 \end{Bmatrix}\mathbf{mV}_n, \text{ independently of } \omega$$

Since

$$F(t) = ma(t)$$

$$\mathbf{F}(t)\mathbf{F}^{\mathrm{T}}(t) = \mathbf{ma}(t)\mathbf{a}^{\mathrm{T}}(t)\mathbf{m} = \mathbf{m}\begin{Bmatrix} a_y a_y & a_y a_x & a_y a_x \\ a_x a_y & a_x a_x & a_x a_x \\ a_x a_y & a_x a_x & a_x a_x \end{Bmatrix}\mathbf{m}$$

Therefore,

$$\sigma_{M_B}^2 = \sum_{m=1}^{3}\sum_{n=1}^{3}\left[B_m B_n S_{P_m P_n}\int_{-\infty}^{\infty} H_m(-\omega)H_n(\omega)\mathrm{d}\omega\right]$$

By inserting Eq. (5.22), the integral can be evaluated, whereby

$$\sigma_{M_B}^2 = \sum_{m=1}^{3}\sum_{n=1}^{3}\frac{4\pi\lambda B_m B_n S_{P_m P_n}}{M_m M_n(\omega_m + \omega_n)[(\omega_m - \omega_n)^2 + 4\lambda^2\omega_m\omega_n]}$$

With well-separated frequencies and low damping, the cross terms can be neglected to yield

$$\sigma_{M_B}^2 \simeq \sum_{n=1}^{3}\frac{\pi}{2\lambda}\frac{B_n^2 S_{P_n P_n}}{M_n^2\omega_n^3}$$

and upon substitution

$$\sigma_{M_B}^2 \simeq \frac{\mu L^2}{\lambda}\sqrt{\frac{EI}{\mu}}(9.3 + 3.2C_0)S_0$$

If the excitations $a_x(t)$ and $a_y(t)$ are completely uncorrelated, then $S_{a_x a_x} = C_0 S_0 = 0$ and $C_0 \equiv 0$. If the base accelerations on the other hand are fully correlated, then $a_y(t) = \alpha a_x(t)$ where α is a constant. In this case,

$$S_{a_y a_y} = \alpha^2 S_{a_x a_x}, \quad S_{a_x a_y} = \alpha S_{a_x a_x}$$

$$0.5S_0 = \alpha^2 S_0, \qquad C_0 S_0 = S_0 \Rightarrow C_0 = \alpha = \pm\sqrt{2}/2$$

Thus the possible range of values for C_0 is $-1/\sqrt{2} \leqslant C_0 \leqslant 1/\sqrt{2}$, and therefore

$$7.0\frac{(\mu L)^2}{\lambda}S_0\sqrt{\frac{EI}{\mu}} \leqslant \sigma_{M_B}^2 \leqslant 11.6\frac{(\mu L)^2}{\lambda}S_0\sqrt{\frac{EI}{\mu}}$$

5.2 COMPLEX LINEAR SYSTEMS. A GENERAL APPROACH

The analysis of complex engineering structural systems first of all requires that a sensible mathematical model for the structure, which correctly describes its dynamical behaviour, be conceived. In many cases, relatively simple models will do. For instance, in the case of tall buildings, it is often sufficient to imagine the total mass of the structure concentrated in the floors of the building. Thus, the degrees of freedom are restricted to the movement of the mass centre of each floor times the number of floors. Usually, only horizontal translation along the main axis of the building and rotation about its vertical axis are considered. The total number of degrees of freedom is therefore three times the number of floors (roof included), Fig. 5.4. An example of this kind of analysis is shown in the following, where a mathematical model for an N-storey tall, uniform building is shown, and the input–output or response relations for a random earthquake type of excitation of the base of the building are derived. The presentation follows closely a scheme proposed by Yang *et al.* [239].

A more precise modelling of engineering structures is made possible by the finite element technique (FEM). The finite element models offer a powerful technique for obtaining both statical and dynamical information about the behaviour of structures under dynamical loads. Actually, a statical FEM analysis would probably be required to calculate the stiffnesses in the lumped mass system of Fig. 5.4. The advantage of the FEM technique is that it can easily handle complex structures, which are formed by a mixture of structural elements such as massive walls, columns, beams, plates and other structural parts. The structure is modelled by forming a nodal system, which defines the spatial configuration of the structure, with all the structural elements connected at the various nodes. For instance, a wall is subdivided into small shear elements connected at the nodes of the boundaries of the elements. Similarly, the columns, beams and plates are subdivided into elements, and the whole structure is tied together at the nodes of the global system. The mass of the elements is concentrated at the nodes thus forming a consistent mass matrix for the entire structure, and each node has the number of degrees of freedom appropriate for the element in question. The element stiffness is first computed locally for each element, and a global stiffness matrix for the entire structure is then constructed from all the element stiffnesses. Powerful commercial program packages are available for finite

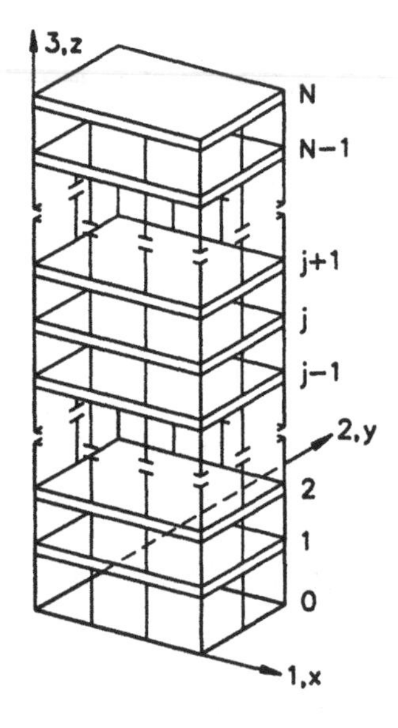

Fig. 5.4 *A tall N-storey building subjected to earthquake excitation, reproduced from [239] by permission of ASCE*

element analysis of complex structures, (see for instance ALGOR [6]). The FEM program packages can handle a variety of structural elements and deal with all sorts of connectivity and boundary conditions. Most program packages offer numerous auxiliary routines for treating frequency analysis and normal mode calculations, time history analysis through direct integration, random excitation etc.

As for the actual response calculations of a structural system subjected to dynamical excitation, the classical normal mode approach to obtain the response characteristics of a linear multi degree of freedom system was discussed in the preceding section. Actually, the normal mode method is only convenient for a certain type of problems and in many cases other methods of solutions have to be sought. Also, with the computing power available in even small desktop computers, the original reason for breaking the system up into its normal modes, that is, seek the response for each mode separately and then sum up the individual modal contributions, has lost some of its appeal. Direct integration of the equations of motion is now a perfectly feasible task, and with the introduction of the fast Fourier transform, FFT, in the mid-1960s by Cooley and Tukey, [39], methods involving the frequency domain offer a very attractive approach. In this case, the equations of motion of the system are formed as usual and the complex frequency response is obtained. The impulse response function is then recovered as the Fourier transform of the frequency response applying the FFT algorithm and the system response characteristics are obtained through numerical integration of the time domain solution.

This approach can be described as follows. Form the equations of motion of the system,

$$L[\mathbf{Y}(t)] = \mathbf{X}(t) \tag{5.27}$$

where $\mathbf{X}(t)$ is a random excitation vector and $\mathbf{Y}(t)$ is the response vector of the system describing any set of quantities of interest such as deflections, stresses, internal forces or any other set of response characteristics that are of interest in a particular problem. The complex frequency response matrix of the system is obtained by applying a simple harmonic excitation vector,

$$\mathbf{X}(t) = \mathbf{A}_0 \mathrm{e}^{\mathrm{i}\omega t} \tag{5.28}$$

where $\mathbf{A}_0$ is a vector of arbitrary constants. If the system Eqs. (5.27) are linear, then

$$\mathbf{Y}(t) = \mathbf{Y}_0 \mathrm{e}^{\mathrm{i}\omega t} \qquad \text{or} \qquad \mathbf{Y}_0 = L^{-1}[\mathbf{A}_0] = \mathbf{H}(\omega)\mathbf{A}_0$$

and

$$\mathbf{H}(\omega) = \mathbf{A}_0^{-1}\mathbf{Y}_0 \tag{5.29}$$

The time domain solution is then given by the Duhamel integral

$$\mathbf{Y}(t) = \int_0^t \mathbf{h}(t-u)\mathbf{X}^{\mathrm{T}}(u)\mathrm{d}u \tag{5.30}$$

where $\mathbf{h}(t)$ is the impulse response matrix and is defined as the Fourier transform of the corresponding frequency response matrix, that is

$$\mathbf{h}(t) = \frac{1}{2\pi}\int_{-\infty}^{\infty} \mathbf{H}(\omega)\mathrm{e}^{\mathrm{i}\omega t}\mathrm{d}\omega \tag{5.31}$$

Once the frequency response matrix has been obtained through the above procedure, the impulse response matrix is either obtained through direct integration of Eq. (5.30) or, as already stated, in many cases more easily computed as the fast Fourier transform (see Chapter 7, Eq. (7.31)), whereby ($\mathbf{h}(t) = \{h_{ij}(t)\}$)

$$h_{ij}^{r} = \frac{1}{N} \sum_{k=0}^{N-1} H_{ij}^{k} \gamma^{-kr} \tag{5.32}$$

5.2.1 *Analysis of Tall Buildings Subjected to Random Base Excitations*

In Fig. 5.4, the layout of an N-storey uniform building is shown. The columns in the building are considered to be massless elastic members that have to withstand the vertical and horizontal forces induced by a random excitation vector applied at the base of the building, Fig. 5.5. No other resisting elements in the building are considered, such as load-bearing shear walls, and the entire mass of the building is assumed to correspond to N equal masses, concentrated in the mass centre of each floor that does not coincide with the centre of torsion of the elastic resisting forces, Fig. 5.6.

First let the base excitation be defined. It consists of a random base vector

$$\mathbf{G}^{\mathrm{T}}(t) = \{G_1(t), G_2(t), G_3(t)\}$$

which has two horizontal acceleration components $\{G_1(t), G_2(t)\}$ and one rotational acceleration component $G_3(t)$, Fig. 5.5. The random acceleration components are of the kind $G_i(t) = \psi(t)X_i(t)$ where the $X_i(t)$ form are a set of stationary stochastic processes with zero mean values and power spectral densities $S_{X_i}(\omega)$. The factor $\psi(t)$ is a deterministic envelope function of time. Thus an earthquake type acceleration signal can be constructed by correctly defining the spectral densities of the stationary excitation and by applying suitable amplitude modulating functions to provide the time history characteristics of the earthquake. The way of creating non-stationary signals from stationary stochastic processes, applying deterministic envelope or amplitude modulating functions, will be further discussed in Chapter 6, where an example of generating earthquake acceleration signals will also be presented.

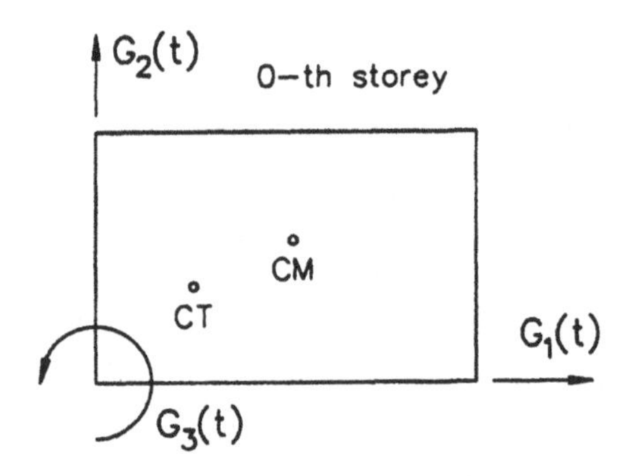

Fig. 5.5 *Random base excitation, reproduced from [239] by permission of ASCE*

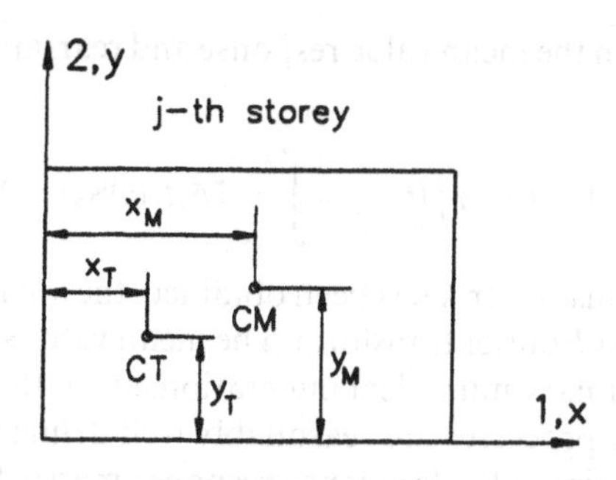

Fig. 5.6 *The mass and torsion centres*

The spectral densities and the autocorrelation functions of the stationary processes $\{X_i(t)\}$ can conveniently be set up in a matrix form for the consequent calculations:

$$\mathbf{S}_X(\omega) = \begin{Bmatrix} S_{11} & S_{12} & S_{13} \\ S_{21} & S_{22} & S_{23} \\ S_{31} & S_{32} & S_{33} \end{Bmatrix} \tag{5.33}$$

and

$$\mathbf{R}_X(\tau) = \int_{-\infty}^{\infty} \mathbf{S}_X(\omega)\mathrm{e}^{\mathrm{i}\omega\tau}\mathrm{d}\omega = \begin{Bmatrix} R_{11} & R_{12} & R_{13} \\ R_{21} & R_{22} & R_{23} \\ R_{31} & R_{32} & R_{33} \end{Bmatrix} \tag{5.34}$$

The mean values and the mean square values of any response vector $\mathbf{Y}(t)$ are then obtained as follows. Firstly, the mean value of all response quantities is necessarily zero since the underlying stationary excitation processes have zero means. As a second step, the covariance kernel of the response, Eq. (5.30), is formed:

$$E[\mathbf{Y}(t)\mathbf{Y}^{\mathrm{T}}(s)] = \int_0^t \int_0^s \mathbf{h}(t-u)E[\mathbf{G}(u)\mathbf{G}^{\mathrm{T}}(v)]\mathbf{h}^{\mathrm{T}}(s-v)\mathrm{d}u\,\mathrm{d}v \tag{5.35}$$

Introducing the envelope function, $\psi(t)$, the covariance kernel of the excitation, $E[\mathbf{G}(u)\mathbf{G}^{\mathrm{T}}(v)]$ is written as $\psi(u)\psi(v)E[\mathbf{X}(u)\mathbf{X}^{\mathrm{T}}(v)] = \psi(u)\psi(v)\mathbf{R}_X(u-v)$. It is therefore convenient to define the auxiliary matrices,

$$\mathbf{M}(t,\omega) = \int_0^t \mathbf{h}(u)\psi(t-u)\mathrm{e}^{-\mathrm{i}\omega u}\mathrm{d}u, \qquad \mathbf{M}^{\mathrm{T}*}(t,\omega) = \int_0^t \mathbf{h}^{\mathrm{T}}(u)\psi(t-u)\mathrm{e}^{\mathrm{i}\omega u}\mathrm{d}u \tag{5.36}$$

which only depend on the system parameters through the impulse response matrix and the envelope function of the excitation. Equation (5.35) can now be written as

$$E[\mathbf{Y}(t)\mathbf{Y}^{\mathrm{T}}(s)] = \int_0^t \int_0^s \mathbf{h}(t-u)\psi(u)\left[\int_{-\infty}^{\infty} \mathbf{S}_X(\omega)\mathrm{e}^{\mathrm{i}\omega(u-v)}\mathrm{d}\omega\right]\mathbf{h}^{\mathrm{T}}(s-v)\psi(v)\mathrm{d}u\,\mathrm{d}v$$

By choosing $t = s$ to obtain the mean value response and rearranging terms, the final result is

$$E[\mathbf{Y}(t)\mathbf{Y}^{\mathrm{T}}(t)] = E[\{Y_i^2(t)\}] = \int_{-\infty}^{\infty} \mathbf{M}(t,\omega)\mathbf{S}_X(\omega)\mathbf{M}^{\mathrm{T}*}(t,\omega)\mathrm{d}\omega \tag{5.37}$$

So once the impulse response matrix has been obtained, the auxiliary matrices $M(t, \omega)$ can be computed using the fast Fourier transform. The mean value square is then given by Eq. (5.37), which involves ordinary numerical integration, given that the spectral densities of the underlying stationary processes are reasonably well defined. It is therefore clear that the main work lies in obtaining the frequency response matrix for any response quantity that is to be studied, which requires that the complete set of the equations of motion of the system be established.

In Fig. 5.7, the internal forces acting on the jth floor of the building are shown. The horizontal deflections and the rotation of the mass centre C_M are denoted by

$$u_j^M = u_j - y_M\,\theta_j; \quad v_j^M = v_j + x_M\theta_j; \quad \theta_j^M = \theta_j \tag{5.38}$$

whereas the deflections and rotation with respect to the torsion centre C_{T} are given by

$$u_j^{\mathrm{T}} = u_j - y_{\mathrm{T}}\,\theta_j, \quad v_j^{\mathrm{T}} = v_j + x_{\mathrm{T}}\theta, \quad \theta_j^{\mathrm{T}} = \theta_j \tag{5.39}$$

The shear forces and torsional moment acting on the jth storey are denoted as follows:

From above: U_j^+, V_j^+, Q_j^+

From below: U_j^-, V_j^-, Q_j^-

Let m be the mass concentrated in the jth floor and I the mass torsional moment with respect to the torsion centre C_{T}. The dynamic equilibrium is then stated through the following set of equations:

$$\begin{aligned} U_j^+ &= U_j^- + m(\ddot{u}_j - y_M\ddot{\theta}_j) + c\dot{u}_j \\ V_j^+ &= V_j^- + m(\ddot{v}_j + x_M\ddot{\theta}_j) + c\dot{v}_j \\ Q_j^+ &= Q_j^- + I\ddot{\theta} + m(\ddot{u}_j - y_M\ddot{\theta}_j)(y_T - y_M) - m(\ddot{v}_j + x_M\ddot{\theta}_j)(x_T - x_M) \end{aligned} \tag{5.40}$$

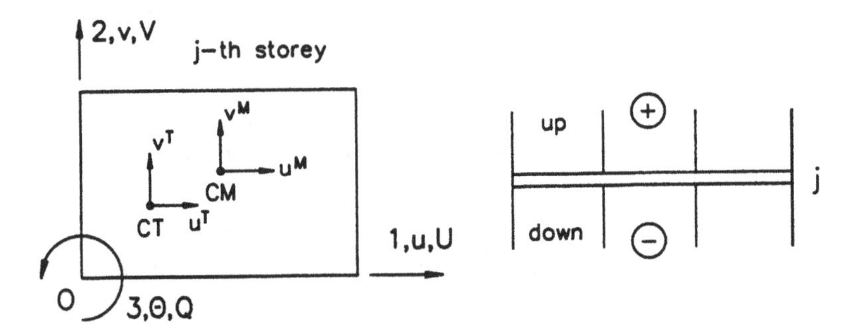

Fig. 5.7 *Deflections and forces acting on the jth storey*

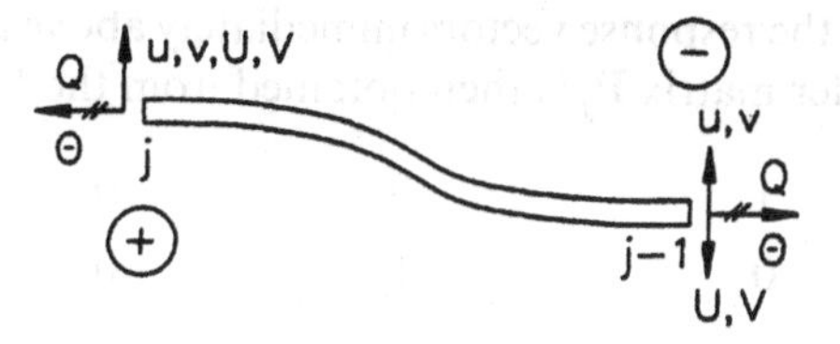

Fig. 5.8 *Elastic forces in the jth storey*

where c is the coefficient of damping. The solution of the above equations is then subject to the boundary conditions

$$\begin{aligned} &U_N^+ = V_N^+ = Q_N^+ = 0 \quad \text{(at the top in free space)} \\ &u_0 = G_1(t), \qquad v_0 = G_2(t), \qquad \theta_0 = G_3(t) \qquad \text{(at the base)} \end{aligned} \tag{5.41}$$

The shear forces and torsional moments are transmitted between the storeys through the load bearing columns, which are considered to be massless elastic members as already stated. The relation between the forces at two different levels is indicated in Fig. 5.8, whereby the following relations between the response quantities are easily obtained:

$$\begin{aligned} &U_j^- = U_{j-1}^+, \qquad V_j^- = V_{j-1}^+, \qquad Q_j^- = Q_{j-1}^+ \\ &u_j = u_{j-1} + \frac{1}{k_{x,j}} U_{j-1}^+ + y_T \frac{1}{k_{T,j}} Q_{j-1}^+ \\ &v_j = v_{j-1} + \frac{1}{k_{y,j}} V_{j-1}^+ - x_T \frac{1}{k_{T,j}} Q_{j-1}^+ \\ &\theta_j = \theta_{j-1} + \frac{1}{k_{T,j}} Q_{j-1}^+ \end{aligned} \tag{5.42}$$

in which $k_{x,j}, k_{y,j}, k_{T,j}$, are the elastic spring constants for the deflection and the torsion respectively.

The frequency response of the entire system can now be obtained. Let the base excitation be the harmonic vector, $G^T(t) = \{a_1, a_2, a_3,\}e^{i\omega t}$, where $\{a_i\}$ are three arbitrary constants. Then

$$\begin{aligned} &U = \bar{U} e^{i\omega t}, \qquad V = \bar{V} e^{i\omega t}, \qquad Q = \bar{Q} e^{i\omega t} \\ &u = \bar{u} e^{i\omega t}, \qquad \bar{v} e^{i\omega t}, \qquad \bar{\theta} = \bar{\theta} e^{i\omega t} \\ &(\bar{u}^+ = \bar{u}^-, \qquad \bar{v}^+ = \bar{v}^-, \qquad \bar{\theta}^+ = \bar{\theta}^-) \end{aligned} \tag{5.43}$$

With the response vector for any one storey, $\mathbf{Z}^T = \{\bar{u}, \bar{v}, \bar{\theta}, \bar{Q}, \bar{V}, \bar{U}\}$, the equations of motion can be established in a matrix form.

$$\mathbf{Z}_j^+ = \mathbf{P}_j \mathbf{Z}_j^- \tag{5.44}$$

This equation connects the response vectors immediately above and below the jth floor of the building. The transfer matrix $\mathbf{P}_j$ is then obtained from the Eqs. (5.40):

$$P_j = \begin{Bmatrix} 1 & 0 & 0 & 0 & 0 & 0 \\ 0 & 1 & 0 & 0 & 0 & 0 \\ 0 & 0 & 1 & 0 & 0 & 0 \\ -m\omega^2(y_T - y_M) & m\omega^2(x_T - x_M) & P_{43} & 1 & 0 & 0 \\ 0 & -m\omega^2 + ci\omega & -m\omega^2 x_M & 0 & 1 & 0 \\ -m\omega^2 + ci\omega & 0 & m\omega^2 y_M & 0 & 0 & 1 \end{Bmatrix} \tag{5.45}$$

in which $P_{43} = \omega^2(m[y_M(y_T - y_M) + x_M(x_T - x_M)] - I)$

Another similar relation connects the forces within each storey or

$$\mathbf{Z}_j^- = \mathbf{L}_j \mathbf{Z}_{j-1}^+ \tag{5.46}$$

where the Eqs. (5.42) have been put into a matrix form and the matrix $\mathbf{L}_j$, which describes the static equilibrium conditions for the massless elastic members in each storey, is given below, Eq. (5.48). Obviously,

$$\mathbf{Z}_j^+ = \mathbf{P}_j \mathbf{Z}_j^- = \mathbf{P}_j \mathbf{L}_j \mathbf{Z}_{j-1}^+ = \mathbf{T}_j \mathbf{Z}_{j-1}^+ \tag{5.47}$$

where $\mathbf{T}_j = \mathbf{P}_j \mathbf{L}_j$ is the transfer matrix across one storey, Fig. 5.9. Now,

$$\mathbf{L}_j = \begin{Bmatrix} 1 & 0 & 0 & y_T k_T^{-1} & 0 & k_x^{-1} \\ 0 & 1 & 0 & -x_T k_T^{-1} & k_y^{-1} & 0 \\ 0 & 0 & 1 & k_T^{-1} & 0 & 0 \\ 0 & 0 & 0 & 1 & 0 & 0 \\ 0 & 0 & 0 & 0 & 1 & 0 \\ 0 & 0 & 0 & 0 & 0 & 1 \end{Bmatrix} \tag{5.48}$$

$$\mathbf{Z}_N^+ = \mathbf{T}_N \mathbf{T}_{N-1} \cdots \mathbf{T}_2 \mathbf{T}_1 \mathbf{T}_0 \mathbf{Z}_0^+ = \mathbf{W} \mathbf{Z}_0^+ \tag{5.49}$$

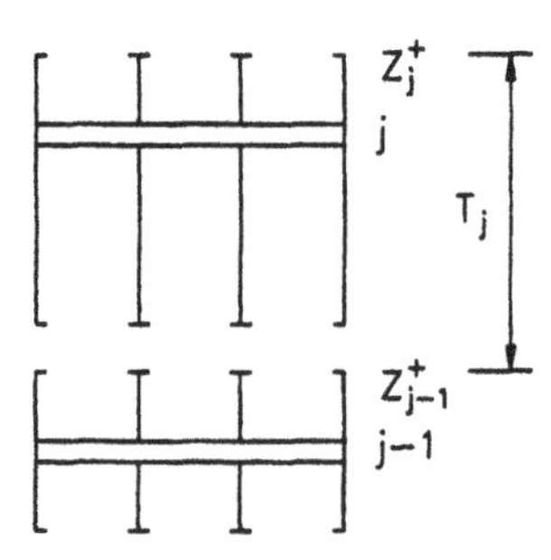

Fig. 5.9 *The transfer matrix across one storey*

and the boundary conditions are given by

$$\bar{U}_N^+ = \bar{V}_N^+ = \bar{Q}_N^+ = 0$$

$$\bar{u}_0 = a_1, \qquad \bar{v}_0 = a_2, \qquad \bar{\theta}_0 = a_3$$

so that

$$\mathbf{Z}_N^+ = \begin{Bmatrix} \bar{u} \\ \bar{v} \\ \bar{\theta} \\ 0 \\ 0 \\ 0 \end{Bmatrix}_N = \mathbf{W} \cdot \begin{Bmatrix} a_1 \\ a_2 \\ a_3 \\ \bar{Q} \\ \bar{V} \\ \bar{U} \end{Bmatrix}_0 \tag{5.50}$$

For so-called periodic buildings, that is, those for which the stiffness and mass distribution of all storeys is the same, obviously,

$$\mathbf{T}_1 = \mathbf{T}_2 = \cdots \mathbf{T}_N \qquad \text{so} \qquad \mathbf{W} = \mathbf{T}^N \tag{5.51}$$

The frequency response of the building is now completely determined by Eqs. (5.44)–(5.50). The internal forces in the first storey are given by

$$\begin{Bmatrix} \bar{Q}_0 \\ \bar{V}_0 \\ \bar{U}_0 \end{Bmatrix} = (-1) \begin{Bmatrix} w_{44} & w_{45} & w_{46} \\ w_{54} & w_{55} & w_{56} \\ w_{64} & w_{65} & w_{66} \end{Bmatrix} \begin{Bmatrix} a_1 w_{41} + a_2 w_{42} + a_3 w_{43} \\ a_1 w_{51} + a_2 w_{52} + a_3 w_{53} \\ a_1 w_{61} + a_2 w_{62} + a_3 w_{63} \end{Bmatrix} \tag{5.52}$$

and the frequency response vectors $\mathbf{Z}_j^+(\omega)$ by

$$\mathbf{Z}_j^+(\omega) = \begin{Bmatrix} \bar{u}_j \\ \bar{v}_j \\ \bar{\theta}_j \\ \bar{Q}_j \\ \bar{V}_j \\ \bar{U}_j \end{Bmatrix} = \mathbf{T}_j \cdot \mathbf{T}_{j-1} \cdots \mathbf{T}_2 \cdot \mathbf{T}_1 \begin{Bmatrix} a_1 \\ a_2 \\ a_3 \\ \bar{Q}_0 \\ \bar{V}_0 \\ \bar{U}_0 \end{Bmatrix} \tag{5.53}$$

A numerical example showing the earthquake response analysis of an eight storey periodical building using the above method of analysis has been presented by Yang *et al.* [239]. They calculate the amplitudes of the frequency response and the non-stationary r.m.s. values of the base shear forces and the roof displacements for a horizontal random acceleration component in the x-direction of the building. The acceleration signal is modelled as a white noise signal filtered by a special ground motion filter and then amplitude modulated using a special modulating function with the characteristics of a strong earthquake (for explanation of this method, see Example 6.3).

5.2.2 *FEM Analysis of a Tall Building Subjected to Earthquake-type Excitation*

The author of this book has been involved in several projects dealing with the analysis of tall buildings subjected to earthquake motion. One of those projects is the structural design of an office tower in the west part of Mexico City by Correa Hermanos C.V. de S.A. In a commercial centre on the road to Toluca, the seventeen-storey Pabellon tower is under construction, which is within Zone I in the seismic risk classification of the city area, [37]. It is a reinforced concrete building, which is founded on Bentonite piles through 10–30 metres deep soft sedimentary layers to reach a load bearing gravel deposit. The total floor area of the tower is 15000 m^2. The structural system is based on an interior shear wall core, which forms the elevator and staircase shaft, the rear facade of the building, which is a load bearing shear wall with window openings, and a ductile framing system, which consists of columns and floor beams. The columns are spaced eight metres apart in both directions. The main floor beams (0.3×0.75 m) are placed over the column rows, forming an $8 \times 8\,m^2$ grid. In addition, secondary beams (0.25×0.60 m) are placed between the main beams parallel to the shorter direction of the cross section of the tower, whence the 12 cm floor slabs are $4 \times 8\,m^2$.

The tower was subjected to static and dynamical analysis using a large finite element analysis program ALGOR, which is a multipurpose commercial and educational software package from ALGOR Inc., [6]. In order to set up the finite element model (FEM), a layered drawing of the structural system of the tower was first drawn up using the AUTOCAD graphical program by AUTODESK Inc., [33]. The shear walls were configured in one layer, the plate elements in another, the beams and columns in a third layer and the boundary elements, which model the connections between the superstructure of the tower and the pile heads, were placed in a fourth layer. Using various decoder routines, which are a part of the ALGOR software, the different layers of the AUTOCAD drawing can be converted into a complete finite element model of the structure with optimum numbering of nodes, and numbering and classification of all structural elements into different element groups. Figures 5.10–5.12 show the finite element model of the tower obtained in this manner.

The FEM system for the tower has 1767 nodal points, 2064 beam elements, 1562 shear elements (walls and plates) and 98 boundary elements because of the pile connections. The rank of the global stiffness matrix is 7383, which is also the total number of independent equations. The bandwidth of the matrix is 518, which is a crucial number for the calculation speed. As soon as the mass distribution of all elements together with the material stiffnesses has been supplied to the program, a static analysis for external static loads and the life loads on the floors (2.5 KN/m^2) can be performed, and at the same time a frequency analysis giving the fundamental frequencies and the normal mode vectors of the structure is carried out. In Table 5.1 the first eight natural frequencies of the model structure are shown. The fundamental frequency is 2.72 rad/s giving a fundamental natural period of 2.31 seconds. It is interesting to note that earthquake records in the Valley of Mexico seem to have a great deal of power concentrated about the 0.5 Hz frequency. The Pabellon tower is close enough to this spectral peak whereby some kind of quasi-resonance can not be excluded.

For the subsequent response analysis, there are two options. Firstly, a classical modal analysis can be performed. Secondly, the program allows for a direct numerical integra-

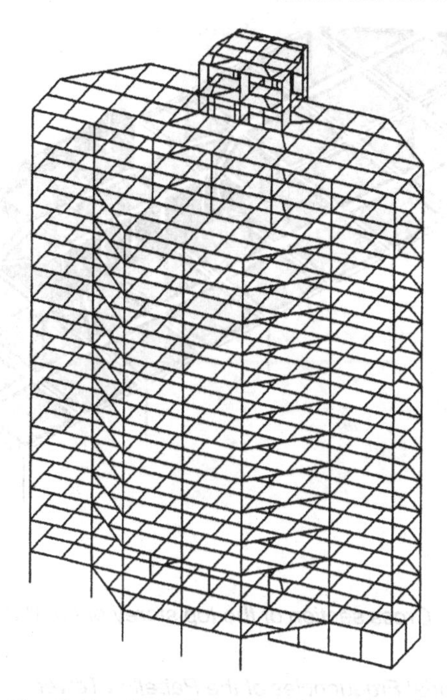

Fig. 5.10 *Pabellon tower: front elevation*

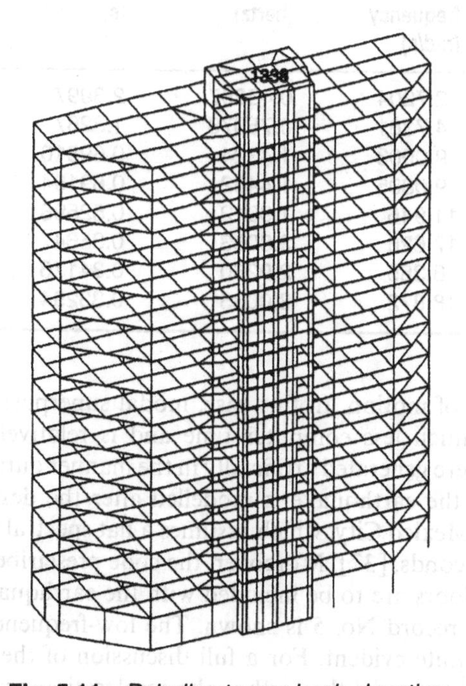

Fig. 5.11 *Pabellon tower: back elevation*

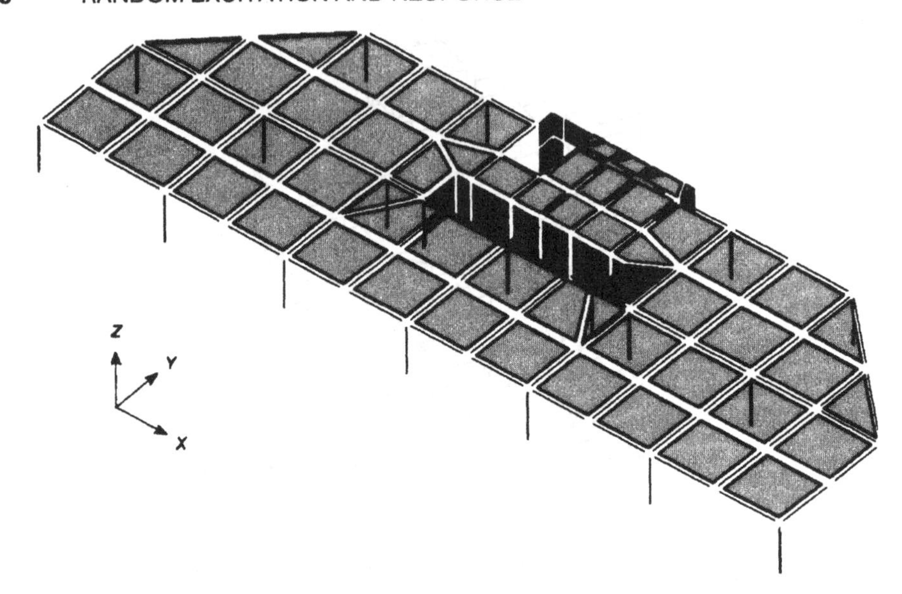

Fig. 5.12 *Cross section of the top storey of the Pabellon tower*

Table 5.1 *Modal Frequencies of the Pabellon Tower*

Natural Frequencies				
Mode number	Circular frequency (rad/s)	Frequency (hertz)	Period (s)	Tolerance
1	2.7204	0.43297	2.3097	1.2001E − 16
2	4.7361	0.75378	1.3267	0.0000E + 00
3	9.0003	1.4324	0.69810	1.7543E − 16
4	9.0805	1.4452	0.69194	3.4469E − 16
5	11.946	1.9012	0.52598	1.5934E − 15
6	17.626	2.8053	0.35647	2.6151E − 09
7	18.385	2.9260	0.34176	1.4165E − 08
8	18.872	3.0035	0.33294	2.3495E − 06

tion of the equations of motion. In this case, modal superposition was chosen, mainly because it requires much less computer time and is relatively simple. Five different earthquake records were generated artificially in the manner outlined in Example 6.3. The spectral character of the earthquake is modelled after the design spectra given in the Earthquake Code of Mexico City, which assumes a flat spectral peak for Zone I between periods 0.4 and 0.8 seconds, [37]. Moreover, the code prescribes that about 72% of the full-life loads on all floors are to be included with the earthquake analysis. In Fig. 5.13, synthetic earthquake record No. 5 is shown. The low-frequency content of the record ($T \approx 0.6$ seconds) is quite evident. For a full discussion of the evolutionary stochastic process, which is used to model the earthquake acceleration, see Example 6.3.

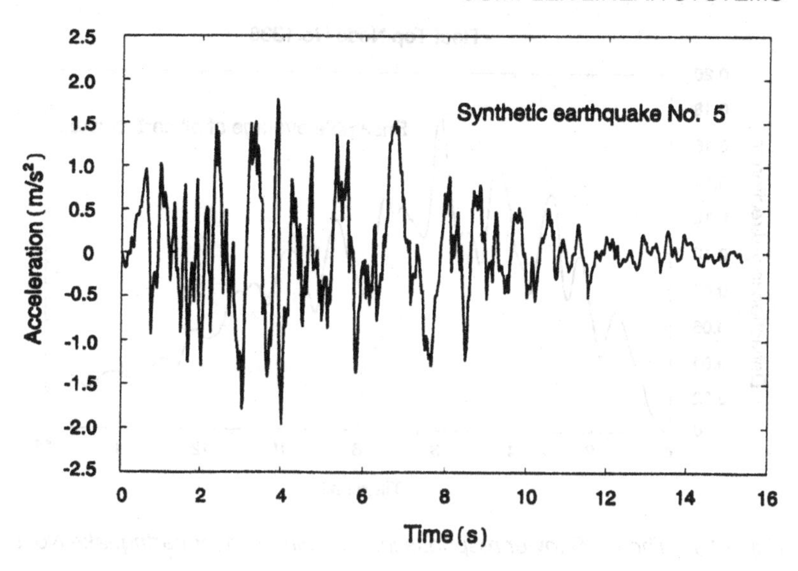

Fig. 5.13 *An artificially generated earthquake acceleration process*

The Pabellon tower has been analysed for five earthquake excitation records acting at the base of the structure in the y-direction. For a more complete analysis, the x-direction should be investigated also, and the resulting deflections from both direction superposed. In the present case, only the response in the y-direction of Node No. 1338, which is the middle of the top of the tower (see Fig. 5.11), was studied. In Fig. 5.14 the response for earthquake record No. 5 is depicted. The response very clearly reflects the fundamental

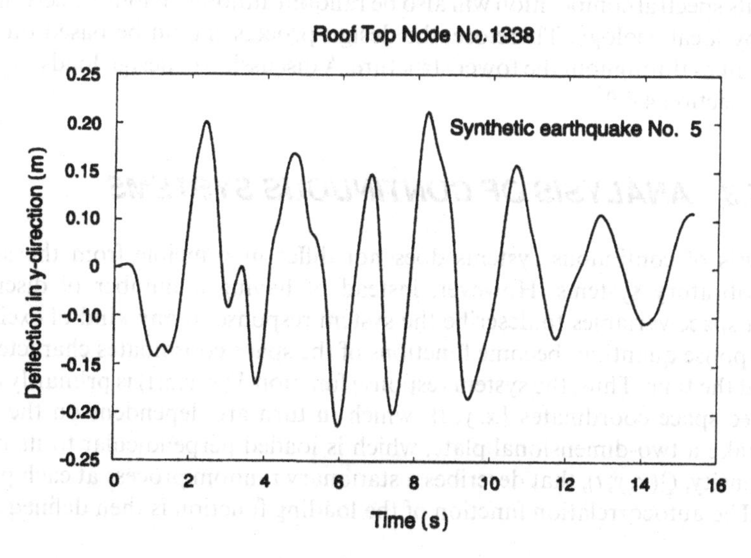

Fig. 5.14 *The tower response in the y-direction for earthquake No. 5*

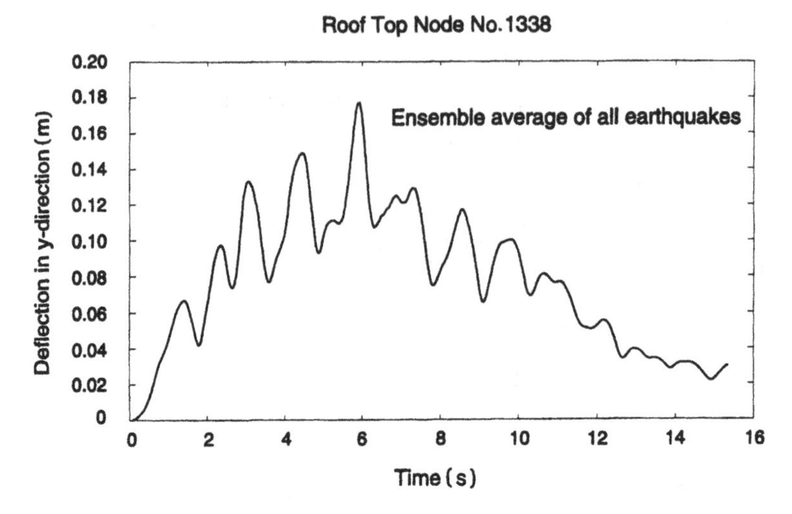

Fig. 5.15 *The RMS tower response in the y-direction for earthquake No. 5*

period of the structure in the vicinity of 2.3 s. Finally, the root mean square response of Node No. 1338 for all five records is shown in Fig. 5.15. The trace is still very much affected by the fundamental period and ensuing first mode contributions. The maximum response of about 0.18 m comes at about 6 s from the onset of the earthquake. For a larger number of artificial records, the first mode effects would slowly vanish, leaving a relatively smooth r.m.s. response in the range 0–16s. Probably, the smoothed maximum response would come down to about 0.14 m. It should be borne in mind that the design of the tower should not be dependent on one single earthquake. The earthquake process is by nature a random event and its spectral composition will also be random although it follows certain patterns enforced by local geology. Therefore, the design process should be based on the r.m.s. response values throughout the tower structure. A discussion of design loads vs. overloads is given in Section 4.4.2.

5.3 ANALYSIS OF CONTINUOUS SYSTEMS

The analysis of continuous systems does not differ in principle from the analysis of multiple vibratory systems. However, instead of having a number of discrete time-dependent space variables to describe the system response to any kind of excitation, all system response quantities become functions of the space coordinates characterizing the system and the time. Thus, the system response function $Y(x, y, z; t)$ is primarily a function of the three space coordinates $\{x, y, z\}$, which in turn are dependent on the time. For instance, take a two-dimensional plate, which is loaded perpendicular to its plane with a load intensity, $Q(x, y; t)$, that describes a stationary random process at each point (x, y), Fig. 5.16. The autocorrelation function of the loading function is then defined as

$$R_Q(x, y; \tau) = E[Q(x, y; t)Q(x, y; t + \tau)] \tag{5.54}$$

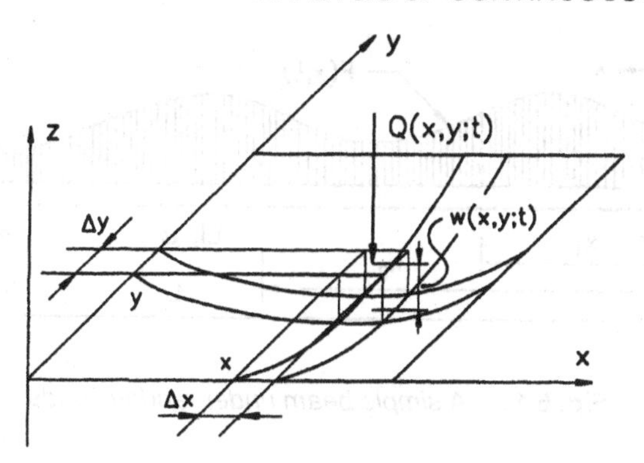

Fig. 5.16 *A plate loaded perpendicular to its plane*

and the spectral density is obtained by integration as

$$S_Q(x, y; \omega) = \frac{1}{2\pi}\int_{-\infty}^{\infty} R_Q(x, y; \tau)e^{-i\omega\tau}d\tau \tag{5.55}$$

The response quantities are obtained in a similar manner as described in the preceding sections. For instance, the vertical deflection of the plate is obtained from the plate equation

$$L[W(x, y; t)] = Q(x, y; t) \tag{5.56}$$

where the equation operator $L[*]$ represents the r.h.s. of the equation

$$\mu(x, y)\frac{\partial^2 w}{\partial t^2} - \frac{1}{D}\left(\frac{\partial^4 w}{\partial x^4} + \frac{\partial^4 w}{\partial x^2 \partial y^2} + \frac{\partial^4 w}{\partial y^4}\right) = 0$$

in which $\mu(x, y)$ is the mass distribution and D is the plate bending stiffness. The solution is given by the standard form.

$$w(x, y; t) = \int_0^t h(x, y; t-u)Q(x, y; u)du \tag{5.57}$$

where the impulse response function $h(x, y; t)$ is a function of both the space coordinates and time.

Example 5.2 In Fig. 5.17, a uniform simply supported elastic beam with mass μ per unit length and elastic stiffness EI is shown. The beam carries a stationary, random load $F(x, t)$ with a given cross-spectrum $S_F(x_1, x_2; \omega)$ between the load intensities at two different positions x_1 and x_2 along the span of the beam:

$$S_F(x_1, x_2; \omega) = \begin{cases} \dfrac{q^2}{\omega_c L^2}\sin^2\left(\dfrac{\pi x_1}{L}\right)\sin^2\left(\dfrac{\pi x_2}{L}\right) & |\omega| < \omega_c \\ 0 & \omega_c < |\omega| \end{cases}$$

where q and ω_c are given constants.

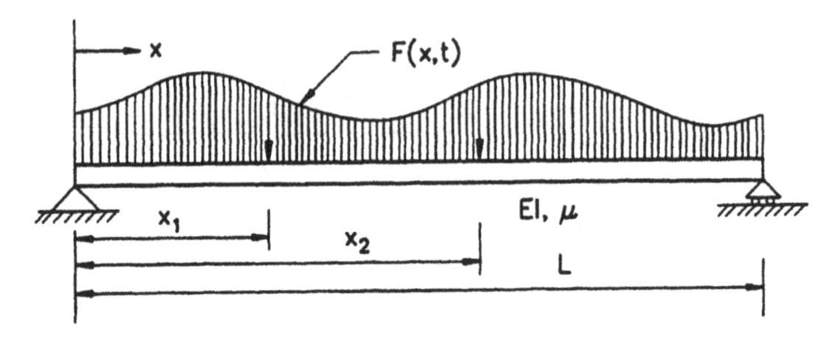

Fig. 5.17 *A simple beam under random loads*

1. Find the autocorrelation function $R_F(\tau)$ of the total force $F(t)$ acting on the beam, where

$$F(t) = \int_0^L F(x, t)\mathrm{d}x$$

2. Find the mean square value $E[F^2]$.

Solution The autocorrelation function of the total load, $F(t)$, is

$$E[F(t)F(t+\tau)] = E\left[\int_0^L F(x_1, t)\mathrm{d}x_1 \int_0^L F(x_2, t+\tau)\mathrm{d}x_2\right]$$

$$= \int_0^L \int_0^L E[F(x_1, t)F(x_2, t+\tau)\mathrm{d}x_1\,\mathrm{d}x_2 = \int_0^L \int_0^L R_F(x_1, x_2; \tau)\mathrm{d}x_1\,\mathrm{d}x_2$$

Now the autocorrelation function for the load $R_F(x_1, x_2; \tau)$ at two different positions x_1 and x_2 is obtained as the Fourier transform of the given spectral density, that is,

$$R_F(x_1, x_2; \tau) = \int_{-\infty}^{\infty} S_F(x_1, x_2; \omega)\mathrm{e}^{\mathrm{i}\omega\tau}\mathrm{d}\omega$$

so

$$R_F(x_1, x_2; \tau) = \frac{q^2}{\omega_c L^2} \int_{-\omega_c}^{\omega_c} \sin^2\left(\frac{\pi x_1}{L}\right) \sin^2\left(\frac{\pi x_1}{L}\right)\mathrm{e}^{\mathrm{i}\omega\tau}\mathrm{d}\omega$$

$$= \frac{2q^2}{L^2} \sin^2\left(\frac{\pi x_1}{L}\right) \sin^2\left(\frac{\pi x_1}{L}\right) \frac{\sin\omega_c\tau}{\omega_c\tau}$$

Therefore,

$$R_F(\tau) = E[F(t)F(t+\tau)]$$

$$= \frac{2q^2}{\pi^2} \int_0^L \int_0^L \sin^2\left(\frac{x_1\pi}{L}\right) \sin^2\left(\frac{x_2\pi}{L}\right)\mathrm{d}x_1\,\mathrm{d}x_2 = \frac{q^2}{2} \frac{\sin(\omega_c\tau)}{\omega_c\tau}$$

and finally, the mean square value of the total force,

$$E[F_2] = R_F(0) = q^2/2$$

5.3.1 *Tall, Slender Structures Subjected to Random Wind Loads*

In Example 2.18, the basic assumptions for the treatment and interpretation of turbulent gusty wind as a random phenomenon were introduced. It was shown how the wind velocity was represented as a stochastic process with a given experimentally derived spectral density. The analysis of tall slender structures excited by stochastic wind loads gives a good example of the way continuous systems behave in an random environment.

In Fig. 5.18, the various aspects of the behaviour of different kind of structures when subjected to turbulent wind loads is shown. The quasi-stationary part of the wind load, that is, due to the slowly varying mean wind speed or the standard 10 minute reference constant wind speed, produces a quasi-static response of the structure through the along-wind pressure and suction. The turbulent part of the wind, i.e. the wind gusts and whirls due to all kinds of obstructions and disturbances of the wind flow, will cause a random buffeting of a structure that can best be described as a stochastic dynamic excitation. The random dynamic response of structures such as tall buildings, towers, masts etc. can be divided into the along-wind response component, which is dominant for bluff structures, and the across-wind response component, which is due to wake-induced vibrations such as eddy shedding and vortices. Certain structures, such as suspended cables, high power tension lines and suspension bridges may suffer self-induced vibration and galloping effects, the most famous case being the failure of the Tacoma bridge in 1940. Flutter of aircraft wings and similar airborne structures is an other example.

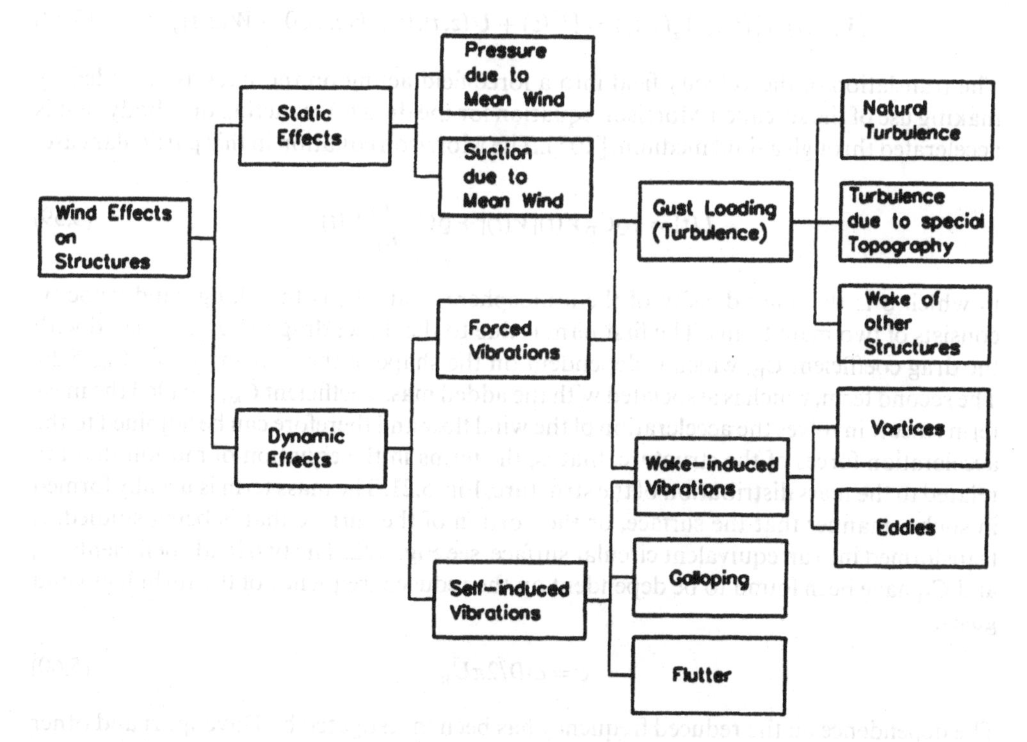

Fig. 5.18 *Wind effects on structures. A schematic diagram, (after Sigbjörnsson, [194])*

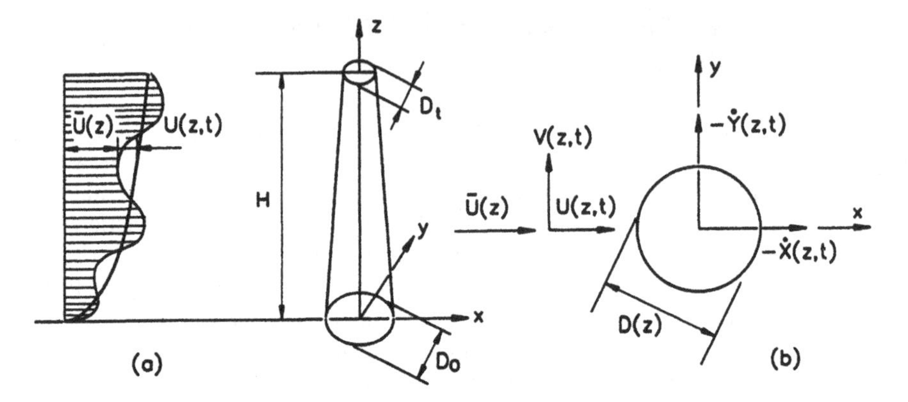

Fig. 5.19 *A tapered chimney stack subjected to wind loads*

To illustrate the problem of describing the wind effects on tall slender structures that behave as continuous vibratory systems, the case of tall chimney stacks will be studied, [194]. In Fig. 5.19, the geometrical data of a tapered chimney stack are given together with the wind loading conditions. The velocity profile of the mean wind speed is given by $\bar{U}(z) = \bar{U}_R(z/z_R)^\alpha$ and the wind velocity field surrounding the stack is therefore given by the three-dimensional velocity vector

$$\{V_X(z,t), V_y(z,t), V_z(z,t)\} = \{\bar{U}(z) + U(z,t), 0 + V(z,t), 0 + W(z,t)\} \tag{5.58}$$

The translation of the velocity field into a force field acting on the stack is provided by making use of the so-called Morrison equation for the drag forces acting on a body that is accelerated through a fluid medium, [105]. The Morrison equation in this particular case,

$$L(t) = \tfrac{1}{2}\varrho C_D V(t)|V(t)| + \varrho C_M \frac{A_0}{D_0}\dot{V}(t) \tag{5.59}$$

in which ϱ is the mass density of the atmosphere, and $V(t)$ is the along-wind velocity, consists of two main terms. The first term is due to the direct drag forces associated with the drag coefficient C_D, which is dependent on the shape of the resisting body, Fig. 5.20. The second term, which is associated with the added mass coefficient C_M, is called the mass term since it involves the acceleration of the wind flow and therefore can be adjoined to the acceleration forces of the structure, that is, the terms in the equation of motion that are related to the mass distribution of the structure, Fig. 5.21. The mass term is usually formed in such a manner that the surface, or the portion of the surface that is being studied, is transformed into an equivalent circular surface, see Fig. 5.22. The two load coefficients C_D and C_M have been found to be dependent on the reduced frequency of the turbulent wind gusts,

$$\xi = \omega D/2\pi\bar{U}_R \tag{5.60}$$

The dependence on the reduced frequency has been investigated by Davenport and other investigators, [51]. Experimental results are shown in Figs. 5.20 and 5.21. The wind

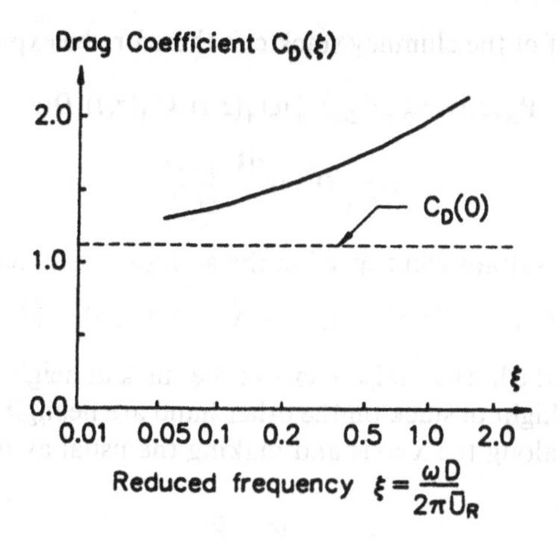

Fig. 5.20 *The drag coefficient, C_D, (after Davenport, [51]), reproduced by permission of the Institution of Civil Engineers*

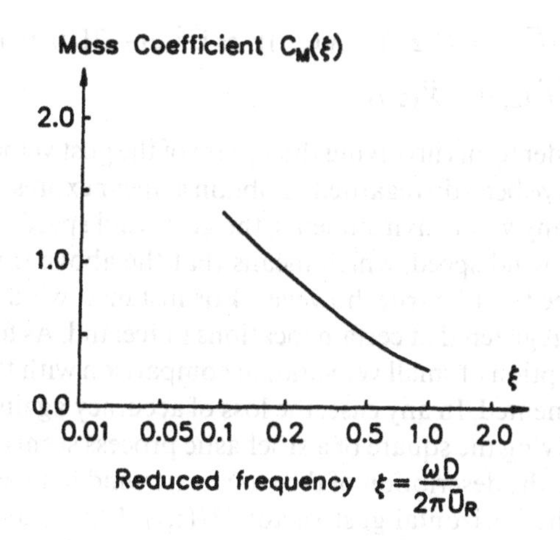

Fig. 5.21 *The mass coefficient C_M, (after Davenport, [51]), reproduced by permission of the Institution of Civil Engineers*

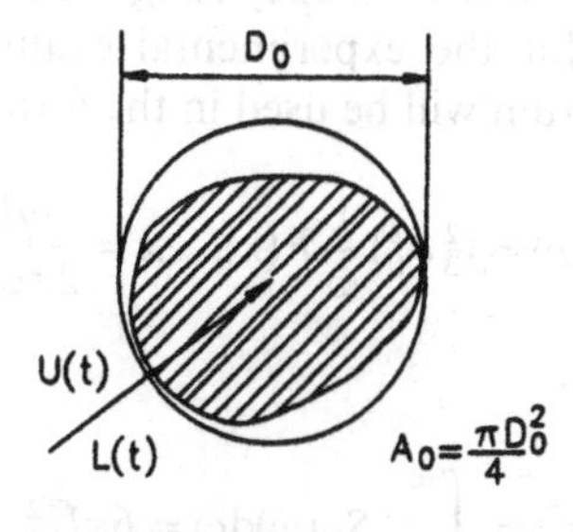

Fig. 5.22 *The equivalent load area of the added mass term*

loading per unit height of the chimney stack can therefore be expressed as follows:

$$\mathbf{P}_A(z,t) = \tfrac{1}{2}\varrho C_D(z,\xi)\mathbf{U}_A(z,t)|\mathbf{U}_A(z,t)|D(z) + \varrho\frac{\pi}{4}D^2(z)\frac{\mathrm{d}\mathbf{U}_A(z,t)}{\mathrm{d}t} \tag{5.61}$$

in which the relative resultant wind speed in the along-wind direction is

$$\mathbf{U}_A^{\mathrm{T}}(z,t) = \{\bar{U}(z) + U(z,t) - \dot{X}(z,t),\ V(z,t) - \dot{Y}(z,t)\} \tag{5.62}$$

and $\{\dot{X}(z,t),\ \dot{Y}(z,t)\}$ is the horizontal velocity of the stack at height z. The influence of the wind forces along the hight of stack on the other hand are negligible. With the dominant direction of the wind along the x-axis and making the usual assumption that the pitch angle

$$\tan\theta = \frac{V - \dot{Y}}{\bar{U} + U - \dot{X}} \sim \theta \tag{5.63}$$

is small, Fig. 5.17, the velocity and acceleration terms in Eq. (5.61) can be simplified as follows:

$$\begin{aligned} \mathbf{U}_A(z,t)|\mathbf{U}_A(z,t)| &\approx (\bar{U}(z) + U(z,t) - \dot{X}(z,t))^2 \approx \bar{U}^2(z) + 2\bar{U}(z,)U(z,t) - 2\bar{U}(z)\dot{X}(z,t) \\ \dot{\mathbf{U}}_A(z,t) &= \dot{U}(z,t) - \ddot{X}(z,t) \end{aligned} \tag{5.64}$$

in which the higher-order terms involving the square of the gust velocity minus the relative velocity of the stack have been disregarded to obtain a linear expression at a certain cost of loss of accuracy. In many wind environments, the gust wind speeds are found to be about one third of the mean wind speed, which means that the above simplification is around 90% accurate. This need not be true, however. For instance, wind gusts about twice the mean wind have been registered at certain locations in Iceland. As for the relative velocity of the stack, the assumption of small velocities in comparison with the mean wind speed is probably better documented. In any case, the loss of accuracy against having to deal with a nonlinear term involving the square of a stochastic process seems more than justifiable.

Finally, to complete the description of the stochastic wind load acting on the stack, the spectral densities of the horizontal gust vector $\{U(z,t),\ V(z,t)\}$ are briefly discussed. In Example 2.17, the height invariant power spectral density for horizontal gustiness in the along-wind direction, the so-called Davenport spectrum, was introduced, Eq. (2.93). In Fig. 5.23, three different proposals for the horizontal along-wind spectrum or longitudinal spectrum are presented, that is, those of Davenport *et al.* (compiled by Sigbjörnsson [194]). They differ mainly in the low-frequency range and as such mirror different wind climates that form the basis for the experimental evaluation of the spectra. In the following, the Davenport spectrum will be used in the form

$$S_U(\omega) = (\tfrac{2}{3}\sigma_U^2)\frac{1}{\omega}F(x), \quad x = \frac{\omega L}{2\pi\bar{U}_R} \tag{5.65}$$

in which the mean square value

$$\sigma_U^2 = \int_{-\infty}^{\infty} S_U(\omega)\mathrm{d}\omega = 6\kappa\bar{U}_R^2 \tag{5.66}$$

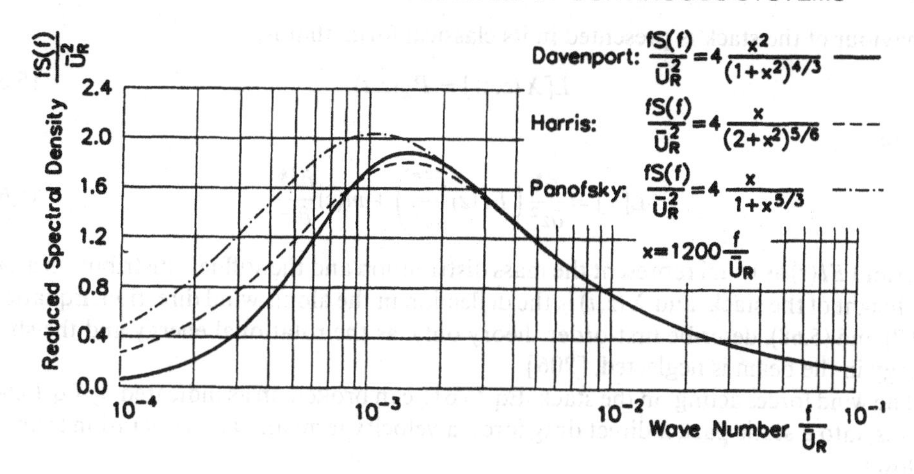

Fig. 5.23 *The longitudinal spectrum of horizontal gustiness, [194]*

where κ is a surface roughness parameter ($\kappa = (0.05, 0.015, 0.005)$) and L is the length scale of the turbulence, $L \approx 1200$ m. In Fig. 5.24, another version of the gust spectrum is shown from the Kennedy Space Centre in Florida (compiled by Sigbjörnsson, [194]). There, the across-wind or the lateral gust spectrum is also shown, indicating the much stronger turbulence effect in the along-wind direction.

Making use of the above assumptions and information about the horizontal gustiness and the drag forces induced by a turbulent wind flow around an obstruction, the equations of motion for the stack are now formed, and consequently the input–output relations for the stochastic processes obtained. The equation of motion governing the vibration

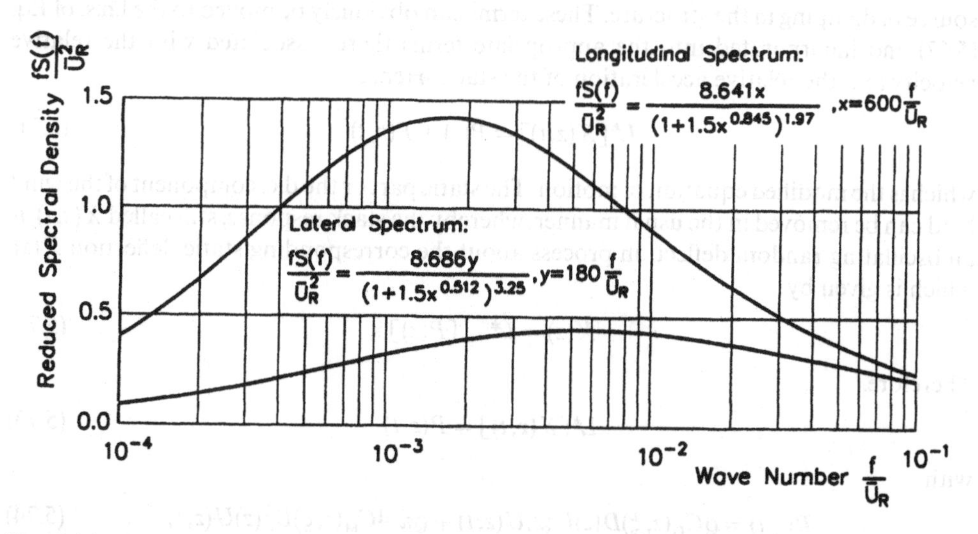

Fig. 5.24 *Longitudinal and lateral gust spectra at 18 m height at the Kennedy Space Centre in Florida, [194]*

behaviour of the stack is presented in its classical form, that is,

$$L[X(z,t)] = P_A(z,t) \tag{5.67}$$

where

$$L[*] = \frac{\partial *}{\partial z^2}\left(EI(z)\frac{\partial^2 *}{\partial z^2}\right) + m(z)\frac{\partial^2 *}{\partial t^2} \tag{5.68}$$

$m(z)$ and $EI(z)$ as usual represent the mass distribution and the stiffness distribution along the height of the stack, and $X(z,t)$ is the deflection in the along-wind direction. Equations (5.67) and (5.68), describe first-order theory only, as the rotational energy and the shear energy in the beam is neglected, [206].

The wind forces acting on the stack, Eq. (5.61), can broken up as indicated by Eq. (5.64), that is, into a static part, a direct drag force, a velocity term and an acceleration terms as follows:

$$L[X(z,t)] = \bar{P}(z) + P(z,t) - C^{\text{aero}}\, X(z,t) - M^{\text{aero}} X(z,t)$$

where

$$\bar{P}(z) = \tfrac{1}{2}\varrho C_{\text{D}}(z,0)D(z)\bar{U}^2(z) \tag{5.69}$$

is the static part of the wind loading force and

$$\begin{aligned} M^{\text{aero}}(z) &= \varrho\pi/4C_{\text{M}}(z,\xi)D^2(z) \\ C^{\text{aero}}(z) &= \varrho C_{\text{D}}(z,\xi)D(z) \end{aligned} \tag{5.70}$$

where $C^{\text{aero}}(z)$ and $M^{\text{aero}}(z)$ are the coefficients associated respectively with the relative velocity of the stack in the r.h.s. forcing function, the so-called aerodynamical damping term, and the added mass term or the aerodynamical mass. The added mass term is usually very small, whereas the aerodynamical damping can be quite large, and is often the main source of damping in the structure. These terms can obviously be moved to the l.h.s. of Eq. (5.67) and incorporated into the appropriate terms there, associated with the relative velocity and the relative acceleration of the stack. Hence

$$L^*[X(z,t)] = \bar{P}(z) + P(z,t) \tag{5.71}$$

which is the modified equation of motion. The static part or the d.c. component of the wind load can be removed in the usual manner, whereby the stack response, still called $X(z,t)$, is an oscillating random deflection process about the corresponding static deflection $\bar{X}(z)$, which is given by

$$\bar{X}(z) = L^{*-1}[\bar{P}(z)] \tag{5.72}$$

Therefore,

$$L^*[X(z,t)] = P(z,t) \tag{5.73}$$

with

$$P(z,t) = \varrho C_{\text{D}}(z,\xi)D(z)\bar{U}(z)U(z,t) + \varrho\pi/4C_{\text{M}}(z,\xi)D^2(z)\dot{U}(z,t) \tag{5.74}$$

as the forcing function or the input process. The complicated transformation of wind

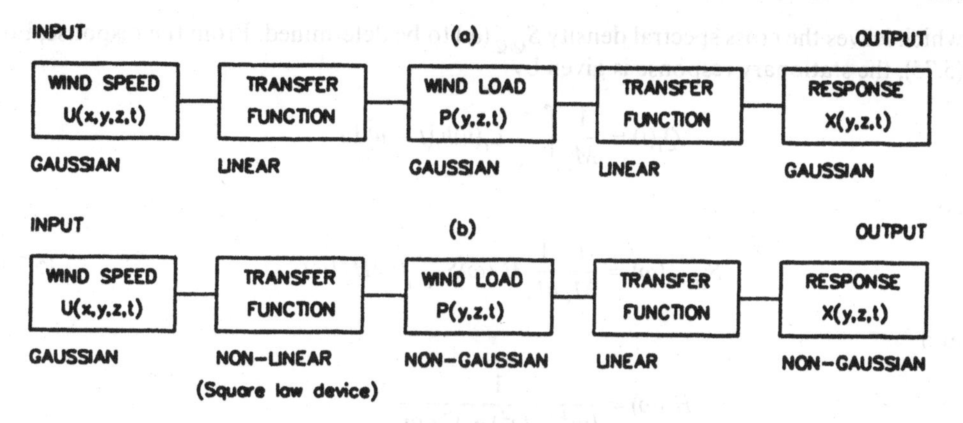

Fig. 5.25 *Schematic input–output relations (after Sigbjörnsson, [194])*

speeds into wind loading and consequently structural response is shown schematically in Fig. 5.25. It is noteworthy that in the case of linear transfer functions, the original Gaussian probability law assumed for the wind speed distribution will be preserved all the way to the corresponding deflection process of the stack, which is an inherent quality of the Gaussian or normal stochastic process (cf. Eq. 6.24). For nonlinear transfer functions, the Gaussian behaviour is of course lost.

Now, the random deflection response of the stack will be evaluated using the normal-mode approach. Then

$$X(z,t) = \sum_{i=1}^{\infty} \phi_i(z)Q_i(t) \tag{5.75}$$

where $\phi_i(z)$ is the ith normal mode shape function and $Q_i(t)$ is the ith generalized response, given by the equation,

$$\ddot{Q}_i(t) + 2\lambda_i\omega_i\dot{Q}_i(t) + \omega_i^2 Q_i(t) = L_i(t)/M_i \tag{5.76}$$

where ω_i is the ith normal mode frequency, λ_i the ith damping ratio and

$$M_i = \int_0^H m(z)\phi_i^2(z)\mathrm{d}z, \quad L_i = \int_0^H P(z,t)\phi_i^2(z)\mathrm{d}z \tag{5.77}$$

are the ith generalized mass and the ith generalized force respectively.

The mean value of the response has already been dealt with, Eq. (5.72). In order to find the mean square response $E[X^2(z,t)]$, the power spectral density of the response process, Eq. (5.75), is sought. Starting with the autocorrelation function

$$E[X(z,t)X(z,t+\tau)] = \sum_{i=1}^{\infty}\sum_{j=1}^{\infty} \phi_i(z)\phi_j(z)E[Q_i(t)Q_j(t+\tau)]$$

the power spectral density is

$$S_X(z,\omega) = \sum_{i=1}^{\infty}\sum_{j=1}^{\infty} \phi_i(z)\phi_j(z)S_{Q_iQ_j}(\omega) \tag{5.78}$$

which leaves the cross spectral density $S_{Q_iQ_j}(\omega)$ to be determined. From the response, Eq. (5.75), the stationary response is given by

$$Q_i(t) = \frac{1}{M_i}\int_{-\infty}^{\infty} L_i(u)h_i(t-u)\mathrm{d}u$$

so

$$S_{Q_iQ_j}(\omega) = \frac{1}{M_i}\frac{1}{M_j}H_i(\omega)S_{L_iL_j}(\omega)H_j^*(\omega) \tag{5.79}$$

and

$$H_i(\omega) = \frac{1}{(\omega^2-\omega_i^2)+\mathrm{i}2\lambda_i\omega_i}$$

From Eq. (5.78), the cross-spectrum for the generalized force is obtained as

$$S_{L_iL_j}(\omega) = \int_0^H\int_0^H \phi_i(z_m)\phi_j(z_n)S_{P_mP_n}(\omega)\mathrm{d}z_m\,\mathrm{d}z_n \tag{5.80}$$

and through Eq. (5.74), the cross spectral density for the direct wind loads, introducing the velocity profiles shown in Fig. 5.19, can be obtained, Eq. (5.81), in which the cross-spectrum $S_{U_mU_n}(\omega)$, Eq. (2.95), together with the coherence function $Coh_{mn}(\omega)$, Eq. (2.92), and the phase angle $\Phi_{mn}(\omega)$, Eq. (2.93), are already known from Example 2.17. The spatial separation r_{mn} is equivalent to $|z_m - z_n|$:

$$\begin{aligned} S_{P_mP_n}(z_m, z_n, \omega) = \varrho^2\bar{U}_R^2\left(\frac{z_m}{z_R}\frac{z_n}{z_R}\right)^{\alpha}&[C_\mathrm{D}(z_m,\xi_m)C_\mathrm{D}(z_n,\xi_n) \\ &+\frac{\pi^4}{4}\xi_m\xi_nC_\mathrm{M}(z_m,\xi_m)C_\mathrm{M}(z_n,\xi_n)] \\ &\times D(z_m)D(z_n)S_{U_mU_n}(\omega), \quad \xi_n = \frac{\omega D(z_n)}{2\pi\bar{U}_R} \end{aligned} \tag{5.81}$$

By combining Eqs. (2.91), (2.92) and (2.94) with Eq. (5.81) and inserting the result into Eq. (5.80), the following expression for the cross spectrum of the wind loads is obtained:

$$\frac{S_{L_iL_j}}{(\frac{1}{2}\varrho A\bar{U}_R^2)} = 4|F_{ij}^u(\omega)|\frac{S_U(\omega)}{\bar{U}_R^2} \tag{5.82}$$

The function, $F_{ij}^U(\omega)$, introduced in Eq. (5.82), is the so-called the aerodynamic admittance Eq. (5.83), which transforms wind speeds into wind loads, that is,

$$\begin{aligned} |F_{ij}^U(\omega)|^2 = \frac{1}{A^2}\int_0^H\int_0^H\left[\frac{z_mz_n}{z_R^2}\right]^{\alpha}&[C_\mathrm{D}(z_m,\xi_m)C_\mathrm{D}(z_n,\xi_n) \\ &+\frac{\pi^4}{4}C_\mathrm{M}(z_m,\xi_m)C_\mathrm{M}(z_n,\xi_n)]D(z_m)D(z_n) \\ &\times\exp\left[-\frac{b\omega r_{mn}}{\bar{U}_R}\right]\cos\left[\frac{c\omega r_{mn}}{\bar{U}_R}\right]\phi_i(z_m)\phi_j(z_n)\mathrm{d}z_m\,\mathrm{d}z_n \end{aligned}$$

where $A = \int_0^H D(z)dz$ and the constants b and c are descriptive of the turbulence conditions as already discussed in Example 2.17.

The aerodynamic admittance $F_{ij}^U(\omega)$ obviously controls the transfer of wind speeds into wind loads according to the model prescribed by Morrison's Eq. (5.59) with subsequent linearization depending on the structure being analysed. In Figs. 5.26–5.28, the aerody-

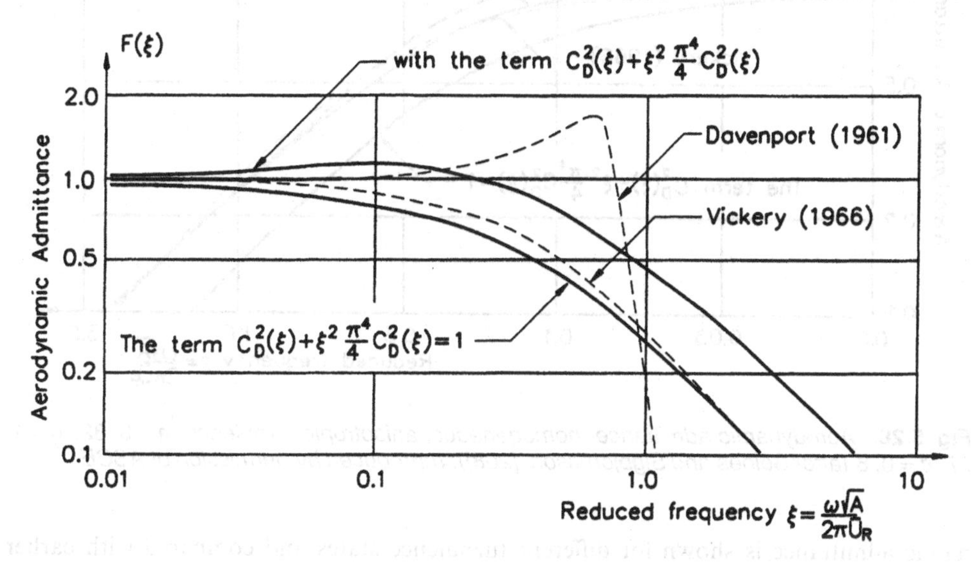

Fig. 5.26 *Aerodynamic admittance: homogeneous, isotropic turbulence, a = b = 1, 27, c = 0, (after Solnes and Sigbjörnsson, [208]), reproduced by permission of ASCE*

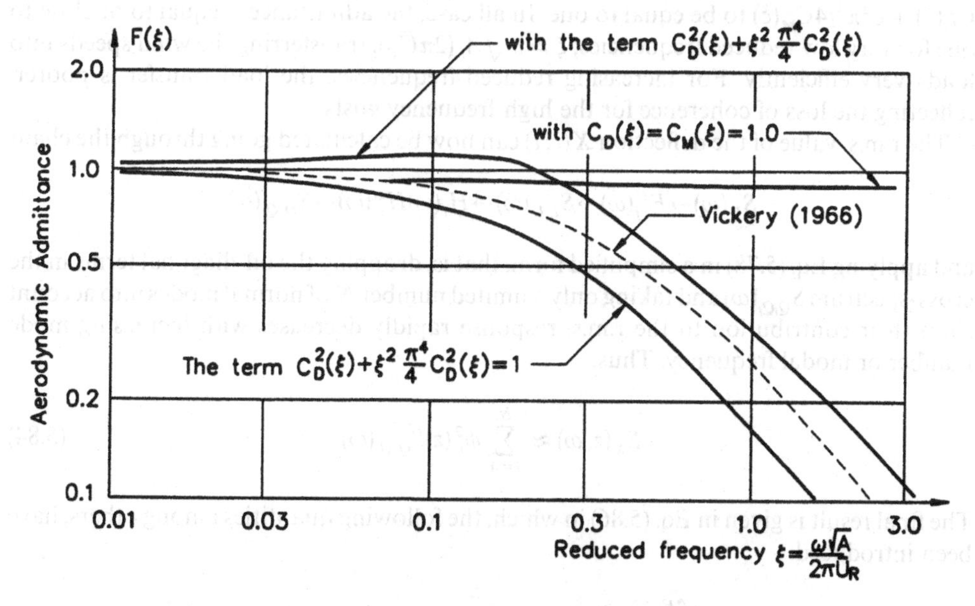

Fig. 5.27 *Aerodynamic admittance: homogeneous, isotropic turbulence, a = b = 1, 27, c = 0, 8, (after Solnes and Sigbjörnsson, [208]), reproduced by permission of ASCE*

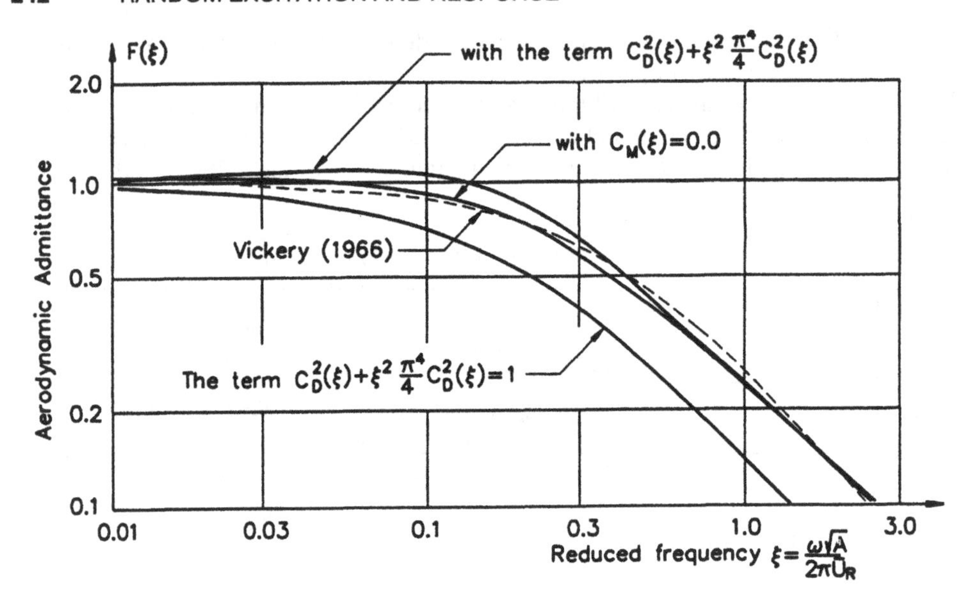

Fig. 5.28 *Aerodynamic admittance: homogeneous, anisotropic turbulence,* $a = 3, 82$, $b = 1, 27$, $c = 0, 8$ *(after Solnes and Sigbjornsson, [208]), reproduced by permission of ASCE*

namic admittance is shown for different turbulence states and compared with earlier results reported by Davenport, [51] and Vickery, [229]. The aerodynamic admittance is computed in two different ways, either using the experimental results for the drag and added mass coefficients shown in Figs. 5.20 and 5.21 or presuming the term $C_D^2(\xi) + \xi^2\pi^4/4C_M^2(\xi)$ to be equal to one. In all case, the admittance is equal to or close to one for very low reduced frequencies, $\xi = \omega\sqrt{A}/(2\pi\bar{U}_R)$, transferring the wind speeds into loads very efficiently. For increasing reduced frequencies, the load transfer is poorer, reflecting the loss of coherence for the high-frequency gusts.

The r.m.s. value of the deflection $X(z, t)$ can now be calculated going through the chain

$$S_U(\omega) \rightarrow F_{ij}^U(\omega) \rightarrow S_{L_iL_j}(\omega) \rightarrow H_i(\omega)H_j^*(\omega) \rightarrow S_{Q_iQ_j}(\omega)$$

and applying Eq. (5.78) in a simplified form, that is, dropping the off-diagonal terms in the cross-spectrum $S_{Q_iQ_j}(\omega)$ and taking only a limited number N of normal modes into account since their contribution to the r.m.s. response rapidly decreases with increasing mode number or modal frequency. Thus,

$$S_X(z, \omega) \approx \sum_{j=1}^{N} \phi_i^2(z) S_{Q_iQ_i}(\omega) \tag{5.84}$$

The final result is given in Eq. (5.86) in which, the following quantities among others, have been introduced.

$$A = \int_0^H D(z)\mathrm{d}z, \quad \sigma_U^2 = 6\kappa\bar{U}_R^2, \quad I_U^2(z_R) = \sigma_U^2/\bar{U}_R^2 \tag{5.85}$$

where $I_U(z)$ is called the intensity of the turbulence.

$$\mathrm{RMS}[X(z,t)] = \sigma^2[X(z,t)] = \sum_{i=1}^{N} \phi_i^2(z) \int_{-\infty}^{\infty} S_{Q_iQ_i}(\omega)\mathrm{d}\omega$$

$$= (\varrho A \overline{U}_R)^2 \int_{-\infty}^{\infty} S_U(\omega) \left[\sum_{i=1}^{N} \phi_i^2(z) \frac{|H_i(\omega)|^2}{M_i^2} |F_{ii}^U(\omega)|^2 \right] \mathrm{d}\omega$$

$$\approx (\tfrac{1}{2}\varrho A \overline{U}_R^2) \frac{I_U^2(z_R)}{\sigma_U^2} \sum_{i=1}^{N} \left(\frac{\phi_i(z)}{\omega_i^2 M_i} \right)^2 \frac{\pi \omega_i}{\lambda_i} |F_{ii}^U(\omega_i)|^2 S_U(\omega_i) \tag{5.86}$$

Having resolved the response of the stack in the along-wind direction, $X(z,t)$, the response in the across-wind direction, $Y(z,t)$, will next be addressed. From Fig. 5.19, it is evident that the small pitch angle θ, Eq. (5.63), will produce a load component in the across-wind or y-direction equal to

$$T(z,t) = P(z,t)\tan\theta \approx P(z,t)\theta \tag{5.87}$$

The analysis of the across-wind response follows a more or less similar path as for the along-wind response. The equations are complicated, however, by the fact that the across-wind response is coupled to the along-wind response through Eq. (5.87). The final result can be represented as follows:

$$\mathrm{RMS}[Y(z,t)] = \sigma^2[Y(z,t)] = (\tfrac{1}{2}\varrho A \overline{U}_R^2) \frac{I_U^2(z_R)}{\sigma_U^2}$$

$$\times \sum_{i=1}^{N} \left(\frac{\phi_i(z)}{\omega_i^2} \frac{(z)}{M_i} \right)^2 \frac{\pi \omega_i}{\lambda_i} |F_{ii}^V(\omega_i)|^2 S_V(\omega_i) \tag{5.88}$$

in which all quantities have the same meaning as before except that $S_V(\omega)$ is now the spectrum of lateral gustiness, Fig. 5.23, and $F_{ii}^V(\omega)$ is the aerodynamic admittance taking into account the complicated coupling of the motion. Sometimes, the lateral response due to the gusts is just estimated as a small portion, for instance one third, of the along-wind response. This may be acceptable as another source of forces producing motion or responses in the across-wind direction, that is, the lift forces, (cf. Eq. (5.89)), can be of more importance.

The airflow around an obstacle can produce a very complicated pattern of vortexes or eddies, which are rythmically released at the edges of the obstacle, [58], [194]. This eddy shedding will produce forces lateral to the direction of the airflow, the so-called lift forces, which can cause a movement of the obstacle in the across-wind direction. In Fig. 5.29, the flow pattern about a round object like the stack with the lift forces $K(z,t)$ is shown. The eddy shedding will depend heavily on the Reynolds number of the flow, $\mathrm{Re} = U(z,t)D(z)/\nu$. The Reynolds number is defined as the ratio between the inertia force and the viscous force acting on a particle in the flow. Thus,

$$\mathrm{Re} = \frac{\text{inertia force}}{\text{viscous force}} = \frac{\varrho l^3 a}{l^2 \tau} = \frac{ul}{\nu} = \frac{U(z,t)D(z)}{\nu} \tag{5.89}$$

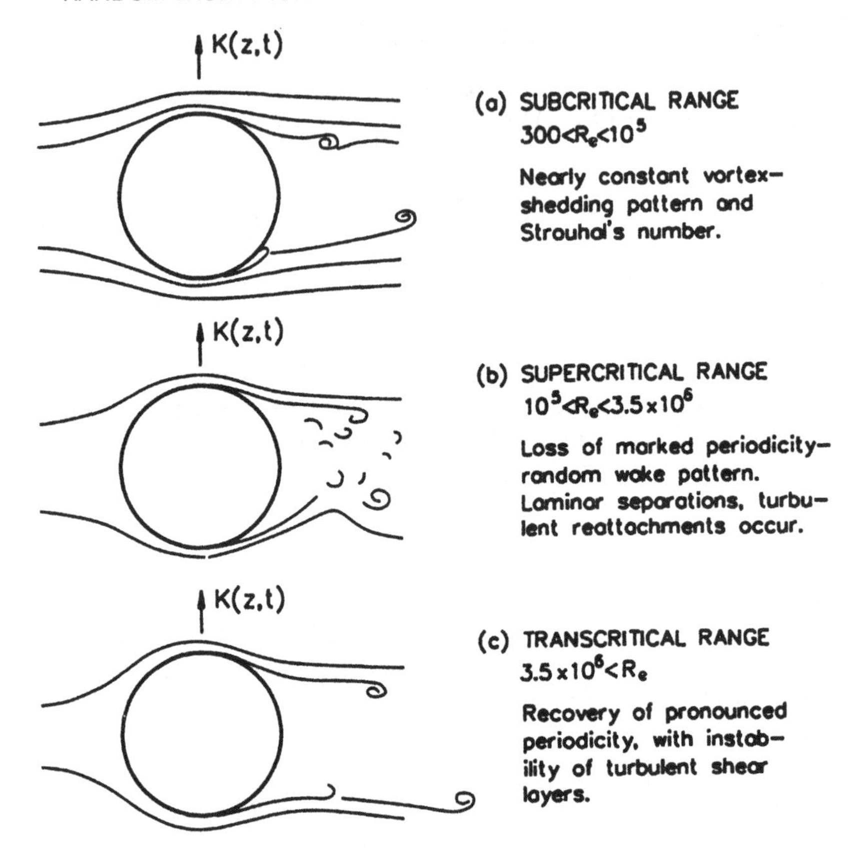

Fig. 5.29 *The eddy shedding pattern for different conditions of flow, [194]*

in which a is the acceleration of the particle, l is a characteristic length, in this case selected as the diameter of the stack, τ is the shear stress in the flow, $u = U(z, t)$ is the velocity and ν is the kinematic viscosity of air. Reynolds, who formulated his model law in 1883, (see [58]), thus disregards the action of the gravity forces, but as the viscous forces are fundamental for the airflow around the surface of structure, the magnitude of Reynolds number will be decisive in determining the flow pattern. For stacks and other civil engineering structures, the Reynolds number will almost without exception fall into the so-called transcritical range where the eddy shedding is fairly periodic, producing systematic lift forces. The drag coefficient C_D is also found to be highly dependent on the Reynolds number, Fig. 5.30 (see e.g., Scruton and Rogers, [187], also reproduced in [58]).

The frequency of the shedding pattern f_s is governed by yet another physical number, Strouhal's number S_t, after the Czech engineer Strouhal who studied this phenomenon over one hundred years ago, [58]. The eddy shedding frequency f_s, which is found by dividing the velocity v_l of the eddies in the so-called von Kármán eddy lane (after Theodore von Kármán, who was a pioneer in this field of research) by the distance l_h between two eddies in the lane, Fig. 5.29, is

$$f_s = v_l/l_h = S_t U(z, t)/D(z) \tag{5.90}$$

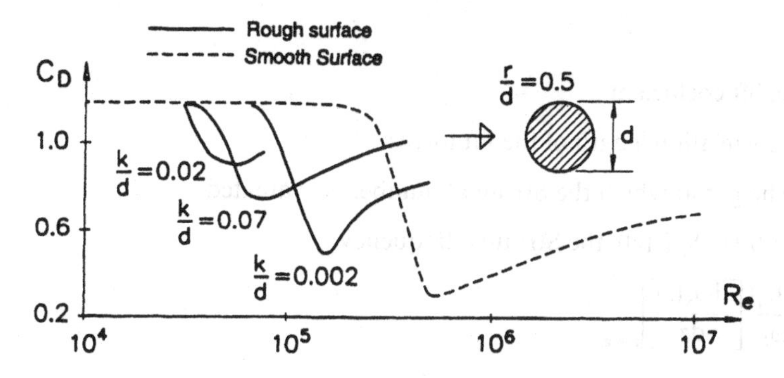

Fig. 5.30 *Dependence of the drag coefficient C_D on Reynolds number and the surface roughness of a cylinder, (after Scruton and Rogers, [187]), reproduced by permission of The Royal Society*

This frequency is found to be proportional to the free field wind velocity divided by the diameter of the cylinder and the factor of proportionality is the Strouhals number. The Strouhals number is found to be dependent on the Reynolds number, the cross section form of the obstacle, including the roughness and the turbulence in the velocity field, Fig. 5.31.

For normal damped structures such as the stack, the maximum excitation takes place when the dominant eddy sheddying frequence f_s is close to the fundamental frequency of the structure, f_0. The critical free field wind velocity U_{cr} is therefore

$$U_{cr} = f_0 \, D/S_t \tag{5.91}$$

which is often very low, with ensuing risk of fatique in a structure prone to across-wind excitation in a frequent wind speed environment.

Taking the above problem into consideration, the mean square response for the across wind direction was shown by Sigbjörnsson to have the following form, [194],

$$\mathrm{RMS}[Y(z,t)] \approx \sigma^2[Y(z,t)] = \sum_{i=1}^{N} \phi_i^2(z) \left[\frac{\frac{1}{2} \varrho C_L \overline{U_R^2} D(z_R)}{M_i^2 \omega_i^2} \right]^2 \frac{\pi \delta \phi_i^2(z_i)}{2\omega_i^* \lambda_i(v_i)^{-1}} \tag{5.92}$$

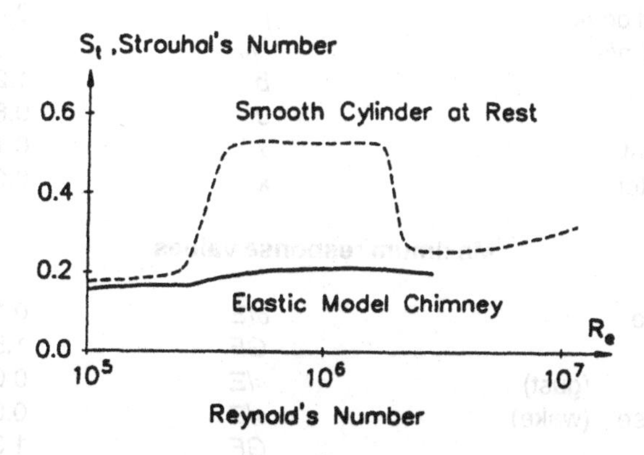

Fig. 5.31 *Strouhals number as a function of Reynolds number, (Wooton and Scruton, 1970, [237]), reproduced by permission of CIRIA*

in which

C_L is the lift coefficient

δ is the correlation length of the lift forces

z_i is the height at which the Strouhal number is evaluated

$\omega_S(z) = (2\pi \bar{U}_R S_t/D(z))$, the Strouhal frequency

$$\omega_i^* = \frac{D(z_i)}{\omega_i}\left[\frac{d\omega_S(z)}{dz}\right]_{z=z_i}$$

and

$\nu_i = F(\phi_i(z_i), D(z_i), H, \omega_i^*)$ which is a complicated function of these quantities.

Sigbjörnsson used the above relations to evaluate the mean square response of two stacks in an open country environment, one 200 m high and another 80 m high, [194]. The pertinent data are given in Table 5.2. The longitudinal spectrum of Davenport, Eq. (2.58), is used for the along-wind direction and the across-wind response due to gusts is estimated

Table 5.2 *Numerical data and relevant response parameters of two stacks in an open country environment (after Sigbjörnsson [194])*

	Symbol	Stack I	Stack II
Height (m)	H	200	80
Base diameter (m)	D_1	14.0	12.0
Tip diameter (m)	D_0	5.0	9.3
Natural frequency (r.p.s.)	ω_1	2.3	6.28
Critical damping ratio	λ_1	0.02	0.02
Drag coefficient	C_D	1.0	1.0
Added mass coefficient	C_M	0.0	0.0
Lift coefficient	C_L	0.1	0.11
Strouhal number	St	0.25	0.25
Correlation length	δ	0.6	0.6
Width of the spectral peak	β	0.2	0.2
Mean wind velocity (m/s)	$\bar{U}_{10}$	30	30
Decay constant	b	1.27	1.27
Phase constant	c	0.8	0.8
Wind profile exponent	α	0.16	0.16
Roughness parameter	κ	0.005	0.005
Maximum response values			
Along wind response	σ/Ξ	0.162	0.162
	GF	1.56	1.60
(gust)	σ/Ξ	0.094	0.092
Across wind response (wake)	σ/Ξ	0.029	0.914
	GF	1.35	3.44
Resulting dynamic response factor *GF*		1.60	3.70

to be $1/\sqrt{3}$ of the longitudinal response. The aerodynamic admittance $F_{ii}^{U}(\omega)$ is approximated by its first generalized component, as the response of the first normal mode carries the bulk of the information with the contribution of higher normal modes rapidly decreasing.

Finally, the expected maximum deflection of the stacks is estimated using the overload design factor, which was presented in Chapter 4, Eq. (4.63). The maximum deflection is then given by

$$\Xi_{max}(z) = GF \times \text{ (the static deflection in the } x \text{ or } y \text{ direction)}$$

where GF the 'gust factor' takes over the role of the overload factor given by Eq. (4.87), or

$$GF = 1 + g(\nu\Delta T, \varepsilon)\sigma[X \text{ or } Y]/E[X \text{ or } Y) \tag{5.93}$$

The gust factor, Eq. (5.93), contains the peak function $g(\nu\Delta T, \varepsilon)$ shown in Fig. 4.20. For narrow band response, the frequency is $\nu \approx \omega_i/2\pi$, the first mode frequency, and $\Delta T = 10$ minutes, which is the usual reference time for measuring wind speed mean values.

6 Some Fundamental Stochastic Processes

In the preceding sections, a brief description of the probability mechanism of stochastic processes has been given and their classification discussed. In this respect, the frequency composition and time-domain correlation of stationary processes played a dominant role, whereas the probability laws governing the processes were shoved into the background. However, the probability distribution describing a stochastic process is the fundamental information and other derived knowledge about the behaviour of the process is often secondary. Since the stochastic process in any kind of an application is a mathematical model for a certain given physical system that behaves and changes in accordance with the laws of probability, the goal is to seek a one-to-one correspondence between the states of the physical system and the stochastic process chosen for the modelling. This is far from being easy since the two things are essentially different, that is, to transfer the sometimes ill-defined probability laws and mechanical behaviour of the physical world to the orderly behaviour of elements of the mathematical world. The stochastic processes being treated have either been continuous random functions of time or discrete random sequences, which can be represented by irregular point events in time or space, in which case, the probability law is that of discrete variables. Sometimes, the stochastic process or the physical system that is being modelled is best described by a combination of both continuous and discrete processes. An example of such processes is the compound Poisson process, as will be seen in the following. Another way of describing the stochastic process is: (a) by its history (sample functions) and (b) by its dynamics or ensemble statistics. If a stochastic process is considered to be ergodic and has an invariant measure, for instance a quasi-stationary mean value, it has already been shown that any complete physical realization of the process, contains its dynamics up to the order of the invariant measure. In the literature, stochastic processes are characterized in a great number of ways and many different and often confusing expressions are in use, [50].

In the following sections, the probability laws and behaviour of certain fundamental stochastic processes, which frequently show up in the study of physical systems, will be briefly discussed. The probability laws of Gauss, Poisson and Markov play an important role in this context. Other types of stochastic processes will also be mentioned such as martingales and certain non-linear Markov processes. Gaussian and Poisson processes will be discussed in some detail, and a few examples with practical applications will be shown. This will be followed by a short presentation of Markov processes with a few examples given. Then, a short overview of martingale processes will be given, and the interrelationship between the different probability mechanisms sought explained. Finally, a typical non-Markovian process, the so-called Boltzmann process, will be introduced.

The Boltzmann process has been applied in such different fields as continuum mechanics, biology and earthquake risk analysis [50], [193].

6.1 GAUSSIAN OR NORMAL RANDOM PROCESSES

Consider a stochastic process $\{X(t), t\in T\}$. If for any t, the probability density Eq. (2.1) is the Gaussian density,

$$p(x,t)=\frac{1}{\sigma(t)\sqrt{2\pi}}\exp\left[-\frac{(x-\mu(t))^2}{2\sigma^2(t)}\right] \tag{6.1}$$

the process is a Gaussian random process. Clearly,

$$E[X(t)]=\mu(t)$$

and

$$E[X^2(t)]=\sigma^2(t)+\mu^2(t) \tag{6.2}$$

A unique property of the Gaussian probability law is that a knowledge of the two first central moments, μ_X and σ_X, completely determines the characteristic function (see Example 1.15, Eq. (1.83))

$$\varphi_X(\omega)=\exp\left[\mathrm{i}\mu_X\omega-\frac{\sigma_X^2}{2}\omega^2\right]$$

and hence any probabilistic assertion concerning the variable X. Therefore, if for any $t\in T$ the two moments are known, the entire information pyramid, (Fig. 2.4), can be constructed.

Example 6.1 Consider a Gaussian random variable X with mean μ and variance σ^2. Find the moments and central moments of any order n.

Start by studying the standardized variable $U=(X-\mu)/\sigma$ with mean 0 and variance 1. Clearly,

$$E[U^n]=\frac{1}{\sqrt{2\pi}}\int_{-\infty}^{\infty}u^n\exp\left[-\frac{u^2}{2}\right]du$$

For n odd, u^n is an odd function of u, whereas the density is an even function. Therefore, the central moments are obtained as follows:

$$E[U^n]=0 \qquad \text{for } n \text{ odd}$$

For n even,

$$E[U^n]=\sqrt{\frac{2}{\pi}}\int_{-\infty}^{\infty}u^n\mathrm{e}^{-u^2/2}du=\sqrt{\frac{2}{\pi}}\times 2^{(n-1)/2}\int_{-\infty}^{\infty}\left(\frac{u^2}{2}\right)^{n-1/2}\mathrm{e}^{-u^2/2}\,\mathrm{d}\left(\frac{u^2}{2}\right)$$

$$= \sqrt{\frac{2}{\pi}} \times 2^{(n-1)/2} \int_{-\infty}^{\infty} (x)^{(n-1)/2} \mathrm{e}^{-x} \mathrm{d}x = \frac{1}{\sqrt{\pi}} 2^{n/2} \Gamma\left(\frac{n-1}{2} + 1\right)$$

$$= \frac{1}{\sqrt{\pi}} 2^{n/2} \Gamma\left(\frac{n+1}{2}\right), \{\Gamma(\alpha+1) = \alpha\Gamma(\alpha), \Gamma(\tfrac{1}{2}) = \sqrt{\pi}\}$$

So

$$E[(X-\mu)^n] = \sigma^n E[U^n] = \begin{cases} 0 & n = \text{odd} \\ \pi^{-1/2} \sigma^n 2^{n/2} \Gamma\left(\frac{n+1}{2}\right) & n = \text{even} \end{cases} \tag{6.3}$$

As for the moments, form the $(n-2)$th moment:

$$E[X^{n-2}] = \int_{-\infty}^{\infty} x^{n-2} p(x) \mathrm{d}x$$

and by integrating by parts:

$$\int_{-\infty}^{\infty} x^{n-2} p(x) \mathrm{d}x = \left[\frac{x^{n-1}}{n-1} p(x)\right]_{-\infty}^{\infty} - \int_{-\infty}^{\infty} \frac{x^{n-1}}{n-1} \left(\frac{x-\mu}{\sigma^2}\right) p(x) \mathrm{d}x$$

$$= \frac{1}{\sigma^2 (n-1)} \left[\int_{-\infty}^{\infty} x^n p(x) \mathrm{d}x - \mu \int_{-\infty}^{\infty} x^{n-1} p(x) \mathrm{d}x\right]$$

therefore

$$E[X^n] = \mu E[x^{n-1}] + \sigma^2 E[X^{n-2}] \tag{6.4}$$

which is a most convenient formula for calculating the higher-order moments.

6.1.1 Multivariate Gaussian Variables

The nth-order probability density (nth row in the information pyramid, Fig. 2.4) describing the n random variables $X(t_1), X(t_2), \ldots, X(t_n)$ is the multivariate normal density. Start by assuming that $\{X(t_i)\}$ have zero mean values and variances equal to one. Then

$$p_X(x_1, x_2, \ldots, x_n) = \frac{1}{(\sqrt{2\pi})^n} \exp[-\tfrac{1}{2}(x_1^2 + x_2^2, \ldots, x_n^2)] = \frac{1}{(\sqrt{2\pi})^n} \exp[-\tfrac{1}{2}\mathbf{x}^{\mathrm{T}}\mathbf{x}] \tag{6.5}$$

where $\mathbf{x}^{\mathrm{T}} = \{x_1, x_2, \ldots, x_n\}$ (as before, vectors and matrices are denoted by bold face letters). Next, consider a set of normal variables $\mathbf{Y}^{\mathrm{T}} = \{Y_1, Y_2, \ldots, Y_n\}$ that have the central moments

$$\mu_{11}(t_i, t_j) = \mu_{ij} = E[(Y_i - \mu_i)(Y_j - \mu_j)] = \varrho_{ij}\sigma_i\sigma_j \tag{6.6}$$

in which (the short hand form μ_{ij} will be used whenever there is no possible confusion with the higher central moments Eq. (1.71)) the mean values and the standard deviations of

$\mathbf{Y}$ respectively are

$$\boldsymbol{\mu}^{\mathrm{T}} = \{\mu_1, \mu_2, \ldots, \mu_n\} \qquad \text{and} \qquad \boldsymbol{\sigma}^{\mathrm{T}} = \{\sigma_1, \sigma_2, \ldots, \sigma_n\} \tag{6.7}$$

and the correlation coefficients are represented by the matrix $\{\varrho_{ij}\}$. Now, the following transformation is made:

$$\mathbf{Y} = \mathbf{AX} + \boldsymbol{\mu} \tag{6.8}$$

where $\mathbf{A}$ is an $n \times n$ matrix to be determined. By forming the central moments,

$$\mathbf{D} = E[(\mathbf{Y} - \boldsymbol{\mu})(\mathbf{Y} - \boldsymbol{\mu})^{\mathrm{T}}] = E[\mathbf{AX}(\mathbf{AX})^{\mathrm{T}}] = \mathbf{A}E[\mathbf{XX}^{\mathrm{T}}]\mathbf{A}^{\mathrm{T}} = \mathbf{AA}^{\mathrm{T}} \tag{6.9a}$$

so

$$\mathbf{D} = \mathbf{AA}^{\mathrm{T}} = \begin{Bmatrix} \mu_{11} & \mu_{12} & \cdots & \mu_{1n} \\ \mu_{21} & \mu_{22} & \cdots & \mu_{2n} \\ \cdots & & & \\ \mu_{n1} & \mu_{n2} & \cdots & \mu_{nn} \end{Bmatrix} = \begin{Bmatrix} \sigma_1^2 & \varrho_{12}\sigma_1\sigma_2 & \cdots & \varrho_{1n}\sigma_1\sigma_n \\ \varrho_{21}\sigma_2\sigma_1 & \sigma_2^2 & \cdots & \varrho_{2n}\sigma_2\sigma_n \\ \cdots & & & \\ \varrho_{n1}\sigma_n\sigma_1 & \varrho_{n2}\sigma_n\sigma_2 & \cdots & \sigma_n^2 \end{Bmatrix} \tag{6.9b}$$

The joint probability density of $\mathbf{Y}$ can now be obtained by a simple transformation, (Eq. (1.85)).

$$p_{\mathbf{Y}}(\mathbf{y}) = |\mathbf{J}|\, p_{\mathbf{X}}(\mathbf{A}^{-1}\mathbf{y} - \mathbf{A}^{-1}\boldsymbol{\mu})$$

where the Jacobian is given by

$$\mathbf{J} = \det\left\{\frac{\partial \mathbf{A}^{-1}\mathbf{y}}{\partial \mathbf{y}}\right\} = \det \mathbf{A}^{-1}, \qquad |\mathbf{J}| = |\det \mathbf{A}^{-1}| = \frac{1}{|\det \mathbf{A}|}$$

Therefore,

$$\begin{aligned} P_{\mathbf{Y}}(\mathbf{y}) &= \frac{1}{|\det \mathbf{A}|} p_{\mathbf{X}}(\mathbf{A}^{-1}(\mathbf{y} - \boldsymbol{\mu})) \\ &= \frac{1}{(\sqrt{2\pi})^n |\det \mathbf{A}|} \exp[-\tfrac{1}{2}(\mathbf{y} - \boldsymbol{\mu})^{\mathrm{T}}(\mathbf{A}^{-1})^{\mathrm{T}}\mathbf{A}^{-1}(\mathbf{y} - \boldsymbol{\mu})] \\ &= \frac{1}{(\sqrt{2\pi})^n \sqrt{|\det \mathbf{D}|}} \exp[-\tfrac{1}{2}(\mathbf{y} - \boldsymbol{\mu})^{\mathrm{T}}\mathbf{D}^{-1}(\mathbf{y} - \boldsymbol{\mu})] \end{aligned} \tag{6.10}$$

In Eq. (6.10), the n random variables $\{Y_i = Y(t_i)\}$ are statistically dependent with correlation coefficients as follows:

$$\varrho_{ij}(t_i, t_j) = \frac{E[(X_i - \mu_i)(X_j - \mu_j)]}{\sqrt{E[(X_i - \mu_i)^2]\,E[(X_j - \mu_j)^2]}} \tag{6.11}$$

If the random process $X(t)$ is stationary, the correlation coefficients are given by

$$\varrho_{ij}(t_i, t_j) = \frac{R_X(t_i - t_j) - \mu_X^2}{\sigma_X^2} \tag{6.12}$$

Now for non-singular correlation matrices $\mathbf{D}$, a simple transformation

$$\mathbf{X} = \mathbf{BY}, \qquad \mathbf{Y} = \mathbf{B}^{-1}\mathbf{X}, \qquad \boldsymbol{\mu}_X = \mathbf{B}\boldsymbol{\mu}_Y \tag{6.13}$$

is possible, which breaks down the statistical dependence between the variables $\mathbf{Y}$ and produces n independent random variables $\mathbf{X}^T = \{X_1, X_2, \ldots, X_n\}$. Inserting Eq. (6.13) into Eq. (6.5):

$$p_X(\mathbf{X}) = \frac{1}{(\sqrt{2\pi})^n \sqrt{|\det \mathbf{D}|}} \exp[-\tfrac{1}{2}(\mathbf{x} - \boldsymbol{\mu}_X)^T (\mathbf{B}^{-1})^T (\mathbf{D}^{-1})(\mathbf{B}^{-1})(\mathbf{x} - \boldsymbol{\mu}_X)] \, |\mathbf{J}_n| \tag{6.14}$$

where J_n is the Jacobian of the transformation. Clearly, $|\mathbf{J}_n| = |\det \mathbf{B}^{-1}|$. Obviously, the transformation matrix is the matrix of eigenvectors of the matrix $\mathbf{D}$ in order to produce a diagonal matrix with zero off-diagonal elements. Since $\mathbf{B}$ is the matrix of the eigenvectors of the symmetric matrix $\mathbf{D}$,

$$(\mathbf{D} - \lambda_i \mathbf{I})\mathbf{B}_i = \mathbf{0} \tag{6.15}$$

where $\mathbf{B}_i$ is the ith eigenvector (a column in $\mathbf{B}$), $\mathbf{I}$ the unit matrix and λ_i is the corresponding ith eigenvalue, that is, the ith latent root of the equation

$$\det(\mathbf{D} - \lambda \mathbf{I}) = 0 \tag{6.16}$$

The eigenvectors also satisfy the orthogonality condition

$$\mathbf{B}_j^T \mathbf{D} \mathbf{B}_i = \lambda_i \mathbf{B}_j^T \mathbf{B}_j = \begin{cases} 0 & \text{if } i \neq j \text{ and } \lambda_i \neq \lambda_j \\ \lambda_i & \text{if } i = j \end{cases}$$

That is, the normalized eigenvectors satisfy the condition $\mathbf{B}^T\mathbf{B} = \mathbf{I}$ or $\mathbf{B}^T = \mathbf{B}^{-1}$ and therefore

$$|\mathbf{J}_n| = |\det \mathbf{B}^{-1}| = 1$$

and

$$\mathbf{B}^{-1}\mathbf{D}^{-1}\mathbf{B} = (\mathbf{BDB}^{-1})^{-1} = (\mathbf{B}^T\mathbf{DB})^{-1} = \boldsymbol{\Lambda}^{-1} \tag{6.17}$$

where $\boldsymbol{\Lambda}^{-1}$ is the diagonal matrix of the eigenvalues,

$$\boldsymbol{\Lambda} = \{\lambda_1, \lambda_2, \ldots, \lambda_n\} \tag{6.18}$$

which, introduced into Eq. (6.14), gives the uncoupled density

$$p_X(\mathbf{x}) = \frac{1}{(\sqrt{2\pi})^n \sqrt{|\det \boldsymbol{\Lambda}|}} \exp[-\tfrac{1}{2}(\mathbf{x} - \boldsymbol{\mu}_X)^T (\boldsymbol{\Lambda}^{-1})(\mathbf{x} - \boldsymbol{\mu}_X)] \tag{6.19}$$

or written out element by element

$$p_{X_1,\ldots,X_n}(x_1, x_2, \ldots, x_n) = \frac{1}{(\sqrt{2\pi})^n \sqrt{\lambda_1 \lambda_2, \ldots, \lambda_n}} \exp\left[-\frac{1}{2}\sum_{j=1}^{n} \frac{1}{\lambda_i}(x_i - \mu_i)^2\right]$$

Also,

$$p_{X_1,\ldots,X_n}(x_1,\ldots,x_n)=\prod_{j=1}^{n}\frac{1}{(\sqrt{2\pi\lambda_j})}\exp\left[-\frac{(x_j-\mu_j)^2}{2\lambda_j}\right]=\prod_{j=1}^{n}p_{X_j}(x_j) \tag{6.20}$$

The random variables $\{X_j\}$ are thus independent and each normally distributed with mean μ_j and variance λ_j.

The above derivation is dependent on the covariance matrix **D** being non-singular, that is det $\mathbf{D}\neq 0$. If D is singular of rank $r<n$, exactly r latent roots are non-zero and the rest is zero or

$$\lambda_1=\lambda_2=\cdots=\lambda_{n-r}=0,\ \lambda_i\neq 0 \qquad \text{for} \quad i>n-r$$

By Eq. (6.20), λ_i is the variance of the random variable X_i. A latent root equal to zero means that the variance is zero or $X_i\,(\lambda_i=0)$ is a deterministic constant equal to the mean value μ_i. Therefore, the joint probability density Eq. (6.20) will in terms of the probability distribution show delta functions at the controversial roots, that is,

$$p_{X_1,\ldots,X_n}(x_1,\ldots,x_n)=\prod_{j=1}^{n-r}\delta(x_i-\mu_i)\prod_{i=n-r+1}^{n}p_{X_i}(x_i) \tag{6.21}$$

Example 6.2 Consider a stationary Gaussian process $X(t)$ and let $U(t)=\mathrm{d}X(t)/\mathrm{d}t$ be its derivative process. The autocorrelation function for $X(t)$ is given as

$$R_X(\tau)=a^2\exp(-b^2\tau^2/2)+b^2$$

Find the joint probability density function for $X(t)$ and $U(t)$.

Solution Since the derivative process is obtained by a linear operation on $X(t)$ (differentiation), it is also a stationary Gaussian process. The joint probability distribution is then given by Eq. (6.8). To compute the D matrix, the variances and covariances have to be determined.

$$\lim_{\tau\to\infty} R_X(\tau)=b^2=\mu_X^2$$

$$\sigma_X^2=R_X(0)-\mu_X^2=a^2+b^2-b^2=a^2$$

$$\mu_U=E[U]=E[\dot{X}]=\mathrm{d}E[X]/\mathrm{d}t$$

$$\sigma_U^2=R_U(0)=-R_X''(0)=a^2b^2$$

$$\varrho_{XU}=\frac{E[XU]-\mu_X\mu_U}{\sigma_X\sigma_U}=0$$

since

$$E[XU]=\frac{1}{2}\frac{\mathrm{d}E[X^2]}{\mathrm{d}t}=\frac{1}{2}\frac{\mathrm{d}(a^2+b^2)}{\mathrm{d}t}=0.$$

Now

$$\mathbf{D}=\begin{Bmatrix} a^2 & 0 \\ 0 & a^2b^2 \end{Bmatrix}, \qquad \det(\mathbf{D})=a^4b^2, \qquad \mathbf{D}^{-1}=\begin{Bmatrix} a^{-2} & 0 \\ 0 & a^{-2}b^{-2} \end{Bmatrix}$$

$$\{\mathbf{x}-\boldsymbol{\mu}^{\mathrm{T}}\}\mathbf{D}^{-1}\{\mathbf{x}-\boldsymbol{\mu}\}=\{(x-b),u\}\begin{bmatrix} a^{-2} & 0 \\ 0 & a^{-2}b^{-2} \end{bmatrix}\begin{bmatrix} x-b \\ u \end{bmatrix}=\frac{(x-b)^2}{a^2}+\frac{u^2}{a^2b^2}$$

so

$$p(x,u)=\frac{1}{2\pi a^4b^2}\exp\left[-\frac{(x-b)^2}{a^2}-\frac{u^2}{a^2b^2}\right]$$

For a Gaussian random process $X(t)$ with a zero mean, consider the joint moments of the higher order of the vectors of the amplitudes $\{X_1=X(t_1), X_2=X(t_2),\ldots,X_n=X(t_n)\}$. Then the nth joint moment is $E[X_1X_2,\ldots,X_n]$ and $E[X(t)]=0$ for all t. It can be shown (see e.g. [122]), that for a Gaussian, stationary process,

$$E[X_1X_2,\ldots,X_n]=\sum_k E[X_{r_1}X_{r_2}X_{r_3},\ldots,X_{r_{n-2}}]\,E[X_kX_j] \tag{6.22}$$

where the summation over k does not include $j=k$ and $r_1,r_2,\ldots,r_{n-2}$ are the numbers left when k and j are excluded; k and j run through all possible combinations of pairs in $(1,2,\ldots,n)$, but each pair (k,j) or (j,k) is counted only once.

For $n=3$,

$$E[X_1X_2X_3]=E[X_3]\,E[X_1X_2]=0$$

For $n=4$,

$$E[X_1X_2X_3X_4]=\sum_{k=2,3,4}E[X_{r_1}X_{r_2}]\,E[X_kX_1]$$

In general

$$\left.\begin{aligned} &E[X_1X_2,\ldots,X_{2m+1}]=0 \\ &E[X_1X_2,\ldots,X_{2m}]=\sum_N E[X_jX_k]\,E[X_rX_s] \end{aligned}\right\} \tag{6.23}$$

where the last summation is to be taken over all the different ways $2m$ elements can be grouped in two pairs. The total number of two pairs, thus obtained, is N, which is then the number of terms in the summation Eq. (6.23). Now, the total permutations of $(2m)$ numbers are $(2m)!$ and $N<(2m)!$ since the sum does not include identical terms. Secondly, for each term in the sum, permutations of the m factors result in identical ways of breaking up the $(2m)$ elements. Thirdly, since the pairs $E[X_jX_k]=E[X_kX_j]$ are equivalent and should be counted only once, the number of terms is reduced by 2^3. Therefore,

$$N=\frac{(2m)!}{(2m-4)!2^3}=\frac{2m(2m-1)(2m-2)(2m-3)}{8} \tag{6.24}$$

If the condition $E[X(t)] = 0$ is relaxed, the relations Eqs. (6.22) and (6.23) are still valid for the corresponding central moments. The above relations are unique for the Gaussian process. Any process for which the above relations can be shown to be true is therefore a Gaussian process.

It has been mentioned that any linear operation performed on a Gaussian process, including that of differentiation and integration, yields another Gaussian process. This statement is now easily proved with the aid of the above obtained relations Eq. (6.23). In fact, consider a random Gaussian process $X(t)$, which as an input signal is passed through a linear time-independent filter (see Eq. (3.2)). The output process $Y(t)$ will be given by the general expression

$$Y(t) = \int_{a(t)}^{b(t)} X(u)\, h(u, t) du \tag{6.25}$$

which also covers all linear operations performed either on $X(t)$ or $Y(t)$. The mean value at any particular time instant $t = t_i$ is

$$E[Y(t)] = \mu_Y(t_i) = \int_{a(t)}^{b(t)} E[X(u)]\, h(u, t) du = \int_{a(t)}^{b(t)} \mu_X(u)\, h(u, t) du \tag{6.26}$$

Now form the n-multiple integral

$$(Y_1 - \mu_Y(t_1))(Y_2 - \mu_Y(t_2)) \cdots (Y_n - \mu_Y(t_n))$$
$$= \int_{a(t_1)}^{b(t_1)} \cdots \int_{a(t_n)}^{b(t_n)} (X(u_1) - \mu_X(u_1)) \cdots (X(u_n) - \mu_X(u_n))\, h(u_1, t_1) \cdots h(u_n, t_n) du_1 \cdots du_n$$

Taking the expectation of both sides, and introducing the nth joint central moment or cumulant $\kappa_n[Y_1 Y_2 \cdots Y_n] = E[(Y_1 - \mu_{y_1}) \cdots (Y_n - \mu_{Y_n})]$,

$$\kappa_n[(Y_1 Y_2 \cdots Y_n)]$$
$$= \int_{a(t_1)}^{b(t_1)} \cdots \int_{a(t_n)}^{b(t_n)} \kappa_n[X(u_1) \cdots (X(u_n))]\, h(u_1, t_1) \cdots h(u_n, t_n) du_1 \cdots du_n \tag{6.27}$$

Obviously, since $\kappa_n[X_1 X_2 \cdots X_n]$ satisfies the conditions Eqs. (6.20) or (6.21), then by Eq. (6.27), the nth cumulant Y also does so. Hence $Y(t)$ is a Gaussian random process.

6.1.2 Stationary Gaussian Processes

From the discussion of spectral composition of stochastic processes in Section 2.5, it was made clear that any covariance stationary stochastic process with a zero mean has the discrete spectral representation

$$X(t) = \sum_{i=-\infty}^{\infty} e^{i\omega_j t} \Delta\Phi(\omega_j) \tag{2.68}$$

where $\Phi(\omega)$ is a non-differentiable random function the increments of which satisfy the orthogonality condition

$$E[\Delta\Phi^*(\omega_1)(\Delta\Phi(\omega_2))] = S(\omega_1)\,\delta(\omega_1 - \omega_2)\Delta\omega_1\Delta\omega_2 \tag{2.70}$$

where $S(\omega)$ is the power spectral density function of the process and $\delta(\omega)$ is Dirac's delta function. The above discrete spectral representation of a covariance stationary stochastic process can be used to interpret stochastic processes that are formed as finite discrete series of harmonic components with random phases or both random phases and frequencies. This offers a simple method for generating sample functions of such processes (see Clough and Penzien, [37], and Langen and Sigbjörnsson, [119]).

Consider a random process, which has the form

$$X(t) = \sum_{k=1}^{N} C_k \cos(\omega_k t + \phi_k) \tag{6.28}$$

where C_k are real positive constants to be determined, $\omega_k = (k - \frac{1}{2})\Delta\omega$ is the kth frequency ($\Delta\omega$ is a real unit of frequency, a measure of the desired or possible frequency resolution), and ϕ_k are random phase angles, independent and uniformly distributed on the interval $[0, 2\pi]$. The mean value of $X(t)$ is given by the expression

$$E[X(t)] = \int_0^{2\pi} \sum_{k=1}^{N} C_k \cos(\omega_k t + \phi_k)\frac{1}{2\pi}\mathrm{d}\phi_k = 0 \tag{6.29}$$

whereby the above condition for a zero mean value is satisfied. The autocorrelation function is given by the expression

$$\begin{aligned} R_X(\tau) &= E[X(t+\tau)X(t)] \\ &= \int_0^{2\pi}\int_0^{2\pi} \sum_{j=1}^{N}\sum_{k=1}^{N} C_j C_k \cos(\omega_j(t+\tau) + \phi_j)\cos(\omega_k t + \phi_k)\frac{1}{(2\pi)^2}\mathrm{d}\phi_j \mathrm{d}\phi_k \\ &= \frac{1}{2}\sum_{k=1}^{N} C_k^2 \cos(\omega_k \tau) = R_X(\tau) \end{aligned} \tag{6.30}$$

which shows that the process is covariance stationary. Thus the stochastic process Eq. (6.28) satisfies both requirements for having a spectral representation. Moreover, the process tends to a covariance stationary Gaussian process as N becomes large, because it is the sum of a large number of random contributions that are identically distributed and thus falls under the central limit theorem (see Eq. (1.121a). Therefore, for a sufficiently large number of harmonic components N, the process is a stationary Gaussian process. The process Eq. (6.28) is ergodic with respect to the mean and the second moments, since time averaging of a sample function $x(t)$ yields the following results:

$$\begin{aligned} \overline{x(t)} &= \lim_{T\to\infty} \frac{1}{2T}\int_{-T}^{T} x(t)\mathrm{d}t \\ &= \lim_{T\to\infty} \frac{1}{2T}\sum_{k=1}^{N}\int_{-T}^{T} C_k \cos(\omega_k t + \phi_k)\mathrm{d}t = 0 \end{aligned} \tag{6.31}$$

and

$$R_X(\tau, T) = \overline{x(t+\tau)\,x(t)} = \lim_{T\to\infty} \frac{1}{2T}\int_{-T}^{T} x(t+\tau)\,x(t)\mathrm{d}t$$

$$= \lim_{T\to\infty} \frac{1}{2T}\sum_{j=1}^{N}\sum_{k=1}^{N} C_j C_k \int_{-T}^{T} \cos(\omega_j(t+\tau)+\phi_j)\cos(\omega_k t+\phi_k)\mathrm{d}t$$

$$= \sum_{k=1}^{N} C_k^2 \cos(\omega_k \tau) \tag{6.32}$$

which is the same as previously obtained for the ensemble averages Eqs. (6.29) and (6.30).

The average power of the process at any time instant is equal to the mean square value, which is given by Eqs. (6.30) and (2.76) as

$$R_X(0) = E[X(t)^2] = \frac{1}{2}\sum_{k=1}^{N} C_k^2 = \int_{-\infty}^{\infty} S_X(\omega)\mathrm{d}\omega \tag{6.33}$$

Therefore, if the power spectral density $S_X(\omega)$ of the process is known or can be constructed, the process can be represented as a large number of harmonic components having amplitudes that are obtained by discretizing the above integral (see Fig. 2.11). By Eq. (6.33), the discrete amplitudes C_k are given by

$$C_k^2 = 2S_X(\omega_k)\Delta\omega \tag{6.34}$$

and the process Eq. (6.28) becomes

$$X(t) = \sum_{k=1}^{N} \sqrt{(2S_X(\omega_k)\Delta\omega)}\cos(\omega_k t + \phi_k) \tag{6.35}$$

The autocorrelation function given by Eq. (6.30) is then written as

$$R_X(\tau) = \sum_{k=1}^{N} S_X(\omega_k)\Delta\omega\cos(\omega_k\tau) \tag{6.36}$$

If N tends to infinity and $\Delta\omega \to 0$, the above sum converts to a Riemann integral, and the autocorrelation function assumes the classical form (see Eq. (2.79)):

$$R_X(\tau) = \int_{-\infty}^{\infty} S_X(\omega)\cos(\omega\tau)\mathrm{d}\omega \tag{6.37}$$

This shows that when the amplitudes are selected by Eq. (6.34), the autocorrelation and spectral density functions converge to the desired continuous functions for $N \to \infty$.

Sample functions of the stationary Gaussian process Eq. (6.35) are easily generated by sampling the random phases from the uniform distribution and adding up a large number of harmonic terms with the amplitudes given by the power spectral density function. It should be noted that the sample functions will be periodic with the sampling period $T_0 = 2\pi/\Delta\omega$. Let the highest frequency in the signal be $f_{\max}$. The lowest frequency in the signal is related to the observation interval or the record length T, that is, $f_{\min} = 1/T$, whereby $N = f_{\max}/f_{\min}$. However, in order to prevent unacceptable frequency distortion in the signal, the number of points or rather the sampling frequency $f_0 = 1/T_0$ has to be selected as twice the maximum frequency (for a fuller discussion see Section 7.2.1). Keeping the sampling frequency $\Delta\omega = 2\pi f_0$ constant, the number of terms needed to reflect the highest frequencies present in the power spectral density without unacceptable distortion, that is, $S_X(\omega) \approx 0$ for $\omega > \omega_{max}$, is therefore given by

$$N \geqslant 2f_{max}/f_{min} = 2f_{max}T \tag{6.38}$$

The requirement of a sampling frequency $f_0 \geqslant 2f_{max}$ can in certain cases be in conflict with the necessity of using smaller frequency intervals to cover narrow peaks in the shape of the power spectral density.

Instead of only one random variable associated with the process Eq. (6.28), namely the phase ϕ, it is natural to assume that the amplitudes are also random variables, that is, have to be drawn from some probability distribution. Therefore, consider the harmonic series process

$$X(t) = \sum_{k=1}^{N} C_k \cos(\omega_k t + \phi_k) \tag{6.39}$$

where, as before, the phase angles ϕ_k are independent and uniformly distributed random variables on the interval $[0, 2\pi]$ and the amplitudes C_k are sampled from a prescribed probability distribution with the probability density $p_C(c)$ having a mean value equal to zero and a variance σ_C^2. The mean value is now given by the expression

$$E[X(t)] = \int_0^{2\pi}\int_{-\infty}^{\infty}\sum_{k=1}^{N} p_C(c_k)\cos(\omega_k t + \phi_k)\frac{1}{2\pi}\mathrm{d}\phi_k\,\mathrm{d}c_k = 0 \tag{6.40}$$

since the variables C_k and ϕ_k are independent and the order of integration and summation can be interchanged. The autocorrelation function is

$$\begin{aligned} R_X(\tau) &= E[X(t+\tau)X(t)] \\ &= \sum_{j=1}^{N}\sum_{k=1}^{N} \mathrm{Cov}[C_j C_k]\,\mathrm{Cov}[\cos(\omega_j(t+\tau)+\phi_j)\cos(\omega_k t + \phi_k)] \\ &= \tfrac{1}{2}\sigma_C^2 \sum_{k=1}^{N} \cos(\omega_k \tau) = R_X(\tau) \end{aligned} \tag{6.41}$$

so the only change is that the amplitudes are replaced by the variance. The process is still covariance stationary and by the same argument as before, it tends to a Gaussian process for sufficiently large N. Testing for ergodicity, the temporal mean value of a sample function $x(t)$ is still equal to zero but the temporal autocorrelation Eq. (6.30) remains unchanged. The process is therefore no longer ergodic with respect to the second moments, which makes it less attractive.

Finally, discarding probabilistic amplitudes, the frequencies ω_k may also be considered to be random variables with a probability density to be modelled after the desired spectral density of the process. Therefore, consider the stochastic process

$$X(t) = C\sum_{k=1}^{N} \cos(\omega_k t + \phi_k) \tag{6.42}$$

where the amplitude is now a real constant C, ω_k are independent random variables with a probability distribution shaped after the desired spectral density of the process, and ϕ_k as before are independent random phases uniformly distributed on the interval $[0, 2\pi]$. Given the power spectral density $S_X(\omega)$ of the process, the probability distribution of the frequencies can be defined as follows:

$$F_\Omega(\omega) = G_X(\omega)/\sigma_X^2 \tag{6.43}$$

where

$$G_X(\omega) = \int_{-\infty}^{\omega} S_X(u)\mathrm{d}u \tag{6.44}$$

By Eqs. (2.76) and (6.44), $G_X(\infty) = \sigma_X^2$ provided that the mean value of $X(t)$ is zero. Sample random frequencies are therefore easily drawn from the above distribution by generating a random number y from the uniform distribution $[0, 1]$, whereby

$$F_\Omega(\omega_k) = P[\Omega \leqslant \omega_k] = y_k \qquad \text{or} \qquad \omega_k = F_\Omega^{-1}(y_k) \tag{6.45}$$

that is, ω_k is determined by solving the equation

$$y_k = \frac{1}{\sigma_X^2}\int_{-\infty}^{\omega_k} S_X(u)\mathrm{d}u \tag{6.46}$$

The probability density on the other hand is given by the expression

$$p_\Omega(\omega) = S_X(\omega)/\sigma_X^2 \tag{6.47}$$

For spectral density functions with sharp peaks, this automatically concentrates the random frequencies about those peaks.

As before, the mean value and autocorrelation of the process need to be evaluated. Starting with the mean value,

$$E[X(t)] = C\sum_{k=1}^{N}\int_{-\infty}^{\infty}\int_0^{2\pi}\cos(\omega_k t + \phi_k)p_\Omega(\omega_k)\frac{1}{2\pi}\mathrm{d}\phi_k\mathrm{d}\omega_k = 0 \tag{6.48}$$

which shows that the mean value is zero as had already been assumed. The autocorrelation is given by the expression

$$\begin{aligned} R_X(\tau) &= E[X(t+\tau)X(t)] \\ &= C^2\sum_{j=1}^{N}\sum_{k=1}^{N}\int_{-\infty}^{\infty}\int_{-\infty}^{\infty}\int_0^{2\pi}\int_0^{2\pi}\cos(\omega_j(t+\tau)+\phi_j)\cos(\omega_k t+\phi_k) \\ &\quad\times\frac{1}{(2\pi)^2}p_\Omega(\omega_j)p_\Omega(\omega_k)\mathrm{d}\phi_j\mathrm{d}\phi_k = \tfrac{1}{2}C^2\sum_{k=1}^{N}\int_{-\infty}^{\infty}\cos(\omega_k\tau)p_\Omega(\omega_k)\mathrm{d}\omega_k \\ &= \frac{C^2N}{2\sigma_{X^2}}\int_{-\infty}^{\infty}S_X(\omega)\cos(\omega\tau)\mathrm{d}\omega = R_X(\tau) \end{aligned} \tag{6.49}$$

in which the expression for the probability density Eq. (6.47) has been used. Thus the process is at least covariance stationary, and as before, it is Gaussian by strength of the central limit theorem. Obviously the autocorrelation Eq. (6.49) assumes the proper form by choosing the real constant C to be equal to

$$C = \sigma_X \sqrt{\frac{2}{N}} \tag{6.50}$$

Thus, for each discrete frequency component the amplitude is given by the above expression.

To test for ergodicity, the temporal mean value and autocorrelation functions are formed.

$$\overline{x(t)} = \lim_{T \to \infty} \frac{C}{2T} \sum_{k=1}^{N} \int_{-T}^{T} \cos(\omega_k t + \phi_k) dt = 0 \tag{6.51}$$

and

$$\begin{aligned} R_X(\tau, T) &= \overline{x(t+\tau)\, x(t)} \\ &= \lim_{T \to \infty} \frac{C^2}{2T} \sum_{j=1}^{N} \sum_{k=1}^{N} \int_{-T}^{T} \cos(\omega_j (t+\tau) + \phi_j) \cos(\omega_k t + \phi_k) dt \\ &= \frac{C^2}{2} \sum_{k=1}^{N} \cos(\omega_k \tau) \end{aligned} \tag{6.52}$$

which shows the process to be ergodic respect to the first and second moments. This version of a Gaussian process has been used to simulate ocean wave heights by Borgman, [21].

The above discussion has shown that it is comparatively simple to construct stationary Gaussian processes using the spectral representation, which was introduced in Section 2.5. A common denominator is that the number of harmonic terms in the spectral representation has to be large enough to warrant the normalcy of the probability law, that is, to produce a Gaussian process. Another common requirement is that the random phases of each harmonic term have to be uniformly distributed between 0 and 2π in order to ensure stationarity. If the distribution of the random phases were, say, defined on the interval $[0, \theta]$ with $\theta < 2\pi$, the time could no longer be made to disappear through integration in the expressions for the mean values and the autocorrelation. Another important feature of stationary Gaussian processes is that sample functions are easily generated for many practical purposes.

Example 6.3 Generation of Gaussian process artificial earthquakes
A simple method of producing artificial strong motion earthquake records is to generate sample functions of Gaussian white noise processes and pass them through appropriate filters that may provide the output process with the spectral characteristics normally associated with a strong motion earthquake at a given location. The Fourier spectra of strong motion earthquake records show irregular and rapid oscillations with often several dominant frequency peaks, which are indicative of the geological character of

the subsoil of the site where the earthquake is recorded (see Section 6.2.4 for further discussion). Also, since the earthquake motion is essentially an evolutionary process, the filtered output sample functions have to be amplitude-modulated to resemble the time evolution of the real motion, that is, show a build-up phase, a strong motion phase and a coda or attenuating tail. One of the earlier attempts to describe earthquake accelerograms as sample functions of stochastic processes was made by Bycroft, [27]. Bogdanoff *et al.*, [18], and Housner and Jennings, [85]. Jennings *et al.*, [95], further developed this technique by applying filters and amplitude modulation, which by now has become a routine exercise in simulation of synthetic or artificial earthquake records. An innovative practical approach is presented by Clough and Penzien, [37]. A more traditional approach using the Gaussian process Eq. (6.28) is presented in the following.

Consider a Gaussian white noise process $\{X(t), -\infty < t < \infty\}$ with an intensity S_0 (cf. Ex. 2.14) and a mean value $E[X] = 0$. In order to shape the constant power spectrum after the conditions prevailing in surface earthquake motions, Kanai, [98], and Tajimi [218], suggested that the frequency response function of a simple inverted pendulum with a fundamental frequency ω_g and the ratio of critical damping λ_g could be used as a suitable filter. Denoting the absolute acceleration of the pendulum $Y(t)$, the equaticn of motion is

$$\ddot{Y}(t) + 2\lambda_g\omega_g(\dot{Y}(t) - \dot{X}(t)) + \omega_g^2(Y(t) - X(t)) = 0 \tag{6.53}$$

where $X(t)$ is the earthquake motion (displacement) at the base of the pendulum. The frequency response function is given by

$$|H_g(\omega)|^2 = \frac{1 + 4\lambda_g^2(\omega/\omega_g)^2}{(1 - (\omega/\omega_g)^2)^2 + 4\lambda_g^2(\omega/\omega_g)^2} \tag{6.54}$$

which is supposed to furnish an ideal characterization of the subsoil at the site being investigated. The filter parameters are the fundamental frequency ω_g, which acts as a main characteristic for the vibratory behaviour of the ground, also called the predominant frequency of the ground, and the damping ratio λ_g, which is also a main characteristic and describes the hardness of the subsoil. High values of λ_g are thus representative for soft subsoil of the diluvium type, whereas low values of λ_g are representative of hard ground, i.e. rock sites or hard gravel subsoil. For rock sites or hard gravel sites, the values suggested are $\omega_g = 15.6$ rad/s and $\lambda_g = 0.6$, [98]. The power spectral density of the output acceleration process is then given by (see Eq. (3.52))

$$S_{\ddot{Y}}(\omega) = |H_g(\omega)|^2 S_0 \tag{6.55}$$

which can be used as a basis for constructing a stationary Gaussian process according to Eq. (6.28). Obviously, S_0 is a measure of the mean square acceleration peak at $\omega = \omega_g$.

By Eq. (6.55), the low frequency portion of the filtered output is prominent as $|H_g(0)| = 1$. The thus defined earthquake spectrum Eq. (6.54) would therefore impart a significantly low frequency portion to the earthquake motion, which is in contradiction to what is found in strong motion records. It is also a matter of interpretation whether the white noise input process should be an acceleration process or a displacement process. For an acceleration process, the frequency response function needs to be divided by the frequency square, and the spectrum by $1/\omega^4$, which would result in further problems and an instability at very low frequencies. For a displacement input process, the output velocity and displacement spectra are obtained by dividing Eq. (6.55) by ω^2 and ω^4 respectively, which means that the velocity and the displacement become unbounded at $\omega = 0$. In order to truly model the earthquake in realistic terms, the synthetic earthquake signal has therefore to be cleansed of unrealistically low-frequency components. This can be done by passing the signal through a suitable highpass filter. Clough and Penzien, [37], propose the following filter, which greatly attenuates the very low-frequency components, given by the frequency response function

$$|H_h(\omega)|^2 = \frac{(\omega/\omega_h)^4}{(1-(\omega/\omega_h)^2)^2 + 4\lambda_h^2(\omega/\omega_h)^2} \tag{6.56}$$

The fundamental frequency and the damping ratio (ω_h, λ_h) of this filter are selected to ensure the desired frequency content of the earthquake signal. For rock sites, appropriate values are $\omega_h = 4$ rad/s and $\lambda_h = 0.5$, [119]. Thus, the power spectral density of the earthquake acceleration process is given by

$$S_{\ddot{Y}}(\omega) = |H_g(\omega)|^2 \, |H_h(\omega)|^2 \, S_0 \tag{6.57}$$

In Fig. 6.1, the two frequency response functions are shown, and the resulting power spectral density for the filtered output is shown in Fig. 6.2 ($S_0 = 1$).

This part of the exercise has produced a stationary Gaussian earthquake acceleration process with the desired frequency content. The next step is to apply an amplitude modulating function to more closely simulate the situation in real earthquakes. Therefore, let

$$\ddot{U}(t) = A(t)\,\ddot{Y}(t) \tag{6.58}$$

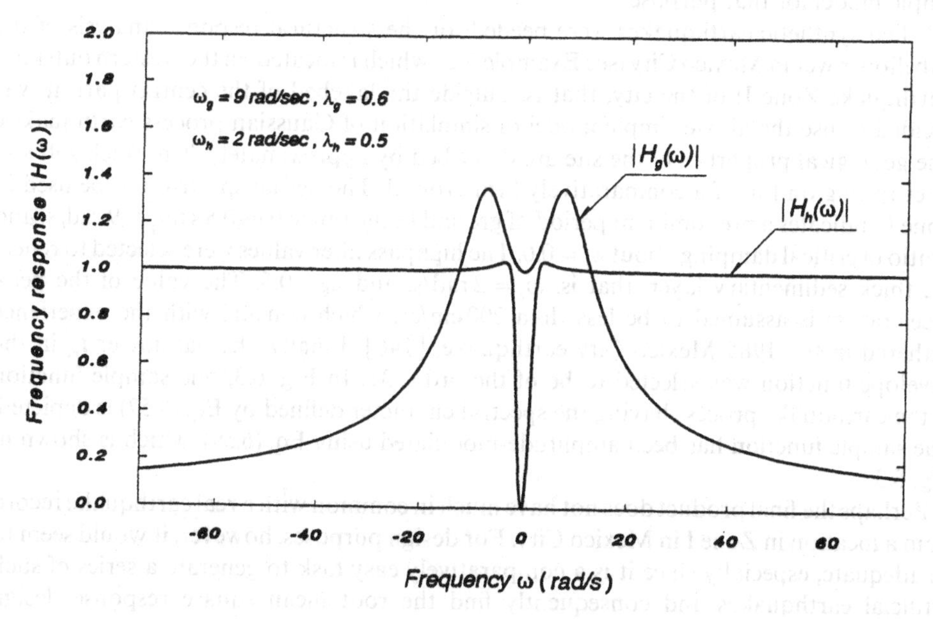

Fig. 6.1 *Filter response functions*

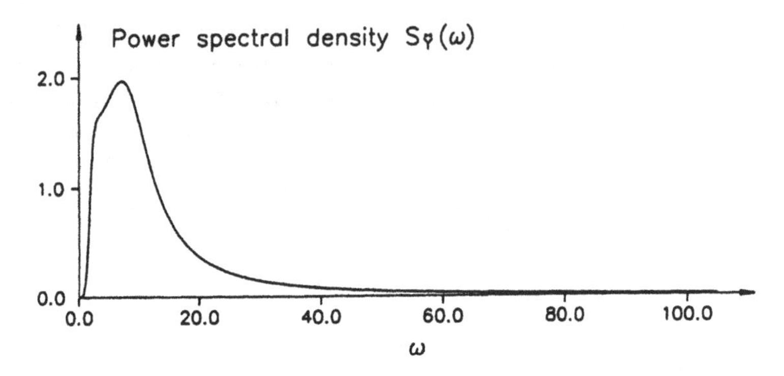

Fig. 6.2 *The power spectral density of an artificial earthquakes process*

where $\ddot{U}(t)$ is the final non-stationary amplitude modulated earthquake, and $A(t)$ is an envelope function of the type shown in Fig. 2.22. Many such functions have been suggested by various authors (see for instance Jennings *et al.*, [95]). Hsu and Bernard, [86], have suggested the following simple model:

$$A(t) = (t/t_m)\exp[1 - t/t_m] \tag{6.59}$$

where t_m is the time the earthquake motion is at its maximum. Actually, the choice of envelope functions does not seem much to affect the results of standard response analysis of structures subjected to earthquake type excitations, which warrants the use of the above simple model for that purpose.

When synthetic earthquakes were needed for the structural response analysis of the Pabellon tower in Mexico City (see Example 5.2), which is located on the western outskirts (earthquake Zone I) of the city, that is, outside the lakebed of the central part, it was decided to use the above simple model of simulation of Gaussian process earthquakes. The geological properties of the site are described by approximately 30 m thick soft soil layer resting on top of a comparatively firm ground. The design spectrum to be used in Zone I, indicates a predominant period of ground in the range 0.6–0.8 s ($\omega_g \approx 9$ rad/s) and a ratio of critical damping about $\lambda_g = 0.6$. The highpass filter values were selected to reflect the thick sedimentary layer, that is, $\omega_h = 2$ rad/s, and $\lambda_h = 0.5$. The value of the peak acceleration is assumed to be less than 200 cm/s^2, which concurs with the experience gathered in the 1985 Mexico City earthquake, [146]. Finally, the parameter t_m in the envelope function was selected to be of the order 3 s. In Fig. 6.3, one sample function of the earthquake process having the spectral character defined by Eq. (6.57) is depicted. The sample function has been amplitude-modulated using Eq. (6.59), which is shown in Fig. 6.3.

Perhaps the final product does not have much in common with a real earthquake record from a location in Zone I in Mexico City. For design purposes, however, it would seem to be adequate, especially since it is a comparatively easy task to generate a series of such artificial earthquakes and consequently find the root mean square response design parameters of a structure in the zone (see Section 5.2.2).

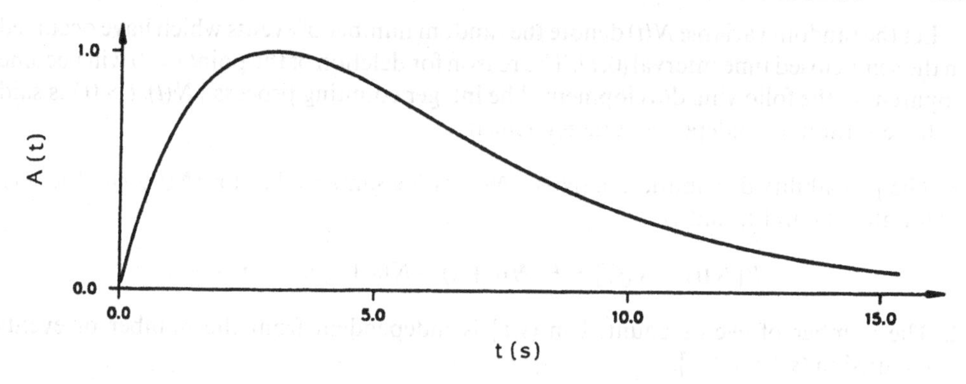

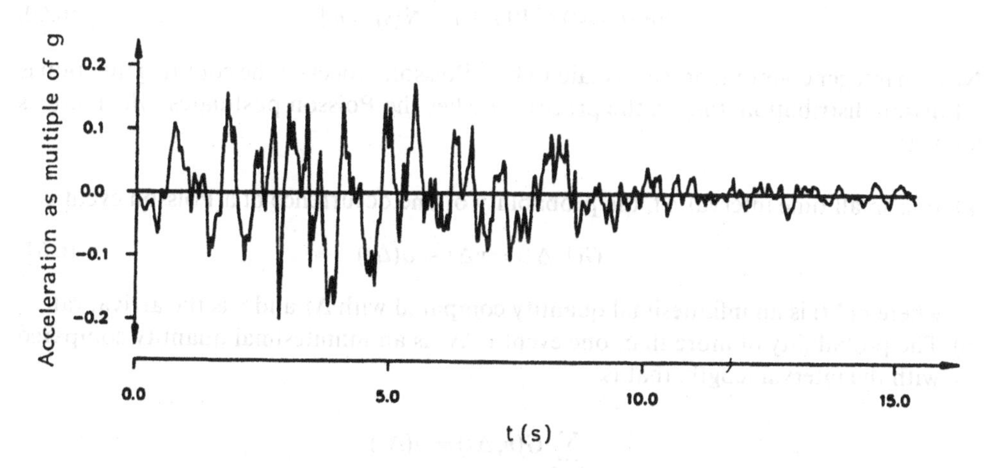

Fig. 6.3 *A synthetic amplitude-modulated earthquake acceleration process*

Finally, a FORTRAN 77 program was written to handle the entire simulation process. The program is available to interested readers, who can download it by accessing the web site: http://www.hi.is/~solnes/.

6.2 THE POISSON PROCESS

The Poisson process $\{N(t), t \in T\}$, where $N(t)$ is an integer, is a special case in connection with more general counting processes, which arise in the problems concerning the counting of events in the course of time. The probability law of certain rare events such as the occurrence of an earthquake, of an accident at a road intersection or the arrival of a public transport vehicle (at least when one is waiting it happens as a rare event) is often described by a Poisson counting process.

Let the random variable $N(t)$ denote the random number of events which have occurred in the semi-closed time interval $(0, t]$. The reason for deletion of the point $t = 0$ will become apparent in the following development. The integer counting process $\{N(t), t > 0\}$ is said to have stationary, independent increments if

1. The probability distribution of $(N(t) - N(s))$ is the same as that for $(N(t+\tau) - N(s+\tau))$ for all $s < t$ and τ, that is

$$P[N(t) - N(s)] = P[N(t+\tau) - N(s+\tau)], \qquad \tau > -t.$$

2. The number of events counted in $(s, t]$ is independent from the number of events counted in $(s+\tau, t+\tau]$.

Introducing the counting function $G(n, (t-s))$, the probablity of n counts in a time interval $(s, t]$ is

$$G(n,(t-s)) = P[N(t) - N(s) = n] \tag{6.60}$$

Now an integer counting process is said to be a Poisson process if the counting function is a Poisson distribution, that is, the process satisfies the Poisson postulates, which are as follows:

(a) In a small time interval Δt, the probability of one occurrence of a Poisson event is

$$G(1, \Delta t) = v\Delta t + o(\Delta t) \tag{6.61}$$

where $o(\Delta t)$ is an infinitesimal quantity compared with Δt and v is the arrival rate.

(b) The probability of more than one event in Δt is an infinitesimal quantity compared with the interval length, that is,

$$\sum_{n=2}^{\infty} G(n, \Delta t) = o(\Delta t)$$

(c) The process has independent increments, that is, the count or number of occurrences in non-overlapping intervals is statically independent.

It should be noted that the Poisson process can be non-stationary, which means that the arrival rate v is a function of time $v(t)$. If the independent increments are stationary, the arrival rate v must be a constant.

To find the probability distribution for the Poisson process, that is, the counting function $G(n, t)$, note that the probability of no arrival/occurrence in the time interval $(0, t+\Delta t]$ is the product of probabilities of no events in both $(0, t]$ and $(t, t+\Delta t]$, that is,

$$G(0, t+\Delta t) = G(0, t)G(0, \Delta t)$$

and the probability of exactly n events in the $(0, t+\Delta t]$ is

$$\begin{aligned} G(n, t+\Delta t) &= G(n, t)G(0, \Delta t) + G(n-1, t)G(1, \Delta t) + \sum_{i=2}^{n} G(n-i, t)\, G(i, \Delta t) \\ &= G(n, t)G(0, \Delta t) + G(n-1, t)G(1, \Delta t) + o(\Delta t) \end{aligned}$$

Form

$$G(0,t+\Delta t)-G(0,t)=G(0,t)(G(0,\Delta t)-1)=-G(0,t)\,G(1,\Delta t)$$
$$=-G(0,t)\,v(t)-o(\Delta t)\,G(0,t)$$

or

$$\frac{G(0,t+\Delta t)-G(0,t)}{\Delta t}=-vG(0,t)-G(0,t)\frac{o(\Delta t)}{\Delta t}$$

and letting $\Delta t\to 0$,

$$\frac{dG(0,t)}{dt}=-vG(0,t)$$

whereby

$$G(0,t)=C\,e^{-vt}=e^{-vt} \tag{6.62}$$

since $C=G(0,0)=1$. Also,

$$G(n,t+\Delta t)-G(n,t)=-G(n,t)+G(n,t)\,G(0,\Delta t)+G(n-1,t)\,G(1,\Delta t)+o(\Delta t)$$
$$=(-G(n,t)+G(n-1,t))\,G(1,\Delta t)+o(\Delta t)$$
$$=v(t)(-G(n,t)+G(n-1,t))+o(\Delta t)$$

so for $\Delta t\to 0$,

$$\frac{dG(n,t)}{dt}+vG(n,t)=vG(n-1,t) \tag{6.63}$$

The general solution of Eq. (6.63) is of the form

$$G(n,t)=F_n(t)\,e^{-vt}$$

where $f_n(t)$ satisfies the equation

$$\frac{df_n(t)}{dt}=vf_{n-1}(t) \tag{6.64}$$

for $n=1, f'_1(t)=vf_0(t)=v$ or $f_1(t)=vt+C_1=vt$, since $G(n,0)=G(1,0)=0$. Therefore,

$$f_2(t)=\frac{(vt)^2}{2},\qquad f_3(t)=\frac{(vt)^3}{3\cdot 2}\qquad\text{and}\qquad f_n(t)=\frac{(vt)^n}{n!}$$

or

$$G(n,t)=\frac{(vt)^n}{n!}e^{-vt} \tag{6.65}$$

As shown in Example 1.26, for $P[X=k]=e^{-\lambda}(\lambda^k/k!)$, $E[X]=\mathrm{Var}[X]=\lambda$, whereby

$$E[N(t)]=vt,\ \mathrm{Var}[N(t)]=vt \tag{6.66}$$

and the constant ν is the expected mean arrival rate. Also, the variation coefficient

$$V_N = \frac{\sigma_N}{\mu_N} = \frac{\sqrt{\nu t}}{\nu t} = \frac{1}{\sqrt{\nu t}} \tag{6.67}$$

so statistical stability is obtained, that is, when $V_N \to 0$, for very large times.

For a non-stationary Poisson process, i.e. a process having non-stationary increments, the mean arrival rate is a function of time, $\nu = \nu(t)$. By analogous reasoning it is found that

$$G(n,t) = \exp\left[-\int_0^t \nu(u)\mathrm{d}u\right] \int_0^t \int_0^{u_n} \cdots \int_0^{u_1} \nu(u_n)\nu(u_{n-1})\cdots\nu(u_1)\mathrm{d}u_1\cdots\mathrm{d}u_n$$

or

$$G(n,t) = \frac{1}{n!}\left[\int_0^t \nu(u)\mathrm{d}u\right]^n \exp\left[-\int_0^t \nu(u)\mathrm{d}u\right] \tag{6.68}$$

Example 6.4 Consider a time interval of length T. Given that exactly one Poisson event takes place in T, find the probability that it occurs between t and $t + \Delta t$ (see Fig. 6.4).

Solution

$$P[A_2 = \text{'one event in } \Delta t\text{'} \mid A_1 = \text{'one event in } T\text{'}]$$

$$= P[A_2 \mid A_1] = \frac{P[A_1 \cap A_2]}{P[A_1]} = \frac{G(0,t)\,G(1,\Delta t)\,G(0, T - t - \Delta t)}{G(1,T)}$$

$$P[A_2 \mid A_1] = \frac{\mathrm{e}^{-\nu t}(\nu \Delta t\, \mathrm{e}^{-\nu \Delta t})\,\mathrm{e}^{-\nu(T-\Delta t - t)}}{\nu T \mathrm{e}^{-\nu T}}$$

$$= \frac{\Delta t}{T} = p(t \mid 1 \in T)\Delta t$$

Therefore, the conditional probability density is

$$p(t \mid 1 \in T) = 1/T \tag{6.69}$$

the uniform density, independent of time t. In other words, one event is equally likely any place in T. This can be generalized for n events taking place in T. The arrival time of each

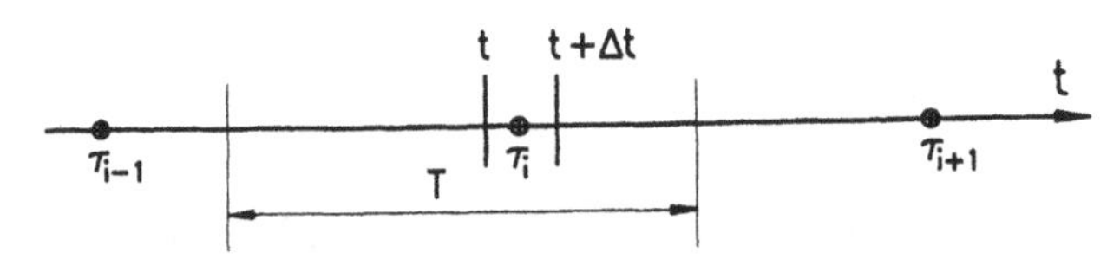

Fig. 6.4 *Arrival times τ_i of Poisson events*

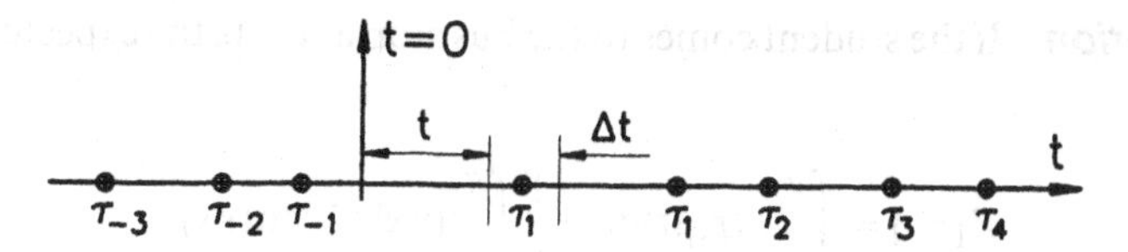

Fig. 6.5 *Poisson inter-arrival times*

event is uniformly distributed in T, $(p(t|n \in T) = 1/T)$, and the arrival times are statistically independent. This is the simplest model for waiting times and queuing theory.

Example 6.5 The arrival of a series of Poisson events is clocked on a time axis (Fig. 6.5). If the zero time axis is placed arbitrarily, what is the probability distribution of the first arrival time ('the first occurrence problem')? Also, find the probability distribution of the inter-arrival times, that is, the waiting time for the next Poisson event to take place.

Solution

$$P[t < \tau_1 \leqslant t + \Delta t] = p_{\tau 1}(t)\Delta t = G(0, t)\,G(1, \Delta t) = \mathrm{e}^{-\nu t}(\nu \Delta t)\,\mathrm{e}^{-\nu \Delta t} = \nu\,\mathrm{e}^{-\nu(t+\Delta t)}\,\Delta t$$

Letting $\Delta t \to 0$

$$p_{\tau_1}(t) = \begin{cases} \nu \mathrm{e}^{-\nu t} & t \geqslant 0 \\ 0 & t < 0 \end{cases} \tag{6.70}$$

Let the inter-arrival time $T_i = \tau_i - \tau_{i-i}$. In the same manner

$$P[t < T \leqslant t + \Delta t] = p_T(t)\Delta t = \nu\,\mathrm{e}^{-\nu(t+\Delta t)}\,\Delta t$$

and

$$p_T(t) = \nu \mathrm{e}^{-\nu t} \tag{6.71}$$

Also,

$$P[T \leqslant t] = 1 - P[T > t] = 1 - G(0, t) = 1 - \mathrm{e}^{-\nu t}$$
$$p_T(t) = d(P[T \leqslant t])/dt = \nu \mathrm{e}^{-\nu t}$$

so the distribution of the inter-arrival times is the same as that of the first arrival time. This example also demonstrates clearly that the process is memory-less. The arrival times of new events are totally independent on the previous event.

Example 6.6 A student arrives at a bus stop of the Poisson bus company. The bus schedule says that the buses run every 10 minutes on the average. What is the expected waiting time for the next bus to arrive, and what is the probability that the student has to wait 15 minutes or more for the next bus to arrive?

Solution If the student comes to the bus stop at $t = 0$, the expected waiting time is $E[\tau_1]$,

$$E[\tau_1] = \int_{-\infty}^{\infty} t p_{\tau_1}(t) dt = \frac{1}{\nu} \int_0^{\infty} (\nu x) e^{-\nu x} d(\nu x)$$

$$= \left[-\frac{1}{\nu} (\nu x) e^{-\nu x} \right]_0^{\infty} + \frac{1}{\nu} \int_0^{\infty} e^{-u} du = \frac{1}{\nu}$$

The mean arrival rate of the buses is one bus every 10 minutes, $\nu = 1/10$, and the expected waiting time is $1/\nu = 10$ minutes.

Secondly, calling the waiting time T,

$$P[T > 15] = G(0, 15) = e^{-\nu 15} = e^{-15/10} = 0.223$$

so there is about 22% probability that the student has to wait more than 15 minutes.

6.2.1 Compound Poisson Processes

The Poisson counting process is an extremely useful tool in such problems when a single stochastic variable or a random process is generated by a superposition of random quantities. For instance, consider a family of random variables $\{Y_i\}$ which are independent and identically distributed (IID variables). Then, the random sum

$$X(t) = \sum_{i=1}^{N(t)} Y_i \tag{6.72}$$

where $N(t)$ is a stationary Poisson counting process, is called a compound Poisson process.

This kind of superposition of Poisson events has been widely used in many applications. The only drawback is the inherent quality of the Poisson process that it is memoryless. For instance, using the Poisson process to describe the occurrences of large earthquakes, which have an average recurrence time equal to, say, T, the Poisson arrival time of the next big earthquake is independent of the occurrence of the last earthquake, even if almost T years have passed since the last earthquake (cf. Chapter 8). The Poisson process takes no notice of things that happened in the past. However, an important property of the Poisson process, called the superposition property found by Khinchine, [106], states that if an array of generalized Poisson processes are generated by a single source, it can be expected that the parameters are identically distributed throughout the system. The superposition of all such individually generated processes will tend to a pure Poisson process. Therefore, even if there were a discrepancy in the arrival times for any individual process, the superposition process would have purely Poisson arrival times.

Example 6.7 Consider the compound Poisson process

$$X(t) = \sum_{i=1}^{N(t)} (-1)^{N(\tau_i)} 1(t - \tau_i) \tag{6.73}$$

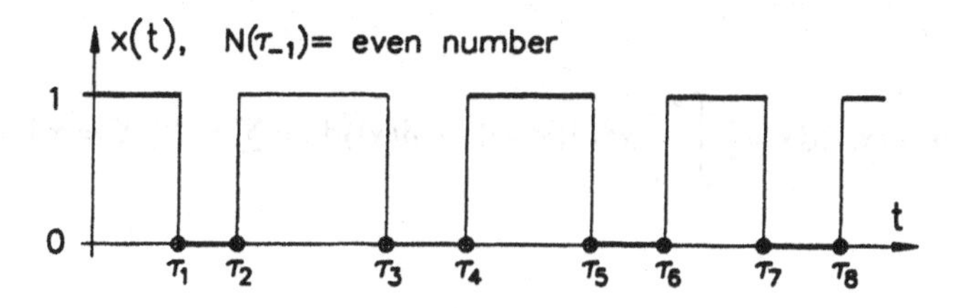

Fig. 6.6 *Operation of an on-off switch*

where $1(t)$ is Heaviside's step function. A typical sample function can have the form shown in Fig. 6.6. This situation arises when a switch is either on (+ 1) or off (0) at random times. As the observation of the process is started at a random time $t = 0$, it is equally likely that the switch is on (as in Fig. 6.6) or off corresponding to $x(0) = 0$. The times t_i are Poisson arrival times in accordance with the stationary Poisson counting process $N(t)$ with intensity ν (mean arrival rate). Is the switching process $X(t)$ a stationary process? Find the mean value and variance of $X(t)$. Given that the process is stationary also find the autocorrelation function.

Solution (1) At time t, $X(t)$ is either 1 or 0 (the probability that $X(t)$ is exactly changing at this instance is equal to zero). The probability that $X(t) = 1$ is

$$P[X(t) = 1] = P[X(0) = 1]\,P[\text{'}N(t) = \text{an even number'}] + P[X(0) = 0]\,P[\text{'}N(t) = \text{an odd number'}]$$

Now since $P[X(0) = 1] = P[X(0) = 0] = \frac{1}{2}$

$$P[X(t) = 1] = \tfrac{1}{2}(G(0, t) + G(2, t) + G(4, t) + \cdots + \tfrac{1}{2}(G(1, t) + G(3, t) + G(5, t) + \cdots)$$

as

$$P[N(t) = n] = G(n, t)$$

Therefore

$$P[X(t) = 1] = \tfrac{1}{2} \sum_{n=0}^{\infty} G(n, t) = \tfrac{1}{2}$$

Thus at all times, $X(t)$ is equally distributed between 1 and 0 so $p(x, t)$, the probability density, does not depend on t. The process at least is first-order stationary and

$$p(x, t) = \tfrac{1}{2}\delta(x - 1) + \tfrac{1}{2}\delta(x)$$

(2)

$$E[X] = \int_{-\infty}^{\infty} x p(x, t)\mathrm{d}x = \tfrac{1}{2}\int_{-\infty}^{\infty} x(\delta(x - 1) + \delta(x))\mathrm{d}x = \tfrac{1}{2}$$

also

$$E[X^2]=\int_{-\infty}^{\infty} x^2 p(x,t)\mathrm{d}x=\tfrac{1}{2}\int_{-\infty}^{\infty} x^2(\delta(x-1)+\delta(x))\mathrm{d}x=\sum x^2 P[X=x]=\tfrac{1}{2}\cdot 1+\tfrac{1}{2}\cdot 0=\tfrac{1}{2}$$

so

$$\mathrm{Var}[X]=\tfrac{1}{2}-\tfrac{1}{4}=\tfrac{1}{4}$$

(3) Form

$$X(t)\,X(t+\tau)=\begin{cases}1 \text{ if } X(t)=1 \text{ and there is an even number of changes in } \tau\\ 0 \text{ if } X(t)=1 \text{ and there is an odd number of changes in } \tau\\ 0 \text{ if } X(t)=0 \text{ and there is an even number of changes in } \tau\\ 0 \text{ if } X(t)=0 \text{ and there is an odd number of changes in } \tau\end{cases}$$

The random variable $Z=X(t)\,X(t+\tau)$ can assume the above four values. Therefore, $p(z)=P[Z=1]\,\delta(z-1)+P[Z=0]\,\delta(z)$ and

$$\begin{aligned}E[Z]&=\int_{-\infty}^{\infty} zp(z)\mathrm{d}z=1\cdot P[Z=1]=P[X(t)=1]\,P[\text{`}N(\tau)=\text{even'}]\\ &=\tfrac{1}{2}(G(0,t)+G(2,t)+G(4,t)+\cdots)=\tfrac{1}{2}\mathrm{e}^{-\nu\tau}\left(1+\frac{(\nu\tau)^2}{2!}+\frac{(\nu\tau)^4}{4!}+\cdots\right)\\ &=\tfrac{1}{2}\mathrm{e}^{-\nu\tau}\cosh\tau=R_X(\tau)\end{aligned}$$

The process is covariance stationary since the autocorrelation function is only a function of the time difference.

Another class of compound or filtered Poisson processes is constructed from the superposition of random pulses arriving at random Poisson times. In fact, consider the random process $\{X(t),\, t\in T\}$ where

$$X(t)=\sum_{i=0}^{N(t)} w(t,\tau_i,Y_i) \tag{6.74}$$

in which

1. $\{N(t),\, t>0\}$ is a Poisson process with intensity ν.
2. Y_i is a sequence of independent and identically distributed (IID) random variables with the probability density $p_Y(y)$.
3. τ_i are Poisson arrival times with arrival rate ν.
4. $w(x,y,z)$ is a three-valued shape function called the impulse response function or the shape function of the process.

An intuitive interpretation of Eq. (6.74) is that at the time instant τ_i, a Poisson type event takes place, which is the arrival of a time signal or pulse $w(t,\tau_i,Y_i)$. Y_i represents a measure of the amplitude of the signal, which is a random quantity, and $w(t,\tau,y)$ is the time history

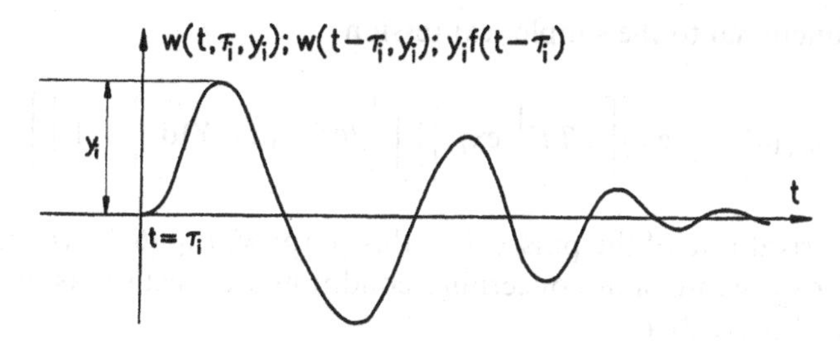

Fig. 6.7 *The impulse response function*

of the signal $X(t)$ is therefore the value at time t of the sum of all such signals that have occurred or arrived in the time interval $(0, t]$.

The impulse response function is typically of the form shown in Fig. 6.7. It is generally defined as a causal function of time, having backwards-oriented memory, that is, $w(t, \tau, y) = 0$ for $t < \tau$, or the pulse can not influence the process until it has arrived. Secondly, the shape function is most often a function of the real time or the time difference only, $\theta = t - \tau$, whereby $w(t, \tau, y) = w(\theta, y)$. For causal shape functions, the Poisson counting process $N(t)$ as the upper limit of the sum Eq. (6.74) can either be replaced by plus infinity $(+\infty)$ or by a large time T through $N(T)$, since for arrival times larger than τ, $w(t, \tau, y) \equiv 0$. Thus,

$$X(t) = \sum_{i=0}^{N(T)} w(t, \tau_i, Y_i), \qquad T \gg t \tag{6.75}$$

The following facts should be noted: (i) the number of pulses in T is random, that is, $P[N(T) = m] = G(m, T)$, (ii) each arrival time is a random variable with the conditional probability density $p(\tau | m \in T) = 1/T$ (see Example 6.4), (iii) each amplitude Y_i is a random variable with the common distribution or density $p_Y(y)$, (iv) all τ_i and Y_i are mutually independent.

The next step is to obtain the mean value and covariance kernel of the random process Eq. (6.75). For this purpose, introduce the so-called characteristic functional of a stochastic process, Eq. (2.35), that is,

$$\varphi_X(\theta(t)) = E\left[\exp\left\{\mathrm{i} \int_0^T \theta(t) X(t) \mathrm{d}t\right\}\right] \tag{2.35}$$

which was shown to be a natural extension of the multivariate characteristic function of a sequence of random variables, Eq. (1.111). This approach has been described by Lin, [122], among others. If the sum Eq. (6.75) is introduced into the expression Eq. (2.35),

$$\varphi_X(\theta(t)) = E\left[\exp\left\{\mathrm{i} \int_0^T \theta(t) \left[\sum_{j=0}^{N(T)} w(t, \tau_j, Y_j)\right] \mathrm{d}t\right\}\right] \tag{6.76}$$

The next task is to show how the sum in Eq. (6.76) can be broken up to reduce the characteristic functional to the simpler expression

$$\varphi_X(\theta(t)) = \exp\left[\nu T E\left[\exp\left\{i\int_0^T \theta(t)\,w(t,\tau,Y)\,dt\right\} - 1\right]\right] \tag{6.77}$$

where ν is the arrival rate of the pulses. For this purpose, Eq. (6.76) is first rewritten by using a well-known expression concerning conditional expectations, namely $E[Y] = E[E[Y|X]]$, see Eq. (1.63), or

$$\begin{aligned}\varphi_X(\theta(t)) &= E\left[E\left[\exp\left\{i\int_0^T \theta(t)\left[\sum_{j=0}^{N(T)} w(t,\tau_j,Y_j)\right]dt\right\}\middle| N(T)=n\right]\right]\\ &= E\left[\int_{-\infty}^{\infty} \exp\left\{i\int_0^T \theta(t)\left[\sum_{j=0}^{u} w(t,\tau_j,Y_j)\right]dt\right\} p_N(u)\,du\right]\end{aligned}$$

where $p_N(u) = G(n,T)\,\delta(u-n)$. Therefore,

$$\begin{aligned}\varphi_X(\theta(t)) &= E\left[\sum_{n=0}^{\infty} G(n,T)\exp\left\{i\int_0^T \theta(t)\left[\sum_{j=0}^{n} w(t,\tau_j,Y_j)\right]dt\right\}\right]\\ &= \sum_{n=0}^{\infty} G(n,T)\,E\left[\exp\left\{i\int_0^T \theta(t)\left[\sum_{j=0}^{n} w(t,\tau_j,Y_j)\right]dt\right\}\right]\end{aligned} \tag{6.78}$$

The expectation term in Eq. (6.78) can be simplified as follows:

$$\begin{aligned}E\left[\exp\left\{i\int_0^T \theta(t)\left[\sum_{j=0}^{n} w(t,\tau_j,Y_j)\right]dt\right\}\right] &= E\left[\prod_{j=0}^{n}\exp\left\{i\int_0^T \theta(t)\,w(t,\tau_j,Y_j)\,dt\right\}\right]\\ &= \prod_{j=0}^{n} E\left[\exp\left\{i\int_0^T \theta(t)\,w(t,\tau_j,Y_j)\,dt\right\}\right]\end{aligned} \tag{6.79}$$

since all τ_j and Y_j are independent.

In Eq. (6.79), each exponential term is now expanded into infinite series, that is,

$$\begin{aligned}&E\left[\exp\left\{i\int_0^T \theta(t)\,w(t,\tau_j,Y_j)\,dt\right\}\right]\\ &\quad = 1 + \sum_{m=1}^{\infty}\frac{(i)^m}{m!}E\left[\left\{\int_0^T \theta(t)\,w(t,\tau_j,Y_j)\,dt\right\}^m\right] = 1+\alpha\end{aligned} \tag{6.80}$$

The value of α in Eq. (6.80) depends on neither the individual arrival times τ_j nor the amplitudes Y_j because

$$\begin{aligned}&E\left[\left\{\int_0^T \theta(t)\,w(t,\tau_j,Y_j)\,dt\right\}^m\right]\\ &= \int_0^T\cdots\int_0^T \theta(t_1)\cdots\theta(t_m)\,E[w(t_1,\tau_j,Y_j)\,w(t_2,\tau_j,Y_j)\cdots w(t_m,\tau_j,Y_j)]\,dt_1\,dt_2\cdots dt_m\end{aligned}$$

$$= \int_0^T \cdots \int_0^T \theta(t_1)\cdots\theta(t_m)\left\{\int_{-\infty}^{\infty}\int_0^T w(t_1,\tau,y)\cdots w(t_m,\tau,y)\,p_Y(y)\,p(\tau|m\in T)\mathrm{d}y\,\mathrm{d}\tau\right\}\mathrm{d}t_1\cdots\mathrm{d}t_m$$

$$= \frac{1}{T}\int_0^T \cdots \int_0^T \theta(t_1)\cdots\theta(t_m)\int_{-\infty}^{\infty} p_Y(y)\int_0^T w(t_1,\tau,y)\cdots w(t_m,\tau,y)\mathrm{d}\tau\,\mathrm{d}y\,\mathrm{d}t_1\cdots\mathrm{d}t_m \quad (6.81)$$

which shows that the variables τ_j and Y_j are no longer present after performing the integrations in Eq. (6.81). In fact the whole expression Eq. (6.81) only depends on T and m, and is independent of the number j. By Eq. (6.80),

$$\alpha = E\left[\exp\left\{\mathrm{i}\int_0^T \theta(t)\,w(t,\tau_j,Y_j)\,\mathrm{d}t\right\} - 1\right] \quad (6.82)$$

which inserted into Eq. (6.79) gives

$$E\left[\exp\left\{\mathrm{i}\int_0^T \theta(t)\left[\sum_{j=0}^{n} w(t,\tau_j,Y_j)\right]\mathrm{d}t\right\}\right] = \prod_{j=0}^{n}(1+\alpha) = (1+\alpha)^n$$

that can be brought into Eq. (6.78) to yield

$$\varphi_X(\theta(t)) = \sum_{n=0}^{\infty} \mathrm{e}^{-\nu T}\frac{(\nu T)^n}{n!}(1+\alpha)^n = \mathrm{e}^{-\nu T}\mathrm{e}^{\nu T(1+\alpha)} = \mathrm{e}^{\nu T\alpha} \quad (6.83)$$

or

$$\varphi_X(\theta(t)) = \exp\left[\nu T E\left[\exp\left\{\mathrm{i}\int_0^T \theta(t)\,w(t,\tau,Y)\mathrm{d}t\right\} - 1\right]\right] \quad (6.77)$$

which is the desired result.

A general expansion of the characteristic functional into series yields the moments of $X(t)$, that is,

$$\varphi_X(\theta(t)) = 1 + \mathrm{i}\int_0^T \theta(t)\,E[X(t)]\mathrm{d}t + \frac{\mathrm{i}^2}{2!}\int_0^T\int_0^T \theta(t_1)\theta(t_2)\,E[X(t_1)X(t_2)]\mathrm{d}t_1\,\mathrm{d}t_2 + \cdots \quad (6.84)$$

and by forming and expanding the log-characteristic functional

$$\psi_X(\theta(t)) = \log_e \varphi_X(\theta(t)) = \mathrm{i}\int_0^T \theta(t)\kappa_1[X(t)]\mathrm{d}t + \frac{\mathrm{i}^2}{2!}\int_0^T\int_0^T \theta(t_1)\theta(t_2)\kappa_2[X(t_1)X(t_2)]\mathrm{d}t_1\,\mathrm{d}t_2 + \cdots \quad (6.85)$$

the cumulant functions are obtained. Now by Eqs. (6.85), (6.83) and (6.82)

$$\psi_X(\theta(t)) = \nu T\alpha = \nu T\sum_{m=1}^{\infty}\frac{(\mathrm{i})^m}{m!}E\left[\left\{\int_0^T \theta(t)\,w(t,\tau_j,Y_j)\mathrm{d}t\right\}^m\right]$$

so by Eqs. (6.81) and (6.85), the mth term in the above sum can be written as

$$\frac{i^m}{m!}\int_0^T \cdots \int_0^T \theta(t_1)\cdots\theta(t_m)\left[\nu\int_{-\infty}^{\infty} p_Y(y)\int_0^T w(t_1,\tau,y)\cdots w(t_m,\tau,y)\,d\tau\,dy\right]dt_1\cdots dt_m$$

using Eq. (6.81). Therefore, the mth cumulant is given by

$$\kappa_m[X(t_1)\cdots X(t_m)] = \nu\int_{-\infty}^{\infty}\int_0^T p_Y(y)w(t_1,\tau,y)\cdots w(t_m,\tau,y)d\tau\,dy \tag{6.86}$$

which is the same as

$$\kappa_m[X(t_1)\cdots X(t_m)] = \nu\int_0^{\min(t_1,\ldots,t_m)} E[w(t_1,\tau,Y)\cdots w(t_m,\tau,Y)]d\tau \tag{6.87}$$

since $w(t_i,\tau,y)\equiv 0$ for $\tau > t_i$ and the expectation is only with respect to Y.

For impulse response functions of the form

$$w(t,\tau,y) = yw(t,\tau) \tag{6.88}$$

the mth cumulant simplifies to (see Eq. (6.86))

$$\begin{aligned}\kappa_m[X(t_1)\cdots X(t_m)] &= \nu\int_{-\infty}^{\infty} p_Y(y)y^m\int_0^T [w(t_1,\tau)\cdots w(t_m,\tau)]d\tau\,dy\\ &= \nu E[Y^m]\int_0^{\min(t_1,\ldots,t_m)} w(t_1,\tau)\cdots w(t_m,\tau)d\tau\end{aligned} \tag{6.89}$$

The mean value function, covariance kernel and the variance of the compound Poisson process

$$X(t) = \sum_{i=0}^{N(t)} Y_i w(t,\tau_i) \tag{6.90}$$

are therefore given by

$$E[X(t)] = \mu_X(t) = \nu\mu_Y\int_0^t w(t,\tau)d\tau \tag{6.91}$$

$$K_X(t_1,t_2) = \nu E[Y^2]\int_0^t w(t_1,\tau)\,w(t_2,\tau)d\tau \tag{6.92}$$

and

$$\sigma_X^2(t) = \nu E[Y^2]\int_0^t w^2(t,\tau)d\tau \tag{6.93}$$

Example 6.8 Consider the random process

$$X(t) = \sum_{i=1}^{N(t)} A_i f(t-\tau_i)$$

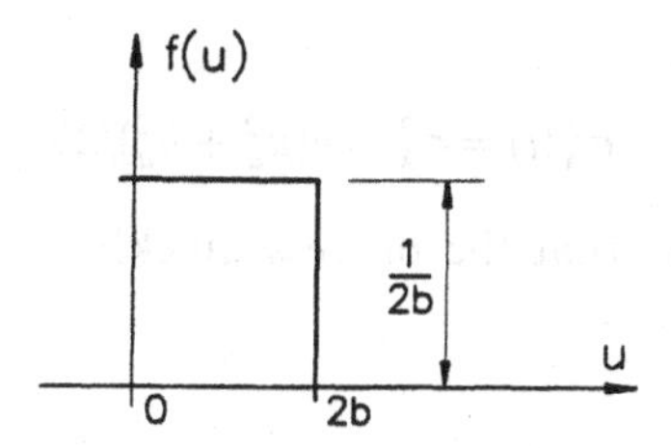

Fig. 6.8 *An ideal square wave*

where $N(t)$ is a stationary Poisson counting process with intensity ν and $\{A_i\}$ are independent random amplitudes taken from a common distribution with mean value μ_a and variance σ_a^2. The function $f(u)$ is an ideal square wave as shown in Fig. 6.8. Find the mean value function, the variance and the covariance kernel of the process.

Solution The mean value is given by Eq. (6.91):

$$E[X(t)] = \nu\mu_a \int_0^t f(t-\tau)\mathrm{d}\tau = \begin{cases} \dfrac{\nu\mu_a}{2b}t & t < 2b \\ n\mu_a & t \geqslant 2b \end{cases}$$

The covariance kernel is given by Eq. (6.92). For $t < s$

$$K_X(t,s) = \nu E[A^2] \int_0^t f(t-\tau) f(s-\tau)\mathrm{d}\tau$$

The integral is evaluated by introducing the time difference $u = s - t$ and the transformation $\theta = t - \tau$, see Fig. 6.9. For $u \geqslant 2b$, the integral is obviously zero, otherwise the value of the integral $\int f(\theta) f(\theta + u)\mathrm{d}\theta$ is the area of the shaded column times $1/2b$.

Now,

$$E[A^2] = \mu_a^2 + \sigma_a^2$$

so

$$K_X(t,s) = \Gamma_X(u) = \begin{cases} \dfrac{\nu(\mu_a^2 + \sigma_a^2)}{4b^2}(2b - u) & u < 2b \\ 0 & u \geqslant 2b \end{cases}$$

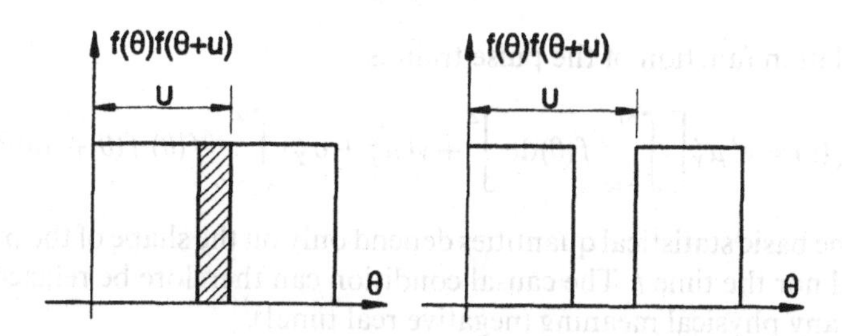

Fig. 6.9 *The integral* $\int (f(\theta) f\theta + u)\mathrm{d}\theta$

and

$$\sigma_X^2(t) = \sigma_X^2 = \nu(\mu_a^2 + \sigma_a^2)/2b$$

This interesting result shows that the process quickly becomes covariance stationary (for $t \geqslant 2b$).

6.2.2 Stationary Pulse Trains

In most physical problems, the impulse response functions (Fig. 6.7) are not only causal, ($f(t-\tau) \equiv 0$, for $\tau > t$), but also decaying such that ($f(0) \to 0$ for $t - \tau = \theta \to \infty$. In this case, if the time t is large enough, the compound Poisson process, Eq. (6.91), is stationary since the upper limit in the integrals, Eq. (6.89) and Eqs. (6.91)–(6.93), can be replaced by plus infinity. The random pulse train,

$$X(t) = \sum_i Y_i f(t - \tau_i) \tag{6.94}$$

that has a sufficient large number of terms (t is large) has the mth cumulant function equal to

$$\kappa_m[X(t_1)\cdots X(t_m)] = \nu E[Y^m] \int_{-\infty}^{\infty} f(t_1 - \tau)\cdots f(t_m - \tau)\mathrm{d}\tau \tag{6.95}$$

The mean value is

$$E[X(t)] = \nu \mu_Y \int_{-\infty}^{\infty} f(\theta)\mathrm{d}\theta \tag{6.96}$$

The autocovariance function ($u = t_2 - t_1$, $\theta = t_1 - \tau$)

$$\Gamma_X(u) = \nu(\mu_Y^2 + \sigma_Y^2) \int_{-\infty}^{\infty} f(\theta) f(\theta + u)\mathrm{d}\theta \tag{6.97}$$

and the variance

$$\sigma_X^2 = \nu E[Y^2] \int_{-\infty}^{\infty} f^2(\theta)\mathrm{d}\theta \tag{6.98}$$

The autocorrelation function of the pulse train is

$$R_X(u) = \nu^2 \mu_Y^2 \left[\int_{-\infty}^{\infty} f(\theta)\mathrm{d}\theta \right]^2 + \nu(\mu_Y^2 + \sigma_Y^2) \int_{-\infty}^{\infty} f(\theta) f(\theta + u)\mathrm{d}\theta \tag{6.99}$$

It is seen that the basic statistical quantities depend only on the shape of the pulses and not on their arrival nor the time t. The causal condition can therefore be relaxed even if that does not have any physical meaning (negative real time!).

An infinite train of random pulses is often referred to as 'shot noise'. The name stems from the physical description of random emission of electrons from a heated cathode in vacuum tubes. If there is no accumulated space charge, the emission of electrons from the cathode can be shown to be a Poisson event. As early as 1909, Campbell, [29], was able to derive Eqs. (6.96)–(6.99) for the stationary shot noise

$$X(t) = \sum_{-\infty}^{\infty} Y_i f(t - \tau_i) \tag{6.100}$$

which is equivalent to Eq. (6.94) for sufficiently large time t.

Campbell's argument is based upon taking a large interval of time T and observing the process in the interval $(-T/2 < t < T/2)$. Considering the random process Eq. (6.100), the number of pulses in this interval is $N(T)$ with probability $G(n, T)$. Each arrival time in the interval is random with the probability density $p(\tau_i | n \in T) = 1/T$. Each amplitude is random with the common probability density $p_Y(y)$. Then, for a fixed t in $(-T/2, T/2)$,

$$E[X(t)] = \lim_{T\to\infty} E[E[X(t)|N(T)]]$$

$$= \lim_{T\to\infty} \sum_{n=0}^{\infty} G(n, T) \prod_{j=1}^{n} \int_{-\infty}^{\infty} p_Y(y_j) \mathrm{d}y_j \prod_{j=1}^{n} \int_{-T/2}^{T/2} \left[p(\tau_j | n \in T) \sum_{j=1}^{n} y_j f(t - \tau_j) \right] \mathrm{d}\tau_j$$

since the joint density function

$$p(y_1, y_2, \ldots, y_n, \tau_1, \tau_2, \ldots, \tau_n) = \prod_{j=1}^{n} p_Y(y_j) \prod_{j=1}^{n} p(\tau_j | n \in T)$$

as all these variables are mutually independent. Then,

$$E[X(t)] = E\left[\lim_{T\to\infty} \sum_{j=1}^{n} Y_j f(t - \tau_j) \right]$$

$$= \lim_{T\to\infty} \sum_{n=0}^{\infty} G(n, T) \prod_{j=1}^{n} \int_{-\infty}^{\infty} p_Y(y_j) \mathrm{d}y_j \frac{1}{T} \prod_{j=1}^{n} \int_{-T/2}^{T/2} \left(\sum_{j=1}^{n} y_j f(t - \tau_j) \right) \mathrm{d}\tau$$

Now take a typical term $y_j f(t - \tau_j)$ through the double Π multiplication:

$$\int_{-\infty}^{\infty} p_Y(y_1) \mathrm{d}y_1 \cdots \int_{-\infty}^{\infty} p_Y(y_j) y_j \, \mathrm{d}y_j \cdots \int_{-\infty}^{\infty} p_Y(y_n) \mathrm{d}y_n \int_{-T/2}^{T/2} \frac{\mathrm{d}\tau_1}{T}$$

$$\cdots \int_{-T/2}^{T/2} \frac{\mathrm{d}\tau_j}{T} f(t - \tau_j) \cdots \int_{-T/2}^{T/2} \frac{\mathrm{d}\tau_n}{T}$$

$$= \frac{1}{T} \int_{-\infty}^{\infty} y_j p_Y(y_j) \mathrm{d}y_j \int_{-T/2}^{T/2} \mathrm{d}\tau_j \, f(t - \tau_j)$$

$$= \frac{1}{T} E[Y] \int_{-(T/2)+\tau_j}^{(T/2)+\tau_j} f(\theta) \mathrm{d}\theta$$

The whole sum $\sum_{j=1}^{n} y_j f(t-\tau_j)$ thus gives $(n/T)\,E[Y]\int_{-(T/2)+t}^{(T/2)+t} f(\theta)\mathrm{d}\theta$ and

$$E[X(t)] = \lim_{T\to\infty} \sum_{n=0}^{\infty} nG(n,T)\frac{1}{T}E[Y]\int_{(T/2)+t}^{(T/2)+t} f(\theta)\mathrm{d}\theta$$

$$= \lim_{T\to\infty}\left[\frac{1}{T}E[Y]\int_{(T/2)+t}^{(T/2)+t} f(\theta)\mathrm{d}\theta \sum_{n=0}^{\infty} nG(n,T)\right]$$

$$= \lim_{T\to\infty}\left[\frac{1}{T}E[Y]\int_{(T/2)+t}^{(T/2)+t} f(\theta)\mathrm{d}\theta\, E[N(T)]\right] = \nu E[Y)\int_{-\infty}^{\infty} f(\theta)\mathrm{d}\theta$$

since $E[N(T)] = \sum_{n=0}^{\infty} nG(n,T) = \nu T$, that is, the same result as before. In the same manner, Campbell obtained the second moment $E[X(t)X(t+u)]$ which gives the autocorrelation function Eq. (6.99).

Now consider a stationary shot noise with zero mean value, that is, $E[Y]=\mu_Y=0$, and amplitude variance $\sigma_Y^2 = E[Y^2]$. Then the autocorrelation function Eq. (6.99) becomes

$$R_X(u) = \nu\sigma_Y^2 \int_{-\infty}^{\infty} f(\theta)\, f(\theta+u)\mathrm{d}\theta \tag{6.101}$$

The power spectral density function of this process is then given by

$$S_X(\omega) = \frac{1}{2\pi}\int_{-\infty}^{\infty} R_X(u)\mathrm{e}^{-\mathrm{i}\omega u}\mathrm{d}u = \frac{\nu\sigma_Y^2}{2\pi}\int_{-\infty}^{\infty}\int_{-\infty}^{\infty} f(\theta)\, f(\theta+u)\mathrm{e}^{-\mathrm{i}\omega u}\mathrm{d}\theta\,\mathrm{d}u$$

$$= \frac{\nu\sigma_Y^2}{2\pi}\int_{-\infty}^{\infty} f(\theta)\mathrm{e}^{+\mathrm{i}\omega\theta}\mathrm{d}\theta\int_{-\infty}^{\infty} f(\theta+u)\mathrm{e}^{-\mathrm{i}\omega(\theta+u)}\mathrm{d}(\theta+u)$$

$$= \frac{\nu\sigma_Y^2}{2\pi}F(-\omega)\,F(\omega) = \frac{\nu\sigma_Y^2}{2\pi}|F(\omega)|^2 \tag{6.102}$$

It should be noted that the intensity of the power spectrum only depends on the pulse arrival rate and the mean square of the pulse amplitude. The shape of the spectrum on the other hand depends only on the frequency content of the pulses (the pulse shape function).

Example 6.9 A stationary random process is given by

$$X(t) = \sum_{-\infty}^{\infty} Y_i \exp\left[-\frac{(t-\tau_i)^2}{2\omega_0^2}\right]\frac{1}{\omega_0\sqrt{2\pi}}$$

where the amplitudes Y_i are independent Gaussian random variables with zero mean and variance σ^2, and the arrival times τ_i are Poisson-distributed random times with an average rate ν. Find the mean value, the autocorrelation function and the power spectral density of this process.

Solution Since $E[Y]=0$, $E[X(t)]=0$. The autocorrelation function is given by

$$R_X(u)=\nu\sigma^2\int_{-\infty}^{\infty} f(\theta)\,f(\theta+u)\mathrm{d}\theta=\frac{\nu\sigma^2}{\omega_0^2 2\pi}\int_{-\infty}^{\infty} \mathrm{e}^{-\theta^2/(2\omega_0^2)}\,\mathrm{e}^{-(\theta+u)^2/(2\omega_0^2)}\,\mathrm{d}\theta$$

$$=\nu\sigma^2\exp\left[-\frac{u^2}{4\omega_0^2}\right]\left(\frac{1}{\omega_0\sqrt{2}\sqrt{2\pi}}\right)\frac{1}{(\omega_0/\sqrt{2})\sqrt{2\pi}}\int_{-\infty}^{\infty}\exp\left[-\frac{(\theta+u/2)^2}{\omega_0^2}\right]\mathrm{d}\theta$$

$$=\nu\sigma^2\frac{\exp[-u^2/(4\omega_0^2)]}{(\omega_0\sqrt{2})\sqrt{2\pi}}$$

and the power spectral density by

$$S_X(\omega)=\frac{\nu\sigma^2}{2\pi}|F(\omega)|^2$$

Obviously, $F(\omega)=\exp(-\omega_0^2\omega^2/2)$ (the Gaussian density, see Example 1.24), and

$$S_X(\omega)=\frac{\nu\sigma^2}{2\pi}\exp(-\omega_0^2\omega^2)$$

Now letting $\omega_0\to 0$, the impulse response function (a Gaussian density with zero variance) tends to the delta function. Therefore, the shot noise process

$$X(t)=\sum_{-\infty}^{\infty} Y_n\delta(t-\tau_n) \tag{6.103}$$

which is a random superposition of impulses with a random amplitude Y_n, has the mean value equal to zero and the autocorrelation and power spectrum equal to

$$R_X(u)=\lim_{\omega_0\to 0}\nu\sigma^2\frac{\exp[-u^2/(4\omega_0^2)]}{(\omega_0\sqrt{2})\sqrt{2\pi}}=\nu\sigma^2\delta(u) \tag{6.104}$$

$$S_X(\omega)=\lim_{\omega_0\to 0}\frac{\nu\sigma^2}{2\pi}\exp(-\omega_0^2\omega^2)=\frac{\nu\sigma^2}{2\pi} \tag{6.105}$$

Therefore, the process Eq. (6.103) has an impulse-like autocorrelation function and a constant power spectrum which corresponds to that of white noise. The process is therefore equivalent to white noise.

6.2.3 Non-stationary Pulse Trains

Consider compound or filtered Poisson processes of the type Eq. (6.74), where the underlying Poisson counting process is non-homogeneous, that is, has a time-dependent intensity $\nu=\nu(t)$. By comparing Eqs. (6.65) and (6.68), the mean value of the non-homogeneous Poisson process is

$$E[N(t)]=\mu_N(t)=\int_0^t \nu(u)\mathrm{d}u \tag{6.106}$$

where

$$\nu(t) = \frac{d\mu_N(t)}{dt} \qquad \text{or} \qquad \Delta\mu_N(t) = \nu(t)\Delta t$$

that is, on the average, $\Delta\mu_N(t) = \nu(t)\Delta t$ events will occur in a time interval $(t, t + \Delta t]$.

The non-homogeneous Poisson process has the probability function $G(n, t)$ given by Eq. (6.68). The conditional probability density for the arrival time of a Poisson event (pulse) in a time interval $(0, T]$ is therefore, (Ex. 6.4),

$$p(t|n \in T) = \frac{\nu(t)}{\int_0^T \nu(u)du}$$

and by going through the same argument leading to Eq. (6.83), the log-characteristic functional now becomes

$$\psi_X(\theta(t)) = \int_0^T \nu(u)du \sum_{m=1}^{\infty} \frac{i^m}{m!} E\left[\left\{\int_0^T \theta(t)\, w(t, \tau, Y)dt\right\}^m\right]$$

and the mth term in the expansion is

$$\int_0^T \nu(u)du \frac{i^m}{m!} \int_0^T \cdots \int_0^T \theta(t_1)\cdots\theta(t_m)\left[\int_{-\infty}^{\infty} p_Y(y) \int_0^T w(t_1, \tau, y)\cdots w(t_m, \tau, y)\right.$$
$$\left. \times \frac{\nu(\tau)}{\int_0^T \nu(u)du} d\tau\, dy\right] dt_1 \cdots dt_m$$

The integral $\int_0^T \nu(u)du$ cancels out and the mth cumulant function is therefore

$$\kappa_m[X(t_1)\cdots X(t_m)] = \int_{-\infty}^{\infty} \int_0^{\min(t_1,\ldots,t_m)} p_Y(y)\, w(t_1, \tau, y)\cdots w(t_m, \tau, y)\, \nu(\tau)d\tau\, dy \quad (6.107)$$

Actually, the results Eqs. (6.86), (6.89), (6.91)–(6.93) for an underlying homogeneous Poisson process with a constant intensity ν can be generalized by moving the now time-dependent intensity function inside the integral sign.

As a special case, consider a compound non-homogeneous Poisson process, Eq. (6.90), for which the shape function $w(t, \tau) \equiv 1$ for all t and τ and $\{Y_i\}$ are IID random variables, that is,

$$X(t) = \sum_{i=0}^{N(t)} Y_i \quad (6.108)$$

According to the above result, the mean value and the variance are therefore given by

$$E[X(t)] = \mu_X(t) = \mu_Y \int_0^t \nu(\tau)d\tau = \mu_Y \eta(t) \quad (6.109)$$

and

$$\sigma_X^2(t) = E[Y^2]\int_0^t \nu(\tau)d\tau = E[Y^2]\eta(t) \tag{6.110}$$

by introducing the auxiliary function $\eta(t) = \int_0^t \nu(\tau)d\tau$. For a constant intensity ν_0, the mean value and variance are linearly increasing functions of time, i.e.

$$\mu_X(t) = \nu_0 \mu_Y t, \qquad \sigma_X^2(t) = \nu_0 E[Y^2]t \tag{6.111}$$

that is, the process is cumulative by nature, its mean value and variance increases with each incremental event Y_i. Compound non-homogeneous Poisson processes are often classified as birth or death processes according to the behaviour of the intensity function. For instance, a linear birth process is obtained with the intensity function $\nu(t) = \nu_0 t$, that is, the registration of random events (e.g. childbirths) increases linearly with time. The mean value and variance of the linear birth process increases quadratically as time passes since

$$\mu_X(t) = \tfrac{1}{2}\nu_0 \mu_Y t^2, \qquad \sigma_X^2(t) = \tfrac{1}{2}\nu_0 E[Y^2]t^2 \tag{6.112}$$

The intensity function $\nu(t) = \nu_0/t$ on the other hand would produce a death process as the registration of events peters out, and the mean value and variance rapidly diminish as time passes.

Example 6.10 Sobczyk, [60], [204], has used the linear birth process to decribe fatigue crack growth in a material component under load (cf. Section 4.5). Studies of crack growth or crack propagation are essential in the prediction of fatigue life of structural components. Crack propagation in solid materials is a highly complex phenomenon, complicated by the crack growth retardation which normally follows a high overload. A crack growth process is therefore best represented by a suitable stochastic process taking into account the randomness of the phenomenon.

Denote by $L(t)$ the length of the dominant crack in a material component at time t, which has been growing randomly from an initial crack of length L_0 (a real constant) that had sufficient size to start the crack propagation process. Assume now that the crack length $L(t)$ is the cumulative sum of elementary crack increments, that is,

$$L(t) = L_0 + L_1(t) = L_0 + \sum_{i=1}^{N(t)} Y_i \tag{6.113}$$

in which $N(t)$ is an underlying Poisson counting process, and $\{Y_i\}$ are non-negative IID random variables characterizing the magnitudes of elementary crack increments or propagation length in the time interval between two Poisson counts. Moreover, the counting process $N(t)$ is independent of the Y_i. Thus, the crack propagation model is based on the probability distribution of the magnitudes of elementary crack increments and the intensity of crack growth in concordance with the Poisson process.

Of immediate interest is now to derive the probability distribution of the dominant crack length, i.e.

$$p_L(l;t) = \frac{d(P[L(t) \leq l])}{dl} \tag{6.114}$$

For this purpose, form the moment generation function for the cumulative part $L_1(t)$, Eq. (6.113),

$$\varphi_{L_1}(\omega) = E[\mathrm{e}^{\mathrm{i}\omega L_1}] = \int_{-\infty}^{\infty} \mathrm{e}^{\mathrm{i}\omega l} p_{L_1}(l;t)\mathrm{d}l \tag{6.115}$$

whereby the probability density is obtained as the inverse Fourier transform (cf. Eqs. (1.97) and (1.98)). Rather than working on the integral in Eq. (6.115), the expectation term can be evaluated by using the inherent properties of the Poisson counting process and the law of total probability Eq. (1.63), i.e.

$$\varphi_{L_1}(\omega) = E[\mathrm{e}^{\mathrm{i}\omega L_1}] = \sum_{j=0}^{\infty} E[\mathrm{e}^{\mathrm{i}\omega L_1} | N(t) = j]\, P[N(t) = j]$$

The conditional expectation in the above equation is simply $E[\mathrm{e}^{\mathrm{i}\omega(Y_1 + \cdots + Y_j)}] = (E[\mathrm{e}^{\mathrm{i}\omega(Y_i)}])^j$ since the $\{Y_i\}$ are IID variables. Therefore, by Eqs. (6.60) and (6.68),

$$\begin{aligned}\varphi_{L_1}(\omega) = E[\mathrm{e}^{\mathrm{i}\omega L_1}] &= \sum_{j=0}^{\infty} (E[\mathrm{e}^{\mathrm{i}\omega Y_i}])^j \frac{(\eta(t))^j}{j!} \mathrm{e}^{-\eta(t)} \\ &= \mathrm{e}^{-\eta(t)} \sum_{j=0}^{\infty} (\varphi_Y(\omega))^j \frac{(\eta(t))^j}{j!}\end{aligned} \tag{6.116}$$

Based on experiments on the distribution of elementary crack increments, it is assumed that the probability distribution of $\{Y_i\}$ is the exponential distribution of the form

$$p_Y(y) = \begin{cases} \alpha \mathrm{e}^{-\alpha y} & y > 0,\ \alpha > 0 \\ 0 & y \leqslant 0 \end{cases} \tag{6.117}$$

Then, the characteristic function is given by

$$\varphi_Y(\omega) = E[\mathrm{e}^{\mathrm{i}\omega Y}] = \int_0^{\infty} \mathrm{e}^{\mathrm{i}\omega y} \alpha \mathrm{e}^{-\alpha y} dy = \frac{\alpha}{\alpha - \mathrm{i}\omega} \tag{6.118}$$

which inserted into Eq. (6.116) gives the characteristic function

$$\varphi_{L_1}(\omega) = \mathrm{e}^{-\eta(t)} \sum_{j=0}^{\infty} \left(\frac{\alpha}{\alpha - \mathrm{i}\omega}\right)^j \frac{(\eta(t))^j}{j!} \tag{6.119}$$

The inverse Fourier transform of Eq. (6.119) yields the desired probability density, that is,

$$\begin{aligned} p_{L_1}(l;t) &= \mathrm{e}^{-\eta(t)} \sum_{j=0}^{\infty} \frac{(\eta(t))^j}{j!} \frac{1}{2\pi} \int_{-\infty}^{\infty} \left(\frac{\alpha}{\alpha - \mathrm{i}\omega}\right)^j \mathrm{e}^{-\mathrm{i}\omega l} \mathrm{d}\omega \\ &= \mathrm{e}^{-\eta(t) - \alpha l} \sum_{j=0}^{\infty} \frac{(\alpha\eta(t))^j}{j!} \frac{1}{2\pi} \int_{-\infty}^{\infty} \frac{\mathrm{e}^{(\alpha - \mathrm{i}\omega)l}}{(\alpha - \mathrm{i}\omega)^j} \mathrm{d}\omega \end{aligned} \tag{6.120}$$

The integral in Eq. (6.120) can be inverted using Cauchy's integral theorem, (cf. Ex. 3.5). There is one multiple pole at $z = \alpha/\mathrm{i} = -\mathrm{i}\alpha$ so the value of the integral, using the same arguments about integration paths as in Ex. 3.5, is

$$\oint \frac{e^{(\alpha - iz)l}}{(\alpha - iz)^j} dz = \int_{-\infty}^{\infty} \frac{e^{(\alpha - i\omega)l}}{(\alpha - i\omega)^j} d\omega = -2\pi i \operatorname{Res}[z = -i\alpha]$$

Now, the residue at $z = -i\alpha$ is given by

$$\begin{aligned}\operatorname{Res}[z = -i\alpha] &= \lim_{z \to -i\alpha} i(\alpha - iz)\frac{e^{(\alpha - iz)l}}{(\alpha - iz)^j} \\ &= i \lim_{z \to -i\alpha} \frac{e^{(\alpha - iz)l}}{(\alpha - iz)^{j-1}} = i\frac{l^{j-1}}{(j-1)!}\end{aligned}$$

and the last result is obtained by applying L'Hôpital's rule a repeated number of times. The sought after probability density Eq. (6.120) thus becomes

$$p_{L_1}(l; t) = e^{-\eta(t) - \alpha l} \sum_{j=0}^{\infty} \frac{(\alpha\eta(t))^{j+1} l^j}{(j+1)! j!} \tag{6.121}$$

by shifting the initial count from $j = -1$ to $j = 0$. The probability density Eq. (6.121) can be associated with a modified Bessel's function of the first order for which tabulated values exist, [232], that is,

$$J_1(x) = \sum_{j=0}^{\infty} \frac{1}{\Gamma(j+2) j!} \left(\frac{x}{2}\right)^{2k+1}$$

whereby

$$p_{L_1}(l; t) = e^{-\eta(t) - \alpha l} \sqrt{\frac{\alpha\eta(t)}{l}}\, J_1(2\sqrt{\alpha l \eta(t)}) \tag{6.122}$$

is the final closed form version of the probability density of the dominant crack length. The mean value and the variance of the dominant crack length, on the other hand, can be obtained directly from Eqs. (6.109) and (6.110). Thus,

$$E[L(t)] = L_0 + \frac{\eta(t)}{\alpha} \tag{6.123}$$

and

$$\sigma_L^2(t) = \frac{2\eta(t)}{\alpha} \tag{6.124}$$

Finally, in the case the initial crack length L_0 is a random variable rather than a constant, independent of the crack propagation process $L_1(t)$, the probability density of $L(t)$ would be given by (cf. Example 1.21)

$$p_L(l; t) = \int_0^l p_{L_0}(x; t) p_{L_1}(l - x; t) dx \tag{6.125}$$

Various forms for the crack propagation intensity $\nu(t)$ have been proposed. The most obvious choice is to set $\nu(t) = \nu_0$, a constant. Then, the mean value and variance of the dominant crack length in Eqs. (6.123) and (6.124) are linearly increasing with time (cf. Eq. (6.111)). A pure linear birth process would yield a quadratic increase with time according to Eq. (6.112). In the case of periodic loading, Sobczyk (in Elishakoff and Lyon, [60]) proposes that the intensity can be assumed to be

$$\nu(t) = \nu_j = j\nu_0, \qquad j = 1, 2, \ldots, \text{ and } \nu_0 > 0 \tag{6.126}$$

where j denotes the crack growth state at time t. Therefore, the probability of transition Eq. (6.61) from state j to $j + 1$ in the interval $(t, t + \Delta t)$ is proportional to the state j. Finally, the intensity can be modified by taking into account overloading with resulting crack retardation, that is,

$$\nu_j(t) = j(\nu_0 - \nu_{OL}(t; t_1, t_2, \ldots, t_n) \tag{6.127}$$

in which $\nu_{OL}(t; t_1, t_2, \ldots, t_n)$ is the reduction in crack propagation intensity due to n overloads at times $t_1, t_2, \ldots, t_n$. For further discussion of this topic and some numerical results, see Sobczyk and Spencer, [204].

Any stationary random process can be made non-stationary by applying a deterministic amplitude modulating function. Thus, consider the compound Poisson process $X(t)$, Eq. (6.90), and have it multiplied by an amplitude modulating function (also called the envelope function, see Fig. 6.10). Then the random process

$$Y(t) = X(t)\psi(t) = \sum_i A_i \psi(t) w(t, \tau_i) \tag{6.128}$$

is a typical non-stationary process since it is identically zero for negative times and the amplitude shows a rapid build-up in the time interval $(0, t_1)$, is stationary in the time interval (t_1, t_2) and decays for $t > t_2$.

Now by Eqs. (6.91)–(6.93), the first moments of the non-stationary process $Y(t)$ are

$$E[Y(t)] = \psi(t) E[X(t)] = \nu\mu_A \int_0^t \psi(t) w(t, \tau) d\tau \tag{6.129}$$

$$K_Y(t, s) = \nu(\sigma_A^2 + \mu_A^2) \int_0^t \psi(t) w(t, \tau) \psi(s) w(s, \tau) d\tau \tag{6.130}$$

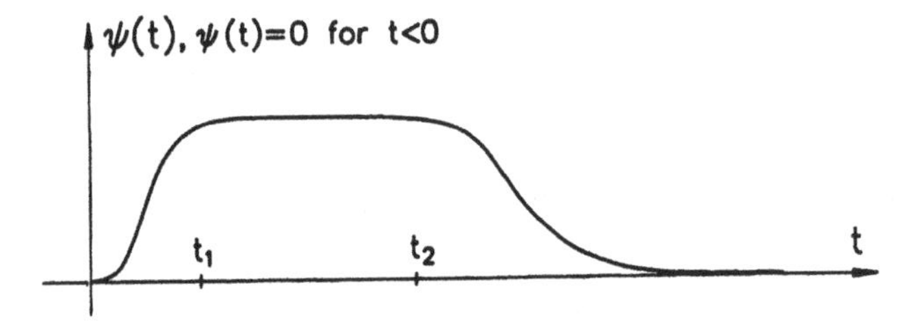

Fig. 6.10 *An amplitude modulating function*

in which the amplitude $\{A_i\}$, as before, are taken from a set of independent random variables. By comparison with Eq. (6.92) it is further evident that

$$K_Y(t,s) = \psi(t)\,\psi(s)\,K_X(t,s) \tag{6.131}$$

that is, the modulating function acts on the statistical moments directly.

A second possibility exists, this is, to apply the modulating function at the time of arrival of each pulse. In this case, the amplitude is modulated by the value of the envelope function at the time of arrival τ instead of applying the envelope function directly, i.e.

$$Y(t) = \sum_i A_i \psi(\tau_i)\,w(t,\tau_i) \tag{6.132}$$

The first two moments are now found to be

$$E[Y(t)] = \nu_0 \mu_A \int_0^t \psi(\tau)\,w(t,\tau)\mathrm{d}\tau \tag{6.133}$$

$$K_Y(t,s) = \nu_0(\sigma_A^2 + \mu_A^2) \int_0^t \psi^2(\tau)\,w(t,\tau)\,w(s,\tau)\mathrm{d}\tau \tag{6.134}$$

where ν_0 is the constant intensity of the Poisson counting process.

An interesting case arises when the mean value $\mu_A = 0$. Then the mean value $E[Y(t)]$ is also zero and the covariance kernel is given by

$$K_Y(t,s) = \sigma_A^2 \int_0^t \nu_0 \psi^2(\tau)\,w(t,\tau)\,w(s,\tau)\mathrm{d}\tau \tag{6.135}$$

Consider another non-homogeneous compound Poisson process with the covariance kernel given by Eq. (6.107), that is,

$$K_Y(t,s) = \int_{-\infty}^{\infty} p_A(a) \int_0^t \nu(\tau)\,w(t,\tau,a)\,w(s,\tau,a)\mathrm{d}\tau\,\mathrm{d}a$$

$$= \sigma_A^2 \int_0^t \nu(\tau)\,w(t,\tau)\,w(s,\tau)\mathrm{d}\tau \tag{6.136}$$

if $w(t,\tau,a) = aw(t,\tau)$. Up to their second moments, the non-homogeneous compound Poisson process with the intensity function

$$\nu(t) = \nu_0 \psi^2(t) \tag{6.137}$$

and the amplitude-modulated homogeneous, compound Poisson process, Eq. (6.132), are equivalent. Their higher order moments, of course, need not be and will not be equal. This interesting result, pointed out by Shinosuka and Sato [189], provides a simple method for generating sample functions of non-stationary, i.e. non-homogeneous processes, numerically.

6.2.4 Earthquake Motion Processes

The above considered non-homogeneous compound Poisson process, essentially a non-stationary pulse train, furnishes a simple model for simulation of random artificial or synthetic earthquake motions. The observed or recorded earthquake motion at a certain location ('building site') may be considered to be a superposition of random wave forms with random amplitudes and phases arriving in chaos according to a non-homogeneous Poisson process with an intensity function $\nu(t)$. It is plausible to assume that the intensity function will be of the form shown in Fig. 6.11, that is, during the initial phase, a few incoming waves are registered but in increasing numbers. Then the arrival rate rapidly builds up, and during the strong motion phase it remains essentially constant. The tail of the earthquake motion or the weak phase is recognized by fewer and fewer waves, the arrival rate drops until the motion fades away.

The study of several factors may provide enough material to construct a so-called representative earthquake amplitude spectrum for the site. This includes the geological properties of the site for which an artificial earthquake, representative of past and future earthquakes, is to be generated, information of the past earthquake history, possibly by actual earthquake records obtained at the site, and other studies, such as microtremor analysis. Let $A(T)$ be a representative earthquake amplitude spectrum for the site, where T is the wave period. $A(T)$ is an averaged and smoothed Fourier amplitude spectrum, which is thought to be the average of all past and future real earthquake spectra, drawn to the same scale (same magnitude). An individual earthquake spectrum (a realization) may appear chaotic and with many peaks. The averaged, smoothed spectrum on the other hand will loose all such peaks and only reflect magnification peaks due to local geology, (see Fig. 6.11), Sólnes, [210].

In Fig. 6.11, a common and often acceptable method of obtaining the frequency distribution of wave periods is indicated. Observing a surface record of an earthquake or a microtremor in a suitable interval of time, a frequency analysis of wave periods is performed by counting the number of cycles in the record with the same period $n(T)$. This number plotted as a function of the period, represented by the broken line in Fig. 6.11, strongly reflects the shape of the corresponding amplitude spectrum. It is therefore possible to obtain the probability distribution of wave periods as

$$F_T(x) = P[T \leqslant x] = \frac{1}{D}\int_0^x A(T)\mathrm{d}T \tag{6.138}$$

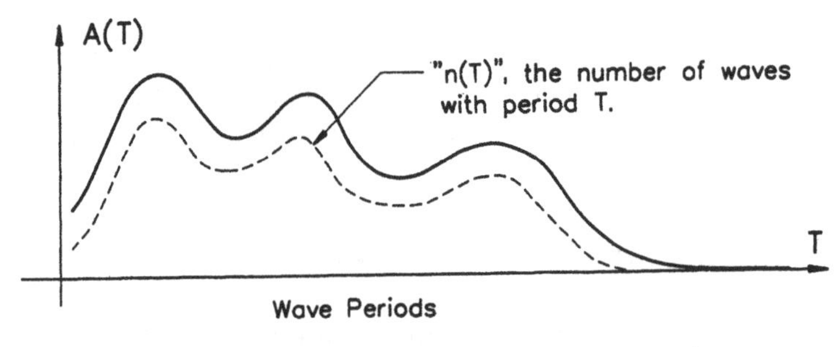

Fig. 6.11 *An averaged smoothed earthquake amplitude spectrum*

where D is a scaling factor such that

$$\frac{1}{D}\int_0^{\text{large } x} A(T)\mathrm{d}T = 1 \tag{6.139}$$

An earthquake motion process is now put together as follows:

$$X(t) = \sum_{i=0}^{N(t)} w(t-\tau_i, A_i, T_i, \Phi_i) \tag{6.140}$$

where τ_i are Poisson arrival times with intensity $v(t)$, A_i a representative random amplitude, T_i the random wave periods, distributed as shown above, and Φ_i are random phase angles, uniformly distributed between 0 and 2π.

A suitable shape function for an earthquake motion process has been suggested to be of the form (see Sólnes, [210]).

$$w(t-\tau, T, \Phi) = A(T)\exp[-f(T, Q)(t-\tau)]\sin\left[\frac{2\pi}{T}(t-\tau)+\Phi\right] \tag{6.141}$$

where $A(T)$ is the amplitude to be drawn from an appropriate Fourier amplitude spectrum, and $f(T, Q)$ is some function of the wave period and the attenuation factor Q of the earthquake waves or the quality factor (see Knopoff, [110], Aki and Richards, [3], and Joyner and Boore, [96]). In earlier studies of the shape function, [210], the attenuation term was assumed to be the same as for body waves in solids, that is, $\exp[-\pi t/TQ]$. This results in a very harsh compression of the high-frequency components in the thus defined surface motion, which does not hold for real measured records. Therefore, it is suggested that the square root of the period T be used instead. Thus, the attenuation term becomes $\exp[-\pi t/(Q\sqrt{T})]$. This modification seems to impart a reasonable frequency content to the signal.

The amplitude spectrum selected for the generation of an artificial earthquakes can be based on average smoothed measurements of earthquake spectral amplitudes, theoretical earthquake spectra or otherwise available spectra representative of the ground displacement, the ground velocity or the surface acceleration according to the mode desired for the site under study, [210]. In the event of lack of measurements or if no provisions for a suitable amplitude spectra are available, a way of constructing a theoretical Fourier amplitude spectrum for a specific site can be based on the work of Hanks and McGuire, [80], and Boore, [19], which use the Brune spectrum, [25], [26] to explain the high-frequency portion of the motion. Bessason, [14], has used this approach to simulate the spectral characteristics of earthquakes in the South Iceland Seismic Zone for generation of synthetic earthquakes according to the Gaussian white noise model of Example 6.3.

A short overview of the theoretical approach to a suitable Fourier amplitude spectrum is as follows. Formally, express the modulus of an amplitude spectrum of the horizontal surface acceleration of an earthquake as

$$|A(\omega, r, M_0)| = |\ddot{X}(\omega, r, M_0)| = C\Gamma(r, r_0)\,\Omega(\omega, M_0)\,P(\omega)\,H_Q(\omega, r)|H_B(\omega)| \tag{6.142}$$

in which r is the distance in kilometres from the site to the hypocentre, M_0 is the seismic moment of the earthquake, C is a scaling factor, Γ is a geometric spreading function, Ω is the source spectral model, P is an amplification factor, H_Q is an attenuation factor and H_B is a lowpass filter. For a full discussion and detailed description of these parameters, the reader is referred to Bessason, [14]. The source spectrum $\Omega(\omega, M_0)$ can be represented by the so-called omega-square law

$$\Omega(\omega, M_0) = M_0 \frac{\omega^2}{1 + (\omega/\omega_0)^2} \tag{6.143}$$

in which the seismic moment M_0 is defined as the stress drop $\Delta\sigma$ during an earthquake times the fault break area (see Section 8.1.1). ω_0 is the corner frequency or cutoff frequency given by

$$\omega_0 = 3.08\,\beta_0 \left(\frac{\Delta\sigma}{M_0}\right)^{1/3} \tag{6.144}$$

where β_0 is the shear wave velocity in the source region, (cf. [79]). The attenuation factor can be expressed as

$$H_Q(\omega) = \exp\left(\frac{-\,\omega r}{2Q(\omega)\beta}\right) \tag{6.145}$$

where β is the wave propagation velocity. $H_Q(\omega)$ is an attenuation term similar to the one already introduced in Eq. (6.141) containing the quality factor Q. As the waves are observed at the site rather than 'following' the waves, $r/\beta = t$ and the attenuation term becomes $\exp[-\pi t/TQ]$. The quality factor was initially assigned a constant value, often of the order 100, (see [110]). More recent studies of the anelastic attenuation of regional seismograph records (see [3] and [96]) indicate a frequency dependence for Northern America given by

$$Q(\omega) = 29.4 \frac{1 + (\omega/(0.6\,\pi))^{2.9}}{(\omega/(0.6\,\pi))^2} \tag{6.146}$$

However, Aki and Richards, [3], point out that Q assumes more or less a constant value equal to 100 in the frequency range 0.0001–10 Hz, which would seem to be an appropriate value for the simulation. Finally, the lowpass filter suggested is a fourth-order Butterworth filter (see Section 3.3.2), that is,

$$|H_B(\omega)| = \frac{1}{\sqrt{1 + (\omega/\omega_c)^8}} \tag{6.147}$$

in which typical values for the cutoff frequency $f_c = \omega_c/2\pi$ are of the order 15–50 Hz. Higher cutoff frequencies indicate harder and more solid rock sites.

All the main ingredients for the earthquake shape functions Eq. (6.141) are thus in place. The main statistical parameters of the process are easily obtained. The mean value of the random process, Eq. (6.140), is given by (cf. Eq. (6.107))

$$E[X(t)] = \int_0^t E[w(t-\tau, T, \Phi)]\, v(\tau)\mathrm{d}\tau$$

and the covariance kernel by

$$K_X(t,s)=\int_0^t E[w(t-\tau,T,\Phi)w(s-\tau,T,\Phi)]\,v(\tau)\mathrm{d}\tau$$

The expectation of the earthquake shape functions is given by

$$E[w(t-\tau,T,\Phi)]=\int_0^\infty \frac{A(x)}{D}\mathrm{d}x\int_0^{2\pi}\frac{1}{2\pi}\mathrm{d}z\int_0^t \exp\left[-\frac{\pi(t-y)}{Q\sqrt{x}}\right]\sin\left(\frac{2\pi}{x}(t-y)+z\right)v(y)\mathrm{d}y \tag{6.148}$$

Since $\int_0^{2\pi}\sin z\,\mathrm{d}z=\int_0^{2\pi}\cos z\,\mathrm{d}z=0$, $E[w(t-\tau,T,\Phi]=0$ and the mean value Eq. (6.148) is zero. The covariance kernel can be obtained in the same manner. It is completely determined when the two functions $A(x)$ and $V(y)$ have been obtained, i.e.

$$\begin{aligned}K_X(t,s)&=\frac{1}{2\pi D}\int_0^\infty A^2(x)\mathrm{d}x\int_0^{2\pi}\mathrm{d}z\int_0^t \exp\left[-\frac{\pi(t+s-2y)}{xQ}\right]\\&\quad\times\sin\left(\frac{2\pi}{x}(t-y+z\right)\sin\left(\frac{2\pi}{x}(s-y)+z\right)v(y)\mathrm{d}y\\&=\frac{1}{2D}\int_0^\infty A^2(x)\exp\left[-\frac{\pi(t+s)}{xQ}\right]\cos\left[\frac{2\pi}{x}(t-s)\right]\int_0^t v(y)\exp\left[\frac{2\pi y}{xQ}\right]\mathrm{d}y\,\mathrm{d}x\end{aligned} \tag{6.149}$$

As the purpose of this study is to obtain sample functions of the earthquake process defined by Eq. (6.140) rather than find an explicit expression for the second moment, the manner in which a sample function is generated will be briefly described. Obviously, a sample function will have the form

$$x(t)=\sum_i A(x_i)\exp\left[-\frac{\pi}{Q\sqrt{x_i}}(t-y_i)\right]\sin\left[\frac{2\pi}{x_i}(t-y_i)+z_i\right] \tag{6.150}$$

where $\{x_i, y_i, z_i\}$ are sampled values of the period, the arrival time and the phase, $\{T_i, \tau_i, \Phi_i\}$. The number of terms in the sum is assumed to be arbitrarily large, depending upon the computer power available. For each term in the above sum Eq. (6.150), three random numbers $\{u_1, u_2, u_3\}$ are generated by sampling the uniform distribution [0, 1]. A sample period is obtained by

$$F_T(x_i)=u_1^{(i)}=\frac{1}{D}\int_0^{x_i}A(T)\mathrm{d}T \tag{6.151}$$

through numerical integration (see also Eq. (6.43)). The amplitude $A(x_i)$ is obtained at the same time. The first arrival time τ_1 and the inter-arrival times $(\tau_i-\tau_{i-1})$ are exponentially distributed, so

$$\left.\begin{aligned}P[\tau_1\leqslant y_1]&=1-\mathrm{e}^{-v(y_1)y_1}=u_2^1\\P[\tau_i-\tau_{i-1}\leqslant t_i]&=1-\mathrm{e}^{-v(t_i)t_i}=u_2^i\end{aligned}\right\} \tag{6.152}$$

or through iteration

$$\left.\begin{aligned} y_1 &= -\frac{1}{v(y_1)}\log_e(1-u_2^{(1)}) \\ t_i &= -\frac{1}{v(y_{i-1})}\log_e(1-u_2^{(i)}), \qquad y_i = y_{i-1}+t_1 \end{aligned}\right\} \tag{6.153}$$

The intensity function can be obtained by noting the covariance equivalence between the amplitude modulated process and the non-homogeneous process using Eq. (6.137), that is,

$$v(t) = v_0\psi^2(t) \tag{6.137}$$

In addition to the envelope functions for strong earthquake motions discussed in Example 6.3, additional envelope functions $\psi(t)$ have been proposed by Amin and Ang, [8], among others, in the form

$$\psi(t) = \begin{cases} \left(\dfrac{t}{t_1}\right)^2 & 0 \leqslant t \leqslant t_1 \\ 1.0 & t_1 \leqslant t \leqslant t_2 \\ e^{-c(t-t_2)} & t > t_2 \end{cases} \tag{6.154}$$

The selection of proper constants t_1, t_2 and c, which completely define $\psi(t)$, has been discussed by Jennings *et al.*, [95], who point out that the envelope function is dependent on the magnitude of the earthquake, the distance from the causative fault and the focal depth. The duration of the strong motion is characterized by the constant t_2 which for the three magnitude values 6, 7 and 8 may be selected of the order 4 s, 15 s and 35 s respectively. The constant t_1 is estimated to be of the order 2–4 s. Finally, c is selected according to the focal distance. In Fig. 6.12, an envelope function for a particular earthquake is shown. The

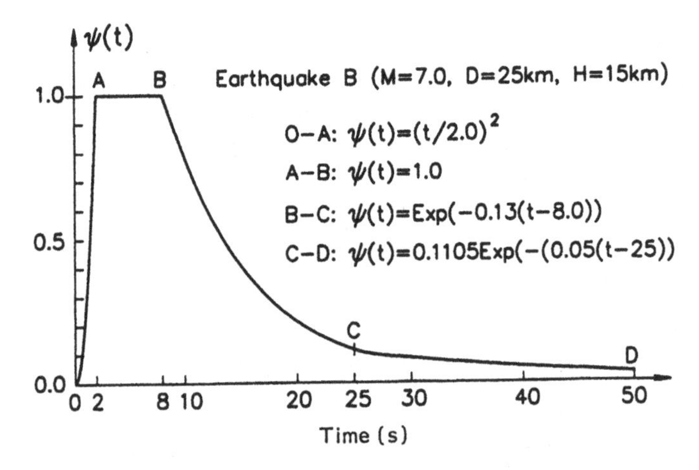

Fig. 6.12 *An amplitude envelope function, (after Jennings et al., [95])*

constant ν_0 in Eq. (6.137) is undetermined but can be selected more or less on a trial and error basis. The frequency content of the earthquake will of course be affected by ν_0, so the selection has to be made according to the desired frequency characteristics. Values as high as 10–30 per second can be assumed to be appropriate. Finally it may be mentioned that it is perhaps more logical to work with the equivalent modulated process, which has the sample functions

$$x(t) = \sum_i A(x_i)\psi(y_i)\exp\left[-\frac{\pi}{Q\sqrt{x_i}}(t - y_i)\right]\sin\left[\frac{2\pi}{x_i}(t - y_i) + z_i\right] \tag{6.155}$$

This makes the determination of sample arrival times easier since in Eq. (6.154), $\nu(t)$ is replaced by a constant ν_0. The sample phase in both versions (Eqs. (6.150) and (6.155)) is $z_i = 2\pi u_3^{(i)}$.

Example 6.11 Generation of synthetic accelerograms The Engineering Research Institute of the University of Iceland operates a strong motion earthquake acceleration monitoring network throughout the zones of major earthquake activity in the country. On the 25 May 1987, a 5.8 magnitude earthquake occurred in Vatnafjöll in the eastern part of the South Iceland Seismic Zone south of the Volcano Hekla. This earthquake was recorded on several of the strong motion accelerographs in the area, which are a part of the strong motion network, and has subsequently been subjected to detailed analysis, [195]. Figure 6.13 shows an accelerogram recorded at the Station of Minni Núpur in the N–S direction with a sampling frequency of 200 Hz.

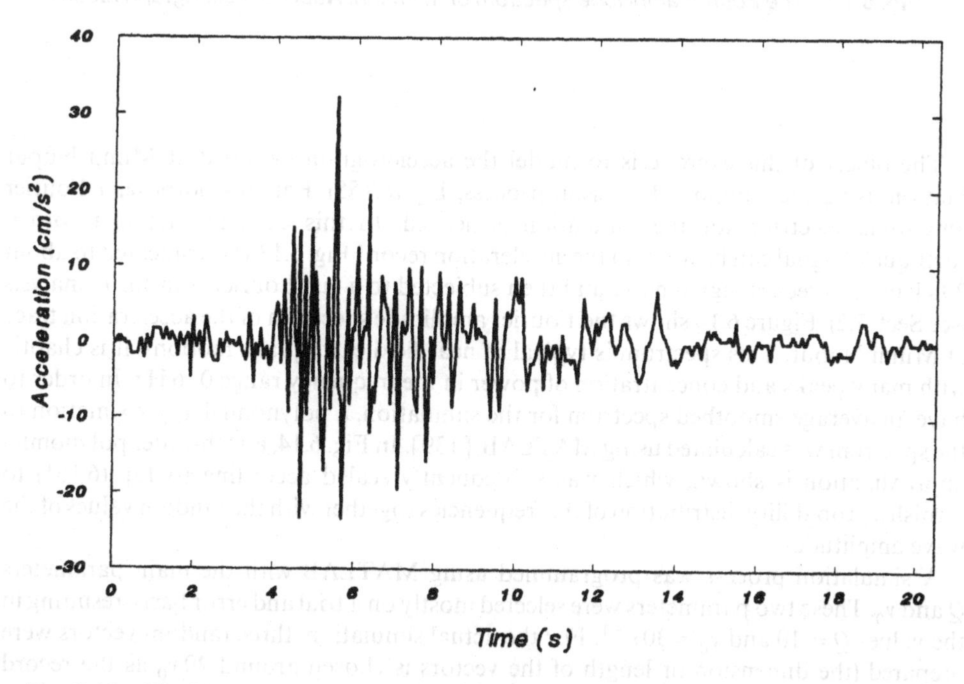

Fig. 6.13 *The Vatnafjöll earthquake of 25 May 1987, station Minni Núpur accelerogram*

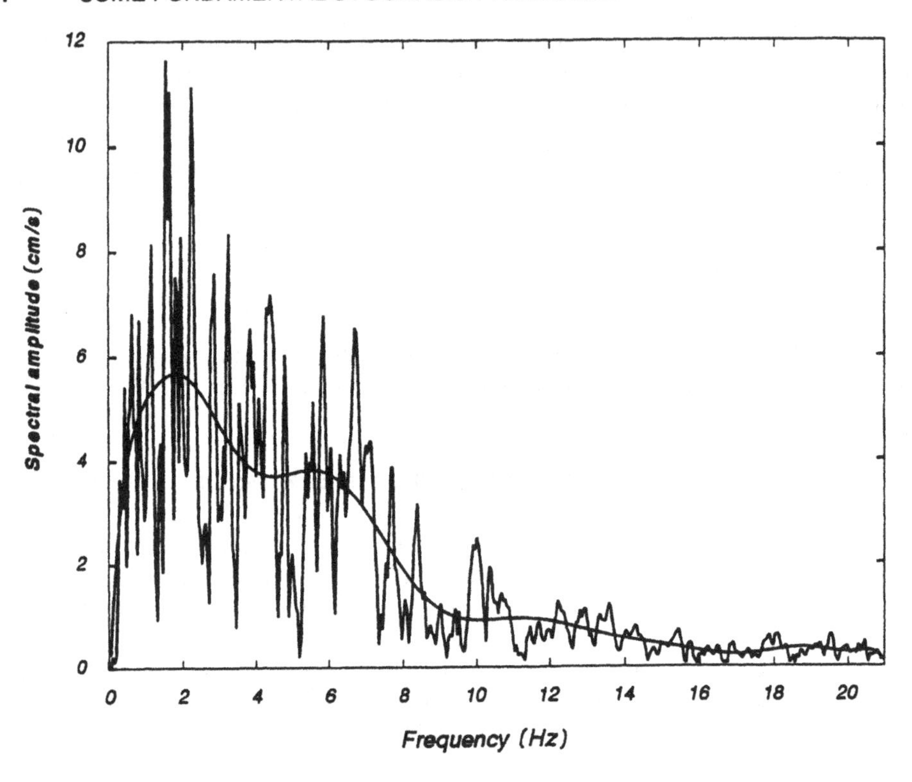

Fig. 6.14 *The Fourier amplitude spectrum of the Minni Núpur accelerograph record*

The object of this exercise is to model the accelerogram obtained at Minni Núpur Station using the compound Poisson process, Eq. (6.155). For this purpose, a Fourier amplitude spectrum for the simulation is needed. In this case, the actual recorded earthquake signal can be used, so the acceleration record Fig. 6.13 was truncated to about 20 s length to reduce signal noise and then subjected to a fast Fourier transform analysis (see Sect. 7.2). Figure 6.14 shows the Fourier amplitude spectrum of the acceleration trace at Minni Núpur. This spectrum is typical of near field earthquake motions; it is chaotic with many peaks and concentration of power in the frequency range 0–6 Hz. In order to have an average smoothed spectrum for the simulation, a polynomial approximation to the spectrum was calculated using MATLAB, [139]. In Fig. 6.14, a 15th-order polynomial approximation is shown, which was subsequently scaled according to Eq. (6.151) to furnish a probability distribution of the frequencies together with the random values of the wave amplitudes.

A simulation process was programmed using MATLAB with the main parameters Q and v_0. These two parameters were selected mostly on a trial and error basis, resulting in the values $Q = 10$ and $v_0 = 30\,\mathrm{s}^{-1}$. For the actual simulation, three random vectors were prepared (the dimension or length of the vectors is chosen around $20\,v_0$ as the record length is to be 20 s:

(a) the wave periods—the vector contains random numbers between 0 and 1;
(b) the arrival times—the vector contains random numbers between 0 and 1;
(c) the phase angles—the vector contains random numbers between 0 and 2π.

The calculation process is now as follows:

1. Simulation of wave periods according to Eq. (6.151). At the same time an amplitude for the shape function is picked from the smoothed average Fourier spectrum (the order 15 polynomial approximation).
2. Simulation of the arrival times according to the scheme in Eqs. (6.152) with v_0 a constant, as the stationary process Eq. (6.155) is being simulated.
3. Each wave is calculated according to Eq. (6.155) and amplitude modulated using the envelope function shown in Fig. 6.12. The parameters of the envelope function were selected at $t_1 = 1.5$ s, $t_2 = 3.0$ s and $c = 0.18$. The thus modulated shape functions are then ordered and summed according to Eq. (6.155) to provide a simulated record. In order to ensure a correct scaling of the amplitude of the simulated record, the standard deviation of the entire records was computed and set equal to the standard deviation of the original record measured at Minni Núpur. Thus, the simulated record has been given a proper scale for the amplitudes. The simulated record thus adjusted is shown in Fig. 6.15. Obviously, it does not look identical to the measured record since a smoothed average spectrum was used for the simulation. The artificial record, however, can be interpreted as being descriptive of earthquakes of magnitude close to 6, which might later be recorded at the same station.

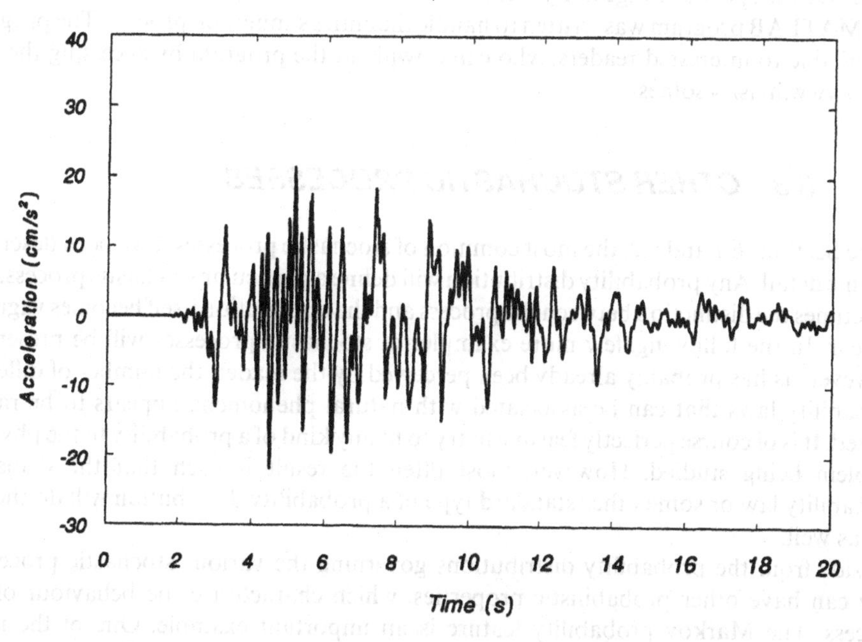

Fig. 6.15 *An artificially generated earthquake accelerogram*

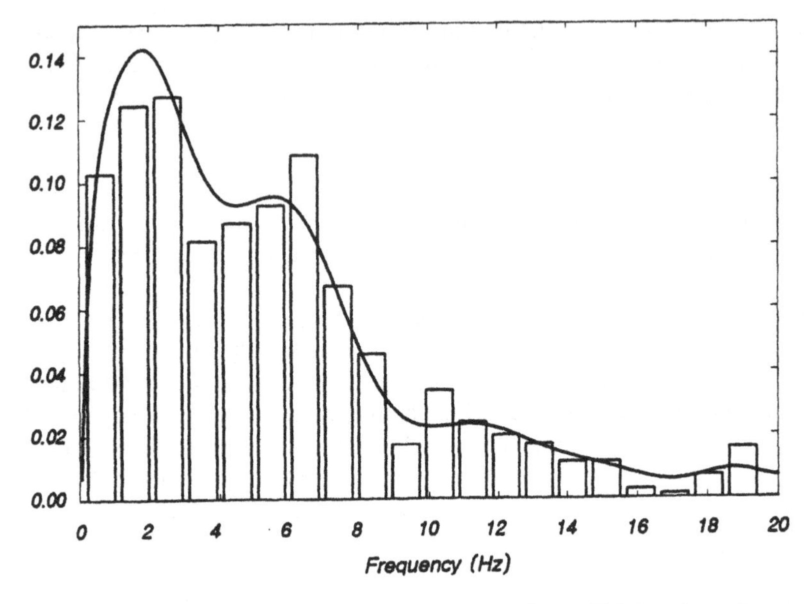

Fig. 6.16 *A histogram of simulated frequencies in the artificial earthquake record*

Finally, the frequency content of the artificial earthquake was compared with the smoothed average Fourier spectrum, which was the basis of the frequency simulation. Figure 6.16 shows a histogram of the simulated frequencies together with the smoothed average Fourier spectrum of the recorded earthquake. Apparently, the simulated frequencies cover the spectral range fairly well.

A MATLAB program was written to handle the entire simulation process. The program is available to interested readers, who can download the program by accessing the web site: //www.hi.is/~solnes/.

6.3 OTHER STOCHASTIC PROCESSES

In the Sections 6.1 and 6.2, the most common of stochastic processes have been described in some detail. Any probability distribution will define a particular stochastic process, and sometimes the distinction between the process and the distribution itself becomes vague or unclear. In the following, few more examples of stochastic processes will be presented. However, as has probably already been perceived by the reader, the number of different probability laws that can be associated with natural phenomena appears to be rather limited. It is of course perfectly feasible to try to fit any kind of a probability to the physical problem being studied. However, most often the result is such that the Gaussian probability law or some other standard type of a probability distribution will do the job just as well.

Aside from the probability distributions governing the various stochastic processes, they can have other probabilistic properties, which characterize the behaviour of the process. The Markov probability feature is an important example. One of the more studied stochastic processes is the Markov process, which was briefly introduced in

Chapter 2 (Example 2.4). The Markov process encompasses a family of processes, which have the common fundamental probability law outlined in Example 2.4. Actually, many of the processes already discussed, like the compound Poisson process and some Gaussian processes, belong to the family of Markov processes even if that fact was not particularly reflected in passing. In the following, an overview of the basis properties of the Markov process will be presented with some examples.

Martingales form a group of stochastic processes, which possess another common interesting probabilistic feature that will be briefly discussed. Finally, there exists a number of stochastic processes with particular features, which have been introduced for special application purposes in various fields such as physics, biology, economy and finance. One such process, the Boltzmann process, will be briefly described.

6.3.1 Markov Processes

The Markov stochastic process has the interesting probabilistic behaviour that its immediate future development is only dependent on its present but not its past performance. The Markov process is basically a memoryless process similar to the Poisson process but its present state depends nevertheless exclusively on the immediate previous state. Markov processes play an important role in many practical applications since many applicable stochastic processes are Markovian in character. Markov processes, which have both continuous state space and parameter set, are sometimes called *diffusion processes.* Discrete Markov processes on the other hand are of two different types (a) discrete both in space and time; (b) discrete in space and continuous in time. Markov processes of the former category are usually called *Markov chains.* Continuous time processes will of course be converted to chains when they are studied at discrete times only. For instance, Poisson type processes are of the last category, since the rare Poisson event is a point event in space but the compounded result $\sum X$ yields a continuous function of time. Thus the compound Poisson process is essentially a Markov process as it is a cumulative process with its future development dependent on its last state only. Concerning Poisson type Markov processes, a further classification is possible according to the rate of occurrence of the Poisson events. If the rate is increasing, that is, more and more events are registered, the process is called a *birth process.* Otherwise it is called a *birth and death process* or a purely *death process* if the rate is decreasing. For the mixed birth and death process, sometimes the form *immigration process* or *emigration process* is used if the difference between the birth rate and death rate is great. Many other classifications exist and some of them will be briefly introduced in the following along the fundamental definitions and properties of the Markov process.

The fundamental probability law of the Markov process is best illustrated by considering the Markov chain, $X(t_0), X(t_1), \ldots, X(t_{n-1}), X(t_n), \ldots$, which is formed by the random amplitudes of a Markov process taken at discrete time instances $t_0, t_1, \ldots, t_{n-1}, t_n, \ldots$, belonging to the index set T of the process. If the observation of the process begins at $t = t_n$, then the probability of the next amplitude $X(t_{n+1})$ is given by

$$P[X_{n+1} \leqslant x_{n+1} | X_n = x_n, X_{n-1} = x_{n-1}, X_{n-2} = x_{n-2}, \ldots, X_0 = x_0] = P[X_{n+1} \leqslant x_{n+1} | X_n = x_n] \tag{2.14}$$

that is, it is only dependent on the present amplitude and not the previous ones.

The conditional probability $P[X(t) \leqslant x | X_0 = x_0]$ of a Markov process $X(t)$ is called the transition probability distribution function. A Markov process is completely characterized by its transition probability distribution and the probability distribution at an initial time t_0. As shown in Example 2.4, the same holds for the special case were the first amplitude $X(t_0)$ is a real constant. The transition probability density of a Markov process is given by

$$p(x|x_0; t, t_0) = \frac{\partial}{\partial x} P[X(t) \leqslant x | X(t_0) = x_0] \tag{6.156}$$

assuming that the usual existence conditions for the derivative are in place. The higher-order joint probability densities of the Markov process can now be obtained as follows. For any kind of a stochastic variable, the joint densities can be expressed in terms of the conditional density and the basic density function, that is,

$$p(x, y) = p(y|x)\, p(x) = p(x|y)\, p(y) \tag{1.60}$$

Thus, the joint density

$$\begin{aligned} p(x_0, x_1, \ldots, x_n; t_0, t_1, \ldots, t_n) = {} & p(x_n|(X(t_{n-1}) = x_{n-1}, \ldots, X(t_1) = x_1, X(t_0) = x_0) \\ & \times p(x_0, x_1, \ldots, x_{n-1}; t_0, t_1, \ldots, t_{n-1}) \end{aligned}$$

which, due to the Markov probability law, Eq. (2.14), is the same as

$$p(x_0, x_1, \ldots, x_n; t_0, t_1, \ldots, t_n) = p(x_n|(X(t_{n-1}) = x_{n-1})\, p(x_0, x_1, \ldots, x_{n-1}; t_0, t_1, \ldots, t_{n-1})$$

Repeating this procedure $n-2$ times,

$$\begin{aligned} p(x_0, x_1, \ldots, x_n; t_0, t_1, \ldots, t_n) = {} & p(x_n|(X(t_{n-1}) = x_{n-1})\, p(x_{n-1}|(X(t_{n-2}) = x_{n-2}) \\ & \times p(x_{n-2}|(X(t_{n-3}) = x_{n-3}) \cdots p(x_1|(X(t_0) = x_0)\, p(x_0; t_0) \end{aligned} \tag{6.157}$$

whereby the nth-order joint probability density of a Markov chain is constructed from the transition density and the initial probability density. Further, using the transition density, the basic probability density $p(x_i; t_i)$ at any time instant t_i can be derived from the initial density $p(x_0; t_0)$. For any stochastic process $X(t)$, the lower order joint densities can be obtained by integrating a higher-order density, (Eq. (1.57). Therefore,

$$p(x_i; t_i) = \int_{-\infty}^{\infty} p(x_i, x_0; t_i, t_0) \mathrm{d}x_0 = \int_{-\infty}^{\infty} p(x_i | X_0 = x_0)\, p(x_0; t_0) \mathrm{d}x_0 \tag{6.158}$$

whereby the basic density $p(x_i; t_i)$ has been obtained through a kind of convolution integral, involving the transition density and the initial density, making use of Eq. (6.157).

If the initial state of a Markov process is known, ($P[X_0 = x_0] = 1$), which is a common situation in many practical applications, the process is completely characterized by its transition density, as made evident by Eq. (6.157). Of particular interest is the case where the transition mechanism is stationary, that is, the transition densities depend only on the time difference $t_i - t_{i-1}$ but not on the individual time instances t_i and t_{i-1}. Markov processes with stationary probability mechanism are said to be homogeneous to stress the fact that they need not be at the same time stationary processes according to the definition of stationarity, which implies among other things an autocorrelation as a function of the time difference only. A strictly stationary Markov process is easily shown to be homogeneous as well, but the converse is not necessarily true. Consider a homogeneous Markov process $\{X(t), t \geqslant 0\}$, which is evaluated at a discrete set of event times t_i, $i = 1, 2, \ldots, n, \ldots$, that are uniformly spaced such that $t_i - t_{i-1} = \tau$ for all i. Then, the one-stage transitional densities are all equal, that is,

$$p(x_i|(X(t_{i-1}) = x_{i-1}) = p(x_j|(X(t_{j-1}) = x_{j-1}) \quad \text{for all } i \text{ and } j \tag{6.159}$$

It is also convenient to introduce the n-stage transition joint density defined as $p(x_i|(X(t_{i-n}) = x_{i-n})$, that is, when the observation of the process jumps n stages from t_{i-n} to t_i. The n-stage transition density can be determined by combining the associated n one-stage transition densities. First apply Eq. (1.57) to a triple joint density, that is,

$$\int_{-\infty}^{\infty} p(x_i, x_{i-1}, x_{i-2}; t_i, t_{i-1}, t_{i-2}) \mathrm{d}x_{i-1} = p(x_i, x_{t-2}; t_i, t_{i-2})$$

Applying Eq. (6.157) and using the expression Eq. (1.60), the two-stage transition density can be obtained as

$$\int_{-\infty}^{\infty} p(x_i|X(t_{i-1}) = x_{i-1}) p(x_{i-1}|X(t_{i-2}) = x_{i-2}) p(x_{n-2}; t_{i-2}) \mathrm{d}x_{i-1}$$

$$= p(x_i|X(t_{i-2}) = x_{t-2}) p(x_{i-2}; t_{i-2})$$

Now, the term $p(x_{i-2}; t_{i-2})$ cancels out in the above equation which finds the final form

$$\int_{-\infty}^{\infty} p(x_i|X(t_{i-1}) = x_{i-1}) p(x_{i-1}|X(t_{i-2}) = x_{i-2}) \mathrm{d}x_{i-1} = p(x_i|X(t_{i-2}) = x_{t-2}) \tag{6.160}$$

giving the two-stage transition density as an integral of the two associated one-stage transition densities. This result is known as the Chapman–Kolmogorov equation for transitional densities of the Markov process. It can easily be extended to cover n-stage transition densities, whereby

$$\int_{-\infty}^{\infty} \cdots \int_{-\infty}^{\infty} \prod_{j=1-n+1}^{i} [p(x_j|X(t_{j-1}) = x_{j-1})] \mathrm{d}x_j = p(x_i|X(t_{i-n}) = x_{t-n}) \tag{6.161}$$

In the case of pure discrete Markov chains, the Chapman–Kolmogorov equation is usually presented in a discrete form of the kind

$$\sum_k p_{i,k}^m p_{k,j}^{n-m} = p_{i,j}^n, \qquad 0 \leqslant m \leqslant n \tag{6.162}$$

in which the states of the process are numbered by integers, that is, integer j refers to the state $X_n = i$, with the m-stage transition probability of moving from state i at time t_{n-m} to state j at time t_n defined as

$$p_{i,j}^{m} = P[X_n = j | X_{n-m} = i] \tag{6.163}$$

and the summation tends over all possible intermediate paths k leading to the nth state.

For diffusion processes, the Kolmogorov backward and forward differential equations of the transition densities of the process play an important role. The Kolmogorov equations can be derived in several different ways, using fundamentally different procedures. In the following, the procedure will mostly be based on Kolmogorov's original presentation as described in Bharucha-Reid's text, [15]. Let $\{X(t), t \in T\}$ be a Markov process with a continuous state space and indexing parameter t. In this case, it is more convenient to define the transition probabilities and densities as follows:

$$p(y|x; t, s) = \frac{\partial}{\partial y} P[X(t) \leqslant y | X(s) = x], \qquad t > s \tag{6.164}$$

and the corresponding Chapman–Kolmogorov Eq. (6.160) is then given by

$$p(y|x; t, s) = \int_{-\infty}^{\infty} p(y|z; t, r)\, p(z|x; r, s) \mathrm{d}z, \qquad t > s, \quad r \in (t, s) \tag{6.165}$$

In order to derive the Kolmogorov equations, the following general assumptions are made pertaining to the process. For any $\delta > 0$,

(a)
$$P[|X(t + \Delta t) - X(t)| > \delta | X(t) = x] = o(\Delta t) \tag{6.166}$$

which simply states that the probability that major changes occur in the process during a small time interval Δt are very small compared to Δt, ($o(\Delta t)/\Delta t \to 0$ as $\Delta t \to 0$).

(b)
$$\begin{aligned}\lim_{\Delta t \to \infty} & \frac{E[X(t + \Delta t) - X(t) | X(t) = x]}{\Delta t} \\ &= \lim_{\Delta t \to 0} \frac{1}{\Delta t} \int_{|y - x| \leqslant \delta} (y - x)\, p(y|x; t + \Delta t, t) \mathrm{d}y \\ &= b(t, x)\end{aligned} \tag{6.167}$$

that is, the limit of the conditional expectation of a change in the process during an infinitesimal time interval Δt exists and is equal to $b(t, x)$, which is known as the drift coefficient of the process.

(c)
$$\begin{aligned}\lim_{\Delta t \to \infty} & \frac{E[(X(t + \Delta t) - X(t))^2 | X(t) = x]}{\Delta t} \\ &= \lim_{\Delta t \to 0} \frac{1}{\Delta t} \int_{|y - x| \leqslant \delta} (y - x)^2\, p(y|x; t + \Delta t, t) \mathrm{d}y \\ &= a(t, x) \geqslant 0\end{aligned} \tag{6.168}$$

that is, the corresponding conditional mean square value or the infinitesimal variance of a change in $X(t)$ exists and is equal to $a(t, x)$, which is known as the diffusion coefficient of the process. In passing, it should be noted that in the case of homogeneous Markov processes, the transition density function depends only on the time difference, that is,

$$p(y|x; t+\Delta t, t) = p(y|x; \Delta t) \tag{6.169}$$

Therefore, both coefficients will become functions of x only, that is, $b(x)$ and $a(x)$.

With the above defined conditions in place, the Kolmogorov backward equation is derived as follows. From Eq. (6.165),

$$p(y|x; t, s-\Delta s) = \int_{-\infty}^{\infty} p(y|z; t, s)\, p(z|x; s, s-\Delta s)\mathrm{d}z$$

in which $s = s - \Delta s$ and $r = s$. Further, noting that $1 = \int_{-\infty}^{\infty} p(z|x; s, s-\Delta s)\mathrm{d}z$

$$p(y|x; t, s) = \int_{-\infty}^{\infty} p(y|x; t, s)\, p(z|x; s, s-\Delta s)\mathrm{d}z$$

Therefore,

$$\begin{aligned}\frac{p(y|x; t, s-\Delta s) - p(y|x; t, s)}{\Delta s} &= \frac{1}{\Delta s}\int_{-\infty}^{\infty} [p(y|z; t, s) - p(y|x; t, s)]\, p(z|x; s, s-\Delta s)\mathrm{d}z \\ &= \frac{1}{\Delta s}\int_{|z-x|>\delta} [p(y|z; t, s) - p(y|x; t, s)]\, p(z|x; s, s-\Delta s)\mathrm{d}z \\ &\quad + \frac{1}{\Delta s}\int_{|z-x|\leqslant\delta} [p(y|z; t, s) - p(y|x; t, s)]\, p(z|x; s, s-\Delta s)\mathrm{d}z\end{aligned} \tag{6.170}$$

On the strength of assumption (a), the first integral can be shown to tend to zero as $\Delta s \to 0$. In fact, since

$$\begin{aligned}\lim_{\Delta s\to 0}\frac{1}{\Delta s}\int_{|z-x|>\delta} p(z|x; s, s-\Delta s)\mathrm{d}z &= \lim_{\Delta s\to 0}\frac{1}{\Delta s}\left[\int_{-\infty}^{-x-\delta} p(*)\mathrm{d}z + \int_{\delta+x}^{\infty} p(*)\mathrm{d}z\right] \\ &= \lim_{\Delta s\to 0}\frac{1}{\Delta s} P[|X(s-\Delta s) - X(s)| > \delta \,|\, X(s) = x] = 0\end{aligned}$$

whereby the first integral of Eq. (6.170) also tends to zero, assuming that the two density functions behave in a normal manner ($p(y|x; t, s)$ can be moved outside the integral). To avoid such speculations, however, the distribution functions can be used instead. This leaves the second integral to be evaluated. For this purpose, the density functions the brackets can be expanded into Taylor series as follows:

$$\begin{aligned}p(y|z; t, s) - p(y|x; t, s) &= (z-x)\frac{\partial p(y|x; t, s)}{\partial x} \\ &\quad + \tfrac{1}{2}(z-x)^2\frac{\partial^2 p(y|x; t, s)}{\partial x^2} + o((z-x)^2)\end{aligned}$$

which inserted into the second integral of Eq. (6.170) gives

$$\frac{p(y|x;t,s-\Delta s)-p(y|x;t,s)}{\Delta s}=\frac{1}{\Delta s}\int_{|z-x|>\delta}[p(y|z;t,s)-p(y|x;t,s)]\,p(z|x;s,s-\Delta s)\mathrm{d}z$$
$$+\frac{1}{\Delta s}\frac{\partial p(y|x;t,s)}{\partial x}\int_{|z-x|\leqslant\delta}(z-x)\,p(z|x;s,s-\Delta s)\mathrm{d}z$$
$$+\frac{1}{2\Delta s}\frac{\partial^2 p(y|x;t,s)}{\partial x^2}\int_{|z-x|\leqslant\delta}(z-x)^2\,p(z|x;s,s-\Delta s)\mathrm{d}z$$
$$+\frac{1}{\Delta s}\int_{|z-x|\leqslant\delta}o((z-x)^2\,p(z|x;s,s-\Delta s)\mathrm{d}z$$

Now, passing to the limit, $\Delta s\to 0$, and using the two assumptions Eqs. (6.167) and (6.168), the above equation simplifies to the backward Kolmogorov equation

$$-\frac{\partial p(y|x;t,s)}{\partial s}=\tfrac{1}{2}a(t,x)\frac{\partial^2 p(y|x;t,s)}{\partial x^2}+b(t,x)\frac{\partial p(y|x;t,s)}{\partial x} \tag{6.171}$$

which is a second-order partial differential equation satisfied by both the transition density function and the corresponding transition distribution function $F(y|x;t,s)$ of the Markov process. (The derivation for the distribution function follows exactly the same procedure. Also, Eq. (6.171) can be integrated over $-\infty<y<\infty$). The backward variables are x and s, whereas the forward variables y and t only enter the equation by the way of the boundary conditions. A Markov process with continuous sample functions is completely defined by the backward equation such that for given $a(t,x)$ and $b(t,x)$, there is a unique and completely determined transition density or transition probability distribution $F(y|x\colon t,s)$, which satisfies Eq. (6.171).

The corresponding *forward Kolmogorov equation*, which is the adjoint equation of the backward equation, is more difficult to derive, and the reader is referred to the specialist literature, for example, Bharucha-Reid, [15]. A comprehensive and lucid derivation based on Kolmogorov's original method is presented in Lin and Cai, [123], where a thorough discussion of the solution of the two equations and their boundary conditions is given. The forward equation has the form

$$\frac{\partial p(y|x;t,s)}{\partial t}=\frac{1}{2}\frac{\partial^2[a(t,y)\,p(y|x;t,s)]}{\partial y^2}-\frac{\partial[b(t,y)\,p(y|x;t,s)}{\partial y} \tag{6.172}$$

in which the forward variables (y,t) are the basic variables and the backward variables (x,s) are essentially constant and enter only through the boundary condition. The forward equation is also known as the Fokker–Planck diffusion equation.

Another alternative approach to obtaining the Fokker–Planck equation was presented by Itô, [89], which showed that a Markov diffusion process $X(t)$ is governed by a stochastic differential equation of the form

$$dX(t)=\mu(x,t)\mathrm{d}t+\sigma(x,t)\,\mathrm{d}W(t) \tag{6.173}$$

in which μ and σ are the drift and diffusion coefficients of the process, and $W(t)$ is an unit

Wiener process describing Brownian motion, which was already introduced in Example 2.5. Actually, Itô's equation has to be interpreted as holding conditionally on $X(t)=x$. Now, the Wiener process is a Gaussian unit ($\sigma=1$) process with stationary independent increments and is therefore a Markov process itself by virtue of Example 2.4. Also, $E[W(t)]=0$ and $E[(W(t)-W(s))^2]=|t-s|$. Therefore, given that $X(t)=x$, the increment $\Delta X(t)$ in a small time interval Δt has the mean value $\mu(x,t)\Delta t$, the variance $\sigma^2(x,t)\Delta t$ and is independent of all previous increments.

The Fokker–Planck equation can now derived as follows. Introducing the moment generating function of the process $X(t)$, (cf. Ex. 1.25),

$$M(\theta;t)=E[\mathrm{e}^{-\theta X(t)}]=\int_{-\infty}^{\infty}\mathrm{e}^{-\theta x}\,p_X(x;t)\mathrm{d}x \tag{6.174}$$

where $p_X(x;t)$ is the probability density function of $X(t)$. Further, the moment generating function of the increment $\Delta X(t)$ conditional to $X(t)=x$ is given by

$$E[\mathrm{e}^{-\theta\Delta X(t)}|X(t)=x]=1-\theta\mu(x,t)\Delta t+\frac{\theta^2}{2!}\sigma^2(x,t)\Delta t+o(\Delta t) \tag{6.175}$$

where the term $o(\Delta t)$ can be assumed to be uniform in x. Now form the partial derivative of $M(\theta;t)$, that is,

$$\frac{\partial M(\theta;t)}{\partial t}=\int_{-\infty}^{\infty}\mathrm{e}^{-\theta x}\frac{\partial p_X(x;t)}{\partial t}\mathrm{d}x \tag{6.176}$$

The partial derivative of $M(\theta;t)$ can also be obtained by forming the expression

$$\begin{aligned}\frac{\partial M(\theta;t)}{\partial t}&=\lim_{\Delta t\to 0}\frac{1}{\Delta t}[M(\theta;t+\Delta t)-M(\theta;t)]\\&=\lim_{\Delta t\to 0}\frac{1}{\Delta t}E[\mathrm{e}^{-\theta X(t+\Delta t)}-\mathrm{e}^{-\theta X(t)}]\\&=\lim_{\Delta t\to 0}\frac{1}{\Delta t}E[\mathrm{e}^{-\theta(X(t)+\Delta X(t))}-\mathrm{e}^{-\theta X(t)}]\end{aligned} \tag{6.177}$$

The right-hand side expectation, $E[\mathrm{e}^{-\theta(X(t)+\Delta X(t))}-\mathrm{e}^{-\theta X(t)}]=E[(\mathrm{e}^{-\theta\Delta X(t)}-1)\mathrm{e}^{-\theta X(t)}]$, can be evaluated using the expression for conditional expectations (sometimes referred to as the law of total probability), Eq. (1.63), that is, $E[E[V|U]]=E[V]$. Denoting

$$V=(\mathrm{e}^{-\theta\Delta X(t)}-1)\,\mathrm{e}^{-\theta X(t)},\ \text{and}\ U=X(t)$$

the expectation is given by

$$\begin{aligned}E[V]=E[(\mathrm{e}^{-\Delta X(t)}-1)\mathrm{e}^{-\theta X(t)}]&=\int_{-\infty}^{\infty}E[V|x]\,p_X(x;t)\mathrm{d}x\\&=\int_{-\infty}^{\infty}E[(\mathrm{e}^{-\Delta X(t)}-1)\mathrm{e}^{-\theta X(t)}|X(t)=x]\,p_X(x;t)\mathrm{d}x\\&=\int_{-\infty}^{\infty}E[(\mathrm{e}^{-\Delta X(t)}-1)|x]\,\mathrm{e}^{-\theta x}\,p_X(x;t)\mathrm{d}x\end{aligned}$$

Now, by Eq. (6.175),

$$E[(e^{-\Delta X(t)}-1)|x]=\int_{-\infty}^{\infty}(-\theta\mu(x,t)\Delta t+\frac{\theta^2}{2!}\sigma^2(x,t)\Delta t+o(\Delta t))e^{-\theta x}p_X(x;t)dx$$

so the partial derivative Eq. (6.177) becomes in the limit ($\Delta t\rightarrow 0$)

$$\frac{\partial M(\theta;t)}{\partial t}=\lim_{\Delta t\rightarrow 0}\frac{1}{\Delta t}\int_{-\infty}^{\infty}(-\theta\mu(x,t)\Delta t+\frac{\theta^2}{2!}\sigma^2(x,t)\Delta t+o(\Delta t))e^{-\theta x}p_X(x;t)dx$$

$$=\int_{-\infty}^{\infty}(-\theta\mu(x,t)+\frac{\theta^2}{2!}\sigma^2(x,t))e^{-\theta x}p_X(x;t)dx$$

Integrating by parts,

$$\frac{\partial M(\theta;t)}{\partial t}=-\int_{-\infty}^{\infty}\frac{\partial}{\partial x}[(\mu(x,t)p_X(x;t)]e^{-\theta x}dx+\int_{-\infty}^{\infty}\frac{\partial}{\partial x}\left[\frac{\theta}{2}\sigma^2(x,t)p_X(x;t)\right]e^{-\theta x}dx$$

$$=-\int_{-\infty}^{\infty}e^{-\theta x}\left[-\frac{\partial}{\partial x}\left\{(u(x,t)p_X(x;t)\right\}+\frac{1}{2}\frac{\partial^2}{\partial x^2}\left\{\sigma^2(x,t)p_X(x;t)\right\}\right]dx \quad (6.178)$$

Finally, by comparing the two expressions for the derivative $\partial M(\theta;t)/\partial t$, Eqs. (6.176) and (6.178), the Fokker–Planck equation, Eq. (6.172) has been obtained. It should be noted that the moment generating function Eq. (6.174) can also be based on the transition probability density. Using the same procedure, the Fokker–Planck equation is then derived for the transition densities, that is, Eq. (6.172).

Example 6.12 The arithmetic mean of n random variables was studied in Section 1.2.11 where it was shown that for large n, the probability distribution tended towards the normal distribution under fairly general conditions. Show that the arithmetic mean of a set of independent and identically distributed variables X_i, $i=1,2,\ldots,n\ldots$ is indeed a Markov chain.

Solution The arithmetic mean for n such variables is given by

$$Y_n=\frac{X_1+X_2+\cdots+X_n}{n}, \qquad n=1,2,\ldots$$

Rearranging the above fraction yields

$$Y_n=\frac{X_1+X_2+\cdots+X_n}{n}=\frac{n-1}{n}\frac{X_1+X_2+\cdots+X_{n-1}}{n-1}+\frac{X_n}{n}$$

$$=\frac{1}{n}[(n-1)Y_{n-1}+X_n]$$

The conditional distribution of Y_n given Y_{n-1} is obviously independent of the values of Y_k for $k<n-1$ and thus fulfils the Markov probability requirement Eq. (2.14). Therefore, the sequence of random variables Y_n, $n=1,2,\ldots$, is a Markov chain.

Example 6.13 Consider a Gaussian stationary random process $\{X(t), -\infty < t < \infty\}$ with zero mean and variance σ_X^2, which has the autocorrelation function

$$R_X(\tau) = \sigma_X^2 e^{-\alpha|\tau|} \tag{6.179}$$

Show that the process is Markovian.

Solution There are several ways to establish the Markov property. The following way is along the lines shown by Breiman, [24]. Consider a Gaussian white noise process $Z(t)$, (cf. Ex. 2.14) with a constant power spectral density S_0. If the white noise process is passed through a first-order linear system, that is,

$$\frac{dX(t)}{dt} + \alpha X(t) = Z(t) \tag{6.180}$$

the output process $X(t)$ is also stationary Gaussian, (cf. Eq. (6.25). Solving the differential equation, the output process is given by

$$X(t) = \int_{-\infty}^{t} e^{-\alpha(t-u)} Z(u) du \tag{6.181}$$

since the impulse response function is simply $e^{-\alpha t}$. Now, the autocorrelation function of $X(t)$ is

$$R_X(r, s) = E[X(r) X(s)] = \int_{-\infty}^{r} \int_{-\infty}^{s} e^{-\alpha(r-u)} e^{-\alpha(r-u)} e^{-\alpha(s-v)} E[Z(u) Z(v)] du\, dv$$

Since $Z(t)$ is a white noise process, $E[Z(u) Z(v)] = 2\pi S_0 \delta(v-u)$. Therefore,

$$R_X(r-s) = 2\pi S_0 \int_{-\infty}^{r} \int_{-\infty}^{s} e^{-\alpha(r-u)} e^{-\alpha(s-v)} \delta(v-u) du\, dv = 2\pi S_0 \int_{-\infty}^{s} e^{-\alpha(r-v)} e^{-\alpha(s-v)} dv$$

$$= 2\pi S_0 e^{-\alpha(r+s)} \int_{-\infty}^{s} e^{2\alpha v} dv = \frac{\pi S_0}{\alpha} e^{-\alpha(r-s)} \tag{6.182}$$

so the autocorrelation function of the output process has the same form as for the given process.

$$R_X(\tau) = \frac{\pi S_0}{\alpha} e^{-\alpha|\tau|}$$

Note that the integration process is dependent on the sign of $r - s$, that is, the difference has to be positive to keep the integrals finite, hence the absolute value. Thus, it has been shown that a stationary Gaussian process with the given autocorrelation function is essentially a white noise process passed through a first-order linear filter.

To show the Markovian property, consider once more the above response equation in a finite differences form, that is,

$$\frac{\Delta X(t)}{\Delta t} + \alpha X(t) = Z(t)$$

which for small time steps is approximated by

$$\frac{X(t_i) - X(t_{i-1})}{t_i - t_{i-1}} + \alpha X(t_i) = Z(t_i)$$

or

$$(1 + \alpha(t_i - t_{i-1}) X(t_i)) = X(t_{i-1}) + Z(t_i)(t_i - t_{i-1})$$

From this last equation it is obvious that any probability assertion about the random variable $X(t_i)$ is only dependent on its last value $X(t_{i-1})$ and the value of $Z(t_i)$. As $Z(t)$ is white noise, it has no memory and is independent of everything that happened before. Therefore, given the entire past history of $X(t)$ up to and including the time t_{i-1}, any probability assertion about its next value $X(t_i)$ is only dependent on its value at t_{i-1}, which is exactly the Markov property. Actually, the above argument has been put forward in an heuristic manner. A more formal proof is offered by Melsa and Sage, [145], where it is shown that if two sequences of independent random variables $\{Y_i\}$ and $\{X_i\}$ have the relationship

$$Y(t_i) = Y(t_{i-1}) + X(t_i)$$

the random sequence $\{Y_i\}$ is a Markov chain. An entirely different and perhaps more mathematically correct proof of the original statement about a Gaussian stationary process with the given autocorrelation function is found in Doob, [55].

Example 6.14 The Ornstein–Uhlenbeck process The Wiener process, Example 2.5, does not pretend to be a proper physical model for Brownian motion in realistic terms. For instance, whereas the change in motion in a very small time interval Δt is of the order $O(\sqrt{\Delta t})$, ($\text{Var}[W] = \sigma^2 \Delta t$), the change of velocity in the same interval is of the order $O(\sqrt{\Delta t}/\Delta t) = O(1/\sqrt{\Delta t})$, which tends of infinity as $\Delta t \to 0$. Thus the Wiener process fails to provide a satisfactory model to describe Brownian motion correctly for small values of t, although it works quite well for moderate to large time intervals. In 1930, Uhlenbeck and Ornstein proposed an alternative model, considering instead the velocity $V(t)$ of a particle in Brownian motion to overcome the above difficulty, [227].

Let $V(t)$ be the velocity of a Brownian particle of mass m, which is suspended in liquid where it is constantly being bombarded by molecular collisions producing a random force $F(t)$ acting on the particle. The other forces acting on the particle will be the drag force exerted by the liquid, which according to Stoke's law is $-c\,V(t)$, where c is the viscous frictional constant depending on the viscosity of the liquid, and the mass and diameter of the particle. According to Newton's second law of motion, the equation of the particle motion can be written as

$$m\Delta V(t) = -c\,V(t)\Delta t + \Delta F(t) \tag{6.183}$$

in which $\Delta F(t)$ is the instantaneous random change in the force acting on the particle due to the molecular bombardment. The random molecular force $F(t)$ is now assumed to be a Wiener process with zero mean value and the variance $\sigma^2 \Delta t$, or defining the unit process $W(t)$, $F(t) = \sigma W(t)$. Dividing Eq. (6.183) by Δt and letting $\Delta t \to 0$, the following differential equation is obtained

$$m \frac{\mathrm{d} V(t)}{\mathrm{d}t} = -c V(t) + \sigma \frac{\mathrm{d}W(t)}{\mathrm{d}t} \tag{6.184}$$

Equation (6.184) is called the Langevin equation. It has a singular term $\mathrm{d}W(t)/\mathrm{d}t$ as the Wiener process does not have continuous stochastic derivatives. This creates difficult interpretation problems, which will not be addressed here (see for example Todorovic, [222]). Instead, consider Eq. (6.115) as the governing equation for a Markov diffusion process. The instantaneous mean value of the Ornstein–Uhlenbeck process (O–U process) is

$$\lim_{\Delta t \to 0} \frac{E[V(t+\Delta t) - V(t) \mid V(t) = v]}{\Delta t}$$

$$= -\frac{c}{m} v + \frac{1}{m} \lim_{\Delta t \to 0} \frac{E[\Delta F]}{\Delta t} = -\frac{c}{m} v$$

and the instantaneous variance is

$$\lim_{\Delta t \to 0} \frac{\mathrm{Var}[V(t+\Delta t) - V(t) \mid V(t) = v]}{\Delta t}$$

$$= -\frac{c}{m} \lim_{\Delta t \to 0} \frac{[v - (c/m)v]^2 (\Delta t)^2}{\Delta t} + \frac{1}{m} \lim_{\Delta t \to 0} \frac{\mathrm{Var}[\Delta F]}{\Delta t} = \frac{\sigma^2}{m}$$

Therefore, the process $\{V(t), t \geqslant 0\}$ is a Markov diffusion process, whose transition probability density $p(u|v; t, s)$ satisfies the Fokker–Planck equation with the forward variables u and t and $a(t, u) = \sigma^2/m$ and $b(t, u) = -cu/m$, i.e.

$$m \frac{\partial p(u|v;t,s)}{\partial t} = \tfrac{1}{2}\sigma^2 \frac{\partial^2 p(u|v;t,s)}{\partial u^2} + c \frac{\partial [up(u|v;t,s)]}{\partial u} \tag{6.185}$$

It is convenient to transform the above equation. Multiplying both sides by the term $e^{i\omega u}$ and integrating the whole equation over the entire range $-\infty < u < \infty$, it can be transformed into an equation involving the characteristic function, which is defined as follows (cf. Eq. (1.97):

$$\varphi(\omega; u, v, t, s) = \int_{-\infty}^{\infty} e^{i\omega u} p(u|v;t,s) \mathrm{d}u \tag{6.186}$$

Further, it is assumed that $V(0) = v_0$, and as $u \to 0$ both $p(u|v;t,s)$ and its derivative $\partial p/\partial u$ tend to zero as well. After the transformation, the left-hand side of Eq. (6.163) simply becomes $\partial\varphi/\partial t$. The first term of the right-hand side is transformed to yield

$$\int_{-\infty}^{\infty} \mathrm{e}^{\mathrm{i}\omega u}\frac{\partial^2 p}{\partial u^2}\mathrm{d}u = \left[\mathrm{e}^{\mathrm{i}\omega u}\frac{\partial p}{\partial u}\right]_{-\infty}^{\infty} - \mathrm{i}\omega\int_{-\infty}^{\infty} \mathrm{e}^{\mathrm{i}\omega u}\frac{\partial p}{\partial u}\mathrm{d}u$$

$$= -\mathrm{i}\omega\left\{[\mathrm{e}^{\mathrm{i}\omega u}p]_{-\infty}^{\infty} - \mathrm{i}\omega\int_{-\infty}^{\infty}\mathrm{e}^{\mathrm{i}\omega u}p\,\mathrm{d}u\right\} = -\omega^2\varphi$$

The second term yields

$$\int_{-\infty}^{\infty} \mathrm{e}^{\mathrm{i}\omega u}\frac{\partial(up)}{\partial u}\mathrm{d}u = [\mathrm{e}^{\mathrm{i}\omega u}(up)]_{-\infty}^{\infty} - \mathrm{i}\omega\int_{-\infty}^{\infty}\mathrm{e}^{\mathrm{i}\omega u}(up)\mathrm{d}u$$

$$= -\omega\frac{\partial}{\partial\omega}\int_{-\infty}^{\infty}\mathrm{e}^{\mathrm{i}\omega u}p\,\mathrm{d}u = -\omega\frac{\partial\varphi}{\partial\omega}$$

The transformed equation then becomes

$$m\frac{\partial\varphi}{\partial t} + c\omega\frac{\partial\varphi}{\partial\omega} = \tfrac{1}{2}\sigma^2\omega^2\varphi \tag{6.187}$$

Equation (6.187) can be simplified further by considering the log-characteristic function $\psi(\omega) = \log_e\varphi(\omega)$, whereby it converts to

$$m\frac{\partial\psi}{\partial t} + c\omega\frac{\partial\psi}{\partial\omega} = -\tfrac{1}{2}\sigma^2\omega^2 \tag{6.188}$$

Equations (6.187) and (6.188) are special cases of a first-order partial differential equation called the Lagrange equation. The solution of Eq. (6.188) with the initial velocity $V(0) = v_0$ and putting $\alpha = c/m$ is given by

$$\psi(\omega; u, v, t, 0) = \mathrm{i}\omega(v_0\,\mathrm{e}^{-\alpha t}) - \frac{\sigma^2}{m^2}\frac{\omega^2}{2}\left(\frac{1}{2\alpha}(1-\mathrm{e}^{-2\alpha t})\right) \tag{6.189}$$

Comparing this result with the characteristic function for a Gaussian random variable, Eq. (1.101), it has been shown that the transition probability density of the O–U process is a Gaussian density with a mean value function

$$\mu(t) = v_0\,\mathrm{e}^{-\alpha t} \tag{6.190}$$

and a variance function

$$\sigma^2(t) = \frac{\sigma^2}{m^2}\frac{1}{2\alpha}(1-\mathrm{e}^{-2\alpha t}) \tag{6.191}$$

Thus it has been shown that the Ornstein–Uhlenbeck process, that is, the velocity $\{U(t), t \geqslant 0\}$ of a particle in Brownian motion, is a Gaussian Markov process. However, unlike the Wiener process it does not possess independent increments. For large $t\,(t\to\infty)$, the distribution of the velocity is normal with zero mean and variance $\sigma^2/(2m^2\alpha)$. Therefore, $U(t)$ is said to be in a statistical equilibrium, i.e. the Cartesian components of the velocity have independent normal distributions with zero mean and common variance. For small $t\,(t \to 0)$, $\mu(t) = v_0$ and $\sigma^2(t) = t\sigma^2/m^2$.

Example 6.15 The O–U process can be obtained as a transformation of the Wiener process. In fact, let $\{W(t),\ t \geqslant 0\}$ be a unit Wiener process with $W(0) = 0$ and variance t. Consider the transformation

$$Y(t) = a(t)\, W(f(t)) \tag{6.192}$$

in which $a(t)$ is a real continuous function of time and $f(t)$ is a real continuous increasing function of time. Actually, this simply implies that the time scale for observing the process has been changed. The expectation of $Y(t)$ is zero as $E[W(f(t))] = 0$. The variance of $Y(t)$ on the other hand is given by

$$\operatorname{Var}[Y(t)] = a^2(t) \operatorname{Var}[W(f(t))] = a^2(t) f(t) \tag{6.193}$$

since $\operatorname{Var}[W(t)] = t$ and $f(t)$ is strictly increasing. The process $Y(t)$ is also a Gaussian diffusion process. In fact, from Eq. (6.192) the differential equation

$$\mathrm{d}Y(t) = a'(t)\, W(f(t))\, \mathrm{d}t + a(t)\, \mathrm{d}W(f(t)) = a'(t)\, W(f(t))\, \mathrm{d}t + a(t) f'(t)\, \mathrm{d}W(f)) \tag{6.194}$$

is formed, and the last term is indicative of the necessary steps that have to be taken when differentiating a function of a function of the main variable. Inserting Eq. (6.192) into Eq. (6.194) yields

$$\mathrm{d}\,Y(t) = \frac{a'(t)}{a(t)}\, Y(t)\mathrm{d}t + a(t)\, d(f'(t)\, W(f)) \tag{6.195}$$

By comparing Eqs. (6.195) and (6.173), it is clear that conditional on $Y(t) = y$, $Y(t)$ has the drift coefficient $\mu(y, t) = y a'(t)/a(t)$ and the diffusion coefficient $\sigma^2(y, t) = a^2(t) f'(t)$. Thus it has been shown that a Gaussian Markov diffusion process can be obtained from the Wiener process by suitable transformation and scaling.

By selecting the functions $a(t) = \mathrm{e}^{-\alpha t}$ and $f(t) = \sigma^2 \mathrm{e}^{2\alpha t}/(2\alpha)$, the transformed Wiener process becomes

$$Y(t) = \mathrm{e}^{-\alpha t}\, W(\sigma^2 \mathrm{e}^{2\alpha t}/(2\alpha)), \qquad \alpha > 0, \qquad -\infty < t < \infty \tag{6.196}$$

The instantaneous mean value and variance of $Y(t)$ respectively are then given by

$$\mu(y, t) = -\alpha y \qquad \text{and} \qquad \sigma^2(y, t) = \sigma^2$$

These values are exactly the same ($m = 1$) as the corresponding values found for the O–U process, Example 6.14, so $Y(t)$ has the structure of a O–U process. The mean value of $Y(t)$, Eq. (6.175), is zero whereas the variance of $Y(t)$ is $\operatorname{Var}[Y(t)] = \mathrm{e}^{-2\alpha t}(\sigma^2 \mathrm{e}^{2\alpha t}/(2\alpha)) = \sigma^2/2\alpha$. The last result indicates that $Y(t)$ is stationary. Conversely, if $Y(t)$ is a stationary O–U process, the Wiener process may be obtained from the transformation

$$W(t) = \sqrt{\frac{\sigma^2}{2\alpha t}}\, Y\!\left(\frac{1}{2\alpha} \log_e\left(\frac{2\alpha t}{\sigma^2}\right)\right) \tag{6.197}$$

6.3.2 Martingales

Markov process have the common fundamental probability property, which relates the future development of the process, conditional to its past values, to the probability of the present value of the process only. Another family of processes possesses a common probability feature, which relates the expectation of the future development of the process, conditional to its past values, to the present value of the process only. Measure of this family of stochastic processes are called *martingales* or martingale processes. The theory of martingale processes, which was developed by Doob, [55], has found wide application in various fields of applied science. An extensive and comprehensive account can be found in Karlin and Taylor, [100], and in Todorovic, [222].

The definition of the martingale property can be stated as follows. Let $\{X(t), t \in T\}$ be a real stochastic process with a bounded expectation, that is, $E[|X(t)|] < \infty$ for all $t \in T$. The stochastic process $X(t)$ is said to be a martingale if, for every $t_1 < t_2 < \cdots < t_n$ in T,

$$E[X(t_n)|X(t_1)=x_1, X(t_2)=x_2, \ldots, X(t_{n-1})=x_{n-1}] = x(t_{n-1}) \tag{6.198}$$

The name martingale stems from the French expression for progressively doubling one's bets in gambling until a win has been secured. In fact, letting $X(t_n)$ describe the fortune of, say, a roulette player at time t_{n-1}, his fortune on the next spin at t_n is on the average his current fortune and not otherwise affected by the previous history.

The stochastic process $X(t)$ is said to be submartingale if

$$E[X(t_n)|X(t_1)=x_1, X(t_2)=x_2, \ldots, X(t_{n-1})=x_{n-1}] \leqslant x(t_{n-1}) \tag{6.199}$$

and finally, it is said to be a supermartingale if

$$E[X(t_n)|X(t_1)=x_1, X(t_2)=x_2, \ldots, X(t_{n-1})=x_{n-1}] \geqslant x(t_{n-1}) \tag{6.200}$$

It follows from the two last equations that in the case of a submartingale,

$$E[X(t_n)] \geqslant E[X(t_{n-1})] \cdots \geqslant E[X(t_2)] \geqslant E[X(t_1)] \tag{6.201}$$

and in the case of supermartingale

$$E[X(t_n)] \leqslant E[X(t_{n-1}] \cdots \geqslant E[X(t_2)] \leqslant E[X(t_1)] \tag{6.202}$$

A clear distinction has to be made between discrete martingale processes and continuous martingales as was the case for Markov processes. The treatment of continuous martingales involves σ-fields and probability spaces, which is beyond the scope of the present text. The reader is referred to Karlin and Taylor, [100] and Todorovic, [222]. Therefore, the following discussion is limited to the discrete martingales only.

Consider two discrete stochastic process $\{X_i, i=1,2,\ldots\}$ and $\{Y_i, i=1,2,\ldots\}$. By definition, $\{X_i\}$ is a martingale with respect to $\{Y_i\}$ if, for $i=1,2,\ldots$

(i) $E[|X_i|] < \infty$

and

$$\text{(ii)} \quad E[X_{i+1}|Y_1=y_1, Y_2=y_2, \ldots, Y_i=y_i] = x_i \tag{6.203}$$

In the literature, generally no distinction is made between the outcome x_i, that is, the real value assumed by the random variable X_i, and the variable itself. The latter condition (ii) is therefore written as

$$\text{(ii)} \quad E[X_{i+1} | Y_1, Y_2, \ldots, Y_i] = X_i$$

This style will be adopted when there is no ground for misunderstanding.

The last definition of a martingale can be extended to cover supermartingales and submartingales as well. Whereas the original definition Eq. (6.198) considers the past history of the process $\{X_i\}$ only, the process $\{Y_i\}$ can contain more information or history up to the ith stage, which is relevant for the outcome X_{i+1}. The past history determines X_i in a sense that X_i is a function of $Y_1, Y_2, \ldots, Y_i$ since by Eq. (6.203),

$$X_i = f(Y_1, Y_2, \ldots, Y_i) = E[X_{i+1} | Y_1 = y_1, Y_2 = y_2, \ldots, Y_i = y_i] \qquad (6.204)$$

The conditional expectation of X_i with respect to itself, that is $E[X_i | X_i = x_i] = x_i$ or,

$$E[f(Y_1, Y_2, \ldots, Y_i) | Y_1 = y_1, Y_2 = y_2, \ldots, Y_i = y_i] = f(y_1, y_2, \ldots, y_i)$$

therefore,

$$E[X_i | Y_1 = y_1, Y_2 = y_2, \ldots, Y_i = y_i] = x_i \qquad (6.205)$$

Applying the relation of conditional expectations or the law of total probability Eq. (1.63) together with Eq. (6.203),

$$E[X_{i+1}] = E[E[X_{i+1} | Y_1 = y_1, Y_2 = y_2, \ldots, Y_i = y_i]] = E[X_i] \qquad (6.206)$$

so that by induction,

$$E[X_{i+1}] = E[X_i] = \cdots = E[X_1] \qquad (6.207)$$

in other words, the mean values of a martingale are all equal.

Any stochastic process with independent increments and zero mean value is a martingale. In fact, since $E[X_n] = 0$ for all n,

$$E[X_n - X_{n-1} | X_1 = x_1, X_2 = x_2, \ldots, X_{n-1} = x_{n-1}] = 0$$

for all $n \geqslant 2$. Therefore, by Eq. (6.205)

$$E[X_n | X_1 = x_1, X_2 = x_2, \ldots, X_{n-1} = x_{n-1}] = x_{n-1}$$

which is the martingale property, or

$$E[X_n | X_{n-1} = x_{n-1}] = x_{n-1} \qquad (6.208)$$

since the process has independent increments. The converse is not true, that is, a martingale does not necessarily have independent increments or zero mean values.

Under quite general conditions, it can be shown that a martingale $\{X_i\}$ converges to a limit random variable X as $i \to \infty$. More precisely, let $\{X_i\}$ be a martingale with respect to $\{Y_i\}$, which has a finite mean value square, that is,

$$E[X_n^2] \leqslant \infty \qquad \text{for all } n \tag{6.209}$$

Then, $\{X_i\}$ converges to a limit random variable X as $i \to \infty$ in the mean square sense and consequently with probability one (cf. Eq. (2.18))

$$\lim_{i \to \infty} E[|X_i - X|] = 0$$

and by Eq. (6.207)

$$E[X_1] = E[X_2] = \cdots = E[X_i] = E[X] \tag{6.210}$$

The proof of the above assertion can be found in Karlin and Taylor, [100], where an extensive discussion of other features of martingale theory with applications is presented.

Example 6.16 Consider a sequence of arbitrary random variables $\{Z_i, i = 1, 2, \ldots,\}$ and a random variable X with bounded mean, that is, $E[|X|] < \infty$. Show that the random sequence $\{X_i\}$ where

$$X_i = E[X|Z_1, Z_2, \ldots, Z_i] \tag{6.211}$$

forms a martingale with respect to $\{Z_i\}$, called Doob's process.

Taking the expectation of the absolute value of both sides,

$$E[|X_i|] = E[|E[X|Z_1, Z_2, \ldots, Z_i]|] \leqslant E[E[|X\|Z_1, Z_2, \ldots, Z_i]]$$

as $|\int_{-\infty}^{\infty} up(u)du| \leqslant \int_{-\infty}^{\infty} |u|p(u)du$. Invoking the total probability law Eq. (1.63),

$$E[|X_i|]] \leqslant E[E[|X\|Z_1, Z_2, \ldots, Z_i]] = E[|X|] < \infty$$

which is the first requirement for a martingale (see Eq. (6.203)). Furthermore,

$$\begin{aligned} E[X_{i+1}|Z_1, Z_2, \ldots, Z_i] &= E[E[X|Z_1, Z_2, \ldots, Z_{i+1}]|Z_1, Z_2, \ldots, Z_i] \\ &= E[X|Z_1, Z_2, \ldots, Z_i] = X_i \end{aligned}$$

where the last manipulation is an obvious extension of the law of total probability (see Eq. (1.63)). This, however, is the second requirement for a martingale, which proves the assertion.

Example 6.17 Consider a sequence of IID random variables $\{Z_i, i = 1, 2, \ldots\}$ with zero mean. Show that the random sequence $\{X_n\}$, which is formed by the moving sum, $X_n = \sum_{i=1}^{n} Z_i$, $n = 1, 2, \ldots$, is a martingale.

Solution Since

$$X_{n+1} = X_n + Z_{n+1}$$

the expectation of X_{n+1}, conditional on its past history, is

$$\begin{aligned}E[X_{n+1}|X_1,X_2,\ldots,X_n] &= E[X_n + Z_{n+1}|X_1,X_2,\ldots,X_n]\\ &= E[X_n|X_1,X_2,\ldots,X_n] + E[Z_{n+1}|X_1,X_2,\ldots,X_n]\\ &= X_n + E[Z_{n+1}] = X_n\end{aligned}$$

because $E[X_n|X_1,X_2,\ldots,X_n] = X_n$ and $E[Z_{n+1}|X_1,X_2,\ldots,X_n] = E[Z_{n+1}|Z_1,Z_1 + Z_2,\ldots,Z_1 + Z_2 + \cdots + Z_n]$, which is equal to $E[Z_{n+1}] = 0$ since the Z_i are independent variables with zero mean.

Example 6.18 Consider a sequence of IID random variables $\{Z_i, i = 1,2,\ldots\}$ with mean value $E[Z_i] = 1$. Show that the random sequence $\{X_n, n = 1,2,\ldots\}$, which is formed by the moving product $X_n = \prod_{i=1}^{n} Z_i$, is a martingale.

Solution Since

$$X_{n+1} = X_n Z_{n+1}$$

the expectation of X_{n+1}, conditional on its past history, is

$$\begin{aligned}E[X_{n+1}|X_1,X_2,\ldots,X_n] &= E[X_n Z_{n+1}|X_1,X_2,\ldots,X_n]\\ &= E[X_n|X_1,X_2,\ldots,X_n]\,E[Z_{n+1}|X_1,X_2,\ldots,X_n]\\ &= X_n E[Z_{n+1}] = X_n\end{aligned}$$

because $X_n = Z_1 Z_2 \cdots Z_n$ and Z_{n+1} are independent, $E[X_n|X_1,X_2,\ldots,X_n] = X_n$ and $E[Z_{n+1}|X_1,X_2,\ldots,X_n] = E[Z_{n+1}] = 1$ using the same argument as in the previous example.

Example 6.19 Martingales play an important role in the analysis of Markov processes. To illustrate this relationship consider an homogeneous Markov chain $Y_1, Y_2,\ldots, Y_n,\ldots$ with the probability of transition from state i to state j (see Eq. (6.163))

$$p_{i,j} = P_{ij} = P[Y_{n+1} = j|\,Y_n = i] \tag{6.212}$$

The transition probabilities form the transition matrix

$$\mathbf{P} = \begin{Bmatrix} P_{11} & P_{12} & P_{13} & P_{14} & \cdots \\ P_{21} & P_{22} & P_{23} & P_{24} & \cdots \\ P_{31} & P_{32} & P_{33} & P_{34} & \cdots \\ \vdots & \vdots & \vdots & \vdots & \cdots \\ P_{n1} & P_{n2} & P_{n3} & P_{n4} & \cdots \\ \vdots & \vdots & \vdots & \vdots & \cdots \end{Bmatrix} \tag{6.213}$$

Now, Eq. (6.213) displays an infinite square array called a stochastic matrix or simply a Markov matrix with non-negative elements (the transition probabilities). Clearly, the sum of the elements in any row is equal to one since $\sum_{j=1}^{\infty} P_{ij} = 1$. For a fixed n, the transition matrix is an $n \times n$ matrix, which has the eigenvalues $\{\lambda_j\}$ and eigenvectors $\mathbf{v}_j = \{v_i(j)\}$ that satisfy the equation

$$(\mathbf{P} - \lambda_i \mathbf{I})\mathbf{v}_i = 0$$

for all i, where $\mathbf{I}$ is the unit diagonal matrix. For any pair $(\lambda_s, \mathbf{v}_s)$, the ith row in the above matrix equation is

$$\sum_{k=1}^{n} P_{ik} v(k) = \lambda v(i) \tag{6.214}$$

The eigenvectors are indirectly functions of the states i and consequently the Markov variables Y_i. Therefore, form the variable

$$X_n = \lambda^{-n} v(Y_n)$$

The random sequence $\{X_n, n = 1, 2, \ldots\}$ is a martingale since

(i) $E[|X_n|] = \lambda^{-n} E[|v(E[|v(Y_n)|])|] < \infty$ because $E[|v(Y_n)|] < \infty$ for all n

and

$$\begin{aligned}
\text{(ii)}\quad E[X_{n+1} | Y_1, Y_2, \ldots, Y_n] &= E[\lambda^{-n-1} v(Y_{n+1}) | Y_1, Y_2, \ldots, Y_n] && (1)\\
&= \lambda^{-n-1} E[v(Y_{n+1}) | Y_1, Y_2, \ldots, Y_n] && (2)\\
&= \lambda^{-n-1} E[v(Y_{n+1}) | Y_n] && (3)\\
&= \lambda^{-n-1} \sum_{j=1}^{n} P_{jY_n} v(j) && (4)\\
&= \lambda^{-n-1}(\lambda v(Y_n)) = \lambda^{-n} v(Y_n) = X_n && (5)
\end{aligned}$$

Step (2) to (3) is due to the Markovian property of Y_i. Step (3) to (4) is taken by forming the conditional expectation of a discrete random variable given $Y_n = y_n$. Step (4) to (5) is taken by invoking Eq. (6.214), which proves the assertion that X_n is a martingale.

6.3.3 The Boltzmann Process

When the effects of any one event can be superimposed on the cumulative effects of all preceding events, a class of linear non-Markovian processes is obtained that is often referred to as Boltzmann processes after Ludwig Boltzmann, 1876, who used similar arguments when studying elastic aftereffects in continuum mechanics. Boltzmann proposed that all strain effects resulting from all previous stress inputs are linearly additive, that is

$$\dot{E}(t) = \sum_i \Delta\sigma_i \dot{\varphi}(t - \tau_i) \tag{6.215}$$

where $\dot{E}(t)$ is the strain rate, $\Delta\sigma_i$ the stress input and $\varphi(t)$ is a memory function also called the creep function (see Lominitz, [129] and [132]). The Boltzmann process is widely applicable as a rate process in physics where the 'rate' normally decays with time.

The Boltzmann process can be applied to pure rate functions, describing the probability of occurrence of any kind of a physically realizable event. Assume that a change of variables has been effected, whereby $\chi(\Delta\sigma) = \Delta\sigma$, and the relation between the rate function $\Lambda(t)$ and the new variable χ is linear. That is, twice the value of χ will produce twice the value of Λ. Let $\gamma(t)$ be the influence function or the 'creep' function, then

$$\Lambda(t) = \Lambda_0 + \sum \chi_i \gamma(t - \tau_i) \tag{6.216}$$

where Λ_0 is a base rate, which prevents the process from dying out. In other words, there is always a small positive probability $\Lambda_0 \Delta t$ of an occurrence of an event, even if no previous event has occurred for a long time. Such variables as χ are known as tags and the process is sometimes called a tagged process.

In the above formulation, the Boltzmann process has been used in biological applications, (see Daley and Vere-Jones, [50]), where it has been called a tagged self-exciting process. The mean rate of activity of the process is given by

$$E[\Lambda] = \frac{\Lambda_0}{\left[1 - \int_0^\infty \gamma(\tau)\mathrm{d}\tau\right]} \tag{6.217}$$

The memory or influence functions are often taken to be of the type $c\exp[-\alpha t]$, that is, to have an exponentially decaying memoty. In this case the above mean value is calculated as

$$E[\Lambda] = \frac{\Lambda_0}{1 - c/\alpha} \tag{6.218}$$

The rate function Eq. (6.216) can easily be adapted to a continuous time series $\chi(t)$ by introducing the convolution integral instead of the sum, that is,

$$\Lambda(t) = \Lambda_0 + \int_0^\infty \chi(\tau)\dot{\gamma}(\tau)\mathrm{d}\tau \tag{6.219}$$

and by introducing the exponential memory function

$$\Lambda(t) = \Lambda_0 - \alpha c \int_0^\infty \chi(\tau)\mathrm{e}^{-\alpha(t-\tau)}\mathrm{d}\tau \tag{6.220}$$

In this last form, the rate function $\Lambda(t)$ is amenable to interpretation and conclusions. For instance, let $\Lambda(t)$ be the arrival rate of large earthquakes, [129], [132]. The probability of having no earthquakes in an interval $(t, t+\tau)$ is then given by (compare with Eq. (6.62))

$$P_0[t, t+\tau] = \exp\left[-\int_t^{t+\tau} \lambda(u)\mathrm{d}u\right] \tag{6.221}$$

The cumulative probability distribution of the forward return period τ, given that an earthquake happened at time t, is then obtained by

$$F_t(\tau) = 1 - \exp\left[-\int_t^{t+\tau} \lambda(u)\mathrm{d}u\right] \tag{6.222}$$

that is, the probability of having an earthquake after time τ from the time t of the last earthquake is one minus the probability of having no earthquake in the time interval $t + \tau$. In Eq. (6.222), the influence of older terms in $\lambda(t)$ decreases rapidly, so it is possible to introduce a cutoff in the numerical calculations, which keeps only a finite number of terms in the sum.

6.3.4 Concluding Remarks

This brief overview of stochastic processes will hopefully have given the reader food for thought. It is outside the framework of this monograph to present a full and exhaustive treatment of stochastic processes, which would by itself alone fill a good sized book. It is hoped, however, that the presentation has been able to clarify the main aspects of the theory of stochastic processes and lay a sufficient background for practical applications. For more details and a mathematically more precise presentation of the theory, a wealth of literature is available. In Daley and Vere-Jones' book on the theory of point processes, [50], there are over 400 references cited. Many of these texts require knowledge of advanced mathematical concepts such as σ algebra of Borel sets, Hilbert and Banach spaces, Lebesgue measure, to name but a few of them, which are beyond the scope of this monograph. However, the treatment as presented herein should be sufficiently rigorous in mathematical terms for all practical purposes.

Many important topics have been left out and the presentation, especially regarding Markov processes and martingales which have been treated only sketchily, has been more in an overview form. For a more exhausitive treatment, texts such as Bharucha-Reid, [15], Cox and Miller, [40], Medhi, [141], Karlin and Taylor, [100] and Lin and Cai [123], are recommended for further reading. The last text, [123], moreover, contains a variety of practical examples of engineering applications of Markov process. Dynkin, [56], and Todorovic, [222], on the other hand, offer a mathematically more advanced and stringent presentation of Markov processes and other related topics. Queuing and branching type of processes have not been mentioned nor renewal processes. A thorough discussion is given, for example, by Medhi, [141] and Nelson, [151]. Of older and sometimes more topic-oriented texts on random processes, the texts of Laning and Battin, [120], Bendat, [12], Doob, [55], Jazwinski, [94], Melsa and Sage, [145] and Parzen, [170], may be considered.

7 Fourier Analysis and Data Processing

Fourier analysis (After the French Mathematician Jean Baptiste Joseph Fourier, 1768–1830, [23]) plays a dominant role in the treatment of vibrations of mechanical systems responding to deterministic or stochastic excitation, and, as has already been seen, it forms the basis of spectral representation of stochastic time series and random signals. The Fourier transformation as a mathematical tool offers a clear and simple method for treating complex analytic problems by converting equations involving the basic variable to corresponding equations in the frequency domain where solutions are often more easily accessible. In probability theory, the Fourier transform is the key to understanding certain probability distributions through their characteristic functions. Finally, the discrete Fourier series forms the basis of signal processing and data manipulation, which has turned into a vast field that concerns most if not all scientific studies.

In the following, a brief overview of Fourier analysis is presented to have the background for application elsewhere in the text readily available. Fourier series representation of periodic functions is introduced and the continous Fourier transform is derived for aperiodic functions. Various convenient relations concerning the Fourier transform are presented and a few examples given to clarify the text. This will be followed by an overview section on signal analysis and data processing. Sampling or discretizing of band-limited realizations of stochastic processes or any other time signals is briefly discussed. Thereafter, the discrete and inverse discrete Fourier transforms will be introduced as a natural extension of the continuous Fourier transform and discrete Fourier series analysis. The fast Fourier transform algorithm is briefly discussed together with various aspects of the analysis of digital signals. Signal convolution and aliasing, and the ramifications of signal distortion due to truncation is presented in some detail. Inverse fltering and windowing of signals is also presented with examples of standard digital filters. This is a highly specialized field, which is undergoing rapid expansion and development in connection with the ever-increasing electronic gadgetry and instruments in use in everyday life. Finally, other related methods, such as the z-transformation and auto regressive moving average (ARMA) models will be mentioned in order to furnish information on recent developments in earthquake signal analysis, especially for generation of synthetic earthquake accelerations.

Some exciting new topics in system analysis and signal processing will not be treated for the sake of space and limitation of topics. Notably, new concepts such as fuzzy systems and neural networks should be mentioned in this respect, [114], [115] and [116].

7.1 FOURIER ANALYSIS

It is assumed that the reader has already acquired a fundamental knowledge of the basic theory of Fourier series and integral transforms. The following presentation is therefore a refresher text to clarify the concepts and the tools to be used. The most basic relations and definitions will be presented together with some practical examples. Formal rigorous mathematical proof of the various statements has been avoided whenever deemed acceptable, and more simple methods of derivation employed. The text is both compact and lacking in details. For more rigorous and detailed presentation, there is a variety of standard textbooks available. For example, Weaver, [233], and Papoulis; [168], offer a thorough and detailed version of Fourier analysis and its many applications with the latter text concentrating on the Fourier integral transform.

7.1.1 Periodic Functions

A function $f(t)$ is called a periodic function with the period $T > 0$ if for all $t, f(t+T) = f(t)$. By induction, $f(t+nT) = f(t)$ where n is any integer number. Therefore, the function repeats itself in any number of intervals of length T. The well-known cosine and sine functions are perfect examples of periodic functions with the period 2π. In Fig. 7.1, more general examples of such functions are shown. The question now arises, can any reasonably well-behaved function $f(t)$, where t is a real variable (for instance the time), that reapeats itself with the period $T > 0$, be expanded into trigonometric series, that is, broken down into its harmonic cosine and sine components. The function $f(t)$ would then be synthetized by adding the harmonic components together.

Let $f(t)$ be a periodic function with the period $T > 0$. Define the corresponding basic circular frequency $\omega = 2\pi/T$by which the function repeats itself. Now, formally write $f(t)$

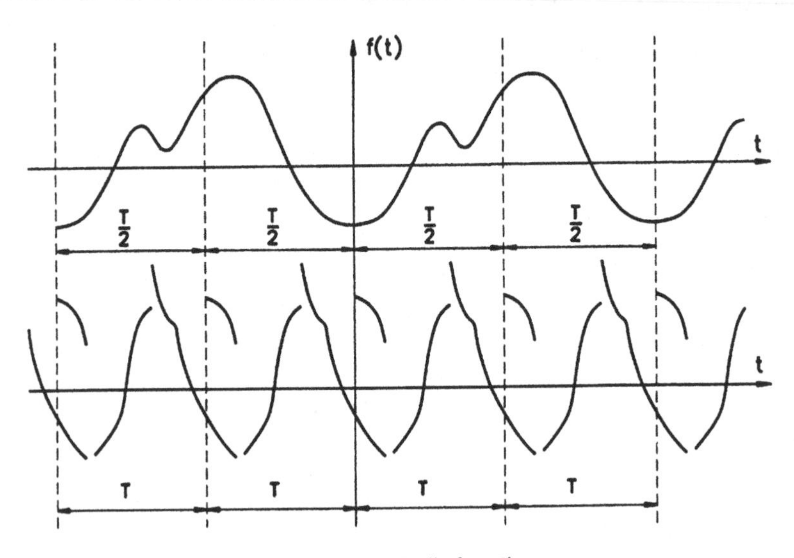

Fig. 7.1 *Periodic functions*

as the trigonometric series:

$$f(t) = \tfrac{1}{2}a_0 + a_1\cos\omega t + b_1\sin\omega t + a_2\cos 2\omega t + b_2\sin 2\omega t + \cdots$$
$$+ a_n\cos n\omega t + b_n\sin n\omega t + \cdots \quad (7.1)$$

in which a_i and b_i are real constants—called the Fourier coefficients—to be determined. For this purpose, the orthogonal properties of the trigonometric functions, shown below, will come in handy:

$$\int_0^T \cos m\omega t \cos n\omega t\, dt = \begin{cases} T & m = n = 0 \\ \dfrac{T}{2} & m = n \neq 0 \\ 0 & m \neq n \end{cases}$$

$$\int_0^T \cos m\omega t \sin n\omega t\, dt = 0 \text{ for all } (m, n)$$

$$\int_0^T \sin m\omega t \sin n\omega t\, dt = \begin{cases} 0 & m = n = 0 \\ \dfrac{T}{2} & m = n \neq 0 \\ 0 & m \neq n \end{cases}$$

By multiplying both sides of Eq. (7.1) with the term $\cos(k\omega t)$ and then integrating over the range $(0, T)$, making use of the above orthogonal properties of the trigonometric functions, the Fourier coefficients a_i are obtained and in the same manner, by multiplying with the term $\sin k\omega t$, the Fourier coefficients b_i are found. Thus,

$$a_k = \frac{2}{T}\int_0^T f(t)\cos k\omega t\, dt$$

$$b_k = \frac{2}{T}\int_0^T f(t)\sin k\omega t\, dt$$

and it has been shown without formal mathematical proof that any reasonably well-behaved function $f(t)$ can be expressed as Fourier series of the form

$$f(t) = \tfrac{1}{2}a_0 + \sum_{k=1}^{\infty} a_k\cos k\omega t + \sum_{k=1}^{\infty} b_k\sin k\omega t \quad (7.1)$$

where the Fourier coefficients $\{a_i, b_i\}$ are given by the relations Eq. (7.2)

It may be asked under what conditions the series Eq. (7.1) converge, that is, reproduce the function $f(t)$. There are several modes of convergence, which can be described as follows:

(i) The Fourier series Eq. (7.1) is said to converge uniformly to $f(t)$ in the interval $[0, T]$ if

(a) $f(t), f'(t)$, and $f''(t)$ exist and are continous for $0 \leqslant t \leqslant T$
(b) $f(t)$ satisfies given boudary conditions

(ii) The Fourier series Eq. (7.1) is said to converge to $f(t)$ in the mean square sense in the interval $[0, T]$ provided only that $f(t)$ is any function for which

$$\int_{-\infty}^{\infty} |f(t)|^2 \mathrm{d}t = c < \infty$$

(iii) The Fourier series Eq. (7.1) is said to converge pointwise to $f(t)$ in the interval of length T if $f(t)$ is a continuous function in $[0, T]$ and $f'(t)$ is piecewise continous in $[0, T]$, that is, is continuous at all except at most finite number of points in $[0, T]$, where it can have a jump discontinuity. The uniform convergence is of course strongest whereas the other two forms assure a certain convergence under a weaker set of assumptions. The statements of existence and convergence of Fourier series will not be addressed further here as the purpose is more to furnish tools to analyse Functions known to have a frequency content, i.e. possess harmonic components. Usually, it is just stated that any function which satisfies the so-called Dirichlet's conditions can be expressed as Fourier series. These conditions are the following:

(a) the functions is periodic, that is, $f(t+T) = f(t)$
(b) the function is bounded, that is, $|f(t)| < \infty$ for all t
(c) the function has at most a finite number of discontinuities within each period
(d) the function has at most a finite number of maxima/minima within each period.

The Dirichlet conditions are sufficient for a periodic function to have a Fourier series representation but not necessary. There are functions which do not satisfy the above conditions but can nevertheless be expressed as Fourier series.

Consider a function $f(t)$ satisfying the Dirichlet conditions that is either an even or odd function of t. For an even function, the Fourier coefficients b_k vanish (see Eq. (7.2)), and for an odd function the a_k vanish. Therefore, the Fourier coefficients are obtained using the simpler form

$$\left.\begin{aligned} a_k &= \frac{4}{T}\int_0^{T/2} f(t)\cos k\omega t\,\mathrm{d}t \quad \text{and} \quad b_k = 0 \quad \text{for} \quad f(t) \text{ even} \\ b_k &= \frac{4}{T}\int_0^{T/2} f(t)\sin k\omega t\,\mathrm{d}t \quad \text{and} \quad a_k = 0 \quad \text{for} \quad f(t) \text{ odd} \end{aligned}\right\} \tag{7.3}$$

The Fourier series Eq. (7.1) can be rewritten using a complex mathematical form by applying the Euler relations

$$\mathrm{e}^{\pm ik\omega t} = \cos k\omega t \pm \mathrm{i} \sin k\omega t$$

or

$$f(t) = \sum_{k=-\infty}^{\infty} c_k \mathrm{e}^{ik\omega t} \tag{7.4}$$

where

$$c_k = \frac{1}{T}\int_{-T/2}^{T/2} f(t)\mathrm{e}^{-k\omega t}\mathrm{d}t \tag{7.5}$$

since

$$\int_{-T/2}^{T/2} e^{-m\omega t} e^{-n\omega t}\,dt = \begin{cases} 0 & m+n \neq 0 \\ T & m+n=0 \end{cases} \tag{7.6}$$

Using the Euler relations Eq. (7.3), and comparing with Eq. (7.2), the Fourier coefficients c_k, a_k and b_k are interrelated as follows:

$$c_k = \begin{cases} \frac{1}{2}(a_k - ib_k) & k>0 \\ \frac{1}{2}(a_{-k} + ib_{-k}) & k<0 \\ \frac{1}{2}a_0 & k=0 \end{cases} \tag{7.7}$$

The Fourier coefficients a_k and b_k give information about the strength of the corresponding harmonic component in the function or signal $f(t)$. For instance, if $x(t)$ is the deflection of a certain elastic structure with a representative spring constant k, the strain energy at any time t is $\frac{1}{2}kx^2(t)$ (also the energy carried by an electric signal where $x(t)$ is the current). The strain energy accumulated during one period T denoted E (in units of $2/k$) is therefore

$$E = \int_0^T x^2(t)\mathrm{d}t = \int_0^T \left(\sum_{-\infty}^{\infty} \sum_{-\infty}^{\infty} e^{im\omega t} e^{in\omega t} c_m c_n \right) \mathrm{d}t$$

upon introducing the Fourier series, Eq. (7.4). Interchanging the order of summation and integration,

$$E = \sum_{-\infty}^{\infty} \sum_{-\infty}^{\infty} c_m c_n \int_0^T e^{im\omega t} e^{in\omega t}\,\mathrm{d}t$$

and applying the orthogonality relations Eq. (7.6),

$$E = T\sum_{-\infty}^{\infty} c_m c_{-m} = T\frac{a_0^2}{4} + 2T\sum_{k=1}^{\infty} \frac{a_k^2 + b_k^2}{4}, \qquad b_0 = 0$$

or

$$E = T\left[\frac{a_0^2}{4} + \frac{a_1^2 + b_1^2}{2} + \cdots + \frac{a_n^2 + b_n^2}{2} + \cdots\right] \tag{7.8}$$

The energy is thus proportional to the sum of the squared amplitudes of the harmonic components in the signal. The terms within the brackets give the energy density per complete period T. A energy density spectrum can therefore be defined by considering the contribution by the kth frequency component, that is,

$$F(k\omega) = F(\omega_k) = \tfrac{1}{2}(a_k^2 + b_k^2) \tag{7.9}$$

The relation Eq. (7.8) is also a statement of the so-called Parseval's theorem, which relates the average energy of the signal over one period to the sum square of the Fourier coefficients, that is,

$$\frac{1}{T}\int_0^T x^2(t)\mathrm{d}t = \frac{a_0^2}{4} + \sum_{n=1}^{\infty} \frac{a_n^2 + b_n^2}{2} = \sum_{n=-\infty}^{\infty} |c_n|^2 \tag{7.10}$$

In Fig. 7.2 the energy density spectrum $F(\omega)$ is plotted for a series of imaginative discrete frequencies.

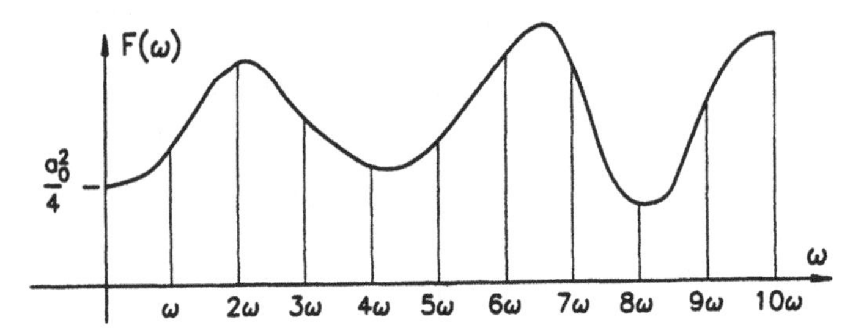

Fig. 7.2 *The energy density spectrum*

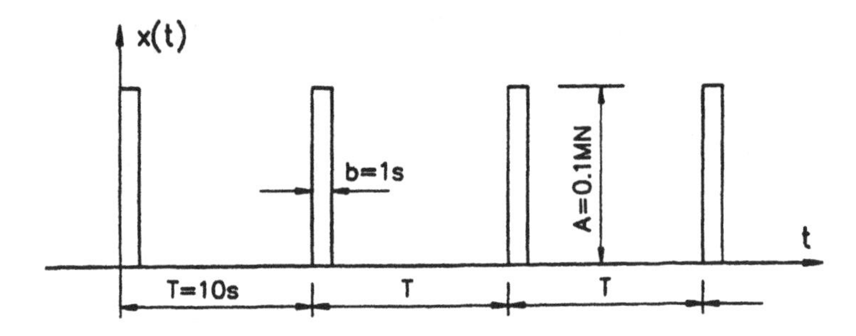

Fig. 7.3 *Pile driving forces*

Example 7.1 A pile hammer produces a rectangular impact load of 1 second duration at the rate of 6 per minute as shown in Fig. 7.3. Obtain the Fourier series of the pile driving load function and its energy density spectrum.

Solution The Fourier coefficients can be obtained directly using Eq. (7.2). Thus,

$$a_0 = \frac{2}{T}\int_0^T x(t)\mathrm{d}t = \frac{2A}{T}$$

$$a_n = \frac{2A}{T}\int_0^1 \cos n\omega t\,\mathrm{d}t = \frac{2A}{T}\frac{\sin n\omega}{n\omega}$$

$$B_n = \frac{2A}{T}\int_0^1 \sin n\omega t\,\mathrm{d}t = \frac{2A}{T}\frac{1}{n\omega}(1 - \cos n\omega)$$

A typical term in the series will be like

$$\frac{2A}{T}\frac{1}{n\omega}\sin n\omega t + \frac{2A}{T}\frac{1}{n\omega}\sin n\omega \cos n\omega t \; - \frac{2A}{T}\frac{1}{n\omega}\cos n\omega t \sin n\omega t$$

$$= \frac{A}{\pi n}\sin n\omega t + \frac{A}{\pi n}\sin(n\omega - n\omega t) = \frac{A}{\pi n}(\sin n\omega t - \sin(n\omega(t-1)))$$

$$= 2\frac{A}{\pi n}\cos\frac{n\omega}{2}\sin\left(n\omega t - \frac{n\omega}{2}\right)$$

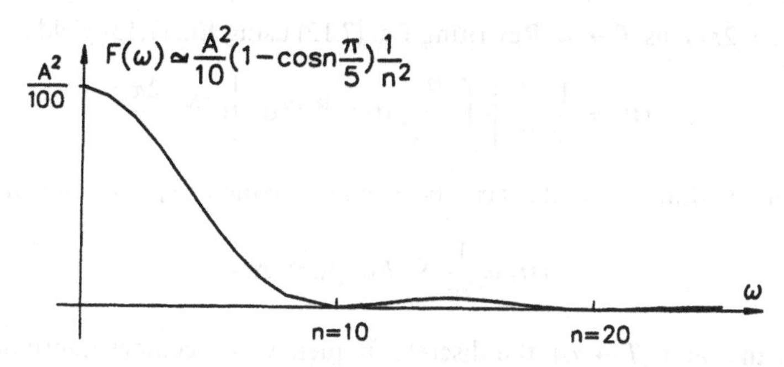

Fig. 7.4 *A discrete fourier spectrum*

or

$$x(t)=\frac{A}{T}+\frac{2A}{\pi}\sum_{n=1}^{\infty}\frac{\cos\frac{n\omega}{2}}{n}\sin\left(n\omega t-\frac{n\omega}{2}\right)$$

The largest contribution to the energy density is due to the first component. The contributions due to the higher-frequency terms rapidly tend to zero, which is reflected in Fig. 7.4.

7.1.2 Aperiodic Functions

Next, consider the type of functions that do not show any pattern of periodicity. Non-periodic functions, i.e. functions that do not repeat themselves systematically in any time interval, are called aperiodic. However, figuratively speaking, one can consider an aperiodic function to have an infinite period, that is, letting $T\rightarrow\infty$, the function is periodic with an infinitely long period which corresponds to the entire real axis $-\infty<t<\infty$. For functions $f(t)$ that satisfy the Dirichlet condition in an extended form, i.e. with an infinite period, and are absolutely integrable, that is,

$$\int_{-\infty}^{\infty}|f(t)|\mathrm{d}t=c<\infty \tag{7.11}$$

it is possible to define the Fourier transformation $F(\omega)$ of $f(t)$. The function $F(\omega)$ is a function of a real variable, a variable frequency ω, and correponds to the Fourier coefficients c_n for the discrete frequencies ω_n in the periodic case, Eq. (7.4). In fact, given an absolutely integrable aperiodic function $f(t)$, which satisfies the Dirichlet condition (most functions of limited duration do), formally write

$$f(t)=\sum_{-\infty}^{\infty}c_k\mathrm{e}^{\mathrm{i}k\,2\pi/t} \tag{7.12}$$

and

$$c_k=\frac{1}{T}\int_{T/2}^{T/2}f(t)\mathrm{e}^{-\mathrm{i}k\,2\pi/T}\mathrm{d}t \tag{7.13}$$

Now put $\Delta\omega = 2\pi/T$ as $T \to \infty$. Rewriting Eq. (7.12) using Eq. (7.13) yields

$$f(t) = \frac{1}{2\pi} \sum_{-\infty}^{\infty} \left[\int_{-T/2}^{T/2} f(t) e^{-ik\Delta\omega t} \, dt \right] e^{ik\Delta\omega t} \frac{2\pi}{T}$$

The expression within the brackets can be given the name $F(\omega_k)$, and putting $\omega_k = k\Delta\omega$,

$$f(t) = \frac{1}{2\pi} \sum_{-\infty}^{\infty} F(\omega_k) e^{i\omega_k t} \Delta\omega \tag{7.14}$$

By going to the limit ($T \to \infty$), the discrete frequency ω_k becomes continuous and the infinite sum converts to the integral

$$f(t) = \frac{1}{2\pi} \int_{-\infty}^{\infty} F(\omega) e^{-i\omega t} \, d\omega \tag{7.15}$$

in which the Fourier transformation function, called the Fourier transform of $f(t)$, is

$$F(\omega) = \int_{-\infty}^{\infty} f(t) e^{-i\omega t} \, dt \tag{7.16}$$

The two functions $f(t)$ and $F(\omega)$ are usually referred to as Fourier transform pairs, which is indicated as follows:

$$f(t) \leftrightarrow F(\omega)$$

The function $f(t)$ can be said to be the time domain representation of certain time-dependent information or signal. The Fourier transform $F(\omega)$ on the other hand is the frequency domain representation of the same information.

The Fourier transform $F(\omega)$ can be broken up into its real and imaginary parts. Obviously, since $f(t)$ is a real function of time,

$$F_R(\omega) = \int_{-\infty}^{\infty} f(t) \cos \omega t \, dt = F_R(-\omega)$$

and

$$F_I(\omega) = -\int_{-\infty}^{\infty} f(t) \sin \omega t \, dt = -F_I(-\omega) \tag{7.17}$$

that is, the real part $F_R(\omega)$ is an even function of ω whereas the imaginary part $F_I(\omega)$ is an odd function of ω.

A physical interpretation of the Fourier transform may be given as follows. By Eq. (7.14), $f(t)$ is dissembled into a series of harmonic components (N is large) such that

$$f(t) \simeq \frac{1}{\pi} \sum_{k=1}^{N} \sqrt{(F_R^2(\omega_k) + F_I^2(\omega_k))} \Delta\omega \cos(\omega_k t + \varphi_k) \tag{7.18}$$

in which

$$\varphi_k(\omega_k) = \arctan\left(\frac{F_I(\omega_k)}{F_R(\omega_k)} \right) \tag{7.19}$$

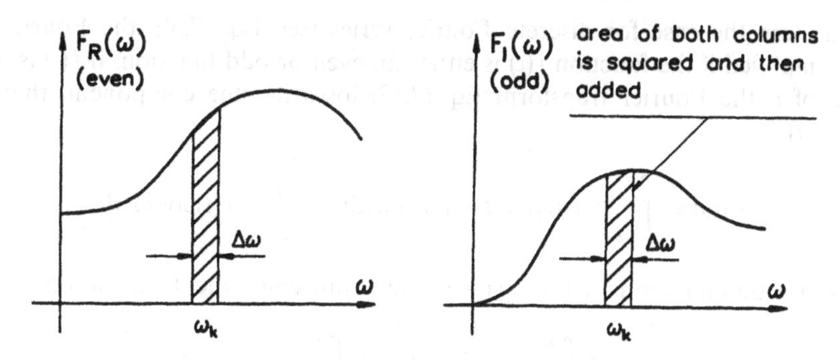

Fig. 7.5 *The continuous Fourier transform*

That is, the lumped area under the Fourier transform at selected frequencies ω_k with equal frequency intervals $\Delta\omega$ adds up to give the amplitude at each selected frequency, Fig. 7.5

Example 7.2 Find the Fourier transform of the function $\exp(-at^2)$, $-\infty < t < \infty, a > 0$.

Solution

$$F(\omega) = \int_{-\infty}^{\infty} e^{-at^2} e^{-i\omega t}\, dt = \int_{-\infty}^{\infty} e^{-[at^2 + i\omega t - \omega^2/4a + \omega^2/4a]}\, dt$$

$$= e^{-\omega^2/4a} \int_{-\infty}^{\infty} \exp\left[-\frac{\left(t+\dfrac{i\omega}{2a}\right)^2}{2\left(\sqrt{\dfrac{1}{2a}}\right)^2}\right] dt = e^{-\omega^2/4a}\sqrt{\frac{1}{2a}}\sqrt{2\pi} = \sqrt{\frac{\pi}{a}}\, e^{-\omega^2/4a}$$

The Fourier transform pairs are shown in Fig. 7.6. Except for the factor $1/\surd(2\pi)$, the time domain function corresponds to a normal or Gaussian density function with zero mean and variance $1/(2a)$. Its Fourier transform is also a Gaussian density function with zero mean, which was already established in the discussion of characteristic functions, Example 1.24.

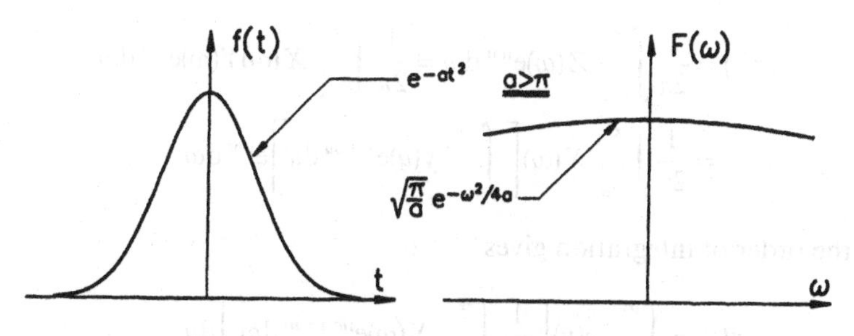

Fig. 7.6 *The normal density. Fourier transformation pairs*

Just as was the case for discrete Fourier series (see Eq. (7.3)), the Fourier transform is simplified if the function $f(t)$ is either an even or odd function. If $f(t)$ is an even function of t, the Fourier transform Eq. (7.17) loses its sine component, that is, for $f(t)=f(-t)$

$$F(\omega)=\int_{-\infty}^{\infty} f(t)(\cos\omega t - \mathrm{i}\sin\omega t)\mathrm{d}t = 2\int_{-\infty}^{\infty} f(t)\cos\omega t\,\mathrm{d}t$$

Similarly, for odd functions $f(t)=-f(-t)$, the cosine component will vanish or

$$F(\omega)=\int_{-\infty}^{\infty} f(t)\mathrm{e}^{-\mathrm{i}\omega t}\mathrm{d}t = 2\int_{0}^{\infty} f(t)\sin\omega t\,\mathrm{d}t$$

The frequency domain representation of an even respectively odd function of time is thus given by the cosine respectively sine transform of the time function, that is,

$$\left.\begin{aligned} C(\omega) &= \int_0^{\infty} f(t)\cos\omega t\,\mathrm{d}t \qquad f(t)\ \text{even} \\ S(\omega) &= \int_0^{\infty} f(t)\sin\omega t\,\mathrm{d}t \qquad f(t)\ \text{odd} \end{aligned}\right\} \tag{7.20}$$

For arbitrary $f(t)$, the cosine and sine transforms correspond to the real and imaginary parts of $F(\omega)$, Eq. (7.17).

7.1.3 Some Important Properties of the Fourier Transform

Given the three Fourier transform pairs

$$x(t)\leftrightarrow X(\omega), \qquad y(t)\leftrightarrow Y(\omega), \qquad z(t)\leftrightarrow Z(\omega)$$

and the frequency domain relation

$$Z(\omega)=X(\omega)Y(\omega) \tag{7.21}$$

a corresponding relation between the three functions $x(t)$, $y(t)$, $z(t)$ in the time domain is sought. Now, through Eq. (7.15)

$$\begin{aligned} z(t) &= \frac{1}{2\pi}\int_{-\infty}^{\infty} Z(\omega)\mathrm{e}^{\mathrm{i}\omega t}\,\mathrm{d}\omega = \frac{1}{2\pi}\int_{-\infty}^{\infty} X(\omega)Y(\omega)\mathrm{e}^{\mathrm{i}\omega t}\,\mathrm{d}\omega \\ &= \frac{1}{2\pi}\int_{-\infty}^{\infty} X(\omega)\left[\int_{-\infty}^{\infty} y(u)\mathrm{e}^{-\mathrm{i}\omega u}\,\mathrm{d}u\right]\mathrm{e}^{\mathrm{i}\omega t}\,\mathrm{d}\omega \end{aligned}$$

Reversing the order of integration gives

$$z(t)=\int_{-\infty}^{\infty} y(u)\left[\frac{1}{2\pi}\int_{-\infty}^{\infty} X(\omega)\mathrm{e}^{\mathrm{i}\omega(t-u)}\,\mathrm{d}\omega\right]\mathrm{d}u$$

or by Eq. (7.14)

$$z(t) = \int_{-\infty}^{\infty} y(u)x(t-u)\,du = \int_{-\infty}^{\infty} x(u)y(t-u)\,du \tag{7.22}$$

The relations Eqs. (7.21) and (7.22) are usually referred to as the convolution theorem and Eq. (7.22) is often presented in the shorthand form $z(t) = x(t) * y(t)$.

The convolution theorem is 'symmetric', that is, given the time domain relation

$$z(t) = x(t) * y(t) \tag{7.23}$$

a relationship corresponding to Eq. (7.20) in the frequency domain can be found. In fact, introducing the inverse transform Eq. (7.15) into Eq. (7.21),

$$Z(\omega) = \frac{1}{2\pi}\int_{-\infty}^{\infty} x(t)\left[\int_{-\infty}^{\infty} Y(\Omega)\mathrm{e}^{\mathrm{i}\Omega t}\,\mathrm{d}\Omega\right]\mathrm{e}^{-\mathrm{i}\omega t}\mathrm{d}t = \frac{1}{2\pi}\int_{-\infty}^{\infty} Y(\Omega)\left[\int_{-\infty}^{\infty} X(t)\mathrm{e}^{-\mathrm{i}(\omega-\Omega)t}\mathrm{d}t\right]\mathrm{d}\Omega$$

or

$$Z(\omega) = \frac{1}{2\pi}\int_{-\infty}^{\infty} Y(\Omega)X(\omega-\Omega)\mathrm{d}\Omega = \frac{1}{2\pi}\int_{-\infty}^{\infty} X(\Omega)Y(\omega-\Omega)\mathrm{d}\Omega \tag{7.24}$$

where Ω is generally a complex variable. Thus, $Z(\omega) = 1/2\pi\, X(\omega) * Y(\omega)$ is the convolution in the frequency domain, also called the complex convolution theorem.

Next, let $Y(\omega) = X^*(\omega)$ be the complex conjugate of the Fourier transform for $x(t)$. Then by Eq. (7.16),

$$X^*(\omega) = \int_{-\infty}^{\infty} x(t)\mathrm{e}^{\mathrm{i}\omega t}\,\mathrm{d}t = \int_{-\infty}^{\infty} x(-t)\mathrm{e}^{-\mathrm{i}\omega t}\,\mathrm{d}t$$

that is, $X^*(\omega)$ and $x(-t)$ are Fourier transform pairs. By Eqs. (7.15) and (7.21),

$$z(0) = \frac{1}{2\pi}\int_{-\infty}^{\infty} X(\omega)X^*(\omega)\mathrm{d}\omega = \frac{1}{2\pi}\int_{-\infty}^{\infty} |X(\omega)|^2\,\mathrm{d}\omega$$

Also by Eq. (7.22),

$$z(0) = \int_{-\infty}^{\infty} x(u)x(-(-u))\mathrm{d}u = \int_{-\infty}^{\infty} x^2(u)\mathrm{d}u$$

Thus Parseval's theorem for the continous transform has been established (cf. Eq. (7.9)), that is,

$$\int_{-\infty}^{\infty} x^2(u)\mathrm{d}u = \frac{1}{2\pi}\int_{-\infty}^{\infty} |X(\omega)|^2\,\mathrm{d}\omega \tag{7.25}$$

If the derivatives of the time domain function $f(t)$ exist, satisfy the extended Dirichlet condition and are absolute integrable, their Fourier transforms are easily obtainable. In fact, by differentiating the inverse transform, Eq. (7.15),

$$\frac{\mathrm{d}f(t)}{\mathrm{d}t}=\dot{f}(t)=\frac{1}{2\pi}\int_{-\infty}^{\infty}F(\omega)(\mathrm{i}\omega)\mathrm{e}^{\mathrm{i}\omega t}\,\mathrm{d}\omega$$

$$\frac{\mathrm{d}^2 f(t)}{\mathrm{d}t^2}=\ddot{f}(t)=\frac{1}{2\pi}\int_{-\infty}^{\infty}F(\omega)(-\omega^2)\mathrm{e}^{\mathrm{i}\omega t}\,\mathrm{d}\omega$$

$$\vdots$$

$$\frac{\mathrm{d}^n f(t)}{\mathrm{d}t^n}=f^{(n)}(t)=\frac{1}{2\pi}\int_{-\infty}^{\infty}F(\omega)(\mathrm{i}\omega)^n\mathrm{e}^{\mathrm{i}\omega t}\,\mathrm{d}\omega$$

Therefore the Fourier transform pairs for the derivatives are the following:

$$\begin{aligned} f(t)&\leftrightarrow F(\omega)\\ \dot{f}(t)&\leftrightarrow \mathrm{i}\omega F(\omega)\\ \ddot{f}(t)&\leftrightarrow -\omega^2 F(\omega)\\ f^{(n)}(t)&\leftrightarrow(\mathrm{i}\omega)^n F(\omega)\end{aligned} \tag{7.26}$$

Example 7.3 In Fig. 7.7, a one-storey elastic frame under the horizontal load $f(t)$ is shown. During the horizontal motion $x(t)$ due to the force $f(t)$, the frame is considered to be massless and purely elastic with a viscous damping coefficient c. Find the time history of the response when

$$f(t)=\begin{cases}0 & t<-T\\ f_0 & -T\leqslant t\leqslant -T\\ 0 & t>T\end{cases}$$

Solution For a static load P, the horizontal deflection δ_P is given by

$$\delta_P=\frac{7}{108}\frac{h^2}{EI}P \qquad \text{or} \qquad P=\frac{108EI}{7h^2}\delta_P=k\delta_P$$

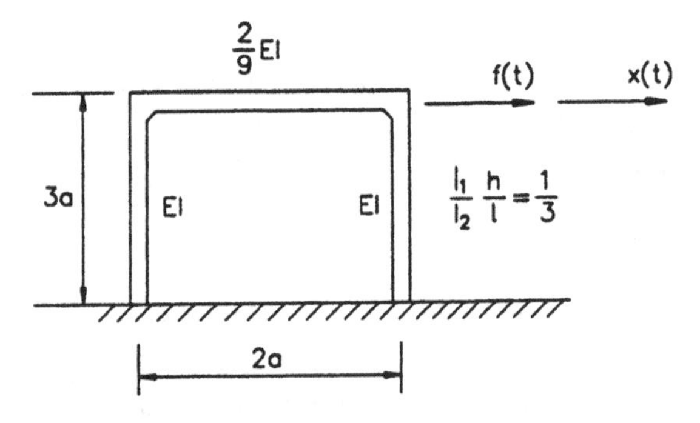

Fig. 7.7 *A massless, elastic frame under dynamic load*

whereby the spring constant k is defined. The equation of motion is then given by

$$(m \sim 0)\ddot{x}(t) + c\dot{x}(t) + kx(t) = f(t)$$

or

$$c\dot{x}(t) + kx(t) = f(t)$$

A frequency domain solution is first sought. By Fourier transforming the whole equation, that is, multiplying both sides by $e^{-i\omega t}$ and integrating from $-\infty$ to $+\infty$, making use of the relations Eq. (7.26),

$$c(i\omega)X(\omega) + kX(\omega) = F(\omega)$$

or

$$X(\omega) = \frac{F(\omega)}{k + i\omega c} = H(\omega)F(\omega)$$

Now

$$F(\omega) = \int_{-\infty}^{\infty} f(t)e^{-i\omega t}\,dt = \int_{-T} f_0 e^{-i\omega t}\,dt = \frac{f_0}{i\omega}(e^{i\omega T} - e^{-i\omega T})$$

and then by the inverse transform Eqs. (7.15), the time history of the response is given by

$$x(t) = \frac{1}{2\pi}\int_{-\infty}^{\infty} X(\omega)e^{i\omega t}\,d\omega = \frac{f_0}{2\pi c}\left[\int_{-\infty}^{\infty} \frac{e^{i\omega(t-T)}}{\omega\left(\omega - i\frac{k}{c}\right)}d\omega - \int_{-\infty}^{\infty} \frac{e^{i\omega(t+T)}}{\omega\left(\omega - i\frac{k}{c}\right)}d\omega\right]$$

$$= \frac{f_0}{2\pi c}[A - B]$$

To recover the time domain representation from the frequency domain representation, the above integral has to be unlocked. For this purpose, elaborate integral tables, [72], may become handy or the more tedious path of Cauchy's integral theorem for analytic functions can be selected (cf. Example 3.5). Choosing to go over the wall where it is highest, study the contour integral $\oint f(z)dz$, where the complex analytic function $f(z)$ is being integrated along a closed contour C. Cauchy's integral theorem then states that the value of the contour integral is equal to the sum of residues at the poles of $f(z)$ within the closed contour C times $2\pi i$. In other words,

$$\oint \frac{e^{iz(t \pm T)}dz}{z\left(z - i\frac{k}{c}\right)} = +2\pi i \sum (\text{Residues at poles within C})$$

The integration path is to be taken anticlockwise. Otherwise the residual sum has to be taken with a negative sign.

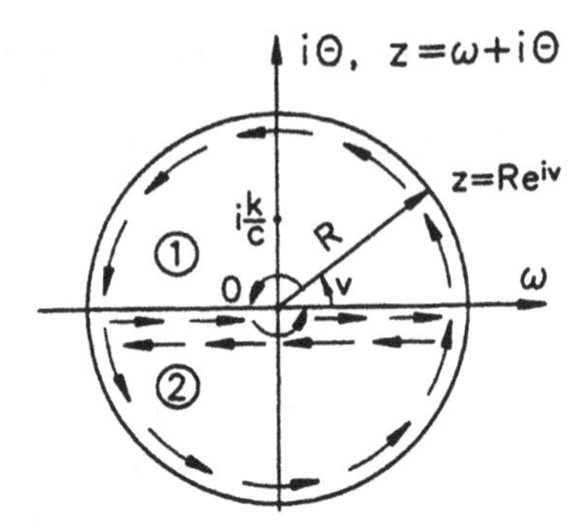

Fig. 7.8 *Contour integration paths*

In Fig. 7.8, the *selected* integration paths are shown. Within the circle of radius R, there are two poles, $z = 0$ and $z = ik/c$. The residues at these poles are

$$\operatorname{Res}(z=0) = \lim_{z\to 0}(z-0)f(z) = \mathrm{i}\frac{c}{k}$$

$$\operatorname{Res}\left(z=\mathrm{i}\frac{c}{k}\right) = \lim_{z\to \mathrm{i}c/k}\left(z-\mathrm{i}\frac{c}{k}\right)f(z) = -\mathrm{i}\frac{c}{k}\mathrm{e}^{k/c(t\pm T)}$$

Now along the ω-axis, the contour integral is equivalent to the sought after integral, that is,

$$\int_{-R}^{R}\frac{\mathrm{e}^{\mathrm{i}\omega(t\pm T)}}{\omega\left(\omega-\mathrm{i}\dfrac{k}{c}\right)}\mathrm{d}\omega \quad \left(\lim_{R\to\infty}\int_{-R}^{R} = A, B\right)$$

Along the great circle contour, path 1:

(I) $$\int_{v=0}^{\pi}\frac{\exp[\mathrm{i}R(\cos v+\mathrm{i}\sin v)(t\pm T)]R\mathrm{e}^{\mathrm{i}v}\,\mathrm{d}v}{(R^2\mathrm{e}^{2\mathrm{i}v}-Rk/c\mathrm{i}\mathrm{e}^{\mathrm{i}v})}$$

and path 2:

(II) $$\int_{\pi}^{2\pi}\frac{\mathrm{e}^{\mathrm{i}R\cos v(t\pm T)}\mathrm{e}^{-R\sin v(t\pm T)}R\mathrm{e}^{\mathrm{i}v}\,\mathrm{d}v}{(R^2\mathrm{e}^{2\mathrm{i}v}-R\,k/c\mathrm{i}\mathrm{e}^{\mathrm{i}v})}$$

Along the small circle contour (about the origin, $z = \varepsilon e^{iv}$), for path 1 (below the ω-axis) and path 2(above the axis) replace R by ε in the two above integrals, now called (III) and (IV). Path 1 includes both poles, whereas path 2 does not include any poles. Therefore,

$$\text{Path 1, } \oint f(z)\,\mathrm{d}z = 2\pi\mathrm{i}\left(\mathrm{i}\frac{c}{k}-\mathrm{i}\frac{c}{k}\mathrm{e}^{-k/c(t\pm T)}\right)$$

$$\text{Path 2, } \oint f(z)\,\mathrm{d}z = 2\pi\mathrm{i}\times 0 = 0$$

Then let $R \to 0$ and $\varepsilon \to 0$ and see what happens. There are four different possibilities:

$$(1)\ t > T, \quad (2)\ t < T, \quad (3)\ t > -T, \quad (4)\ t < -T$$

The integrand in the great circle integrals (I) and (II) is evaluated as:

$$\lim_{R\to\infty} |\text{integrand}| = \lim_{R\to\infty} \frac{e^{-R\sin v(t\pm T)}}{|Re^{iv} - ik/c|} = \begin{cases} 0 \\ \infty \end{cases} \quad \text{(value depends on } t \text{ and the path)}$$

The integrand in the small circle integrals (III) and (IV) is evaluated as

$$\lim_{\varepsilon\to 0} (\text{integrand}) = \lim_{\varepsilon\to 0} \frac{e^{-\varepsilon \sin v(t\pm T)}}{\varepsilon e^{iv} - i\,k/c} = i\frac{c}{k} \quad \text{for all } t$$

Thus the two small circle integrals III and IV both give a contribution $i\pi c/k$ to the two integrals A and B. Since the final value of the time history $x(t)$ is based on the difference $A - B$, the small circle contribution cancels out. Now to evaluate the two integrals A and B for the time domain solution, the integration paths are chosen such that the integrals (II) and (III) vanish in all four cases. That is, for

$$A: \quad \begin{array}{l} \text{For } t > T, \text{ use path 1, } \sin v > 0 \\ \text{For } t < T, \text{ use path 2, } \sin v < 0 \end{array}$$

$$B: \quad \begin{array}{l} \text{For } t > -T, \text{ use path 1, } \sin v > 0 \\ \text{For } t < -T, \text{ use path 2, } \sin v < 0 \end{array}$$

Therefore,

$$A + i\pi\frac{c}{k} = \begin{cases} -2\pi\left(\dfrac{c}{k} - \dfrac{c}{k}e^{-k/c(t-T)}\right) & t > T \\ 0 & t < T \end{cases}$$

$$B + i\pi\frac{c}{k} = \begin{cases} -2\pi\left(\dfrac{c}{k} - \dfrac{c}{k}e^{-k/c(t+T)}\right) & t > -T \\ 0 & t < -T \end{cases}$$

and

$$x(t) = \frac{f_0}{2\pi c}[A - B] = \begin{cases} 0 & t < -T \\ (1 - e^{-k/c(t+T)}) & -T \leqslant t \leqslant T \\ (e^{-k/c(t-T)} - e^{-k/c(t+T)}) = \sin h\left(\dfrac{k}{c}T\right)e^{-k/c\,t} & t > T \end{cases}$$

The response $x(t)$ is shown graphically in Fig. 7.9.

Example 7.4 The delta function ($\delta(t) = \infty$ for $t = 0$, $\delta(t) = 0$ for $t \neq 0$) can be defined as a Gaussian probability density with a zero mean value and a standard deviation σ which tends to zero. That is,

$$f(t) = \frac{1}{\sigma\sqrt{2\pi}} \exp\left(-\frac{t^2}{2\sigma^2}\right), \quad \lim_{\sigma\to 0} f(t) = \delta(t), \qquad -\infty < t < \infty$$

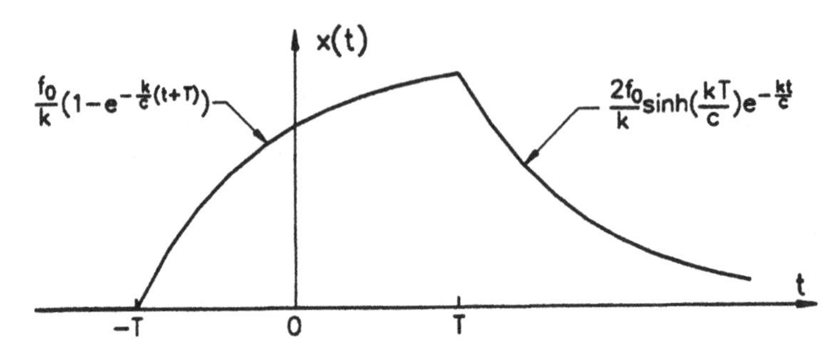

Fig. 7.9 *Time history of the response* $x(t)$

Going to the limit, the delta function has the following properties. Take any reasonably well behaved function $g(t)$ that is smooth in the vicinity of $t = t_0$. Then,

$$\left.\begin{aligned} \lim_{\sigma\to 0} f(t) = \delta(t) &= \begin{cases} \infty & t = 0 \\ 0 & t \neq 0 \end{cases} \\ \lim_{\sigma\to 0}\int_{-\infty}^{\infty} f(t) &= \int_{-\infty}^{\infty} \delta(t)\,\mathrm{d}t = 1 \\ \lim_{\sigma\to 0}\int_{-\infty}^{\infty} f(t-t_0)\,g(t)\,\mathrm{d}t &= \int_{-\infty}^{\infty} g(t)\delta(t-t_0)\,\mathrm{d}t = g(t_0) \end{aligned}\right\} \tag{7.27}$$

where the last of Eqs. (7.27), corresponding to a non-zero mean value t_0, portrays the shifting property of the delta function.

Does the delta function have a Fourier transform? Obviously the function $f(t-t_0)$, $(\mu = t_0)$, does, and it is given by the characteristic function $\varphi(\omega) = \exp[it_0\omega - \sigma^2\omega^2/2]$, (see Example 1.24). Therefore, at least formally, $\delta(t-t_0) = \lim_{\sigma\to 0} f(t-t_0))$

$$\Delta(\omega) = \lim_{\sigma\to 0} \varphi(-\omega) = \mathrm{e}^{-it_0\omega}$$

so

$$\delta(t-t_0) \leftrightarrow \mathrm{e}^{-it_0\omega} \tag{7.28}$$

The time shifting property of the delta function can be used more directly since

$$\Delta(\omega) = \int_{-\infty}^{\infty} \delta(t-t_0)\mathrm{e}^{-it\omega}\,\mathrm{d}t = \mathrm{e}^{-it_0\omega}$$

Next, let $F(\omega) = \delta(\omega - \omega_0)$. What is the time domain representation? By the inverse transform,

$$f(t) = \frac{1}{2\pi}\int_{-\infty}^{\infty} \delta(\omega-\omega_0)\mathrm{e}^{it\omega}\,\mathrm{d}\omega = \frac{1}{2\pi}\mathrm{e}^{it\omega_0}$$

so

$$\mathrm{e}^{it\omega_0} \leftrightarrow 2\pi\delta(\omega-\omega_0) \tag{7.29}$$

Obviously

$$\delta(t) \leftrightarrow 1 \qquad \text{and} \qquad 1 \leftrightarrow 2\pi\delta(\omega) \tag{7.30}$$

The above relations can be given the following physical explanation. A time signal $x(t)$ with a constant value component (a d.c. component) will give rise to an impulse in the Fourier transform at zero frequency. A periodic component in the signal of the type $\exp(\pm i\omega_0 t)$ with a period $T_0 = 2\pi/\omega_0$ will on the other hand give rise to an impulse in the Fourier transform at the controversial frequencies $\omega = \pm\omega_0$. From Eq. (7.30), it can be seen that the delta function can be interpreted as the sum of sinusoids of all frequencies. These are in-phase only at time $t = 0$ or at time $t = t_0$ in the case of Eq. (7.28), giving a very large amplitude, and are out-of-phase at all other frequencies giving zero amplitudes.

Mathematicially speaking, periodic and d.c. components do not belong to the class of absolutely integrable functions for which the Fourier transform is defined. However,by introduction of the delta function such difficulties have been overcome, and discrete Fourier series can now also be handled as continuous Fourier transforms. The analogy with the treatment of discrete and continuous random variables in Chapter 1 is also evident.

Example 7.6 Consider the signal or sample function

$$f(t) = a_0 + \sum_{j=1}^{n} a_j \cos(\omega_j t + \theta_j) + g(t)$$

where a_0 is a constant (a d.c. component), and to a smooth aperiodic function $g(t)$ is added a sum of periodic components. The Fourier transform $F(\omega)$ is sought.

Since $\cos(\omega_j t + \theta_j) = \frac{1}{2}(\exp(+i(\omega_j t + \theta_j))) + (\exp(-(\omega_j t + \theta_j)))$, the Fourier transform consists of a spike at zero frequency and symmetrical spikes at the controversial frequencies $\pm\omega_j$ added to the Fourier transform of the aperiodic function $g(t)$ called $G(\omega)$. That is,

$$F(\omega) = 2\pi a_0 \delta(\omega) + \pi \sum_{j=n}^{n} a_j e^{\theta_j} \delta(\omega - \omega_j) + G(\omega)$$

where $a_j = a_{-j}, \theta_{-j} = -\theta_j$ and $\omega_{-j} = -\omega_j$. The Fourier transform is plotted in Fig. 7.10 to illustrate the behaviour of the separate components.

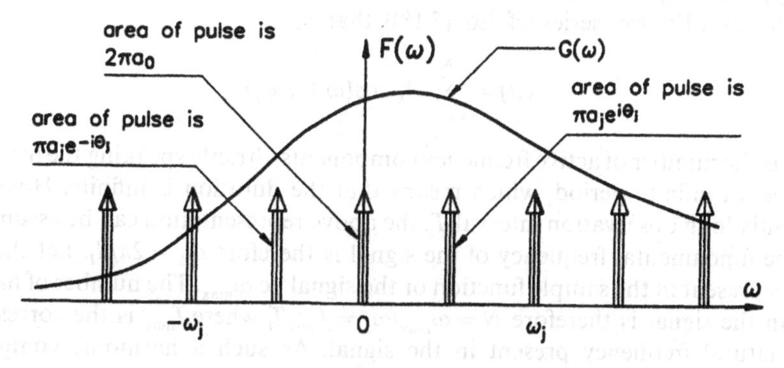

Fig. 7.10 *Composition of the Fourier transform*

Finally, if the function $f(t)$ were periodic with a period T, then by Eqs. (7.4) and (7.29), the continuous Fourier transform is

$$F(\omega) = 2\pi \sum_{k=-\infty}^{n} c_k \delta\left(\omega - k\frac{2\pi}{T}\right)$$

where the coefficients c_k are given by Eq. (7.5).

7.2 DIGITAL SIGNAL ANALYSIS

Digital computational routines are more convenient and much faster than analogue computation. Whereas analogue techniques are based on high technological instrumentation with some fancy circuit gadgetry, digital analyses can be handled by modern desktop computers. After the introduction of the fast Fourier transform (FFT) algorithm by Cooley and Tukey in 1965, [39], data processing and signal analysis using digital techniques has turned into a vast highly specialized field of science. In this section, an overview of digital signal analysis using the discrete Fourier transform will be given without going into much detail. Furthermore, the z-transform and other transformation techniques will be discussed followed by a brief introduction to digital filters, windowing and smoothing of signals. Finally, ARMA models are briefly discussed.

Data processing and signal analysis has taken a giant leap forward during the last few years with ever more powerful computers and processing units available. As in any rapidly growing scientific field, there is a huge amount of literature available describing the various aspects of data processing. Of earlier works, a rather clear and encompassing treatment is for example found in the books by Otnes and Enochsson, [152], and Papoulis, [164]. Stein, [216], offers a fairly extensive treatment of digitalization techniques applied to earthquake signals, and Newland, [152], presents a good overview of digitalization techniques and spectral analysis of random signals. Of more modern texts, Loy, [134], and Cunningham, [49], offer an extensive overview of digital filters. Gardner, [69], and Proakis and Manolakis, [175], give a general presentation of signal analysis and data processing techniques.

7.2.1 Digitized Signals and the FFT Algorithm

Take a typical sample function $x(t)$ of the stochastic process $\{X(t), t \in T\}$, which can be represented as a Fourier series (cf. Eq. (7.18)), that is,

$$x(t) = \sum_{k=1}^{N} A_k \cos(\omega_k t + \varphi_k) \tag{7.31}$$

where N is the number of active frequency components. Strictly speaking the function $x(t)$ must have an infinite period, which means that the duration is infinite. However, for a sufficiently long observation interval T_l, the above representation can be assumed to be valid. The fundamental frequency of the signal is therefore $\omega_1 = 2\pi/T_l$. Let the highest frequency present in the sample function or the signal be $\omega_{\max}$. The number of harmonics present in the signal is therefore $N = \omega_{\max}/\omega_1 = f_{\max} T_l$ where $f_{\max}$ is the corresponding highest natural frequency present in the signal. As such a harmonic component is determined by the two Fourier coefficients Eq. (7.17), the Fourier series representation,

Eq. (7.18), requires that in all $2f_{max}T_l$ coefficients be determined. These can be computed from the sample function record at $2f_{max}T_l$ different time instances or discrete sample times, which are usually chosen by dividing the record length or time duration into equally spaced intervals. Therefore, the function is fed through an analogue-to-digital converter or digitized on a optical digitizing table by sampling the $x(t)$ at regularly spaced time intervals. If the sampling interval is Δ then the discrete values of $x(t)$ at times $t = r\Delta$ are called x_r. Thereby, instead of a continuous time series Eq. (7.18), the following discrete time sequence has been produced,

$$\{x_r\}, \qquad r = 0, 1, 2, \ldots, (N-1), \quad t = r\Delta \tag{7.32}$$

In order to maintain the character of the original signal, that is, a true representation of the frequency content, the sampling frequency $f_s = 1/\Delta$ has to be selected such that dominant frequency components in the signal are not lost. From the above argument, the sampling frequency should be larger than $2f_{max}T_l/T_l\,(\Delta = T_l/(2f_{max}T_l))$ or

$$f_s > 2f_{max} \tag{7.33}$$

Thus, if the maximum frequency in a signal is f_{max}, the sampling frequency for a discrete times sequence representation of the signal is double the maximum frequency. For instance, human speech signals have a maximum audible frequency of 3000 Hz. The sampling frequency for analysing speech signals must therefore at least be f_s = 6000 Hz or the sampling interval Δ be less than 0.17×10^3 seconds. The sampling frequency $f_s = 2f_{max}$ is called the Nyquist frequency. In this manner, any frequency component $f_i < f_{max}$ of the analogue signal is mapped into a discrete time harmonic with a frequency relative to the sampling frequency $n_i = f_i/f_s$, which falls within the range

$$-\tfrac{1}{2} \leqslant n_i \leqslant \tfrac{1}{2} \qquad \text{or} \qquad -\pi \leqslant \omega_i \leqslant \pi \tag{7.34}$$

Obviously, $|\omega| = \pi$ is the highest unique frequency in a discrete signal. If the sampling frequency is selected lower than two times the maximum frequency, there will be frequencies ω_i above π. Such components are replicas or aliases of components within the range Eq. (7.34). Therefore, in order to avoid the problem of aliasing or frequency folding of the discrete signal, the sampling frequency has to be selected higher than the Nyquist frequency in the analogue signal. Thus, all frequency components in the analogue signal are represented in the discrete signal without ambiguity, wherefore the analogue signal can be reproduced from the digital signal without distortion (digital to analogue converter) using suitable interpolation. A suitable interpolation function is defined by the so-called *Shannon sampling theorem*, which can be stated as follows:

If the highest frequency in an analogue signal $x(t)$ is $f_{max} = B$ (the frequency bandwidth), and a discrete time sequence x_r is obtained by sampling with a Nyquist frequency $f_s = 2B = 1/\Delta$, the analogue signal can be exactly reproduced from its sample values by the interpolation formula

$$x(t) = \sum_{k=-\infty}^{\infty} x_k \cdot \frac{\sin 2\pi B(t - k\Delta)}{2\pi B(t - k\Delta)} \tag{7.35}$$

Because of the complexity of the above reconstruction formula and the infinite number of samples needed, the Shannon sampling theorem is of more theoretical than practical

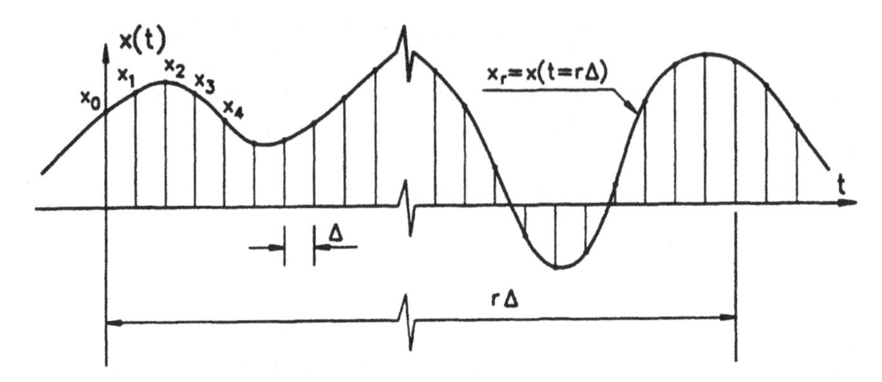

Fig. 7.11 *Sampling of x(t) at regular time intervals*

interest. There are other more practical and amenable methods available, which are for example listed by Proakis and Manolakis, [175].

Of immediate interest is to establish the Fourier transform relations for the discrete time sequences, corresponding to those for the analogue function (cf. Eqs. (7.15) and (7.16). In Fig. 7.11, an example of a discretized sample function $x(t)$ is shown. The function is observed in a time interval of length T, which is subdivided into N equal finite segments, each of length $\Delta = T/N$. Thus, the sampling frequency is $f_s = 1/\Delta$. Without loss of generality the interval length T can be put equal to one (one hour, one day etc.). Now if $x(t)$ and $X(\omega)$ are Fourier transform pairs, the following relations are valid:

$$x(t) = \frac{1}{2\pi}\int_{-\infty}^{\infty} X(\omega)e^{i\omega t}\,d\omega = \int_{-\infty}^{\infty} X(f)e^{i2\pi f t}\,df$$

$$X(\omega) = \int_{-\infty}^{\infty} x(t)e^{-i\omega t}\,dt, \quad \omega = 2\pi f$$

The discretization of the above integrals is straightforward. Let $\omega_k = 2\pi k/T = k\,\delta\omega$, $\delta\omega = 2\pi/T$, then the discrete Fourier transform (DFT), $X(\omega_k) = X_k$, is defined as follows:

$$X_k = \sum_{r=0}^{N-1} x_r e^{i\,2\pi k/T\, r\Delta}\,\Delta = \frac{1}{N}\sum_{r=0}^{N-1} X_k e^{i2\pi kr/N} \tag{7.36}$$

and the corresponding discrete time domain representation or the inverse transform (IDFT) is defined as

$$x_r = \frac{1}{2\pi}\sum_{k=0}^{N-1} X_k e^{-i\,2\pi k/T\, r\Delta}\frac{2\pi}{T} = \sum_{r=0}^{N-1} X_k e^{i2\pi kr/N} \tag{7.37}$$

In both the above relations T has been put equal to one. In many texts, the factor $1/N$ follows the IDFT according to the definition of the Fourier transform. Now for convenience, introduce the weighting kernel $\gamma = \exp[-i2\pi/N]$. Then, Eqs. (7.36) and (7.37) can be written as

$$X_k = \frac{1}{N}\sum_{r=0}^{N-1} x_r\gamma^{kr} \quad \text{and} \quad x_r \sum_{k=0}^{N-1} X_k\gamma^{-kr} \tag{7.38}$$

The computation of the discrete Fourier coefficients X_k, Eqs (7.38), can be organized in a matrix form according to the following computational scheme Eq. (7.39). The fast Fourier transform (FFT) is really a special computational technique for solving the matrix equation,

$$\begin{Bmatrix} X_0 \\ X_1 \\ X_2 \\ \vdots \\ X_{N-1} \end{Bmatrix} = \frac{1}{N} \begin{bmatrix} 1 & 1 & 1 & \cdots & 1 \\ 1 & \gamma & \gamma^2 & \cdots & \gamma^{N-1} \\ 1 & \gamma^2 & \gamma^4 & \cdots & \gamma^{2(N-1)} \\ \vdots & \vdots & \vdots & & \vdots \\ 1 & \gamma^{(N-1)} & \gamma^{2(N-1)} & \cdots & \gamma^{(N-1)(N-1)} \end{bmatrix} \begin{Bmatrix} x_0 \\ x_1 \\ x_2 \\ \vdots \\ x_{N-1} \end{Bmatrix} \tag{7.39}$$

To solve the above equation, N^2 complex multiplications and additions are needed to obtain all X_k. If $N = 2^P$, that is, the number of sampling points is selected as a power of 2 (p is a positive integer), which is a standard requirement for the fast Fourier transform, it means that in all 2^{2P} operations are needed. The FFT algorithm works by partitioning the full sequence $\{x_r\}$ into a number of shorter sequences. The discrete Fourier transform Eq. (7.38) is then obtained for each sequence and the FFT combines these together in a ingenious way to produce the full discrete Fourier transform of the original sequence. The number of mathematical operations is thus greatly reduced. Actually, the FFT only needs Np operations to perform the above calculation compared with direct method which requires N^2 operations. For instance, if $N = 1024, p = 10, N^2 = (2^{10}) \simeq 10^6$ whereas $Np = 2^{10} \times 10 = 10^4$ or the FFT is 100 times faster.

Example 7.6 Given the time history

$$x(t) = a\cos(3\pi t/T) + b\sin(6\pi t/T) \quad 0 < t < T$$

Find the discrete Fourier transform (DFT) by dividing the duration interval into four equal parts.

Solution

$$N = 4, \quad t_0 = 0, \quad t_1 = \frac{T}{4}, \quad t_2 = \frac{T}{2}, \quad t_3 = \frac{3T}{4}$$

$$\gamma = \exp[-2\pi/N] = \exp[-\pi/2] = (-\mathrm{i})$$

$$\gamma^0 = 1, \quad \gamma^1 = -\mathrm{i}, \quad \gamma^2 = -1, \quad \gamma^3 = \mathrm{i}, \quad \gamma^4 = 1, \quad \gamma^5 = -\mathrm{i}$$

$$x_0 = 0, \quad x_1 = a\cos(3\pi/4) + b\sin(3\pi/2) = -a/\sqrt{2} - b,$$

$$x_2 = 0, \quad x_3 = a\cos(9\pi/4) + b\sin(9\pi/2) = a/\sqrt{2} + b$$

$$\begin{Bmatrix} X_0 \\ X_1 \\ X_2 \\ X_3 \end{Bmatrix} = \frac{1}{4} \begin{bmatrix} 1 & 1 & 1 & 1 \\ 1 & -\mathrm{i} & -1 & \mathrm{i} \\ 1 & -1 & 1 & -1 \\ 1 & \mathrm{i} & -1 & -\mathrm{i} \end{bmatrix} \cdot \begin{Bmatrix} 0 \\ -a\dfrac{\sqrt{2}}{2} - b \\ 0 \\ a\dfrac{\sqrt{2}}{2} + b \end{Bmatrix} = \frac{1}{4} \begin{Bmatrix} 0 \\ \mathrm{i}(a\sqrt{2} + 2b) \\ 0 \\ -\mathrm{i}(a\sqrt{2} + 2b) \end{Bmatrix}$$

The inverse transform (IDFT) is also easily obtained, ($\gamma^{-kr} = 1/\gamma^{kr}$),

$$\begin{Bmatrix} x_0 \\ x_1 \\ x_2 \\ x_3 \end{Bmatrix} = \frac{1}{4}\begin{pmatrix} 1 & 1 & 1 & 1 \\ 1 & \mathrm{i} & -1 & -\mathrm{i} \\ 1 & -1 & 1 & -1 \\ 1 & -\mathrm{i} & -1 & \mathrm{i} \end{pmatrix}\begin{Bmatrix} 0 \\ \mathrm{i}(a\sqrt{2}+2b) \\ 0 \\ -\mathrm{i}(a\sqrt{2}+2b) \end{Bmatrix} = \begin{Bmatrix} 0 \\ -a\dfrac{\sqrt{2}}{2}-b \\ 0 \\ a\dfrac{\sqrt{2}}{2}+b \end{Bmatrix}$$

and the original values of the time function are reproduced.

7.2.2 *Dirac's Comb and Aliasing*

The effect of sampling a continuous function $x(t)$ at equally spaced intervals Δ can be represented by a sum of delayed delta functions or spikes as seen in Fig. 7.10. This interpretation is referred to as multiplying or convolving the function $x(t)$ with a special function called Dirac's comb, also called the Shah function, [216], which is defined as follows,

$$\nabla(t;\Delta) = \sum_{n-\infty}^{\infty} \delta(t - n\Delta) \tag{7.40}$$

To study this interesting function, note that it is obviously periodic with period Δ. Therefore, it can be expressed as complex Fourier series. In fact, according to Eqs. (7.4) and (7.5),

$$\nabla(t;\Delta) = \frac{1}{\Delta}\sum_{n=-\infty}^{\infty} \mathrm{e}^{\mathrm{i}\omega_n t} = \frac{1}{\Delta}\sum_{n=-\infty}^{\infty} \mathrm{e}^{\mathrm{i}n\,2\pi/\Delta\,t}, \qquad \omega_n = n\frac{2\pi}{\Delta} \tag{7.41}$$

Since

$$c_n = \frac{1}{\Delta}\int_{-T/2}^{T/2} \nabla(t;\Delta)\mathrm{e}^{-\omega_n t}\,\mathrm{d}t = \frac{1}{\Delta}\int_{-T/2}^{T/2}\sum_{-\infty}^{\infty} \delta(t-n\Delta)\mathrm{e}^{-\omega_n t}\,\mathrm{d}t = \frac{1}{\Delta}\mathrm{e}^{-\mathrm{i}\omega\cdot 0} = \frac{1}{\Delta}$$

as only one delta function, namely $\delta(t)$, is active on the interval $[-T/2, T/2]$. Also, by Eq. (7.28), the continuous Fourier transform of Dirac's comb can be obtained as

$$\nabla(\omega;\Delta) = \sum_{n=-\infty}^{\infty} \mathrm{e}^{-\mathrm{i}n\Delta\cdot\omega} \tag{7.42}$$

which indicates that the Fourier transform of Dirac's comb is another comb, now in the frequency domain. Therefore consider Dirac's comb in the frequency domain, that is,

$$\nabla\left(\omega;\frac{2\pi}{\Delta}\right) = \sum_{-\infty}^{\infty} \delta\left(\omega - n\frac{2\pi}{\Delta}\right) \tag{7.43}$$

which corresponds to a series of spikes, spaced equally, $2\pi/\Delta$ apart, on the frequence axis.

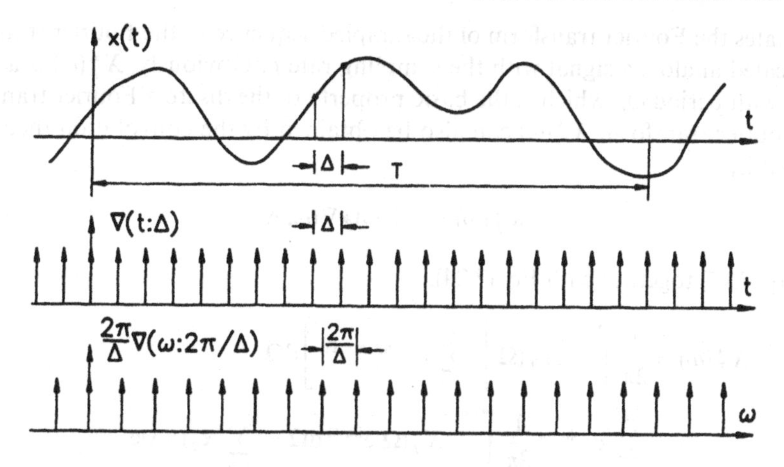

Fig. 7.12 *Sampling with Dirac's comb*

By Eq. (7.29), the inverse Fourier transform is given by

$$\frac{1}{2\pi}\sum_{-\infty}^{\infty} e^{in\, 2\pi/\Delta\, t}$$

and it is the same as the complex series representation of the time domain Dirac's comb (Eq. (7.41)), times $\Delta/2\pi$. Thus the inverse transform of Dirac's comb with a spacing equal to $2\pi/\Delta$ frequency units on the frequency axis is another comb in the time domain spaced at intervals Δ and with an amplitude $\Delta/2\pi$, see Fig. 7.12. In other words,

$$\Delta/2\pi(\nabla(t;\Delta)\leftrightarrow\nabla(\omega{:}\,2\pi/\Delta)) \tag{7.44}$$

are Fourier transform pairs.

Now take a truncated sample function $x_T(t)$, which is available for sampling in the time interval $[0, T]$. By 'convolving' it with Dirac's comb, it is written as

$$x_T^*(t) = x_T(t)\cdot\nabla(t;\Delta) \tag{7.45}$$

whereby the interval length has been subdivided into $N = T/\Delta$ time segments. The new respresentation $x_T^*(t)$ is simply the sequency of sample values $x_r = x_T^*(t = r\Delta)$. Actually, each sample value is a spike of area x_r. Using Eq. (7.41), Eq. (7.45) can be rewritten to give the following relation between the truncated analogue signal and the sampled sequence, that is,

$$x_T^*(t) = f_s \sum_{n=-\infty}^{\infty} x_T(t)\, e^{in\omega_s t}, \qquad \omega_s = 2\pi/\Delta \tag{7.46}$$

Taking the Fourier transform of both sides of Eq. (7.46) yields

$$X_T^*(\omega) = f_s \sum_{n=-\infty}^{\infty} X_T(\omega - n\omega_s) \tag{7.47}$$

which relates the Fourier transform of the sampled sequence to the Fourier transform of the truncated analogue signal with the sampling rate f_s. Obviously, $X_T^*(\omega)$ is a periodic function with period ω_s, which is the basic property of the discrete Fourier transform,

The Fourier transform $X_T^*(\omega)$ can also be obtained by the convolution theorem, Eq. (7.24), that is,

$$X_T^*(\omega) = X_T(\omega) * \nabla(\omega; \Delta)$$

Using Eq. (7.42) together with Eq. (7.24),

$$X_T^*(\omega) = \frac{1}{2\pi}\int_{-\infty}^{\infty} X_T(\Omega)\left[\sum_{-\infty}^{\infty} e^{-in\Delta(\omega-\Omega)}\right] d\Omega$$

$$= \sum_{-\infty}^{\infty} e^{-in\Delta\omega}\frac{1}{2\pi}\int_{-\infty}^{\infty} X_T(\Omega)e^{in\Delta\Omega}\, d\Omega = \sum_{-\infty}^{\infty} x_T(n\Delta)e^{-in\Delta\cdot\omega}$$

$$\sum_{n=0}^{N-1} x_T(n\Delta)e^{-in\Delta\cdot\omega} = \sum_{n=0}^{N-1} x_T(n)e^{-i\,2\pi nk/N} = X_T^*(k) \tag{7.48}$$

since the truncated sample function is zero outside the limits $n=0$ and $n=N-1$, and $T=\Delta N$, whereby $\omega_k = k\,\delta\omega = k2\pi/T = k2\pi/N\Delta$. Except for the factor $1/N$, which now has been moved to the IDFT part, the above result is analogous to the previously obtained form Eq. (7.36). Again it is clear that the Fourier transform $X_T^*(\omega)$ is periodic with period $\omega_s = 2\pi/\Delta$, whereas the Fourier transform of the analogue truncated signal need not be. This of course invokes errors and biasing in the estimates of the true Fourier spectrum of the untruncated signal and other derived quantities such as the spectral density (cf. Chapter 3). For instance, what is the relation of the DFT coefficients X_k to the true Fourier transform of $x(t)$, i.e. $X(\omega)$, outside the sampling interval, that is, k is greater than the number $N-1$, Eq. (7.35). In fact, what happens if one tries to calculate X_k for, say $k > N-1$? For instance, let $k = N + M$. Then,

$$X_{N+M} = \frac{1}{N}\sum_{r=0}^{N-1} x_r e^{-i\,2\pi(N+M)r/N}$$

$$= \frac{1}{N}\sum_{r=0}^{N-1} x_r e^{-i\,2\pi Mr/N}$$

$$= \frac{1}{N}\sum_{r=0}^{N-1} x_r e^{-i\,2\pi+Mr/N} = X_M \tag{7.49}$$

The coefficients X_k therefore just repeat themselves for $k > N-1$, so if the magnitude of the DFT spectrum $|X_k|$ is plotted along a frequency axis, $\omega_k = 2\pi k/(N\Delta)$, the graph will repeat itself periodically as shown in Fig. 7.13 with a period equal to $2\pi/\Delta$. Furthermore, the DFT spectrum is also symmetric about the zero frequency position as by Eq. (7.33), $X_{-k} = X_k^*$, the complex conjugate, so $|X_k| = |X_{-k}|$. The part of the graph which gives a workable estimate of the true spectrum is contained within the frequency range, $|\omega \leqslant \pi/\Delta$. Higher frequencies just show phoney Fourier coefficients that are repetitions of the true spectrum. If there are frequencies higher than π/Δ in the original signal, the calculated spectrum will be distorted by so-called aliasing towards higher frequencies. The

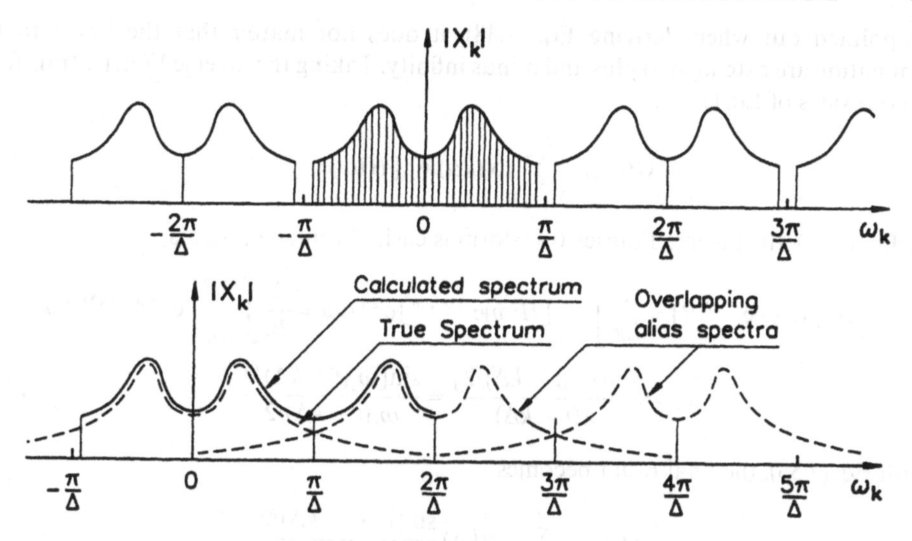

Fig. 7.13 *The aliases of the true Fourier spectrum, (after Newland, [61] and Cunningham, [167]), reproduced by permission of Addison Wesley Longman*

high-frequency components falsely distort the DFT coefficients for frequencies below the folding frequency π/Δ, that is, the overlap area is folded into the principal frequency range. Aliasing can be avoided by either choosing the sampling interval Δ small enough or the sampling frequency f_s (also called the folding frequency) high enough (see Eqs. (7.33) and (7.34)) such that

$$\pi/\Delta > \omega_s \qquad \text{or} \qquad \text{with} \quad f_s = \omega_s/2\pi, \qquad 1/2\Delta > f_s \tag{7.50}$$

Another way is to filter the signal through a so-called antialiasing filter to remove all frequencies above the Nyquist frequency before sampling to produce a discrete signal.

The above discussion furnishes tools for proving the sampling theory Eq. (7.35). Eq. (7.47) relates the Fourier transform of the discretized signal to the analogue signal. If $\omega_s \geqslant 2\omega_{\max}$, (see Eq. (7.33)), the true spectrum $X(\omega)$ can be recovered by convolving $\Delta \cdot X^*(\omega)$ with a lowpass linear filter response function of the kind

$$H(\omega) = \begin{cases} 1 & |\omega| \leqslant \omega_s/2 \\ 0 & |\omega| \geqslant \omega_s/2 \end{cases} \tag{7.51}$$

whereby all the spurious Fourier coefficients are suppressed (see upper part of Fig. (7.13)). Therefore,

$$X(\omega) = \Delta \cdot X^*(\omega) H(\omega) \tag{7.52}$$

and the discrete Fourier transform $X^*(\omega)$ can be replaced by its equivalent form given by Eq. (7.48), whereby,

$$X(\omega) = H(\omega)\Delta \sum_{k=-\infty}^{\infty} x(k\Delta)e^{-ik\Delta\cdot\omega} = \Delta \sum_{k=-\infty}^{\infty} x(k\Delta)H(\omega)e^{-ik\Delta\cdot\omega} \tag{7.53}$$

As pointed out when deriving Eq. (7.48), it does not matter that the limits to the summation are extended to plus and minus infinity. Taking the inverse Fourier transform of both sides of Eq. (7.53),

$$x(t) = \Delta \sum_{k=-\infty}^{\infty} x(k\Delta)\mathscr{F}^{1}[H(\omega)\mathrm{e}^{-\mathrm{i}k\Delta\cdot\omega}] \tag{7.54}$$

By Eq. (7.51), the inverse Fourier transform is easily evaluated, that is,

$$\begin{aligned}\mathscr{F}^{1}[H(\omega)\mathrm{e}^{-\mathrm{i}k\Delta\cdot\omega}] &= \frac{1}{2\pi}\int_{-\infty}^{\infty}[H(\omega)\mathrm{e}^{-\mathrm{i}k\Delta\cdot\omega}]\mathrm{e}^{i\omega t}\,\mathrm{d}\omega = \frac{1}{2\pi}\int_{-\omega_s/2}^{\omega_s/2}\mathrm{e}^{-\mathrm{i}\omega(t-k\Delta)}\,\mathrm{d}\omega \\ &= \frac{\sin[\omega_s(t-k\Delta)/2]}{\pi(t-k\Delta)} = \frac{\sin[\omega_s(t-k\Delta)/2}{\omega_s(t-k\Delta)/2}\end{aligned} \tag{7.55}$$

With Eq. (7.55), the final result becomes

$$x(t) = \sum_{k=-\infty}^{\infty} x(k\Delta)\frac{\sin[\omega_s(t-k\Delta)/2}{\omega_s(t-k\Delta)/2}$$

which is a reconstruction of the signal $x(t)$ from the discrete time sequence, and is the same as the previously introduced sampling theorem Eq. (7.35).

7.2.3 Digital Convolution and Calculation of Spectral Estimates

Since most data is now being handled in digitized form, the operations of convolution and filtering have a slightly different meaning when compared with the same operations for continuous signals, that is, in an analogue form. Following Stein, [75], take for instance two discrete or discretized time signals with a unit sample period Δ; $x(m)$ with M points, $(x(0), x(1), \ldots, x(M-1))$, and $y(n)$ with N points, $(y(0), y(1)), \ldots, y(N-1)$. Now form the convolution in the index domain as

$$z(l) = \sum_{m=0}^{M-1} x(m)y(l-m) \tag{7.56}$$

Since $x(m)$ is zero for m outside the range $(0, M-1)$ and y(n) is zero for n outside the range $(0, N-1)$, the summation is evaluated for each value of $t = 1$ $(\Delta = 1)$ and in all $N + M - 1$ terms in the convolution for $z(l)$ are obtained for $l = 0, 1, \ldots, M + N - 2$. For example, if $M = 4$ and $N = 3$, the $(4 + 3 - 1 = 6)$ terms are as follows:

$$\begin{aligned} z(0) &= x(0)y(0) \\ z(1) &= x/(0)y(1) + x(1)y(0) \\ z(2) &= x(0)y(2) + x(1)y(1) + x(2)y(0) \\ z(3) &= x(1)y(2) + x(2)y(1) + x(3)y(0) \\ z(4) &= x(2)y(2) + x(3)y(1) \\ z(5) &= x(3)y(2) \end{aligned}$$

There are several ways to enact the above multiplications in a simple manner. A relatively simple computational scheme is obtained by arranging the values horizontally as follows:

$$
\begin{array}{llllll}
x(0) & x(1) & x(2) & x(3) & & \\
y(0) & y(1) & y(2) & & & \\
\hline
x(0)y(0) & x(1)y(0) & x(2)y(0) & x(3)y(0) & & \\
 & + & + & + & & \\
 & x(0)y(1) & x(1)y(1) & x(2)y(1) & x(3)y(1) & \\
 & & + & + & + & \\
 & & x(0)y(2) & x(1)y(2) & x(2)y(2) & x(3)y(2) \\
\hline
z(0) & z(1) & z(2) & z(3) & z(4) & z(5)
\end{array}
$$

The above representation can also be written as the matrix product

$$
\begin{Bmatrix} z(0) \\ z(1) \\ z(2) \\ z(3) \\ z(4) \\ z(5) \end{Bmatrix} = \begin{bmatrix} x(0) & 0 & 0 \\ x(1) & x(0) & 0 \\ x(2) & x(1) & x(0) \\ x(3) & x(2) & x(1) \\ 0 & x(3) & x(2) \\ 0 & 0 & x(3) \end{bmatrix} \begin{Bmatrix} y(0) \\ y(1) \\ y(2) \end{Bmatrix} \tag{7.57}
$$

Another scheme is a graphical interpretation of the above multiplications (shown in Fig. 7.14). The operations are carried out by reversing the order of the function $y(n)$ and sliding it past $x(\mathrm{m})$ while performing any non-zero multiplications.

The above formulation shows that the convolution has more terms than either of the two sample functions, which can create some interesting problems both in the index domain and also in the frequency domain. In fact, trying to calculate the Fourier spectrum of the convolution function $z(l)$ from the two spectra, $X(\omega)$ and $Y(\omega)$,

$$
Z(\omega_k) = X(\omega_k)Y(\omega_k), \ \omega_k = k2\pi/T \tag{7.58}
$$

the frequencies can not be the same, since for a unit sampling period, $\Delta = 1$, $T_M = M$ and $T_n = N$, as the two series have different numbers of sample points. To take care of this matter, the shorter series is brought to the same length as the longer series by adding zeroes. Actually this technique is also needed for the FFT routine, which demands that the number of sampling points be an integer power of 2. In many cases the length of available records and the choice of suitable sampling frequency will give a number of sampling points N that need not satisfy this requirement. Rather than reducing N to the nearest lowest integer for which $N = 2^P$ and thereby throwing away data, it is better to extend the series by adding zeroes (zero padding) until N is increased to the next higher power of 2.

Directly related to the above problem is how to calculate the temporal correlation function and the temporal power spectral density, Eqs. (2.37) and (2.99). For the sake of generality, the derivation is carried out for the cross-correlation and cross spectral

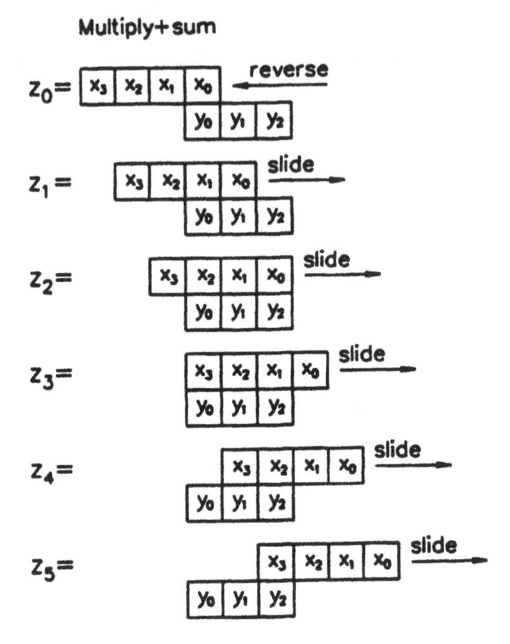

Fig. 7.14 *Digital convolution scheme, (after Stein, [75])*

functions. Therefore consider once more two sample functions of equal length, $x(t)$ and $y(t)$, of the stochastic processes $X(t)$ and $Y(t)$, that have been sampled in the time interval $[0, T]$ with a sample period Δ. The temporal cross-correlation function can obviously only be estimated for time lags τ that are integral multiples of the sampling period. If R_r is the estimate for $R_{XY}(\tau = r\Delta)$, then by Eq. (2.37),

$$R_{XY}(\tau = \Delta r) = \frac{1}{T}\int_0^T x(t)y(t+\tau)\mathrm{d}t = R_r = \frac{1}{N}\sum_{s=0}^{N-1} x_s y_{r+s} \tag{7.59}$$

which can be written in matrix form as

$$\begin{Bmatrix} R(0) \\ R(1) \\ R(2) \\ \vdots \\ R(N-2) \\ R(N-1) \end{Bmatrix} = \frac{1}{N} \begin{pmatrix} y(0) & y(1) & y(2) & \cdots & y(N-2) & y(N-1) \\ y(1) & y(2) & y(3) & \cdots & y(N-1) & 0 \\ y(2) & y(3) & y(4) & \cdots & 0 & 0 \\ \vdots & \vdots & \vdots & \vdots & \vdots & \vdots \\ y(N-2) & y(N-1) & 0 & \cdots & 0 & 0 \\ y(N-1) & 0 & 0 & \cdots & 0 & 0 \end{pmatrix} \cdot \begin{Bmatrix} x(0) \\ x(1) \\ x(2) \\ \vdots \\ x(N-2) \\ x(N-1) \end{Bmatrix} \tag{7.60}$$

Obviously, as the number r increases the estimate for the correlation loses validity and has little to do with the true correlation function, even if ergodicity has been ensured. One way around this problem is to assume that the two sample functions are periodic, such that precisely one cycle of each function is being sampled and the sampled values will therefore repeat themselves according to the scheme

$$x_{r+s} = x_{r+s-N} \quad \text{and} \quad y_{r+s} = y_{r+s-N}$$

In this manner, the zeroes in the above matrix will be filled with functional values. The correlation function Eq. (7.59) is called the circular correlation, $R^c_{XY}(\tau = \Delta r)$. It is peculiar to the two records being analysed, and has the same periodicity, [152]. From Eq. (7.31), its continuous Fourier transform is given by

$$S^c_{XY}(\omega) = 2\pi \sum_{k=-1}^{\infty} S_k \delta\left(\omega - k\frac{2\pi}{T}\right) \quad \text{for} \quad -\frac{\pi}{\Delta} < \omega < \frac{\pi}{\Delta} \tag{7.61}$$

where by Eq. (7.33), S_k is the DFT of the discrete sequence $\{R_r\}$ or

$$S_k = \frac{1}{2\pi}\frac{1}{N}\sum_{r=0}^{N-1} R_r \mathrm{e}^{-\mathrm{i}2\pi kr/N} \tag{7.62}$$

Just as the circular correlation is an estimate of the true correlation function, the circular Fourier transform or Fourier spectrum, $S^c_{XY}(\omega)$, is a kind of an estimate of the cross spectral density, presupposing of course, that the underlying stochastic processes are ergodic. Note that the frequency range within which the circular spectral density is defined is limited by the Nyquist frequency. Also, that the factor 2π, which has to be introduced in the Fourier transform, Eq. (7.62), because of the basic definition of the correlation–spectral density transform relations, Eqs. (2.73) and (2.75), will cancel out.

The circular spectral density can be related directly to the Fourier transform points of the sample functions. By substituting Eq. (7.59) into Eq. (7.62),

$$S_k = \frac{1}{2\pi}\frac{1}{N}\sum_{r=0}^{N-1}\left\{\frac{1}{N}\sum_{s=0}^{N-1} X_s Y_{s+1}\right\}\mathrm{e}^{\mathrm{i}\,2\pi kr/N}$$

Rearranging terms,

$$S_k = \frac{1}{2\pi}\frac{1}{N^2}\sum_{r=0}^{N-1}\sum_{s=0}^{N-1} x_s \mathrm{e}^{\mathrm{i}\,2\pi ks/N}\, y_{s+r}\mathrm{e}^{-\mathrm{i}2\pi k(s+r)/N}$$

$$= \frac{1}{2\pi}\frac{1}{N}\sum_{s=0}^{N-1} x_s \mathrm{e}^{\mathrm{i}\,2\pi ks/N}\left\{\frac{1}{N}\sum_{r=0}^{N-1} y_{s+r}\mathrm{e}^{-\mathrm{i}2\pi k(s+r)/N}\right\} \tag{7.63}$$

Call the variable $s + r$ inside the brackets by the name m. The sum within the brackets is then written as

$$\frac{1}{N}\sum_{m=s}^{(N-1)+s} y_m \mathrm{e}^{-\mathrm{i}2\pi km/N}$$

which is easily recognized as the Fourier coefficient Y_k Eq. (7.33), since the sequence $\{Y_r\}$ is periodic with period N so $y_{N+s} = y_s$. The other sum in Eq. (7.63) is simply the complex conjugate discrete Fourier coefficient X^*_k, so in all,

$$S_k = \frac{1}{2\pi} X^*_k Y_k \tag{7.64}$$

which can be compared with the previously obtained result for the power spectrum of a truncated sample function in an interval of length $2T(T=1)$, Eq. (2.99).

As already mentioned, the above obtained estimates for the correlation functions and the spectral density functions are biased and can only give an estimate of the true values for small time separations. The problem arises from the fact that the two sampled series need to be made equal in length and sampled over the same time interval, whereas to obtain the correlation function, the two series are basically different in time and length. There are several ways to circumvent this problem and in the following one such method is presented that is based on inverse filters, [134], [216].

7.2.4 *Inverse Filters*

In Chapter 3 and 5, it was shown how to find the response of a filter or any black box device by a time convolution, when the input or excitation process and the impulse response were known. The inverse problem is to recover the excitation process when the output process and the impulse response are known. For instance, measuring the acceleration at the top floor, say, of a multi-storey building during an earthquake, one might want to find the base floor excitation to compare it with the free field accelerogram of the earthquake. This problem obviously involves the time convolution of three functions, that is

$$y(t)=\int_{-\infty}^{\infty} x(u)h(t-u)\mathrm{d}u=\int_{-\infty}^{\infty} h(u)x(t-u)\mathrm{d}u=x(t)*h(t) \tag{7.65}$$

which in the frequency domain have the representation

$$Y(\omega)=X(\omega)H(\omega) \tag{7.66}$$

where $y(t)$ is the response, $x(t)$ is the excitation and $h(t)$ is the impulse response function, Eq. (3.37). To recover the excitation function from a measurement of the response $y(t)$, an inverse filter, $f(t)$, is needed such that

$$f(t)*y(t)=x(t) \tag{7.67}$$

From Eq. (7.66), it would seem natural to construct or find the inverse filter by stating that

$$X(\omega)=[1/H(\omega)]Y(\omega)$$

and therefore, $f(t)$ is the inverse Fourier transform of $1/H(\omega)$. If $H(\omega)$ is small, that can sometimes give rise to frequence resolution problems. An alternative way is therefore to design the inverse filter in the time domain. Thus, given the sampled time sequence $y(k)$ and the impulse response function $h(m)$, the excitation $x(n)$ is sought. Now, by the silding technique shown in Fig. 7.13 (Eq. (7.57)), the discretized convolution

$$y(k)=\sum_{m=0}^{M-1} x(m)h(k-m)=\sum_{m=0}^{M-1} h(m)x(k-m) \tag{7.68}$$

is written in a matrix form as follows:

$$
\mathbf{Y}=\mathbf{HX} \text{ or } \begin{Bmatrix} y(0) \\ y(1) \\ \cdot \\ \cdot \\ y(N+M-2) \end{Bmatrix} = \begin{pmatrix} h(0) & 0 & \cdot & 0 & 0 \\ h(1) & h(0) & \cdot & \cdot & 0 \\ \cdot & h(1) & \cdot & 0 & \cdot \\ h(M-1) & \cdot & \cdot & h(0) & 0 \\ 0 & h(M-1) & \cdot & h(1) & h(0) \\ 0 & 0 & \cdot & \cdot & h(1) \\ \cdot & 0 & \cdot & h(M-1) & \cdot \\ 0 & 0 & \cdot & 0 & h(M-1) \end{pmatrix} \cdot \begin{Bmatrix} x(0) \\ x(1) \\ \vdots \\ x(N-1) \end{Bmatrix} \tag{7.69}
$$

where $\mathbf{H}=\{h(ij)\}$ is an $(N+M-1)\times N$ matrix. Since it is not a square matrix, it can not be inverted to give the solution $x(n)$. Therefore an exact solution is not possible. However, a solution may be found by the well-known method of least squares. Introduce the error

$$\boldsymbol{\varepsilon}=\mathbf{Y}-\mathbf{HX} \tag{7.70}$$

and then find the vector $\mathbf{X}$ that minimizes the squared error

$$
\begin{aligned}
\boldsymbol{\varepsilon}^{\mathrm{T}}\boldsymbol{\varepsilon} &= (\mathbf{Y}-\mathbf{HX})^{\mathrm{T}}(\mathbf{Y}-\mathbf{HX}) = (\mathbf{Y}^{\mathrm{T}}-\mathbf{X}^{\mathrm{T}}\mathbf{H}^{\mathrm{T}})(\mathbf{Y}-\mathbf{HX}) \\
&= \mathbf{Y}^{\mathrm{T}}\mathbf{Y}-\mathbf{X}^{\mathrm{T}}\mathbf{H}^{\mathrm{T}}\mathbf{Y}-\mathbf{Y}^{\mathrm{T}}\mathbf{HX}+\mathbf{X}^{\mathrm{T}}\mathbf{H}^{\mathrm{T}}\mathbf{HX} = \mathbf{Y}^{\mathrm{T}}\mathbf{Y}-2\mathbf{X}^{\mathrm{T}}\mathbf{H}^{\mathrm{T}}\mathbf{Y}+\mathbf{X}^{\mathrm{T}}\mathbf{H}^{\mathrm{T}}\mathbf{HX}
\end{aligned} \tag{7.71}
$$

The least square solution is then obtained by differentiating the squared error and equating the result to zero (a minimum). For this purpose, the quadratic form representation of the above equation in an index notation is used, applying the summation convention.

$$\frac{\partial(\varepsilon_i\,\varepsilon_i)}{\partial x_r}=\frac{\partial}{\partial x_r}(y_j\,y_j-2x_j\,h_{jk}\,y_k+x_i\,h_{ji}\,h_{jk}\,x_k)$$

or

$$\frac{\partial(\varepsilon_i\,\varepsilon_i)}{\partial x_r}=0=-2h_{rk}\,y_k+2h_{ji}\,h_{jr}\,x_i \qquad \text{as} \qquad \frac{\partial x_j}{\partial x_r}=\delta_{jr}$$

which in matrix notation is given by

$$\mathbf{H}^{\mathrm{T}}\mathbf{Y}=\mathbf{H}^{\mathrm{T}}\mathbf{HX} \tag{7.72}$$

Now the matrix from $\mathbf{H}^{\mathrm{T}}\mathbf{H}$ constitutes a square matrix and can therefore be inverted provided its determinant is non-zero. Therefore, the final solution for the excitation function $x(n)$ in matrix form is given by

$$\mathbf{X}=(\mathbf{H}^{\mathrm{T}}\mathbf{H})^{-1}\mathbf{H}^{\mathrm{T}}\mathbf{Y} \tag{7.73}$$

and the inverse filter **IF** is given by the expression

$$\mathbf{IF}=(\mathbf{H}^{\mathrm{T}}\mathbf{H})^{-1}\mathbf{H}^{\mathrm{T}} \tag{7.74}$$

Looking more closely at the matrix form $\mathbf{H}^{\mathrm{T}}\mathbf{H}$, some interesting side steps are possible. By Eq. (7.68) it is clear that in the matrix equation Eq. (7.69), $h(i)$ and $x(j)$ can be interchanged provided that $x(j)$ is sufficiently long. Therefore, forming the matrix $\mathbf{X}^{\mathrm{T}}\mathbf{X}$, it can be shown that its elements are directly related to the autocorrelation function of the excitation time series $x(m)$. Now

$$\mathbf{R} = \mathbf{X}^{\mathrm{T}}\mathbf{X} = \left\{ \begin{array}{ccccccccc} x(0) & x(1) & \cdot & x(M-1) & 0 & 0 & 0 & 0 & 0 \\ 0 & x(0) & x(1) & \cdot & x(M-1) & 0 & \cdot & \cdot & 0 \\ \cdot & \cdot & \cdot & \cdot & \cdot & \cdot & \cdot & 0 & \cdot \\ 0 & 0 & 0 & x(0) & x(1) & \cdot & \cdot & x(M-1) & 0 \\ 0 & 0 & \cdot & 0 & x(0) & \cdot & \cdot & \cdot & x(M-1) \end{array} \right\} \tag{7.75}$$

$$\left\{ \begin{array}{ccccc} x(0) & 0 & \cdot & 0 & 0 \\ x(1) & x(0) & \cdot & \cdot & 0 \\ \cdot & x(1) & \cdot & 0 & \cdot \\ x(M-1) & \cdot & \cdot & x(0) & 0 \\ 0 & x(M-1) & \cdot & x(1) & x(0) \\ 0 & 0 & \cdot & \cdot & x(1) \\ 0 & \cdot & \cdot & \cdot & \cdot \\ \cdot & 0 & \cdot & x(M-1) & \cdot \\ 0 & 0 & \cdot & 0 & x(M-1) \end{array} \right\}$$

Carrying out the matrix multiplication, each term in the resulting $\mathbf{R}$ matrix can be written as

$$r_{ij} = \sum_{m=0}^{N+M-1} x(m+(j-i))x(m) = \sum_{m=0}^{M-1} x(m+(j-i))x(m) \tag{7.76}$$

since $x(m) = 0$ for $m < 0$ and $m > M-1$ if $x(m)$ is of length M. Obviously the sum is the auto correlation for the time lag $(j-i)\Delta$. The autocorrelation matrix can therefore be written as

$$\mathbf{R} = \left\{ \begin{array}{ccccc} r(0) & r(1) & r(2) & \cdots & r(N-1) \\ r(1) & r(0) & r(1) & \cdots & r(N-2) \\ r(2) & r(1) & r(0) & \cdots & r(N-3) \\ \vdots & \vdots & \vdots & \cdot & \vdots \\ r(N-1) & r(N-2) & r(N-3) & \cdots & r(0) \end{array} \right\} \tag{7.77}$$

Note that the autocorrelation matrix has only $(N-1)$ independent elements. A matrix with this special kind of structure is called a Toeplitz matrix, [175], [216].

Similarly, the term $\mathbf{X}^T\mathbf{Y}$ is a vector $\mathbf{C}$, whose elements are the cross-correlation terms for the input and output sample functions, $x(i)$ and $y(j)$ for a time lag $k\Delta$, that is,

$$c(k) = \sum_{j=0}^{N+M-1} y(j)x(j+k) \tag{7.78}$$

Now, turning back to the problem of finding the inverse filter, Eq. (7.74), the filter equation can be simplified using the above results Eqs. (7.75)–(7.78). In this manner the correlations are performed on the impulse response function $h(m)$ rather than the excitation $x(m)$, such that

$$\mathbf{IF} = \mathbf{R}^{-1}\mathbf{C} \tag{7.79}$$

which is a system of equations that can be solved comparatively easily since R is a Toeplitz matrix.

The above procedure for solving the time domain convolution Eq. (7.65) for the input rather than the output can be described as an act of deconvolution. It can be found useful if a signal needs to be purified, that is, if effects of the measuring device are to be removed. Also, as the mathematical treatment has shown, the same routines can be applied to obtain the autocorrelation and cross-correlations of the input and output signals.

7.2.5 The z-Transform and Other Transformations

Aside from the Fourier transform there are other, often more convenient, transformations of variables in use in system analysis and signal processing. The Laplace transform, which is widely used by electrical engineers, especially in connection with circuit analysis, can be defined as follows. Let $f(t)$ be a real or complex function of the real variable t. The function $f(t)$ is assumed to satisfy the following three conditions:

1. $f(t)$ is defined on the interval $(0, \infty)$ and $f(t) = 0$ for $t < 0$;
2. $f(t)$ is piecewise continous in any interval $0 < a \leqslant t \leqslant b$, a and b are arbitrary real constants;
3. $f(t)$ is absolutely integrable in a region to the right of $t = 0$.

Introducing the complex variable $s = x + \mathrm{i}y$, the improper integral $\int_0^\infty \mathrm{e}^{-st} f(t)\mathrm{d}t$, if convergent, is called the Laplace transform of the function $f(t)$. The Laplace transform is directly related to the Fourier transform. Consider the positive time function $f(t)$, $t > 0$, which is multiplied by the convergence factor e^{-xt}. Assuming that the modified function $f(t)\mathrm{e}^{-xt}$ is absolutely integrable, that is,

$$\int_0^\infty |f(t)\mathrm{e}^{-xt}|\mathrm{d}t < \infty \tag{7.80}$$

then the Fourier transform exists and is given by

$$\mathscr{F}\{f(t)\mathrm{e}^{-xt}\} = \int_0^\infty f(t)\mathrm{e}^{-xt}\,\mathrm{e}^{-\mathrm{i}\omega t}\mathrm{d}t = \int_0^\infty f(t)\mathrm{e}^{-st}\,\mathrm{d}t \tag{7.81}$$

which is the Laplace transform of $f(t)$, for brevity denoted by

$$F(s) = \mathscr{L}\{f(t)\} \tag{7.82}$$

The Laplace transform $F(s)$ is defined in a complex s-plane with real axis x and imaginary axis y. For values of s along the imaginary axis, for example $s = \mathrm{i}y = \mathrm{i}\omega$, the Laplace transform is equivalent to the Fourier transform of the positive time function, i.e. $F(s) = F(\mathrm{i}\omega)$.

The inverse Laplace transform $\mathscr{L}^{-1}\{f(s)\}$ can be obtained by modifying the inversion Eq. (7.15), which gives the result

$$f(t) = \mathscr{L}^{-1}\{F(s)\} = \frac{1}{2\pi \mathrm{i}} \int_{c-\mathrm{i}\infty}^{c+\mathrm{i}\infty} F(s)\mathrm{e}^{st}\,\mathrm{d}s \tag{7.83}$$

where c is a positive real constant. A word of caution is in order. The inverse Laplace transform need not be unique. In fact, consider two positive time functions $f(t)$ and $g(t)$, defined on the semi-infinite interval $(0, \infty)$, which differ at only a finite number of points. By condition two above, the Laplace integral is unchanged, i.e. the two functions have the same Laplace transform. For all practical purposes, however, the uniqueness of the transformation is well established for all time functions that are continuous (or at least continuous from the same side) at every point in the interval $(0, \infty)$. As for the value of the constant c in the above integral Eq. (7.83), suppose that x_{min} is the minimum value in order to satisfy the condition Eq. (7.80). In that case, the Laplace transform $F(s)$ exists if the real part of s, i.e. x, is greater than x_{min}. Therefore, c must be selected within the region of convergence, that is, where $x > x_{\text{min}}$.

The Laplace transform can be shown to have similar properties to the Fourier transform regarding time and frequency shift, convolution, integration and differentiation. There are many problems in system analysis and signal processing where the Laplace transform is more conveniently applied than the Fourier transform. However, in most cases the Fourier transform can be used just as well. Throughout this book, the Fourier transform is used wherever possible in order not to confuse the issue. For the applicability of the Laplace transform, the reader is referred to the specialist literature, [49].

There are other transforms which relate more directly to digital signal processing, notably the Hilbert transform and the z-transform. Beginning with the z-transform, which is especially useful in applications dealing with discrete time sequences, it can be introduced as follows. Consider the discrete time sequence $x(n)$, $n = \cdots -2, -1, 0, 1, 2, \ldots$. The z-transform of $x(n)$ is then defined as

$$X(z) = \sum_{n=-\infty}^{\infty} x(n)z^{-n} = Z\{x(n)\} \tag{7.84}$$

where z is a complex variable. The z-transform can be either two-sided as above and one-sided if $x(n) = 0$ for $n < 0$. The two-sided version is more general as the choice of the time origin $n = 0$ becomes arbitrary. However, both versions are used and sometimes no distinction is made between the two forms. The z-transform is an infinite power series, so obviously there will be convergence problems. The function $X(z)$ therefore exists only for values of z for which the series Eq. (7.84) is convergent. The set of values of z for which $X(z)$

is finite is called the region of convergence (ROC) for $X(z)$ and should be indicated whenever a z-transform is cited.

In order to investigate the region of convergence for the z-transform, it is convenient to express the complex variable z in a polar form, i.e. $z = re^{i\theta}$. The above version Eq. (7.84) then becomes

$$X(z = re^{i\theta}) \equiv \sum_{n=-\infty}^{\infty} x(n)r^{-n}e^{-i\theta n} \tag{7.85}$$

In the region of convergence, $|X(z)| < \infty$, whereby

$$|X(z)| = |\sum_{n=-\infty}^{\infty} x(n)r^{-n}e^{-i\theta n}| \leqslant \sum_{n=-\infty}^{\infty} |x(n)r^{-n}e^{-i\theta n}|$$

$$= \sum_{n=-\infty}^{\infty} |x(n)r^{-n}| \leqslant \sum_{n=\infty}^{\infty} |x(-n)r^{n}| + \sum_{n=0}^{\infty} |\frac{x(n)}{r^{n}}| \tag{7.86}$$

Hence, the z-transform $X(z)$ is finite if the infinite sequence $x(n)r^{-n}$ is absolutely summable. Breaking the sum up into a non causal ($n < 0$) and causal ($n \geqslant 0$) part as shown in Eq. (7.86), it is clear that the ROC for the first sum consists of all points in the complex plane, which are located within a circle of some radius $r_1 < \infty$, which makes the sum absolutely summable. The ROC of the second sum on the other hand consists of all points that fall outside a circle with a large enough radius $r_2 < \infty$, which makes the sum absolutely summable. Therefore, the ROC for the z-transform is the annulus $r_2 < r < r_1$ in the complex plane.

It is interesting to note that if the unit circle, $|z| = 1$, falls within the ROC, the z-transform evaluated on the unit circle becomes equivalent to the discrete Fourier transform DFT. In fact, consider the discrete time sequence $x(n) = x(n\Delta)$, $n = 0, 1, 2, \ldots, (N-1)$ where N is the number of data points. The z-transform for $z = e^{i\theta}$ ($|z| = 1$) is

$$X(z = e^{i\theta}) \equiv \sum_{n=-\infty}^{\infty} x(n)e^{-i\theta n} \tag{7.87}$$

Instead of using a continuous angle θ, the transform is evaluated at discrete frequencies, i.e. $\theta(k) = 2\pi k/N\Delta$ ($k = 0, 1, 2, \ldots, (N-1)$). Eq. (7.87) then becomes

$$X(k) \equiv \sum_{n=0}^{N-1} x(n)e^{-i\,2\pi kn/N} \tag{7.88}$$

which is equivalent to the DFT as defined by Eq. (7.48). On the other hand, if the ROC for $X(z)$ does not include the unit circle, the Fourier transform of the signal does not exist. Also, the existence of a Fourier transform, defined for finite energy signals, does not necessarily mean that the z-transform of the signal exists.

Example 7.7 Consider the impulse response function of the compound Poisson process, introduced in Example 6.6, in the form

$$x(n) = (-1)^n 1(n), n = 0, 1, 2, 3, \ldots$$

Determine the z-transform of $x(n)$.

Solution The sample functions may be likened to a switch which is either on or off, that is, the signal $x(n)$ consists of an infinite number of samples:

$$x(n) = \{1, -, 1, -1, 1, -1, \ldots\}$$

The z-transform is therefore the power series

$$X(z) = 1 + \left(\frac{-1}{z}\right) + \left(\frac{-1}{z}\right)^2 + \left(\frac{-1}{z}\right)^3 + \left(\frac{-1}{z}\right)^4 + \cdots$$

which is an infinite geometric or quotient series. Recalling that the quotient series

$$1 + q + q^2 + q^3 + q^4 + \cdots = \frac{1}{1-q} \qquad \text{for} \quad |q| < 1$$

it follows that $X(z)$ converges to

$$X(z) = \frac{1}{1 + 1/z} = \frac{z}{1+z} \qquad \text{for} \quad \left|\frac{1}{z}\right| < 1 \text{ or } |z| > 1$$

The z-transform of the infinite signal is a compact and simple complex function with an ROC, which consists of all points outside the unit circle in the complex plane.

Example 7.8 Consider the causal signal $x_1(n) = a^n 1(n), n \geqslant 0$, and the non-causal signal $x_2(n) = a^n 1(-n-1), n < 0$, where a is a positive constant. Together, the two signals consitute an exponentially increasing (decreasing) signal strength with the signal amplitude equal to a^n. Determine the z-transform and the region of convergence for both signals.

Solution From the definition of the z-transform Eq. (7.84),

$$X_1(z) = \sum_{n=-\infty}^{\infty} x_1(n) z^{-n} = \sum_{n=0}^{\infty} a^n z^{-n} = \sum_{n=0}^{\infty} (az^{-1})^n$$

This is a quotient series with the quotient (az^{-1}). For $|az^{-1}| < 1$ or $|z| > a$, the quotient series is convergent with the sum

$$\frac{1}{1 - az^{-1}} = \frac{z}{z-a} = X_1(z)$$

which is the z-transform with a region of convergence outside the circle $r > a$.

The non-causal signal has the z-transform

$$X_2(z) = \sum_{n=-\infty}^{\infty} x_2(n) z^{-n} = \sum_{n=-\infty}^{-1} -a^n z^{-n} = -\sum_{m=1}^{\infty} (a^{-1}z)^m$$

$$= -\frac{z}{a}\left(1 + \frac{z}{a} + \left(\frac{z}{a}\right)^2 + \cdots\right) = -\frac{z}{a} \cdot \frac{1}{1 - z/a} = \frac{z}{z-a}$$

in which $m = -n$. Obviously the two z-transform $X_1(z)$ and $X_2(z)$ are equal whereas the ROC is now the area within the circle $r < a$.

It is interesting to note that in the above example the two z-transform for two different signals are equal, that is, $X_1(z) = X_2(z)$. Therefore, the z-transform need not be unique in the sense that different signals have different z-transform. Thus, a closed form expression for the z-transform does not uniquely specify the signal in the index domain. However, the ROC of the two transforms is different whence this ambiguity can be resolved by always specifying the region of convergence. Therefore, it can be stated that a discrete time sequence $x(n)$ is uniquely specified by its z-transform $X(z)$ and the ROC of $X(z)$. Another interesting fact resulting from Example 7.8 is that a causal signal will have an ROC consisting of the area outside a circle of some radius r_1 whereas the ROC for a non-causal signal is the area within a circle of some radius r_2.

As with other integral transforms, the determiniation of the inverse transform is often desirable, that is, given the function $X(z)$ how can the sequence $x(n)$ be determined? Formally, the inverse transform is written as follows:

$$x(n) = Z^{-1}\{X(z)\} \tag{7.89}$$

There are several ways to derive a suitable inversion formula for Eq. (7.89). Cauchy's integral theorem, which has been used extensively throughout the text of this book, offers a relatively simple and straightforward answer. In fact, multiplying both sides of the transform Eq. (7.84) by z^{k-1} and integrating both sides over a closed contour within the ROC of $X(z)$, which encloses the origin, the following result is obtained:

$$\oint_C X(z)z^{k-1}\,dz = \oint_C \sum_{n=-\infty}^{\infty} x(n)z^{k-1-n}\,dz \tag{7.90}$$

The closed contour C inside the ROC of $X(z)$ is to be taken in a counterclockwise direction. Since the infinite series is convergent on this contour, the summation and integration can be interchanged, whereby

$$\oint_C X(z)z^{k-1}\,dz = \sum_{n=-\infty}^{\infty} x(n)\oint_C z^{k-1-n}\,dz \tag{7.91}$$

The contour integral on the right-hand side is easily handled using Cauchy's integral theorem, that is,

$$\frac{1}{2\pi i}\oint_C z^{k-1}\,dz = \begin{cases} 1 & n = k \\ 0 & n \neq k \end{cases} \tag{7.92}$$

since the integrand has the single pole $z = 0$ within the contour with the residue 1 if $n = k$ and 0 if $n \neq k$. Applying Eq. (7.92) to Eq. (7.91), the right-hand side reduces to $2\pi i x(k)$. The desired inversion formula therefore becomes

$$x(k) = \frac{1}{2\pi i}\oint_C X(z)z^{k-1}\,dz \tag{7.93}$$

Thus, Cauchy's integral theorem offers a closed form inversion formula for retrieving the discrete sequence $x(n)$ from its z-transform. However, it is often advantageous to use a more direct method. Most signals and systems dealt with in practice will have rational z-transforms that are a ratio of two polynomials, that is, are of the form

$$X(z) = \frac{B(z)}{A(z)} \tag{7.94}$$

The inversion of Eq. (7.94) can therefore be carried out using either long division or expansion into partial fractions. Furthermore, it is also possible to expand the function $X(z)$ into power series, that is,

$$X(z) \equiv \sum_{n=-\infty}^{\infty} c_n z^{-n} \tag{7.95}$$

whereby due to the uniqueness of the z-transform given a specified ROC, the time sequence is retrieved by equating the coefficients in the expansion c_n to the signal amplitudes $x(n)$. From Eq. (7.94), it is clear that the function $X(z)$ can have a number of zeros (the roots of $B(z)$) for which $X(z) = B(z) = 0$ as well as a number of poles (the roots of $A(z)$) for which $A(z) = 0$ and $X(z) = \infty$. Both a zero and a pole can occur at either $z = 0$ or $z = \infty$. It is therefore possible to represent the z-transform $X(z)$ graphically by a *pole–zero plot*, which for instance shows the location of zeros by circles (0) and the location of poles by crosses (x). Multiple zeros or poles are indicated by a number close to the corresponding circle or cross. By definition, however, the ROC of a z-transform should not contain any poles.

Example 7.9 The discrete Fourier transform of the unit step function $1(t)$ can be obtained using the z-transform. From Example 7.8, the causal signal $a^n 1(n)$ was found to have the z-transform $z/(z-a)$. Putting $a = 1$, the unit step signal $1(n)$ has the z-transform $X(z) = z/(z-1)$ with an ROC: $|z| > 1$. Now, $X(z)$ has a pole at $z = 1$ but converges for $|z| > 1$. If the $X(z)$ is nevertheless evaluated on the unit circle, except at $z = 1$, the following relation is obtained:

$$X(\omega) = \frac{e^{i\omega}}{e^{i\omega} - 1} = \frac{e^{i\omega/2}}{e^{i\omega/2} - e^{-i\omega/2}} = \frac{e^{i\omega/2}}{2i\sin(\omega/2)} = \frac{e^{i(\omega-\pi)/2}}{2\sin(\omega/2)}, \qquad \omega \neq 2\pi k, \quad k = 0, 1, 2, \ldots$$

as $-i = e^{-i\pi/2}$. Obviously, the Fourier transform is undefined at the zero frequency $\omega = 0$ and for multiples of 2π. At these controversial frequencies, $X(\omega)$ contains impulses of area π. Otherwise, $X(\omega)$ is finite. As the signal is a constant amplitude signal for $n \geqslant 0$, one would expect a zero frequency component, i.e. a d.c. component at all frequencies except at $\omega = 0$. As seen by the above Fourier transform, this is not the case. This is due to the fact that the unit step signal is not a constant for the entire $-\infty < n < \infty$. The signal is abruptly turned on at $n = 0$ causing 'ripples' with frequency components in the range $0 < \omega \leqslant \pi$.

Example 7.10 A linear time invariant system has the transfer function

$$H(z) = \frac{z^3 + z}{(z + 0.5)(z^2 + 0.25)}, \qquad |z| > 0.5$$

Determine and sketch the poles and zeroes of $H(z)$ and find the response function $h(n)$

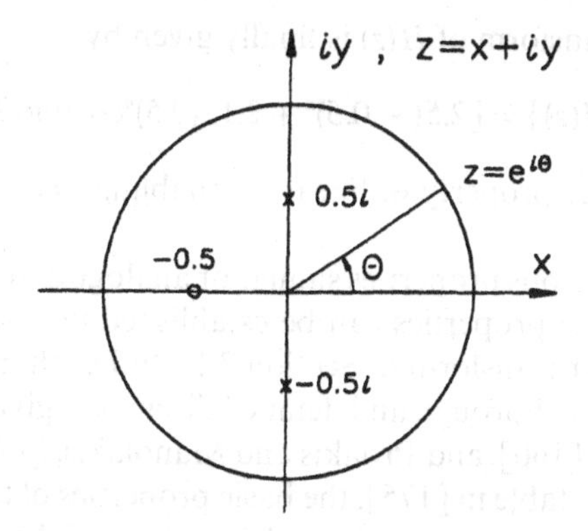

Fig. 7.15 *The poles and zeros of H(z)*

Solution The zeroes of $H(z)$ are $z_1 = -0$, $z_2 = -\mathrm{i}$ and $z_3 = \mathrm{i}$ whereas the poles are $p_1 = -0.5$, $p_2 = 0.5\mathrm{i}$ and $p_3 = -0.5\mathrm{i}$. In Fig. 7.15, a plot of the zeros and poles is shown. A partial fraction expansion of $H(z)/z$ gives

$$\frac{H(z)}{z} = \frac{z^2+1}{(z+0.5)(z-0.5\mathrm{i})(z+0.5\mathrm{i})} = \frac{A}{z+0.5} + \frac{B}{z-0.5\mathrm{i}} + \frac{B^*}{z+0.5\mathrm{i}}$$

where

$$A = (z+0.5)\left.\frac{z^2+1}{(z+0.5)(z^2+0.25)}\right|_{z=-0.5} = 2.5$$

$$B = (z-0.5\mathrm{i})\left.\frac{z^2+1}{(z+0.5)(z^2+0.25)}\right|_{z=0.5\mathrm{i}} = 1.06\mathrm{e}^{-\mathrm{i}\,3\pi/4}$$

$$B^* = 1.06\mathrm{e}^{\mathrm{i}\,3\pi/4}$$

Thus,

$$H(z) = \frac{2.5z}{z+0.5} + \frac{z1.06\mathrm{e}^{-\mathrm{i}\,3\pi/4}}{z-0.5\mathrm{e}^{-\mathrm{i}\pi/2}} + \frac{z1.06\mathrm{e}^{-\mathrm{i}\,3\pi/4}}{z-0.5\mathrm{e}^{-\mathrm{i}\pi/2}}$$

From Example 7.8, it is found that the first fraction has the inverse transform $h_1(n) = 2.5(-0.5)^n(n)$. The second and third fractions are of the type

$$X(z) = \frac{Cz}{z-a\mathrm{e}^{\mathrm{i}\theta}} + \frac{C^*z}{z-a\mathrm{e}^{-\mathrm{i}\theta}}$$

The z-transform of these two fractions is evaluated as follows (again using the result from Example 7.8):

$$x(n) = C(a\mathrm{e}^{\mathrm{i}\theta})^n 1(n) + C^*(a\mathrm{e}^{-\mathrm{i}\theta})^n 1(n) = 2|C|a^n\cos(n\theta + \angle C)1(n)$$

Therefore, the inverse transform of $H(z)$ is finally given by

$$h(n) = Z^{-1}\{H(z)\} = [2.5(-0.5)^n + 2.12(0.5)^n \cos(n\pi/2 - 3\pi/4)]1(n)$$

as the z-transform has the property of linearity, enabling one to solve for each fraction separately.

The z-transform has various properties similar or analogous to those of the Fourier and Laplace transforms. These properties can be established in a similar manner as shown previously for the Fourier transform in Section 7.1. Rather than repeating that exercise, the reader is referred to the thorough and detailed discussion given by Cunningham, [49], Oppenheim and Schafer, [160], and Proakis and Manolakis, [175]. In Table 7.1, which is reproduced after a similar table in [175], the basic properties of the z-transform are listed. The z-transform plays an important role in the analysis of digital signals, the design of digital filters, and in system control. The above presentation is sufficient to give the reader an overview of the z-transform and its potential. For more information, the reader is referred to the specialist literature, [49], [134], [160], [175].

Table 7.1 *The basic properties of the z-transform (from Digital Signal Processing 2/E by Proakis/Manolakis, © 1992. Adapted by permission of Prentice-Hall, Inc., Upper Saddle River, NJ.)*

Property	Time domain	z-Domain	ROC
Notation	$x(n)$	$X(z)$	$r_2 < \|z\| < r_2$
	$x_1(n)$	$X_1(z)$	ROC_1
	$x_2(n)$	$X_2(z)$	ROC_2
Linearity	$a_1x_1(n) + a_2x_2(n)$	$a_1X_1(z) + a_2X_2(z)$	At least the intersection of ROC_1 and ROC_2
Time shifting	$x(n-k)$	$z^{-k}X(z)$	That of $X(z)$, except $z=0$ if $k>0$ and $z=\infty$ if $k<0$
Scaling	$a^n x(n)$	$X(a^{-1}z)$	$\|a\|r_2 < \|z\| < r_2\|a\|$
Time reversal	$x(-n)$	$X(z^{-1})$	$\frac{1}{r_1} < \|z\| < \frac{1}{r_2}$
Conjugation	$x^*(n)$	$X^*(z^*)$	ROC
Real part	$\mathrm{Re}\{x(n)\}$	$\frac{1}{2}(X(z) + X^*(z^*))$	Includes ROC
Imaginary part	$\mathrm{Im}\{x(n)\}$	$\frac{1}{2}(X(z) - X^*(z))$	Includes ROC
Differentiation	$nx(n)$	$-z\frac{dX(z)}{dz}$	$r_2 < \|z\| < r_2$
Convolution	$x_1(n) \otimes x_2(n)$	$X_1(z)X_2(z)$	At least the intersection of ROC_1 and ROC_2
Correlation	$r_{x_1x_2}(l) = x_1(l) \otimes x_2(-l)$	$R_{x_1x_2}(z) = X_1(z)X_2\left(\frac{1}{Z}\right)$	At least the intersection of ROC of $X_1(z)$ and $X_2(z^{-1})$
Initial value	If $x(n)$ is causal	$x(0) = \lim_{z\to\infty} X(z)$	
Multiplication	$x_1(n)x_2(n)$	$\frac{1}{2\pi i}\oint X_1(u)X_2\left(\frac{z}{u}\right)\frac{du}{u}$	At least $r_{li} \cdot r_{2i} < \|z\| < r_{lo} \cdot r_{2o}$
Parseval's relation	$\sum_{n=-\infty}^{\infty} x_1(n)x_2^*(n) = \frac{1}{2\pi i}\oint X_1(u)X_2\left(\frac{1}{u}\right)\frac{du}{u}$		

Finally, a few remarks about the so-called Hilbert transform. When dealing with causal time sequences (positive time only), i.e. $x(n) = 0$ for $n < 0$, certain mathematical relations exist between the discrete Fourier transform of the real and imaginary parts of the sequence, $X_R(\omega)$ and $X_1(\omega)$, which are collectively called Hilbert transforms. Consider a causal time sequence $x(n)$, $n \geqslant 0$. The sequence can be decomposed into an even and an odd sequence, i.e.

$$x(n) = x_e(n) + x_0(n) \tag{7.96}$$

where the even sequence $x_e(n)$ is given by

$$x_e(n) = \tfrac{1}{2}(x(n) + x(-n)) \tag{7.97}$$

and the odd sequence $x_0(n)$ by

$$x_0(n) = \tfrac{1}{2}(x(n) - x(-n)) \tag{7.98}$$

Consequently, for a causal sequence, the even and odd parts are valued as follows:

$$x_e(n) = \begin{cases} x_0(n) & n > 0 \\ x_e(0) & n = 0 \\ -x_0(n) & n < 0 \end{cases} \tag{7.99a}$$

$$x_0(n) = \begin{cases} x_e(n) & n > 0 \\ 0 & n = 0 \\ -x_e(n) & n < 0 \end{cases} \tag{7.99b}$$

The above relations can be written in a shorter form by introducing a kind of two-sided unit step function $u(t)$, that is,

$$u(n) = \begin{cases} 1 & n > 0 \\ 0 & n = 0 \\ -1 & n < 0 \end{cases} \tag{7.100}$$

Also, introducing the delta function $\delta(n)$, ($\delta(n) = 1$ for $n = 0$, $\delta(n) = 0$ for $n \neq 0$), the relations Eqs. (7.99) are written as follows:

$$x_e(n) = x_0(n)u(n) + x_e(0)\delta(n) \tag{7.101a}$$

$$x_0(n) = x_e(n)u(n) \tag{7.101b}$$

Assuming that the sequences $x_e(n)$ and $x_0(n)$ are summable, the Fourier transform exists, and by Eq. (7.17),

$$X_R(\omega) = \mathscr{F}\{x_e(n)\} \quad \text{and} \quad X_I(\omega) = \mathscr{F}\{x_0(0)\} \tag{7.102}$$

Taking the Fourier transform of both sides of Eqs. (7.101), recalling that the Fourier transform of an even and odd function corresponds to the real and imaginary parts (see

Eqs. (7.17) and (7.20)), and using the complex convolution theorem Eq. (7.24) with $\Omega = \mathrm{i}\theta$, yields

$$\left.\begin{aligned} X_R(\omega) &= \frac{\mathrm{i}}{2\pi}\int_{-\infty}^{\infty} X_I(\mathrm{i}\theta)U(\omega - \mathrm{i}\theta)\mathrm{d}\theta + x_e(0) = \frac{\mathrm{i}}{2\pi}\int_{-\pi}^{\pi} X_I(\mathrm{i}\theta)U(\omega - \mathrm{i}\theta)\mathrm{d}\theta + x_e(0) \\ X_I(\omega) &= \frac{\mathrm{i}}{2\pi}\int_{-\pi}^{\pi} X_R(\mathrm{i}\theta)U(\omega - \theta)\mathrm{d}\theta \end{aligned}\right\} \tag{7.103}$$

The function $U(\omega)$ is the discrete Fourier transform of the two-sided unit step function $u(n)$, and the integration limits reflect the definition of the discrete Fourier transform $X(\omega)$ in the interval $-\pi \leqslant \omega \leqslant \pi$. Now, by Eq. (7.48), the discrete Fourier transform $U(\omega)$ is

$$\begin{aligned} U(\omega) = \sum_{n=-\infty}^{\infty} u(n)\mathrm{e}^{-in\omega\Delta} &= -\sum_{n=-\infty}^{-1} \mathrm{e}^{-in\omega\Delta} + \sum_{n=1}^{\infty} \mathrm{e}^{-in\omega\Delta} \\ &= -\sum_{n=1}^{\infty} \mathrm{e}^{in\omega\Delta} + \sum_{n=1}^{\infty} \mathrm{e}^{-in\omega\Delta} \\ &= \frac{-1}{1-\mathrm{e}^{-\mathrm{i}\omega\Delta}} + \frac{1}{1-\mathrm{e}^{-\mathrm{i}\omega\Delta}} = -\mathrm{i}\cot\frac{\omega\Delta}{2} \end{aligned} \tag{7.104}$$

whereby for $\phi = \omega\Delta$

$$U(\omega - \mathrm{i}\theta) = \mathrm{i}\cot\frac{(\phi - 0)}{2} \tag{7.104a}$$

Therefore, the following relations exist between the real and imaginary parts of the Fourier transform of the causal sequence $x(n)$:

$$\begin{aligned} X_R(\phi) &= \frac{1}{2\pi}\int_{-\pi}^{\pi} X_I(\theta)\cot\left(\frac{\phi-\theta}{2}\right)\mathrm{d}\theta + x_e(0) \\ X_I(\phi) &= \frac{1}{2\pi}\int_{-\pi}^{\pi} X_R(\theta)\cot\left(\frac{\phi-\theta}{2}\right)\mathrm{d}\theta \end{aligned} \tag{7.105}$$

where $\phi = \omega\Delta$. Thus, the imaginary part of the discrete Fourier transform of a causal sequence is uniquely related to the real part of the transform, whereas the real part is uniquely related to the imaginary part plus a constant. The above relations Eqs. (7.105) are called the discrete Hilbert transforms of the causal sequence $x(n)$.

The Hilbert transform plays an important role in the design of frequency selective filters and modulation of signals. For instance, a Hilbert transformer, which is defined by the frequency response function

$$H(\mathrm{i}\omega\Delta) = \begin{cases} -\mathrm{i} & 0 \leqslant \omega\Delta < \pi \\ \mathrm{i} & -\pi \leqslant \omega\Delta < 0 \end{cases} \tag{7.106}$$

is an ideal all-pass filter that changes the phase of the signal at its input by 90°. The impulse response function of the Hilbert transformer is given by

$$h(n)=\frac{1}{2\pi}\int_{-\pi}^{\pi} H(\omega\Delta)e^{in\omega\Delta}d(\omega\Delta)=\frac{1}{2\pi}\left(\int_{-\pi}^{0} ie^{in\omega\Delta}d(\omega\Delta)-\int_{0}^{\pi} ie^{in\omega\Delta}d(\omega\Delta)\right)$$

$$=\begin{cases}\dfrac{2}{\pi n}\sin^2\left(\dfrac{\pi n}{2}\right) & n\neq 0\\[2ex] 0 & n=0\end{cases}\tag{7.107}$$

The impulse response function is of infinite duration and is non-causal as $h(n)\neq 0$ for $n<0$. Therefore, realizable approximations of the Hilbert transformer need to be constructed.

Using Eq. (7.107) and Eq. (7.56), the corresponding functional relationship between the real and imaginary sequences $x_R(n)$ and $x_I(n)$ in the time domain can be obtained. The time convolution is simply given by

$$x_R(n)=\sum_{m=-\infty}^{\infty} x_I(n-m)\frac{\sin^2(m\pi/2)}{(m\pi/2)} \qquad m\neq n$$

$$x_I(n)=\sum_{m=-\infty}^{\infty} x_R(n-m)\frac{\sin^2(m\pi/2)}{(m\pi/2)} \qquad m\neq n \tag{7.108}$$

Eqs. (7.108) constitute the Hilbert transform relationships between the signals $x_R(n)$ and $x(n)$.

The Hilbert transformer can among other things be used to reduce the sampling rate required for a real bandpass signal $x_R(n)$. It is a useful tool in filter design, and has other implications of interest, which are treated in more detail in the specialist literature.

7.2.6 An Overview of Digital Filters

Using digital filters offers a systematic approach to the analysis of discrete time sequences or digital signals. By the way of suitable filtering of the signal, undesirable noise or frequency components can be removed, and the signal can be shaped and modified according to prescribed requirements. Also, the errors committed, and the frequency distortion of a signal that has been truncated for numerical or digital applications, can be estimated. In digital signal analysis, distinction is made between so-called infinite impulse response filters (IIR), an example of which is the Hilbert transformer, and finite impulse response filters (FIR). IIR filters can be designed by beginning with an analogue filter and using a mapping to transform the s-plane (the Laplace transform plane) to the z-plane (the z-transform plane), that is, convert the filter frequency function $H(s)$ to the z-transform $H(z)$. FIR filters, on the other hand, are causal, that is, they are non-recursive in the sense that the filter output is formed by a linear combination of the past input values and the current or present input value only. For this reason, FIR filters are stable since there is no feedback from the output to the input. A bonded input thus produces a bonded output. However, FIR filters can also be designed as recursive filters by cheating a little or by adapting suitable IIR filters.

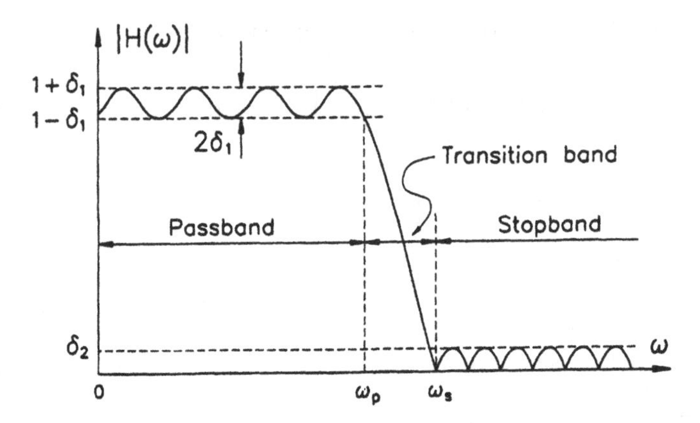

Fig. 7.16 *Characteristics of causal filters, (from Digital Signal Processing 2/E by Proakis/Manolakis, © 1992. Adapted by permission of Prentice-Hall, Inc., Upper Saddle River, NJ.)*

Although the frequency response characteristics of IIR filters may be desirable, they are not absolutely necessary in most practical situations. It is possible to design causal or FIR filters that approximate ideal filters as closely as desired. In particular, some ripples can be tolerated both in the passband and the stopband, Fig. 7.16. The transition of the frequency response from passband to stopband defines the transition band or the transition region of the filter. The band-edge frequency ω_P defines the edge of the passband, while the band-edge frequency ω_s defines the beginning of the stopband. The width of the transition band is $\omega_s - \omega_p$, whereas the bandwidth of the filter is simply ω_p. The ripples in the passband are denoted by δ_1 so the magnitude of the frequency response $|H(\omega)|$ varies between the limits $1 \pm \delta_1$. The ripples in the stopband are denoted by δ_2. To graphically depict the frequency response of most filters, it is necessary to apply a logarithmic scale to accommodate the large dynamic range of the filter magnitudes. A decibel scale is mostly used, whereby the ripples in the passband are $20 \log_{10} \delta_1$ dB, and the ripples in stopband are $20 \log_{10} \delta_2$dB. The filter parameters are therefore selected according to the passband and stopband ripples tolerated, and the desired passband and stopband frequencies. In the following, a brief overview of analogue continuous filters, their relation to IIR filters as a design tool, and a short discussion of FIR filters will be presented with some examples. For a more complete discussion of digital filters, the reader is referred to the many textbooks on this subject.

An analogue filter may be defined by the transfer function $H(s)$, which can be obtained by taking the Laplace transform of the filter convolution equation

$$y(t) = \int_0^\infty x(u)h(t-u)\mathrm{d}u \tag{7.109}$$

where $h(t)$ is the impulse response function of the filter. By the convolution theorem for Laplace transforms (equivalent to the convolution theorem for Fourier transforms),

$$Y(s) = H(s)X(s) \tag{7.110}$$

The filter transfer function is therefore formed as a fraction of two functions, which in many cases can be approximated by polynomials. Thus, using a partial fractions expansion,

$$H(s) = \frac{Y(s)}{X(s)} = \frac{Y(s)}{(s-p_1)(s-p_2)\cdots(s-p_n)}$$

$$= \frac{R_1}{(s-p_1)} + \frac{R_1}{(s-p_2)} + \cdots + \frac{R_1}{(s-p_n)} \quad (7.111)$$

where p_i are the poles of $H(s)$, and the constant denominators R_i are given by the expression

$$R_i = (s-p_i)F(s)|_{s=p_i} \quad (7.112)$$

for single poles. A multiple pole p_k of order m will give rise to m terms in the sum Eq. (7.111), i.e.

$$\frac{R_{k1}}{(s-p_k)^m} + \frac{R_{k2}}{(s-p_k)^{m-1}} + \cdots + \frac{R_{km}}{(s-p_k)} \quad (7.113)$$

where the denominator constants have to be derived through differentiation, that is,

$$R_{km} = \frac{1}{(m-1)!}\frac{\mathrm{d}^{m-1}}{\mathrm{d}s^{m-1}}[(s-p_k)^m F(s)]|_{s-p_k} \quad (7.114)$$

The inverse transform of Eq. (7.111) is easily obtained using Laplace transform tables, [232], or direct derivation, which yields

$$\mathscr{L}^{-1}\{H(s)\} = \mathscr{L}^{-1}\left\{\sum_{i=1}^{n}\frac{R_i}{s-p_i}\right\} = \sum_{i=1}^{n} R_i \mathrm{e}^{p_i t} = h(t) \quad (7.115)$$

By Eq. (7.113), a multiple pole p_k will result in m terms in the inverse transform Eq. (7.115), the typical term given by

$$\mathscr{L}^{-1}\left\{\frac{R_{k1}}{(s-p_k)^m}\right\} = \frac{R_{k1}}{(m-1)!} t^{m-1}\mathrm{e}^{p_k t} \quad (7.116)$$

This whole exercise goes to show that the impulse response function $h(t)$ will be unbounded for poles in the right half of the s-plane, that is, for $x > 0$ ($s = x + iy$), as the exponential term $\mathrm{e}^{p_k t}$ goes to infinity. Thus, the filter becomes unstable. In order to have a stable filter, the poles of the filter transfer function $H(s)$ must lie in the left half of the s-plane in order to produce exponentially decaying impulse response functions $h(t)$. Analogue filters have already been discussed in another context in Section 3.3.2, where the Butterworth filter was briefly explained.

Next, consider the counterpart of the filter convolution Eq. (7.109) in the index domain for discrete signals and filters (see Eq. (7.68)), that is,

$$y(n) = \sum_{m=0}^{M-1} h(m)x(n-m) \quad (7.68)$$

The corresponding relation in the z-domain is given by (see Table 7.1)

$$Y(z) = H_d(z)X(z) \tag{7.117}$$

giving a filter transfer function $H_d(z) = Y(z)/X(z)$, which is comparable to the filter transfer function $H_a(s)$ of the analogue filter. The frequency response of the digital filter is found by setting $z = e^{i\omega\Delta}$ (cf. Eq. (7.48)), whereby

$$H_d(\omega) = \left.\frac{Y(z)}{X(z)}\right|_{z = e^{i\omega\Delta}} \tag{7.118}$$

which is the transfer function for a sinusoidal input of frequency ω. Likewise, the frequency response of the analogue filter $H_a(s)$ is found by setting $s = e^{i\omega}$.

The above discussion shows that a desired recursive digital filter (IIR), having the characteristics of a certain analogue filter, can be designed by applying a suitable transformation of the analogue transfer function $H_a(s)$ to obtain the corresponding digital transfer function $H_d(z)$. This means that a mapping of the left half of the s-plane to the z-plane is required. From Eqs. (7.48) and (7.95), it seems plausible to define the transformation

$$z = e^{\Delta s} \qquad \text{or} \qquad s = \frac{1}{\Delta}\log_e z \tag{7.119}$$

Let $s = \sigma + i\omega$ and $z = x + iy = re^{i\theta}$. By Eq. (7.119),

$$re^{i\theta} = e^{\Delta s} = e^{\Delta\sigma}\, e^{i\omega\Delta} \tag{7.120}$$

whereby

$$r = e^{\Delta\sigma}, \qquad \theta = \omega\Delta = 2\pi\omega/\omega_s \tag{7.121}$$

where ω_s is the sampling frequency. Thus, the left half of the s-plane is mapped into the interior of a unit circle in the z-plane ($-\infty < \sigma < 0 \rightarrow 0 < r < 1$) and the right half of the s-plane is mapped into the exterior of the unit circle. From Eq. (7.121) it is also seen that a horizontal strip of width ω_s in the left half of the s-plane, say, between $-\omega_s/2$ and $\omega_s/2$ completely fills the interior of the unit circle, and strips of the same width between, say, $k\omega_s$ and $(k+1)\,\omega_s$, $k = \pm 1, \pm 2, \ldots$, are mapped on top of each other. This is in concordance with the aliasing effect shown in Fig. 7.13.

The design of recursive filters is thus carried out by seeking suitable approximations to the transformation $H_a(s)$ to $H_d(z)$, using the above mapping. There are several methods in use, which are extensively treated in the specialist literature (see for instance [49], [160] and [175]). In the following, only one such method will be discussed, namely the bilinear transformation. The so-called bilinear transformation, which is an algebraic transformation between the variables s and z, avoids the problem of aliasing by transforming the entire $i\omega$ axis in the s-plane to one revolution of the unit circle in the z-plane. Therefore, $-\infty < \omega < \infty$ maps onto $-\pi < \theta < \pi$, which indicates that the transformation between continous time and discrete time frequency variables must be nonlinear, resulting in a warping of the frequency axis. The bilinear transformation may be introduced as follows. Consider a continuous time signal $y(t)$, which is the output of a simple analogue filter with the transfer function

$$H_a(s) = \frac{Y(s)}{X(s)} = \frac{a}{s+a} \tag{7.122}$$

which corresponds to the differential equation

$$y'(t) = -ay + ax \tag{7.123}$$

where a is a constant. Using the sampling interval Δ, the value $y(k)$ at $t = k\Delta$ is obtained from the previous value $y(k-1)$ using the relation

$$y(k) = y(k-1) + \Delta y \tag{7.124}$$

The increment Δy can be evaluated by different approximative methods using the differential equation Eq. (7.123). The bilinear transformation uses a trapezoidal approximation to Δy, that is, the area of a trapezoidal column with the sides $-ay + ax$, evaluated at $(k-1)\Delta$ and $k\Delta$ respectively. Therefore,

$$\Delta y = y(k) - y(k-1) = [(-ay(k) + ax(k)) + (-ay(k-1) + ax(k-1))]\Delta/2 \tag{7.125}$$

This results in the difference equation

$$(1 + a\Delta/2)y(k) = (1 - a\Delta/2)y(k-1) + a\Delta[x(k-1) + x(k)]/2 \tag{7.126}$$

Taking the z-transform of both sides,

$$(1 + a\Delta/2)Y(z) = (1 - a\Delta/2)z^{-1}Y(z) + a\Delta[z^{-1}X(z) + X(z)]/2$$

giving the transfer function

$$H_d(z) = \frac{(a\Delta/2)(1 + z^{-1})}{1 + (a\Delta/2) - (1 - a\Delta/2)z^{-1}} = \frac{a}{\frac{2}{\Delta}\cdot\frac{z-1}{z+1} + a} \tag{7.127}$$

Comparing Eqs. (7.122) and (7.127) it is seen that the coordinate substitution

$$s = \frac{2}{\Delta}\frac{z-1}{z+1} \tag{7.128}$$

gives the digital transfer function $H_d(z)$ from the analogue transfer function $H_a(s) = a/(s+a)$. The transformation Eq. (7.128) is called the bilinear transformation because of the two linear functions of z in the expression.

To obtain the properties of the bilinear transformation, Eq. (7.128) is solved for z giving

$$z = \frac{1 + s\Delta/2}{1 - s\Delta/2} = \frac{1 + \sigma\Delta/2 + i\omega\Delta/2}{1 - \sigma\Delta/2 - i\omega\Delta/2} \tag{7.129}$$

For $\sigma < 0$, it follows from Eq. (7.129) that $|z| < 1$ for any value of ω, that is, if $H_a(s)$ has a pole in the left half of the s-plane, its image in the z-plane is within the unit circle.

Therefore, causal stable continuous-time filters will result in causal stable discrete-time filters. Furthermore, let $s = i\omega$. Then,

$$z = \frac{1 + i\omega\Delta/2}{1 - i\omega\Delta/2} \tag{7.130}$$

From Eq. (7.130) it follows that $|z| = 1$ for all values of s on the $i\omega$-axis, which shows that the $i\omega$-axis is mapped onto the circumference of the unit circle in the z-plane. Finally, the following relation exists between the continuous time frequency ω and the digital frequency ω_d. Eq. (7.128) evaluated on the unit circle in the z-plane, $z = e^{i\omega_d\Delta}$, gives

$$s = \sigma + i\omega = \frac{2}{\Delta} \cdot \frac{e^{i\omega_d\Delta} - 1}{e^{i\omega_d\Delta} + 1} = i\frac{2}{\Delta} \tan\frac{\omega_d\Delta}{2} \tag{7.131}$$

which shows that $\sigma = 0$ and

$$\omega = \frac{2}{\Delta} \tan\frac{\omega_d\Delta}{2} \tag{7.132}$$

or

$$\omega_d = \frac{2}{\Delta} \arctan\frac{\omega\Delta}{2} \tag{7.133}$$

The properties of the bilinear transformation are summarized as follows. From Eq. (7.133), the range of frequencies $0 < \omega < \infty$ is mapped into $0 < \omega_d < \pi$. The bilinear transformation avoids the problem of aliasing as it maps the entire imaginary axis of the s-plane onto the circumference of the unit circle in the z-plane. This is obtained at the price of severe frequency distortion as made evident by Eq. (7.132). The transformation Eq. (7.133) is often referred to as prewarping of the continuous frequency ω.

A few basic analogue filters are mostly used as tools for designing IIR filters. Besides the Butterworth filter (see Eq. (3.61) there are two other filters, which are commonly used, namely the Chebyshev filters I and II and elliptic filters. The Chebyshev filter has the gain function

$$|H(\omega)|^2 = \frac{1}{1 + C_N(\omega/\omega_c)} \tag{7.134}$$

where $C_N(x)$ is a Nth-order Chebyshev polynomial defined as

$$C_N(x) = \begin{cases} \cos(N \cos^{-1} x) & |x| \leqslant 1 \\ \cosh(N \cosh^{-1} x) & |x| > 1 \end{cases} \tag{7.135}$$

The elliptic filter on the other hand has the gain function

$$|H(\omega)|^2 = \frac{1}{1 + U_N(\omega/\omega_c)} \tag{7.136}$$

where $U_N(x)$ is a Jacobian elliptic function of order N. For a discussion of these filters and other kinds of analogue filters required for more specific purposes, the reader is referred to the specialist literature.

Example 7.11 A simple lowpass digital filter $H_d(z)$ is to be designed applying the bilinear transformation, using a Butterworth filter as an analogue prototype. The digital filter is to have the following specifications:

$$0.9 \leqslant |H_d(e^{i\omega})| \leqslant 1 \qquad \text{for} \qquad 0 \leqslant \omega_d \leqslant 0.6\pi/\Delta$$

$$|H_d(e^{i\omega})| \leqslant 0.2 \qquad \text{for} \qquad 0.8\pi/\Delta \leqslant \omega_d \leqslant \pi/\Delta$$

which indicates that the filter is essentially a lowpass filter with the corner frequency about $0.3f_s$ where f_s is the sampling frequency. The -3 dB frequency $\omega_n = 0.6790$. The corresponding analogue filter must therefore satisfy the following conditions:

$$0.9 \leqslant |H(\omega)| \leqslant 1 \qquad \text{for} \qquad 0 \leqslant \omega \leqslant (2/\Delta)\tan(0.3\pi)$$
$$|H_d(e^{i\omega})| \leqslant 0.2 \qquad \text{for} \qquad (2/\Delta)\tan(0.4\pi) \leqslant \omega < \infty$$

using the frequency warping Eq. (7.132). Now, a Nth-order analogue Butterworth filter has the gain function

$$|H(\omega)|^2 = \frac{1}{1 + (\omega/\omega_c)^{2N}} \tag{3.61}$$

where ω_c is the cutoff frequency. In accordance with the above conditions, the gain function has to satisfy the following two inequalities, that is,

$$|H(2\tan(0.3\pi))| \geqslant 0.9 \qquad \text{and} \qquad |H(2\tan(0.4\pi))| \leqslant 0.2$$

in which the time step Δ has been put equal to one for the sake of convenience. As the task is to determine the two parameters of the analogue Butterworth filter, namely the order N and the cutoff frequency ω_c, which best satisfy the above requirements, the inequality sign can be dropped, and using Eq. (3.61),

$$1 + (2\tan(0.3\pi)/\omega_c)^{2N} = \frac{1}{0.9^2} \qquad \text{and} \qquad 1 + (2\tan(0.4\pi)/\omega_c)^{2N} = \frac{1}{0.2^2}$$

Solving for N,

$$\left(\frac{\tan(0.3\pi)}{\tan(0.4\pi)}\right)^{2N} = \frac{(1/0.9^2) - 1}{(1/0.2^2) - 1} \rightarrow N = 2.8755\ldots$$

whereby $N = 3$. The cutoff frequency is then obtained from the second inequality or

$$(2\tan(0.4\pi)/\omega_c)^6 = \frac{1}{0.2^2} - 1 = 24 \rightarrow \omega_c = 3.624\ldots\text{rad/s}$$

which has as a result that the passband specifications are exceeded and the stopband specifications are met exactly. In other words, with proper prewarping of the frequency, the resulting digital filter will meet the specifications exactly at the stopband edge.

Now, to conclude the filter design, the transfer function of the third-order Butterworth filter has to be examined. Study the analogue transfer function

$$H(s)H(-s) = \frac{1}{1 - s^6}$$

which is equivalent to the third-order Butterworth function Eq. (3.61) with $\omega_c = 1$. The poles are given by the roots of the equation

$$s^6 = 1 = e^{i2k\pi} \quad \text{or} \quad s_k = e^{ik\pi/3}, \qquad k = 0, 1, 2, 3, 4, 5$$

Therefore,

$$s_0 = e^{i0} = 1 \qquad s_1 = e^{i\pi/3} = \tfrac{1}{2} + i\sqrt{3}/2$$

$$s_2 = e^{i2\pi/3} = -1/2 + i\sqrt{3}/2 \qquad s_3 = e^{i\pi} = -1$$

$$s_4 = e^{i4\pi/3} = -1/2 - i\sqrt{3}/2 \qquad s_5 = e^{i5\pi/3} = \tfrac{1}{2} + i\sqrt{3}/2$$

The poles in the left half-plane belong to $H(s)$ whereby

$$H(s) = \frac{1}{(s+1)(s + \frac{1}{2} - i\sqrt{3}/2)(s + \frac{1}{2} + i\sqrt{3}/2)}$$

Introducing $s = s/\omega_c$, the transfer function is written as

$$H(s) = \frac{\omega_i^3}{(s + \omega_c)(s^2 + s\omega_c + \omega_c^2)}$$

Applying the bilinear transformation Eq. (7.128) with $\Delta = 1$,

$$H_d(z) = \frac{\omega_c^3 (z+1)^3}{(2(z-1) + \omega_c(z+1))(4(z-1)^2 + 2(z-1))(z+1)\omega_c + (z+1)^2\omega_c^2}$$

or finally,

$$H_d(z) = \frac{0.347(z+1)^3}{(z + 0.2888)(z^2 + 0.749z + 0.4054)}$$

The magnitude of the frequency response of the discrete time filter is shown in Fig. 7.17. The filter gain at the crucial frequencies, that is, $\omega_i = (0.6, 0.6790, 0.8)$, is $(0.9159, 0.7071, 0.2)$, which verifies the statement that the specifications in the passband are exceeded and are exact for the stopband. The response of the digital filter falls off more quickly than for the continuous filter since the entire $i\omega$-axis has been warped to a circle in the discrete frequency domain. This is particularly obvious for the two corresponding values of $H(\omega)$ at $\omega = \infty$ and $H_d(e^{i\omega_d\Delta})$ at $\omega_d = \pi/\Delta$.

It should be noticed that using the frequency relation Eq. (7.132), the discrete Nth-order Butterworth filter can be obtained by the following simple expression:

$$|H(e^{i\omega_d\Delta})|^2 = \frac{1}{1 + (\tan(\omega_d\Delta/2)/\tan(\omega_c\Delta/2))^{2N}} \tag{1.137}$$

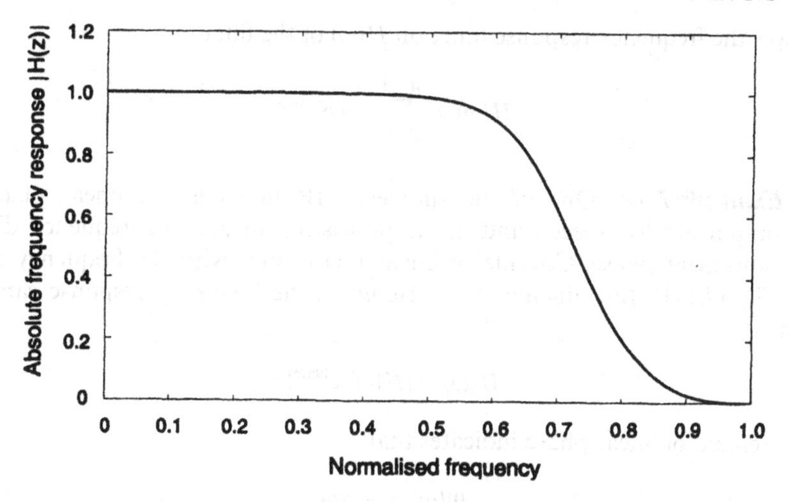

Fig. 7.17 *The frequency response of a discrete time 3rd order Butterworth filter*

The frequency response according to Eq. (7.137) has all the main characteristics of the continuous Butterworth filter. However, the function Eq. (7.137) is periodic with period 2π and falls off more sharply than the continuous filter response. This route to the discrete time filter is not practical, however, as the determination of the pole location is no longer straightforward. Therefore, the method shown in Example 7.11 is to be preferred. Actually, the various signal processing software packages offer direct solutions to the design of digital filters, which does not require the lengthy calculations of Example 7.11.

Turning now to non-recursive filters, the FIR filter has an impulse response that is zero outside of some finite index interval, which essentially produces a finite length record of an infinite process. For a causal FIR filter, the impulse response is given by

$$h(n) = \begin{cases} f(n), 0 \leqslant n \leqslant M-1 \\ 0, n < 0, n > M-1 \end{cases} \tag{7.138}$$

where $f(n)$ is a filter shape function and M is the duration ($T = M\Delta$). Passing of an infinite signal $x(n)$ through the above FIR filter results in the output signal

$$y(n) = \sum_{k=0}^{M-1} f(k)x(n-k) \tag{7.139}$$

Thus, the output at any time index n is simply a weighted linear combination of the input signal samples $x(n)$, $x(n-1), \ldots, x(n-M+1)$ using the "weights" $f(k)$, $k = 0, 1, 2, \ldots, M-1$.

Taking the z-transform of both sides of Eq. (7.139)

$$Y(z) = \sum_{k=0}^{M-1} f(k)X(z)z^{-k} = H(z)X(z) \tag{7.140}$$

which gives the frequency response function $H(\omega)$ of the filter

$$H(\omega)=\sum_{n=0}^{M-1} f(n)e^{-in\omega} \tag{7.141}$$

Example 7.12 One of the simplest FIR filters is the linear phase filter, which is important for speech and music processing to prevent frequency distortion through non-linear phase. Consider a linear phase filter with the frequency response given by Eq. (7.141). Introducing the phase $\Phi(\omega)$, the frequency response can also be written as

$$H(\omega)=|H(\omega)|e^{i\Phi(\omega)} \tag{7.142}$$

The requirement of linear phase indicates that

$$\Phi(\omega)=-\alpha\omega \tag{7.143}$$

where α is a real constant. Thus, Eq. (7.142) becomes

$$H(\omega)=|H(\omega)|e^{-i\alpha\omega} \tag{7.144}$$

which by Eq. (7.141) gives the following relation:

$$\sum_{n=0}^{M-1} f(n)e^{-in\omega}=|H(\omega)|e^{-in\alpha\omega} \tag{7.145}$$

Taking the real and imaginary parts of Eq. (7.145) separately,

$$\sum_{n=0}^{M-1} f(n)\cos n\omega=|H(\omega)|\cos n\alpha\omega \tag{7.146a}$$

and

$$\sum_{n=0}^{M-1} f(n)\sin n\omega=|H(\omega)|\sin n\alpha\omega \tag{7.146b}$$

Multiplying the Eq. (7.146a) by $\sin n\alpha\omega$ and Eq. (7.146b) by $\cos n\alpha\omega$ and subtracting one from the other yields

$$\sum_{n=0}^{M-1} f(n)\cos n\omega \sin n\alpha\omega-\sum_{n=0}^{M-1} f(n)\sin n\omega\cos n\alpha\omega$$

$$=\sum_{n=0}^{M-1} f(n)\sin[(\alpha-n)\omega]=0 \tag{7.147}$$

The only nontrivial solutions to Eq. (7.147) for α and $f(n)$ are obtained by imposing a symmetry condition on the filter, that is,

$$f(n)=f(M-1-n) \tag{7.148}$$

and putting

$$\alpha = \frac{M-1}{2} \tag{7.149}$$

For instance, if $M = 5$, the series Eq. (7.147) becomes

$$\sum_{n=0}^{4} f(n)\sin(2-n)\omega = f(0)\sin 2\omega + f(1)\sin\omega - f(3)\sin\omega - f(4)\sin 2\omega = 0$$

since

$$f(0) = f(5-1-0) = f(4) \quad \text{and} \quad f(1) = f(5-1-1) = f(3)$$

Thus, the filter is symmetric about $f(2)$, which is an essential quality of the linear phase filter. The uniqueness of the solution of Eq. (7.147) is ensured by the uniqueness of the discrete Fourier transform of a given function $f(n)$.

7.2.7 Data Windows and Smoothing

In the previous section, it was shown how IIR filters could be designed from analogue filters, using a suitable transformation. The FIR filter on the other hand is restricted to discrete time implementation only, and as such is a later element in signal processing theory, arriving with the onset of high-speed computers. Throughout the text, truncated sample functions of stochastic processes of infinite duration have often been introduced out of necessity to have a realizable or usable representation of the process. As has already been discussed, the FIR filters provide means to obtain truncated samples of otherwise infinite records. The concept of windowing offers a convenient interpretation of the design and application of FIR filters. In effect, the system described by Eqs. (7.138) and (7.139) acts as a window that views only the most recent M input signal samples in forming the output signal, which is the required truncated form of the original process. Suitable window functions are therefore often applied when constructing a desired FIR filter. In fact, consider a suitable or desired frequency response specification $H_d(\omega)$ and determine the corresponding discrete impulse response $h_d(n)$. From the Fourier transform relations (see Eq. 7.48),

$$H_d(\omega) = \sum_{k=0}^{\infty} h_d(k)\mathrm{e}^{ik\omega\Delta} \tag{7.150}$$

where

$$h_d(n) = \frac{1}{2\pi}\int_{-\pi}^{\pi} H_d(\omega)\mathrm{e}^{in\omega\Delta}\mathrm{d}\omega \tag{7.151}$$

The impulse response function $h_d(n)$ can be of infinite duration and has to be truncated or modified in some way to provide the FIR filter. This can be done by applying a window function $d(n)$ whereby the FIR filter shape function becomes

$$h(n) = h_d(n)d(n) \tag{7.152}$$

where $d(n)$ is for instance a constant in some interval of interest and zero outside of that interval producing the desired truncation effect. Thus, window functions can be applied in different ways to produce different results.

Window functions can be applied both in the time domain (index domain) and in the frequency domain, in which case they are sometimes called lag windows in connection with the correlation function. They are basically nothing but a graphical interpretation of the convolution theorem, Eqs. (7.22) and (7.24). The shape and composition of the window function will greatly affect the results, and various windows have been proposed for solving different kind of problems. In general terms, if the process is viewed through the window $d(t)$, the correlation function will be affected as follows:

$$Y(t) = X(t)d(t) \tag{7.153}$$

and

$$R_Y(\tau) = R_X(\tau)d(t)d(t+\tau) = R_X(\tau)w(\tau) \tag{7.154}$$

where $w(\tau, t) = d(t)d(t+\tau)$ is called the lag window. Even if the lag window and data window thus seem to be directly related, they are often selected independently from one another as different requirements for accuracy and numerical considerations may call for different window applications.

Consider infinite length records of the stochastic process $\{X(t),\ t\in\mathcal{T}\}$, which are 'viewed' or filtered through the *boxcar* window, Fig. 7.18. The operation can be described by multiplying the infinite record $x(t)$ by the boxcar function $d_{T/2}(t)$ to obtain a finite record $y(t)$ of duration T. Thus the boxcar window is in essence nothing but an ideal lowpass filter with a stopband or cutoff frequency $\omega_c = 2\pi/T$. The convolution in the frequency domain is given by

$$Y(\omega) = \int_{-\infty}^{\infty} x(t)d_{T/2}(t)e^{-i\omega t}\,dt$$

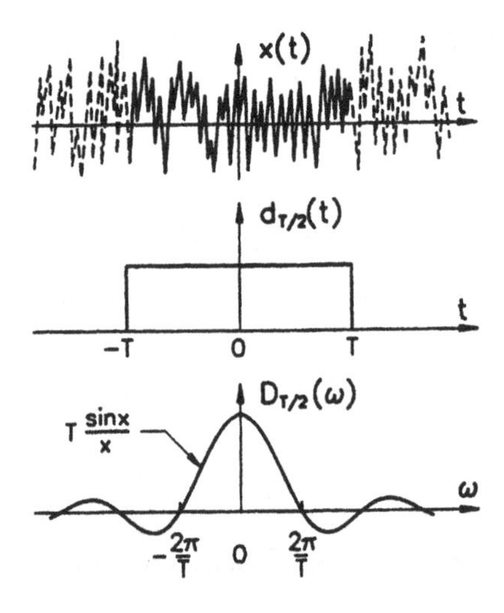

Fig. 7.18 *The boxcar data window*

or

$$Y(\omega) = X(\omega)D_{T/2}(\omega) \tag{7.155}$$

in which the Fourier transform of the data window, the boxcar function, is represented by

$$\begin{aligned} D_{T/2}(\omega) &= \int_{-T/2}^{T/2} d_{T/2}(t)\mathrm{e}^{-\mathrm{i}\omega t}\,\mathrm{d}t \\ &= \int_{-T/2}^{T/2} \cos(\omega t)\mathrm{d}t = T\,\frac{\sin(\omega T/2)}{(\omega T/2)} \end{aligned} \tag{7.156}$$

Because of the large side lobes of the Fourier transform $D_{T/2}(\omega)$ (see Fig. 7.18), power existing in the data at values other than at integral multiples of $2\pi/T$ will be averaged in the values centred at $\omega = 0$. This is partly due to the abrupt cutoff in the data window, and the frequency resolution loss and noise, due to leakage of frequency components outside the filter or window through the side lobes, constitute a part of the undesired effects of the operation.

Example 7.13 To study this phenomenon, usually called the *Gibbs phenomenon*, consider the following exercise where it is desired to construct a FIR filter, which has the frequency response characteristics of an ideal band pass filter, that is,

$$H_d(\omega) = \begin{cases} 1 & \dfrac{\pi}{5\Delta} \leqslant |\omega| \leqslant \dfrac{3\pi}{5\Delta} \\ 0 & \text{otherwise} \end{cases} \tag{7.157}$$

The impulse response function $h_d(n)$ is modified by a discrete version of the one-sided boxcar filter, that is,

$$d(n) = \begin{cases} 1 & 0 \leqslant n \leqslant N-1 \\ 0 & \text{otherwise} \end{cases} \tag{7.158}$$

From Eqs. (7.152), (7.157) and (7.24), the frequency domain convolution is given by

$$H(\omega) = \frac{1}{2\pi}\int_{-\pi/\Delta}^{\pi/\Delta} H_d(\Omega)D(\omega - \Omega)\mathrm{d}\Omega \tag{7.159}$$

which relates the realizable frequency response $H(\omega)$ to the desired response $H_d(\omega)$. To evaluate $H(\omega)$, the frequency response of the window $D(\omega)$ needs to be determined. From Eqs. (7.158) and (7.48),

$$D(\omega) = \sum_{n=0}^{N-1} \mathrm{e}^{\mathrm{i}n\omega\Delta} = \frac{1 - \mathrm{e}^{\mathrm{i}N\omega\Delta}}{1 - \mathrm{e}^{\mathrm{i}\omega\Delta}} = \mathrm{e}^{\mathrm{i}(N-1)\omega\Delta/2}\,\frac{\sin(N\omega\Delta/2)}{\sin(\omega\Delta/2)} \tag{7.160}$$

whereby Eq. (7.159) becomes

$$H(\omega) = \frac{1}{2\pi}\int_{-\pi/\Delta}^{\pi/\Delta} D(\omega - \Omega)\mathrm{d}\Omega = \frac{1}{2\pi}\int_{\omega-\pi/\Delta}^{\omega+\pi/\Delta} D(x)\mathrm{d}x \tag{7.161}$$

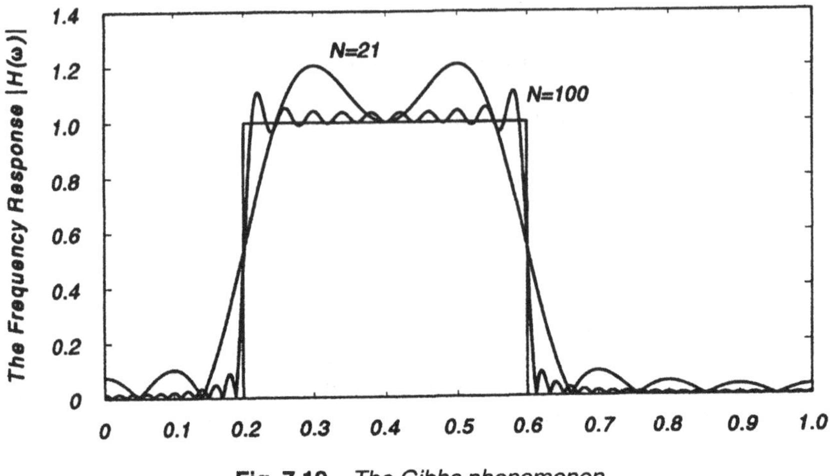

Fig. 7.19 *The Gibbs phenomenon*

in which the integration range is the negative and positive frequency intervals $[\pi/5\Delta, 3\pi/5\Delta]$. Rather than trying to evaluate the integral, the problem can be solved directly by standard mathematical program packages such as the signal toolbox in MATLAB, [139].

In Fig. 7.19, the main results are illustrated for $N = 21$ and 101. The frequency of the ripples in the passband and stopband increases as the number of terms in the Fourier transform, N, increases. The ripples are especially vivid at the cutoff frequency where the discontinuity is about 9%. Actually, regardless of how many terms are used in the Fourier transform, the magnitude of this discontinuity around the cutoff frequency remains the same. This phenomenon is called the Gibbs phenomenon. Approaching an infinite number of terms, the discontinuity will be compressed into a line at the cutoff frequency, which still shows the 9% error. However, it can be stated that the frequency response of the realizable filter truly represents that of the desired filter, as the discontinuity happens at one point only in the frequency range.

To counter-effect some of these problems, which are mostly related to the abrupt changes in frequency at the cutoff frequency, various measures have been suggested. The tapering of the window at its corner frequency or other similar methods can be used. For instance, the boxcar window can be improved by using a cosine taper over 1/10 of each end of the data (see Fig. 7.20). The actual window function, expressed in real time t, is given by the following expression:

$$d_{T/2}(t) = \begin{cases} \cos^2 \dfrac{5\pi t}{T} & -\dfrac{T}{2} \leqslant t < -\dfrac{4T}{10} \\ 1 & -\dfrac{4T}{10} \leqslant t < \dfrac{4T}{10} \\ \cos^2 \dfrac{5\pi t}{T} & \dfrac{4T}{10} \leqslant t < \dfrac{T}{2} \\ 0 & \text{elsewhere} \end{cases} \tag{7.162}$$

The Fourier transform of the tapered cosine window is calculated using the cosine

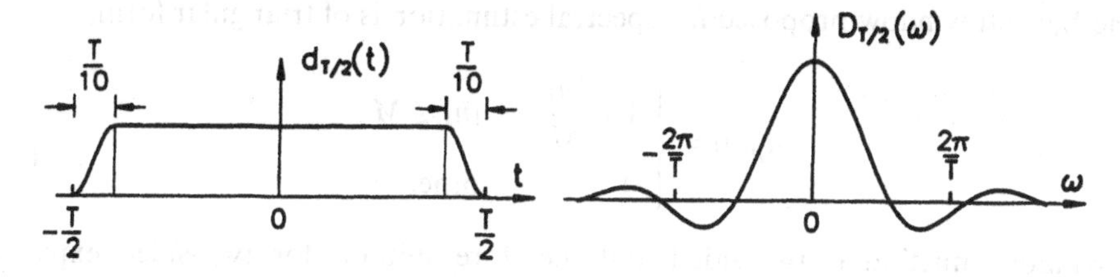

Fig. 7.20 *Cosine taper window and corresponding filter shape*

transform Eq. (7.20), and the result is as follows:

$$D_{T/2}(\omega)=\frac{20\,T}{9}\frac{\sin(9\omega T/10)}{(9\omega T/10)}\cos\omega\frac{T}{10}\frac{(10\pi/T)^2}{(10\pi/T)^2-\omega^2} \tag{7.163}$$

which gives the effective filter shape as shown in Fig. 7.20. It is noted that the two big side lobes are now within the effective frequency range.

The selection of tapering functions sometimes called windowing, has turned into a highly specialized subject and numerous window functions have been suggested. The desirable window characteristics can be described as follows:

(a) The width of the main lobe of the window frequency response should be kept as small as possible, containing as much as possible of the total area or energy.
(b) The side lobes of the frequency response should decrease rapidly as $\omega\Delta$ approaches π.

The above characteristics are contradictory by nature, whence a trade-off between these main goals will be necessary. Of widely used window functions, the Hamming and Bartlett windows may be mentioned. The Hamming window is a modified rectangular boxcar window, which has the following discrete function (see Eq. (7.158))

$$d_H(n)=\begin{cases}1 & -(M-1)\leqslant n\leqslant M-1\\ \dfrac{1}{2} & n=-M, M\\ 0 & \text{otherwise.}\end{cases} \tag{7.164}$$

Taking the z-transform of the Hamming window

$$D_H(z)=\sum_{n=-(M-1)}^{M-1} z^{-n}+\tfrac{1}{2}(z^M+z^{-M}) \tag{7.165}$$

The frequency response of the Hamming window is obtained by letting $z=e^{i\omega T}$. Then, Eq. (7.165) becomes

$$D_H(\omega T)=\frac{\sin(M+\frac{1}{2})\omega T}{\sin(\omega T/2)}-\cos M\omega T=\frac{\sin(M\omega T)\cos(\omega T/2)}{\sin(\omega T/2)} \tag{7.166}$$

The first part of the frequency response is essentially the same as Eq. (7.156), which gives the frequency response of the two-sided boxcar. The added cosine term tends to zero as ωT goes to π and partly smoothens out the ripples around the cutoff frequency. However, often a better smoothing than offered by the modified boxcar window is needed.

The Bartlett window proposed for spectral estimation is of triangular form,

$$d_B(n) = \begin{cases} 1 - \dfrac{|n|}{M} & |n| \leqslant M \\ 0 & \text{otherwise} \end{cases} \tag{7.167}$$

The window function is two-sided and therefore suitable for two-sided expansions truncated at $\pm M$. The window can also be applied as a one-sided window. The z-transform of the Bartlett window is

$$D_B(z) = \sum_{n=-M}^{M} \left(1 - \frac{|n|}{M}\right) z^{-n} \tag{7.168}$$

Setting $z = e^{i\omega T}$, the frequency response is obtained by

$$D_B(\omega T) = \left(\frac{\sin(M\omega T/2)}{M\omega T/2}\right)^2 \tag{7.169}$$

The main lobe of the Bartlett window is in the frequency range $[-4\pi/M, 4\pi/M]$, which is twice the width of the boxcar window (see Fig. 7.18).

A more elaborate real time cosine window has been proposed by von Hann, where

$$d_{\text{VH}}(t) = \begin{cases} \dfrac{1}{2}\left(1 + \cos\dfrac{2\pi t}{T}\right) & -\dfrac{T}{2} \leqslant t \leqslant \dfrac{T}{2} \\ 0 & \text{otherwise} \end{cases} \tag{7.170}$$

Applying the von Hann window to a sample function $x(t)$ is equivalent to using discrete convolution weights $(\frac{1}{4}, \frac{1}{2}, \frac{1}{4})$ for the discrete Fourier transform of $x(t)$ as shown in Example 7.14.

Example 7.14 A sample function $x(t)$ of the stochastic process, $\{X(t), t \in \mathscr{T}\}$ is processed through the von Hann cosine data window $d(t)$, whereby sample functions $y(t)$ of duration T are produced. Find the corresponding spectral window or the effective filter shape $D(\omega)$ and the lag spectral window $W(\omega)$.

Solution Since $d(t)$ is an even function, the cosine transform Eq. (7.21) contains all the spectral information:

$$D(\omega) = 2\int_0^\infty d(t)\cos\omega t\,\mathrm{d}t = \int_0^{T/2} \left(\cos\omega t + \cos\omega t \cos\frac{2\pi t}{T}\right)\mathrm{d}t$$

$$= \int_0^{T/2} \left[\cos\omega t + \frac{1}{2}\left(\cos\left(\omega t + \frac{2\pi t}{T}\right) + \cos\left(\omega t - \frac{2\pi t}{T}\right)\right)\right]\mathrm{d}t$$

$$= \frac{T}{2}\left[\frac{\sin\dfrac{\omega T}{2}}{\dfrac{\omega T}{2}} + \frac{1}{2}\,\frac{\sin\left(\dfrac{\omega T}{2} + \pi\right)}{\dfrac{\omega T}{2} + \pi} + \frac{1}{2}\,\frac{\sin\left(\dfrac{\omega T}{2} - \pi\right)}{\dfrac{\omega T}{2} - \pi}\right]$$

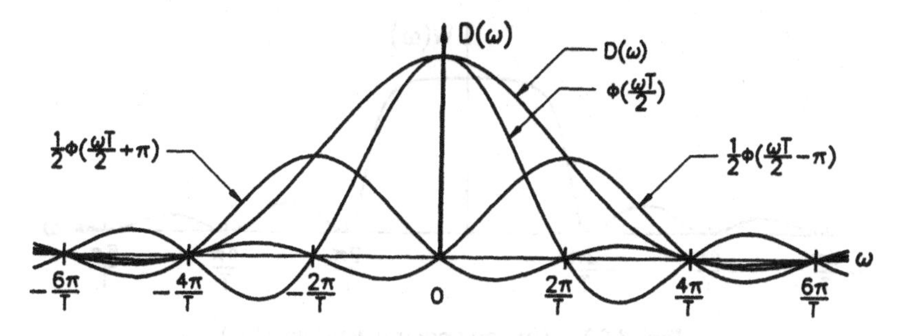

Fig. 7.21 *The effective filter shape D(ω)*

The Fourier transform $D(\omega)$ consists of the real part only, which has simple form in terms of the function $\Phi(x) = (\sin x)/x$, that is (see Fig. 7.21),

$$D(\omega) = (T/2)\left[\Phi(\omega T/2) + \tfrac{1}{2}\Phi(\omega T/2 + \pi) + \tfrac{1}{2}\Phi(\omega T/2 - \pi)\right]$$

The spectral window obviously has the values

$$\{D(-2\pi/T), D(0), D(2\pi/T)\} = T(\tfrac{1}{4}, \tfrac{1}{2}, \tfrac{1}{4})$$

and is zero for all other multiples of $2\pi/T$. Since the discrete Fourier transform of $x(t)$ is evaluated at these frequencies (see Eq. (7.48)), the convolution weights are equivalent to $(\tfrac{1}{4}, \tfrac{1}{2}, \tfrac{1}{4})$ as previously mentioned.

The spectral lag window or simply the spectral window $W(\omega)$ is the Fourier transform of the lag window in the time domain, $w(\tau)$. Most often the spectral window for the PSD functions or the lag window for the autocorrelation functions is selected independently of the data window. In this case however, the lag spectral window $W(\omega)$, is related to the Fourier transform of the data window, that is

$$W(\Omega) = D(\Omega)\, D^*(\Omega)$$

and

$$W(\omega) = |D(\omega)|^2 = D^2(\omega) \text{ since } D(\omega) \text{ is a real function.}$$

Therefore from the previously obtained function $D(\omega)$,

$$W(\omega) = T^2/4\,[\text{sum of nine terms: } a_n \Phi(\omega T/2 + n\pi)\,\Phi(\omega T/2 + m\pi)]$$

where a_n are the convolution weights $\{1, \tfrac{1}{2}, \tfrac{1}{4}\}$ and $n, m = \{-1, 0, +1\}$. The spectral window is shown in Fig. 7.22.

The selection of proper windows for various uses in signal processing has turned into a highly specialized subject. Many more windows have been proposed, and various alternative tapering forms have been studied. The Blackman window is a linear combination of several cosine terms whereas the Kaiser window is based on the ratio of two zero-order Bessel functions. A good overview of window functions and windowing is found in the text of Loy [134], where the properties and performance of 10 window functions are shown.

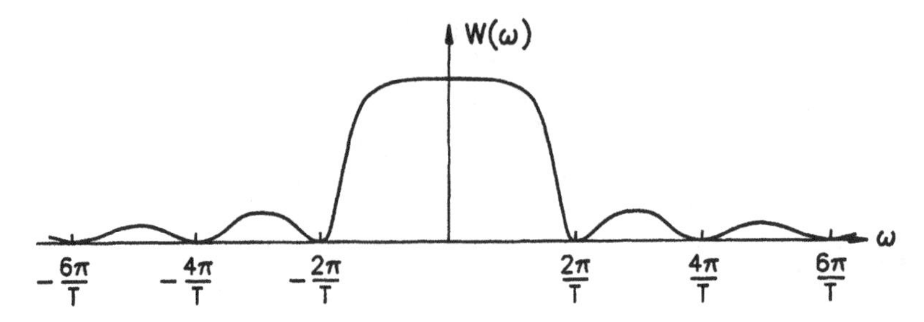

Fig. 7.22 *The lag spectral window W(ω)*

Finally, a few remarks about smoothing and averaging techniques are in order, following the outline presented by Otnes and Enochsson, [164]. Already the idea of zero padding a record to comply with FFT requirement of 2^p-data points and also for convolution calculations where two different records have to be brought up to similar length, has been discussed. Taking a raw data record, the spacing between the points is $\Delta = T/N$, where T is the record length and spacing between the discrete Fourier values is $b = 2\pi/T$. Regarding the boxcar window, the spacing between the first zero crossing on both sides of the main lobe (see Fig. 7.17) is $B_e = 2\pi/T$. When zeroes are added to the sequence, nothing is contributed to the basic shape of the frequency response of the filter, $D_{T/2}(\omega)$, and hence the width of the main lobe is unchanged. However, due to the computational scheme for the discrete Fourier transform points, the spacing is now

$$b = 2\pi/(T + N_z\Delta) \tag{7.171}$$

where N_z is the number of zeroes that has been added. For instance, if a number of zeroes equal to the number of raw data points N is added, the spacing of the Fourier points is halved. This change in the spacing of the Fourier points can lead to some unusual effects, since the convolution with the data window may or may not lead to the desired side lobe cancellation. The problem of modifying the Fourier transform, when an arbitrary number of zeroes has been added to the data record, is therefore equivalent to a filter design.

Frequency averaging is one method of reducing statistical variability and smoothing the spectral estimates. Take r PSD spectral estimate neighbours $S(k)$ and average them as follows:

$$\bar{S}(k) = 1/r\,[S(k) + S(k+1) + \cdots + S(k+r-1)] \tag{7.172}$$

The average spectral value $\bar{S}(k)$ will be distributed as a χ^2 random variable with approximately $2r$ degrees of freedom, Example 2.2. The effective filter shape will be roughly trapezoidal, Fig. 7.23, since adding together triangular shapes (e.g. the main lobe of simple cosine filters) that overlap at the half power points, yields a trapezoid. The effective bandwidth of the new filter shape will be approximately $B_e = rb = r2\pi/T$. The estimate $\bar{S}(k)$ may be considered as representing the midpoing of the frequency interval from kb to $(k+r-1)\,b$, hence the name frequency averaging.

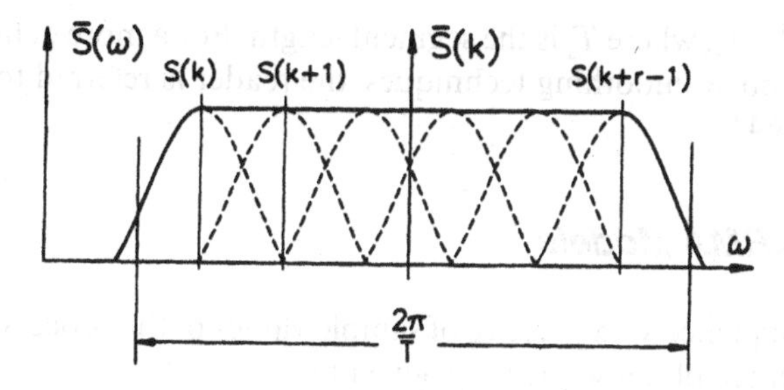

Fig. 7.23 *Effective filter shape after averaging, (after Otnes and Enochsson, [63]), reproduced by permission of John Wiley & Sons, Inc.*

The simple frequency averaging procedure, Eq. (7.172), can be generalized to a more general convolution. For instance,

$$\overline{S_X}(k) = \sum_{i=-m}^{m} a(i) S_X(k-i), \qquad k = 0, 1, 2, \ldots, \frac{N}{2} \tag{7.173}$$

where $a(i)$ are the convolution weights. An example of such smoothing procedures is given by the Hanning window, where the weights are $m = 1$, $a(0) = \frac{1}{2}$, $a(-1) = a(1) = \frac{1}{4}$ with the original spacing of the raw estimates without zero padding is $b = 2\pi/T$. The Hanning window can be interpreted as a special case of binomial smoothing, where the unnormalized weights are

$$\begin{array}{c} 1 \;\; 2 \;\; 1 \\ 1 \;\; 3 \;\; 3 \;\; 1 \\ 1 \;\; 4 \;\; 6 \;\; 4 \;\; 1 \\ \cdots \end{array}$$

In the limit, these weights would be calculated as ordinates of the Gaussian probability density. In general, any recursive filter could be used to smooth a spectrum. However, it should be applied in both directions to avoid the phase shift and retain a symmetric effect.

For long data records, an accepted method for smoothing is to divide the total time history into segments and compute the spectrum for each segment. The average of all such spectra is called an ensemble average, but should not be confused with ensemble average where the average is taken across an ensemble, that is, a large number of available time history representations of the process (cf. Chapter 2). Assuming that the process being studied is stationary, this method appears to be feasible, since the statistics of the various segments will be the same. However, it is also possible to allow for evolutionary process, assuming that the statistics is slowly varying or does not change too rapidly with respect to the segment length. A sensible approach would be to base the segment length on the resolution requirements. That is, if $B_e = b$ has been specified, then the segment length is $2\pi/b$ plus adding zeros to the next power of 2. The spectral window to be used has to have small side lobes to prevent power leakage.

Finally, the two above described methods can be combined, since they are both linear operations in the general sense. Thus, if an ensemble averaging over m segments and then further averaging over r neighbouring spectral estimates is performed, the final spectral estimates obtained will have $n = 2rm$ degrees of freedom in the χ^2 distribution and a resolution $B_e \approx 2\pi/T_s$, where T_s is the segment length. For a more detailed and complete discussion of the above smoothing techniques, the reader is referred to the text of Otnes and Enochsson [164].

7.2.8 ARMA Models

Consider a finite sequence $x_1, x_2, \ldots, x_N$ of sample values of the process $\{x(n), n \geqslant 0\}$. The mean value or average of the sequence is given by

$$\bar{x} = \frac{1}{N} \sum_{k=1}^{N} x_k \tag{7.174}$$

This operation is in effect the same as applying a lowpass filter to the signal, which only passes a single frequency component, i.e. the zero frequency mean. For this purpose, however, all the signal samples must be determined. For very long sequences, this operation loses significance, whereas a running estimate of the mean may be obtained, that is, a moving average. To obtain a moving average of the raw unfiltered data, the one-sided window of length M can be applied with the finite impulse response

$$h(n) = \frac{1}{M+1}, \qquad 0 \leqslant n \leqslant M \tag{7.175}$$

In effect, the window is moved along the data and the output sequence is then given by

$$y(n) = \sum_{k=-\infty}^{\infty} h(k)\, x(n-k) = \frac{1}{M+1} \sum_{k=0}^{M} x(n-k) = \bar{x}_n \tag{7.176}$$

in which $\bar{x}_n$ is the moving average of the sequence at point n. The moving average filter Eq. (7.175) is the essentially the one-sided boxcar window Eq. (7.150) with the Fourier transform or frequency response

$$H(\omega) = \frac{e^{-iM\omega\Delta/2}}{M+1} \frac{\sin((M+1)\omega\Delta/2)}{\sin(\omega\Delta/2)} \tag{7.177}$$

The frequency response magnitude $|H(\omega)|$ is depicted in Fig. 7.24. For visualization purposes, the negative lobes are maintained. The main lobe is from $-2\pi/(M+1)$ to $2\pi/(M+1)$. Introducing the sampling frequency $\omega_s = 2\pi/\Delta$, the main lobe width is $2\omega_s/\Delta$. The moving average operation can therefore be regarded as lowpass filtering with a cutoff frequency $\omega_c = \omega_s/2(M+1)$.

Next, consider an important class of linear time-invariant discrete filters with the convolution equation (cf. Eq. (7.68))

$$y(n) = \sum_{k=0}^{\infty} h(k)\, x(n-k) \tag{7.178}$$

which is defined by the following constant coefficient difference equation

$$y(n) = -\sum_{k=1}^{N} a_k y(n-k) + \sum_{k=0}^{M} b_k x(n-k) \tag{7.179}$$

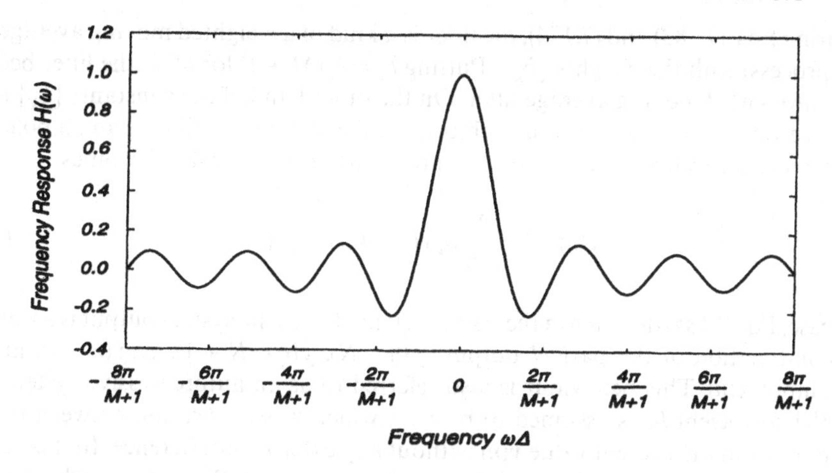

Fig. 7.24 *The frequency response of a moving average filter, $M=10$*

in which $\{a_k\}$ and $\{b_k\}$ are arbitrary real constants. The system described by Eq. (7.179) is also characterized by the z-transform of the impulse response function $h(n)$, which is obtained by taking the z-transform of both sides of the equation using Table 7.1, that is,

$$Y(z) = H(z)X(z) = -\sum_{k=1}^{N} a_k Y(z)z^{-k} + \sum_{k=0}^{M} b_k X(z)z^{-k}$$

whereby

$$H(z) = \frac{Y(z)}{X(z)} = \frac{b_0 + \sum_{k=1}^{M} b_k z^{-k}}{1 + \sum_{k=1}^{N} a_k z^{-k}} = \frac{B(z)}{A(z)} \tag{7.180}$$

The filter characteristics are dependent on the poles and zeroes of $H(z)$. The system is stable and causal if the ROC of $H(z)$ is the exterior of a circle of some radius $r<1$ in the complex plane, including the point $z=\infty$.

The filter equation Eq. (7.180) may be factorized in the form

$$H(z) = \frac{B(z)}{A(z)} = b_0 \frac{\prod_{k=1}^{M}(1-c_k z^{-1})}{\prod_{k=1}^{N}(1-p_k z^{-1})} \tag{7.181}$$

where the c_k are the zeroes and the p_k the poles of $H(z)$. If the constants $\{a_k\}$ are all equal to zero, the filter is an all-zero filter. In this case the linear difference equation becomes

$$y(n) = \sum_{k=0}^{\infty} h(k)x(n-k) = \sum_{k=0}^{M} b_k x(n-k) \tag{7.182}$$

which shows that the filter impulse response function is simply

$$h(k) = \begin{cases} b_k & 0 \leqslant k \leqslant M \\ 0 & \text{otherwise} \end{cases} \tag{7.183}$$

Comparing Eqs. (7.183) and (7.174), the filter is a kind of a weighted moving average filter (a MA process) with the weights $\{b_k\}$. Putting $b_k = 1/(M+1)$ for all k, the filter becomes a pure unweighted moving average filter. On the other hand, if the constants $\{b_k\}$ are all equal to zero for $k \geqslant 1$ while b_0 is an arbitrary real constant, the filter is an all pole filter, and the linear difference equation for the input–output relationship becomes

$$y(n) = -\sum_{k=1}^{N} a_k y(n-k) + b_0 x(n) \tag{7.184}$$

In this case, Eq. (7.184) describes a recursive system, that is, the system output is a weighted linear combination of the past N outputs $y(n-N)$, $y(n-N+1), \ldots, y(n-1)$, and the present input $x(n)$. This behaviour is also referred to as an autoregressive system (AR). Often, the coefficient b_0 is assumed to be zero, which gives a relation between the past values of $y(i)$ and the present value $y(n)$ without any external interference. In this case the filter, Eqs. (7.179) and (7.180), has both finite poles and zeros in the z-plane and the process is called an autoregressive moving average filter (ARMA). The ARMA system really consists of two filtering operation on the raw data, that is, applying an autoregressive filter (AR) and a moving average filter (MA), which is described by Eqs. (7.179) and (7.180).

The ARMA process described by Eq. (7.179) can be extended to cover systems with an exogeneous variable or input signal $u(t)$, which together with a noise signal $w(t)$, usually a zero mean white noise, is input to the ARMA filter. Consider first the system

$$y(n) + \sum_{k=1}^{N_a} a_k y(n-k) = \sum_{k=1}^{N_b} b_k u(n-k) + w(n) \tag{7.185}$$

where the white noise term enters as a direct error in the difference equation. The transfer function of Eq. (7.185) is given by

$$H(z) = \frac{B(z)}{A(z)} + \frac{1}{A(z)} \tag{7.186}$$

The above simple error model Eq. (7.185) can be generalized to allow for more freedom in describing the properties of the disturbance or the error term. For instance, by assuming the error process to be the output of a moving average filter, giving the system equation

$$y(n) + \sum_{k=1}^{N_a} a_k y(n-k) = \sum_{k=1}^{N_b} b_k u(n-k) + w(n) + \sum_{k=1}^{N_c} c_k w(n-k) \tag{7.187}$$

In this form the system is called an ARMAX system to account for the exogenous control variable $u(t)$. For an extensive discussion of ARMAX models, see Ljung, [127], and Söderström and Stoica, [205].

ARMA models play an important role in the analysis of digital signals. They have for instance been applied to simulation of synthetic earthquake acceleration processes with considerable success pioneered by Liu [176]. Since that time, various other researchers have applied ARMA modelling techniques to describe earthquake ground motions as well as to earthquake response analysis and system identification (see for instance Chang *et al.*, [33], Kozin [117], Ólafsson [158], Polhemus and Cakmak [173], Safak [182],

Sigbjörnsson and Ólafsson [196], Snæbjörnsson and co-workers [201], [202] and [203]). In this case, the ARMA filter coefficients are treated as parameters to be determined in order to obtain a desired output signal $y(n)$ from a known input signal $x(n)$. For a comprehensive overview of parametric modelling of time series using ARMA processing, see Box and Jenkins [22].

Consider a Gaussian white noise process $\{W(t), t \in T\}$ with a zero mean and standard deviation σ_W. A sample function of the process is represented by the discrete time sequence $\{w(n)\}, n = 1, 2, \ldots$. Applying a (p, q)-dimensional ARMA model, Eq. (7.179) with $b_0 = 1$, the following relation can be formed

$$y(n) + a_1 y(n-1) + a_2 y(n-2) + \cdots + a_p y(n-p)$$
$$= w(n) + b_1 w(n-1) + b_2 w(n-2) + \cdots + b_q w(n-q) \tag{7.188}$$

in which the AR parameters $\{a_i\}$ and the MA parameters $\{b_j\}$ need to be determined in accordance with the desired characteristics of the output process $y(k)$. It can be convenient to introduce the retrograde operator

$$B^r x_s = x_{s-r}, \; B^\circ = 1 \tag{7.189}$$

Applying Eq. (7.189) to Eq. (7.188), the output value $y(k)$ is obtained as follows:

$$y(n) = \frac{1 + b_1 B^1 + b_2 B^2 + \cdots + b_q B^q}{1 + a_1 B^1 + a_2 B^2 + \cdots + a_p B^p} w(k) = \frac{P_q(B)}{P_p(B)} w(n)$$
$$= \varphi(B) x(n) = (1 + \varphi_1 B^1 + \varphi_2 B^2 + \cdots) w(n) \tag{7.190}$$

in which the polynomials of the retrograde operator B play a similar role to the z-transform (see Eq. (7.180). To ensure the stability of the relationship Eq. (7.190), the denominator polynomial $P_p(B)$ must have all its roots or poles within the unit circle in the complex plane, analogous to the way the ROC of the z-transform $H(z)$ is defined. The poles are found in pairs, that is, for the pole r_i there is a complex conjugate r_i^*. Therefore,

$$P_p(B) = (1 - r_1 B)(1 - r_2 B) \cdots (1 - r_{p/2} B)(1 - r_1^* B)(1 - r_2^* B) \cdots (1 - r_{p/2}^* B) \tag{7.191}$$

The parameters $\{a_i, b_j\}$ in the monic polynomials $P_p(B)$ and $P_q(B)$ (the term monic refers to polynomials with the zeroth coefficient equal to one) are usually arranged in a parametric vector

$$\mathbf{E}^{\mathrm{T}} = \{a_1, a_2, \ldots, a_p, b_1, b_2, \ldots, b_q\} \tag{7.192}$$

with the dimension $(p + q)$. The accuracy of the ARMA (p, q) model obviously depends on the number of parameters (p, q).

There are several methods available for estimation of the ARMA parameters and controlling the errors (see for instance Marple, [138], and Ljung, [127]), which are now a routine part of signal processing software packages. A short overview of error estimation criteria are given in the following. Let $y(k)$ be the measured output and $\hat{y}(k) = \varphi(B) w(k)$ the calculated output. The error $\varepsilon(k)$, which is the difference $(y(k) - \hat{y}(k))$, is to be minimized. Actually, the error function can be treated as the addition to an exogenous function in an ARMAX model, that is,

$$y(n) + \sum_{k=1}^{N_a} a_k y(n-k) = \sum_{k=1}^{N_b} b_k w(n-k) + \varepsilon(n) + \sum_{k=1}^{N_c} c_k \varepsilon(n-k) \tag{7.193}$$

The following assumptions are usually made regarding the error function $\varepsilon(t)$. It should be 'white', i.e. have the characteristics of white noise. Generally, it is assumed that $\varepsilon(t)$ is a zero mean Gaussian white noise independent on the exogenous function $u(t)$. Therefore, one of the criteria for the ARMA model is to test the 'whiteness' of the error function. This can be done by considering the so-called loss function

$$V = \frac{1}{N} \sum_{n-1}^{N} \varepsilon^2(n) \tag{7.194}$$

where N is the total number of data points in the records. The variable $Q = NV$ should follow the chi-square distribution as $\varepsilon(t)$ is a Gaussian variable. The test is usually carried out for a specially constructed variable based on the estimate of the correlation functions, that is,

$$R(i) = \frac{1}{N} \sum_{n=1}^{N} \varepsilon(n)\,\varepsilon(n-i) \tag{7.195}$$

Form the variable

$$P(m) = \sum_{k=1}^{m} R(k) \tag{7.196}$$

where m is of the order $N/5$. Then, the variable

$$Z(m) = \frac{NP(m)}{V} \rightarrow \chi^2(m)$$

Thus, the whiteness of the error is tested by comparing $Z(m)$ with the chi-square variable. It is common to use a 95% confidence level as a reference for accepting or refusing the model.

Based on the loss function V_N, several functions, which give a measure of the error $\varepsilon(k)$, have been proposed. The criterion functions are to be minimized in order to keep the error function within tolerable limits:

(a)

$$V(r) = \sum_{r=1}^{N} \Gamma(s,r)\,\varepsilon^2(s) \tag{7.197}$$

is a criterion function in which $\Gamma(s,r)$ is a weighting factor. This is called the quadratic criterion. For time varying systems, $\Gamma(s,r)$ is selected such that the most recent data points have more weight than the earlier ones.

(b)

$$V(N) = \sum_{r=1}^{N} (-\log_{10} p(\varepsilon, r)) \tag{7.198}$$

is a maximum likelihood criterion function, where N is the total number of points and

$p(\varepsilon, r)$ is the probability density function of $\varepsilon(t)$. The maximum likelihood criterion is an often used method for estimating statistical parameters. It can be defined as follows, [136]. Let $x_1, x_2, \ldots, x_n$ be a sample drawn from a probability density function $p_X(x; \theta)$, where θ is an unknown parameter. Then

$$L = \prod_{i=1}^{n} p_{X_i}(x_i; \theta) \tag{7.199}$$

is the likelihood function. The maximum likelihood estimator (MLE) of θ, say θ^*, is the value that maximizes L, or equivalently the logarithm of L, that is,

$$\frac{d \log L}{d\theta} = \sum_{i=1}^{n} \log p_{X_i}(x_i; \theta) = 0 \tag{7.200}$$

The solutions of Eq. (7.199) have to be admissible, that is, be functions of the sample values $x_1, x_2, \ldots, x_n$ and within the parameter space $\theta \in \Theta$). If the error $\varepsilon(t)$ is a stationary zero mean Gaussian process, it can be shown that the maximum likelihood criterion becomes equivalent to the quadratic criterion Eq. (7.197) (see Ljung, [127]).

To determine the order of the ARMA model, that is, (p, q), there are two main criteria, which were proposed by Akaike, [1] and [2]. They are based on monitoring the decrease in the variance of the error σ_ε^2 as the order of the ARMA model $(p+q)$ increases. The first method is called the final prediction error criterion (FPE), and involves minimizing the function

$$\mathrm{FPE}(p+q) = \frac{N+p+q}{N-(p+q)} V_N \tag{7.201}$$

The second criterion is called the Akaike information criterion (AIC), and is based on minimizing the function

$$\mathrm{AIC}(p+q) = N \log_e V_N + 2(p+q) \tag{7.202}$$

These two criteria are related, and it can be shown that for a large number of data points N and low model order

$$N \log_e \mathrm{FPE}(p+q) \approx \mathrm{AIC}(p+q) \tag{7.203}$$

The actual calculation procedure for the construction of a suitable ARMA model for a given situation is to first try several models of varying model orders, and test the 'whiteness' of the resulting error functions. The next step is to use the Akaike's criteria to determine the minimum model order required to obtain sufficiently accurate results, which is often estimated on the basis of a visual inspection of the original and the ARMA generated records. Finally, an intuitive approach to define the the minimum model order to ensure sufficient accuracy is to examine plots of the location of poles and zeroes in the transfer function $H_{pq}(z)$ of the various models. A clustering of poles and zeroes is supposed to indicate some underlying model order, whereas scattered or distinctive poles and zeroes represent spurious peaks. All the above methods should be used in combination to ensure a good model.

Example 7.15 Ólafsson and Sigbjörnsson of the Engineering Research Institute of Iceland have extensively researched the application of ARMA models in engineering seismology, [10], [101] and [158]. The Vatnafjöll earthquake, which was the subject of Example 6.11, has offered a good opportunity to test the model. In Fig. 7.25, the

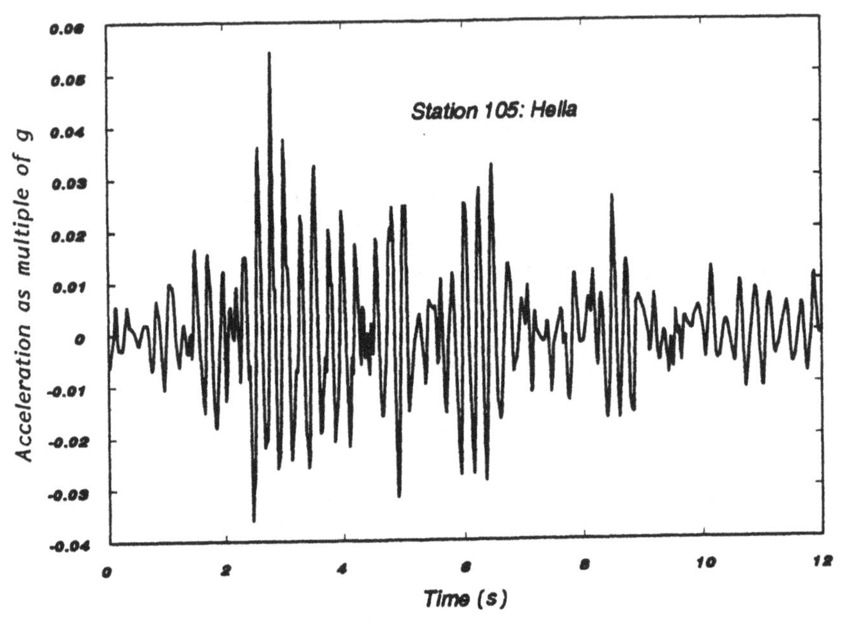

Fig. 7.25 *Vatnafjöll earthquake. An accelerogram measured at Hella*

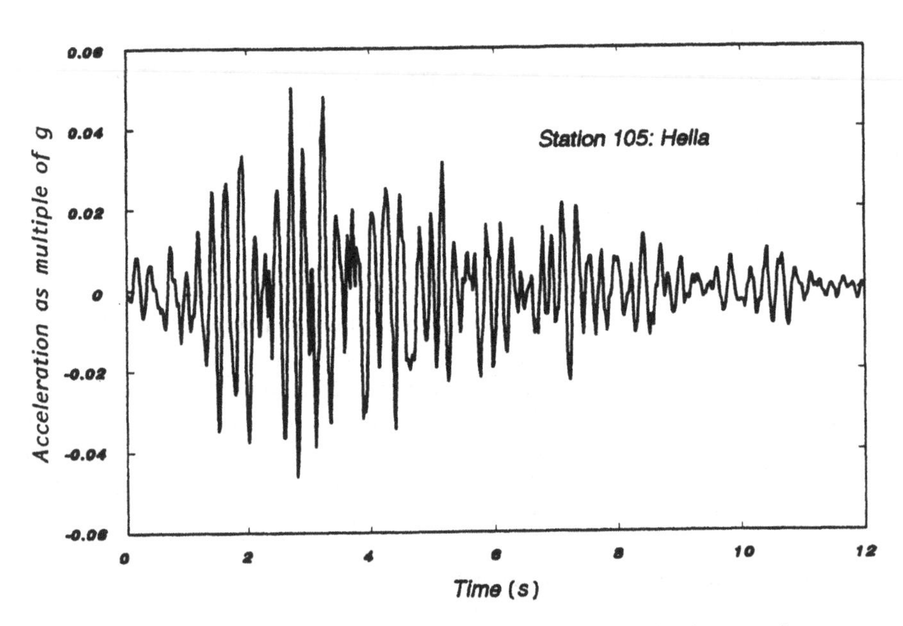

Fig. 7.26 *Vatnafjöll earthquake. An ARMA (4, 1) model record*

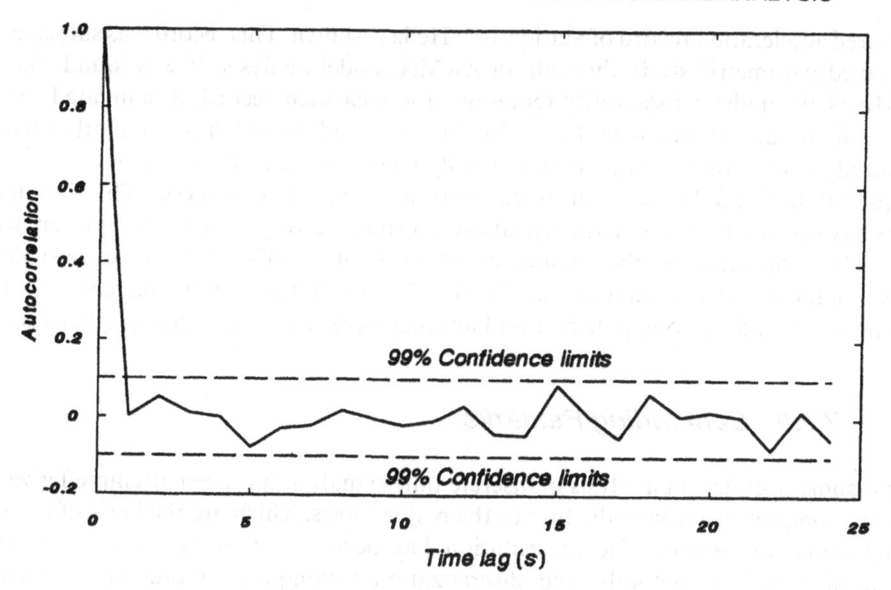

Fig. 7.27 *A correlation test for the whiteness of the error function*

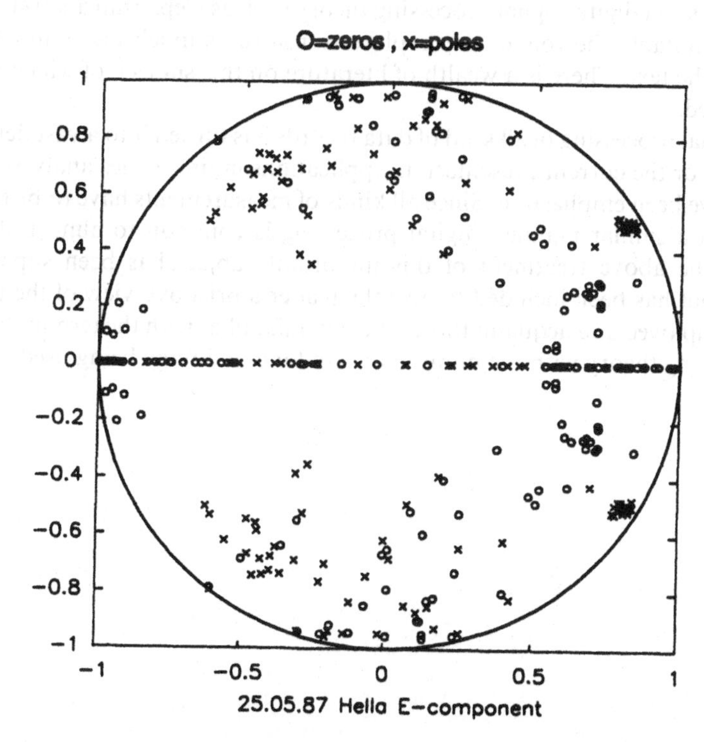

Fig. 7.28 *A pole–zero plot for different model orders. Maximum (7.7)*

measured acceleration record of Station 105: Hella is shown. This record was subjected to a detailed parametric study through an ARMA model analysis. It was found that an ARMA (4, 1) model satisfactorily represents the measured record. A simulated record using this model is shown in Fig. 7.26. The artificial record has been subjected to amplitude modification using a similar modifying function as was used in Example 6.11, Eq. (6.154). In Fig. 7.27, the error function $\varepsilon(t)$ is tested for "whiteness". The confidence limits have been set at 99%. The correlation test shown a very reasonable representation of the white noise autocorrelation function $R(\tau) = R_0\delta(\tau)$. In Fig. 7.28, a pole–zero plot is shown for the 49 different model orders from (1, 2) to (7, 7). The diagram indicates one real zero near -1, and a strong pole pair and another weaker one at higher frequencies.

7.2.9 Concluding Remarks

In this short overview of analysis of discrete time signals, it has been attempted give the reader a comprehensive introduction to the major topics, which are the basis of modern digital signal processing. The presentation has been based on the discrete Fourier transform, sampling methods, and discretization techniques, to thus cover discrete convolutions, the z-transform and digital filters. Topics such as windowing and design of digital filters have been covered very briefly. Finally, ARMA models have been introduced and their application in the analysis of earthquake signals briefly discussed. Many important topics in digital signal processing theory such as Cepstrum analysis have been omitted, and notably the role of the Laplace transform is much more important than indicated by the text. There is a wealth of literature on this subject, of which a few texts have been cited.

Digital signal processing of all kind of data records has turned into a vast field of major importance. For the current presentation, applications in time series analysis and signal treatment have been emphasized. Since all kinds of measurements have to be treated and interpreted in a similar manner, digital processing is common to almost all scientific disciplines. The above treatment of this important subject has been superficial and perfunctory but has been included to give the reader a brief overview of the theory and techniques employed and acquaint those who are unfamiliar with the terminology used in signal processing theory with the phraseology and methodology being used.

8 Earthquake Hazard and Seismic Risk Analysis

It could be possible to write a long essay on the pros and cons of using probability theory in practical applications. There are still people who disregard completely the aspect of uncertainty, which is necessarily an inherent quality of a physical system that is subject to forces and restraints in a natural environment. In engineering, for instance, there are those who look at forces and other kinds of actions that will cause a system to behave in a particular manner as completely deterministic. The behaviour or response of the system can therefore simply be resolved through analysis based on the system and material characteristics, which are elements of a deterministic world. For many generations of civil and mechanical engineers, the forces and actions for which a system had to be designed were predetermined and as such given in manuals or included as elements of specifications or regulation codes. Very few thought much about how these values were defined or had been obtained. Similarly, the material strength and system behaviour were assumed to have deterministic values that could be taken for granted. Many geologists and geophysicists also tend to disregard the randomness of geological processes. They believe that statistics and probability is a necessary evil that has to be applied only very sparingly, or as one geophysicist was heard to say, 'I never use statistics, I only deal with facts', [129]. They would much prefer to uphold the old belief that the processes of evolution and change have been preordained by some obscure deity or almighty power. You can study these processes and learn how to evaluate them, based on past experiences, but the matters will still run their preordained course. Whether an almighty power is in charge of everything or not is a matter beyond the scope of this text. It suffices to be said that such a power might very well have set the things into motion and then let the probability theory take over the management.

In recent decades, however, the way of thinking has changed rapidly and both engineers, material scientists and geophysicists have been busy absorbing new ideas and learning how to readjust their thinking and re-evaluate old truths in terms of rules set by the laws of chance. They have taken to describing natural phenomena, the nature of things, and the environment in which man-made systems have to endure through formal probabilistic methods, which for a long time have been the everyday tools of the pure physicist. Practically all engineering specifications nowadays are based on probabilistic concepts, where the loads are prescribed through their fractile values or by their probability distribution, and the same holds for material strength. Preparation of such specifications or codes of practice is now completely dependent on probabilistic concepts and thinking. As for scientists that are concerned with the interpretation of the complex processes that govern many natural phenomena, probability theory has become the standard tool. Various natural phenomena such as earthquakes, floods, snow avalanches, storms and volcanic eruptions are now categorized and catalogued through statistical and probabilistic methods. The enormous quantities of data that are now being collected in all fields of geophysics, for example, can best be analysed by statistical methods.

In this chapter, which concludes the discussion of the various aspects of the theory of stochastic processes and random vibration, a short overview of earthquake hazard and seismic risk analysis will be presented. In the previous chapters, scattered examples of the evaluation of earthquake magnitudes and accelerations have been given without any clear purpose other than to illustrate the application of various fundamental probabilistic laws. In the following, a more systematic overview of the earthquake risk will be given. Like so many natural phenomena, the earthquake risk has turned into a vast field of science, therefore only a very limited extract of the available material and topics will be covered. The discussion will be divided into two parts, namely the earthquake hazard, which is defined as the evaluation of the random event describing the occurrence of an earthquake in a given region. The earthquake hazard is therefore the assessment of the frequency of such events, their magnitudes and other characteristics, but should not be confused with the other concept of seismic risk, which deals with the risk of having damage and other serious social and economical consequences as the result of an earthquake. These two topics are often confused even if they have a different meaning. For instance, the hazard studies of earthquake-prone regions encompasses the investigation of all earthquakes whether they are likely to cause damage or not. Earthquake hazard studies of uninhabited regions without anthropogenic activity are not likely to involve elements of seismic risk. Earthquake hazard is essentially a probabilistic measure involving future events as well. Hazard maps, which are drawn on the base of historical earthquakes only, can be misleading. For instance, a 5 magnitude intra-plate earthquake occurred in the western part of Iceland in 1974. A seismic risk map for design purposes has been drawn for this region, which has the epicentre of the 1974 earthquake as a focal point. In the future, a similar earthquake may occur 100 km further east or 100 km further west, which would totally distort the current risk map. Risk and hazard maps have to take the possibility of future events into account. They have to be based on the probabilistic nature of these geophysical phenomena. It is not sufficient to apply probabilistic evaluation of the historical data only in order to draw risk maps for future events. This problem will be addressed in Section 8.1.3.

Related to both the above topics, however, is the theory of earthquake prediction, which is advancing very rapidly these years. As the efforts of the numerous scientists working in this field start bearing fruit, it will be possible to expect reasonably accurate estimates of the earthquake hazard in earthquake prone regions. Or as stated by Wallace *et al.*, [231], after the US National Academy of Science National Research Council Panel on Earthquake Prediction: 'An earthquake prediction must specify the expected magnitude range, the geographical area within which it will occur, and the time interval within which it will happen with sufficient precision so that the ultimate success or failure of the prediction can readily be judged. Moreover, scientists should also assign a confidence level to each prediction'. In a recently published monograph on Earthquake Prediction, Lomnitz has surveyed the most recent advances made in this field, [132].

8.1 EARTHQUAKE HAZARD ANALYSIS

8.1.1 Introduction

The frequency and occurrence of large earthquakes is a typical example of geophysical processes that can be adapted to statistical treatment. Even if the ultimate goal might seem elusive, i.e. to be able to predict future earthquakes with some reasonable accuracy based on past events and seismic data, and as such more a part of a spiritual exercise, the point process defined by each event as a point on a time axis is perfectly amenable for statistical analysis. The main statistical parameters may be derived from a set of data points, that is, past earthquake events, and could under certain assumptions fully characterize the dynamics of the process (cf. Chapter 2). In this case, there is no theoretical obstacle against being able to predict the arrival time or occurrence of the next data point within the usual statistical limits. At this stage, however, the interest may be turned rather to the main parameters of the actual event. Namely, such parameters as the location, the magnitude, the maximum intensity observed at certain places, the maximum surface accelerations for a chosen site and many other attributes of an earthquake.

Without going into the extremely complex mechanism of earthquakes or what causes them, the magnitude M since its introduction in the early 1930s by Richter, [178], is perhaps the most important parameter associated with an earthquake. The magnitude M, as the signature of an earthquake, is used in the usual derived sense, that is,

$$M = \log_{10} A \tag{8.1}$$

where M denotes the Richter magnitude of an earthquake, and A is the largest amplitude in microns of the seismograph trace of a torsional motion Wood–Anderson seismograph, which is located 100 kilometres from the instrumental epicentre of the earthquake. To calibrate the Richter Magnitude, usually called the local magnitude of an earthquake M_L, to arbitrary epicentral distances and for different seismographs, the following expression has been suggested:

$$M = a \log_{10} A + b \log_{10} \Delta + c \tag{8.2}$$

where Δ is the epicentral distance to the station in kilometres, and the constants a, b and c are three station parameters, which act as correction factors (to be derived for each seismograph station) to obtain the correct local Richter magnitude of an earthquake according to the original definition, [178]. Many different formulas for the local magnitudes have been suggested. For Japanese near-field earthquakes, Tsuboi, [226], proposed a common formula,

$$M = \log_{10} A + 1.73 \log_{10} \Delta - 0.83 \tag{8.3}$$

and for Californian near field earthquakes, Kanai, [99], proposed

$$M = \log_{10} A + 3 \log_{10} \Delta - 2.92 \tag{8.4}$$

to give an indication of the station values of a, b, c. Later, Gutenberg, [74], defined a teleseismic surface wave magnitude for shallow earthquakes:

$$M_S = \log_{10} A + 1.656 \log_{10} \Delta + 1.818 \tag{8.5}$$

A more recent version for the surface wave magnitude has been given by Vaněk, (see Gutenberg and Richter, [75]), which introduced the amplitude–period ratio (A/T), that is,

$$M_S = \log_{10}(A/T)_{max} + 1.66\Delta + 3.3 \tag{8.6}$$

where $(A/T)_{max}$ is the maximum of all (A/T) values of the wave groups. A similar expression for the magnitude of teleseismic body waves has been adopted (see Aki and Richards, [3] and Freedman, [67]), given by

$$M_B = \log_{10}(A/T)_{max} + Q \tag{8.7}$$

where Q is a function of both epicentral distance and focal depth for eliminating the path effect from the observed amplitude, which was derived empirically by Gutenberg and Richter, [76]. Finally, the moment magnitude M_w, which has become an important measure for earthquake magnitudes, should be mentioned (see Hanks and Kanamori, [79]). The moment magnitude is based on the seismic moment M_0 defined by

$$M_0 = \mu \bar{u} A = (\text{shear modulus}) \times (\text{average slip}) \times (\text{fault area}) \tag{8.8}$$

which furnishes a clear measure of the energy released in an earthquake during a fault break, [3]. Values of M_0 have been measured or assessed to be of the order 10^{30} dyne cm ($\approx 10^{4.3}$ N m), (the 1969 Chile earthquakes and 1964 Alaska earthquake) to about 10^{12} dyne cm ($\approx 10^3$ N m) for microtremors, [3]. The moment magnitude is then defined by the expression

$$M_w = \tfrac{2}{3}\log_{10}(M_0) - 6.033 \tag{8.9}$$

The magnitude of an earthquake, which was originally modelled after the magnitude of light of distant stars, has proved an extremely useful concept in seismology. While the magnitude is a purely instrumental measure of the energy release in an earthquake, an older but often for engineers more useful parameter, describing the surface effects of an earthquake, is the intensity I of an earthquake. Already in 1880–83, de Rossi and Forel described and classified earthquake surface effects, hereunder construction damage, by ten intensity classes designated by Roman letters I–X, [178]. Now, the destruction caused by an earthquake at the surface is directly related to the surface acceleration. Therefore, by 1903, Cancani had determined the values of the acceleration of an earthquake in relation to the de Rossi–Forel scale, and the original intensity scale needed improving. This was done by Mercalli, who added two more classes to the intensity scale, (XI, XII), and through a new description of the classes, obtained better agreement with the acceleration, [10], [178]. The Mercalli scale, as it is known today, was later modified by Wood and Neumann, [236], and the intensity is therefore often designated by MMI. An heuristic connection between the modified Mercalli scale intensities and the probable maximum acceleration in Gals (1 Galilei is 1 cm/s^2), is given in Table 8.1, [211].

In Japan, a separate intensity scale developed by the Japan Meterological Agency has been used, which encompasses 7 classes. Ishimoto has given the following relation between the surface acceleration in Gals and the intensities, Table 8.2 (see Kawasumi, [104]).

Finally, Medvedev *et al.*, [142], proposed a new version of the Mercalli scale in 1964. The new MSK intensity scale has a more precise classification and description of

Table 8.1 *MMI earthquake intensities and corresponding accelerations in Gals, [211]*

MMI	II	III	IV	V	VI	VII	VIII	IX	X	XI	XII
α_{max}	1	1.6	3	10	30	80	170	300	400	500	>500

Table 8.2 *Japanese earthquake intensities and corresponding accelerations in Gals, [104]*

JMAI	I	II	III	IV	V	VI	VII
α_{max}	$0.8 \sim 2.5$	$2.5 \sim 8.0$	$8 \sim 25$	$25 \sim 80$	$80 \sim 250$	$250 \sim 400$	>400

structural damage, incorporating such effects as good or bad workmanship, materials etc. The isoseismal map of an earthquake affected area, drawn using MSK intensities, will therefore locate pockets of bad construction or bad foundation ground.

The intensities and the magnitude of an earthquake are two different things. A low magnitude earthquake can produce large intensities in certain regions due to high accelerations caused by site effects or because of poor construction. Also, a large magnitude earthquake can have comparatively low intensities due to favourable ground conditions and solid construction. However, on the average, when large affected regions are considered, these two quantities are related. To obtain a rough estimate of the magnitude of an earthquake from an isoseismal map, Gutenberg and Richter therefore proposed the following simple relation between the highest Mercalli intensity I and the magnitude M, [75],

$$I = \tfrac{3}{2}(M - I) \tag{8.10}$$

The above relation, however, can only be indicative of the magnitude, knowing the intensity or vice versa.

8.1.2 Magnitude–Frequency Relations

Statistical models for large earthquakes are relatively easy to produce if a reliable earthquake catalogue is available, listing past and recent historical and instrumental earthquakes. A standard model that has been adopted and used by engineering seismologists to categorize the earthquake hazard is based on two simple assumptions (see Lomnitz, [129]): (a) the number of earthquakes which occur in one year in a certain regions is a Poisson random variable (see Example 1.24), and (b) M, the earthquake magnitude of each main earthquake, disregarding foreshocks and aftershocks, is a random variable that is exponentially distributed. Thus, it may be inferred that the probability of having r earthquakes in a certain region in any one year is Poisson distributed, whereby

$$p_r = \frac{n^r e^{-n}}{r!} \tag{8.11}$$

where p_r is the earthquake risk, that is, the probability of having r earthquakes occurring in one year, and n is the mean number of earthquakes in one year. In T years, the hazard is therefore

$$p_r(T) = \frac{(nT)^r e^{-(nT)}}{r!} \tag{8.12}$$

assuming stationary conditions (cf. Chapter 2). The second assumption is that M is a random variable that is exponentially distributed or

$$F(m) = 1 - e^{-\beta m}, \qquad m \geqslant 0 \tag{8.13}$$

This model has been shown to adequately model the modal maxima, the mean return periods and expected number of earthquakes in a region exceeding a given magnitude for a catalogue of large earthquakes. More sophisticated models have been proposed that take into account the aftershock sequences, and make use of Bayesian statistics to evaluate the distribution parameters (see Lomnitz, [129] and Esteva, in [130]).

Now, in the early 1940s, a fundamental discovery was made by Japanese and American seismologists. Without knowledge of one another, Ishimoto and Iida, and Gutenberg and Richter discovered almost simultaneously that the number of earthquakes in a region decreases exponentially with their magnitude. This magnitude–frequency law can be expressed as follows, [75], [178]:

$$\log_{10} N(M) = a - bM \tag{8.14}$$

where $N(M)$ is the number of earthquakes in one year with magnitudes greater than M, and a and b are two parameters that are dependent on the region under study. Obviously, the number of earthquakes in one year that have magnitudes greater than 0 is $N(0) = 10^a$, which gives a measure of the parameter a. Gutenberg and Richter have estimated a to have the approximate value of 8 for the whole world, giving the annual total number of earthquakes with magnitudes greater than 0 as 10^8. The other parameter b, which is the slope of the frequency–magnitude plot, was found by Gutenberg and Richter to have a value of about 0.9 to 1.0 for the whole world. There is indication however, that due to the practical and physically realizable upper limit of earthquake magnitudes of about $M = 9$, the above relation will curve downwards from the straight line close to this upper limit, Fig. 8.1.

Kárnik [101], [102], has studied the above relation for all of Europe and gives the values for a and b throughout the various zones. For the North-Atlantic region, the Mid-Atlantic Ridge and Iceland, the above relation assumes the form

$$\log_{10} N = 4.97 - 0.70\,M \tag{8.15}$$

predicting about six magnitude 6 earthquakes in or around Iceland and the North Atlantic in one single year. This indicates that the curve should have a more pronounced curvature towards the end since earthquakes of this magnitude are not likely to happen more than once every 20 years or so in the region. For the Mediterranean region, the same relation is

$$\log_{10} N = 7.39 - 0.94\,M \tag{8.16}$$

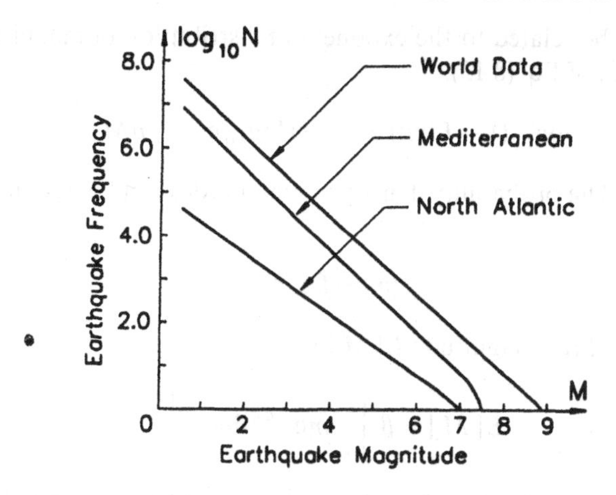

Fig. 8.1 *Earthquake frequency and magnitude relations*

predicting about 56 magnitude 6 earthquakes in one single year, indicating the same problem with the high magnitude end of the curve. In all cases, the high end of the magnitude-frequency relationship shows anomalies, which require modification of the two parameters that appear to become magnitude-dependent for large magnitudes. Actually, many investigators have studied this anomaly and suggested various measures for correction (see for instance Lomnitz-Adler, [133]). Finally, Freedman, [67], has estimated magnitudes of Pacific Rim earthquakes based on the body wave magnitude relation Eq. (8.7). Denoting the estimate by m', she studies a magnitude–frequency relation of the form

$$m' = \alpha + \beta N \tag{8.17}$$

and reports the values for the Pacific region given in Table 8.3.

Now, following Epstein and Lomnitz, [61], the magnitude–frequency relation can be normalized by dividing Eq. (8.14) by $N(0)$ to obtain the frequency distribution of the earthquake magnitudes

$$\log_{10}[1 - F(m)] = \log_{10}(N(m)/N(0)) = -bM \tag{8.18}$$

Table 8.3 *Values of the magnitude–frequency parameters for the Pacific Rim, [67], reproduced by permission of The Seismological Society of America*

Region	α	β
Aleutians	4.15	0.046
Tonga	4.03	0.048
Japan	3.87	0.050
South America	3.87	0.043

This expression can be related to the exponential distribution of magnitudes Eq. (8.13). Taking the logarithm of Eq. (8.13),

$$\log_{10}(1 - F(m)) = -\beta m \log_{10} \mathrm{e} = -bM$$

where $\beta = b/\log_{10} \mathrm{e}$. The probability density of magnitudes can be obtained by differentiation:

$$f(m) = \beta \mathrm{e}^{-\beta m} \tag{8.19}$$

and the mean value of the magnitude $E[M]$ is

$$E[M] = \beta \int_0^\infty m \mathrm{e}^{-\beta m} \, \mathrm{d}m = \frac{1}{\beta} \tag{8.20}$$

Thus, the average magnitude of earthquakes in a region $E[M] = 1/\beta = (\log_{10} \mathrm{e})/b$, which is regionally invariant.

Example 8.1 Frequency and Magnitudes of Californian earthquakes For California, it has been found that $b = 0.85$, $\beta = 2.0$, and $a = 4.97$. Estimate the mean magnitude and the sample size of earthquakes above magnitude 4 in any given year.

Solution The average magnitude is simply $1/\beta = 0.5$ for $M \geqslant 0$. As for earthquakes above magnitude 4.0, it is noteworthy that the exponential density does not change even if the base magnitude is changed. That is, the frequency distribution of earthquakes above magnitude $M_{\min}$ is

$$f(m) = \beta \exp[-\beta(m - M_{\min})] \qquad \text{for } m \geqslant M_{\min}$$

The average magnitude of annual earthquakes above 4.0 is therefore given by

$$E[M] = \beta \int_{M_{\min}}^\infty (m - M_{\min}) \, \mathrm{e}^{-\beta(m - M_{\min})} \, \mathrm{d}m = \frac{1}{\beta} + M_{\min} = 4.5$$

As for the sample size or average number of earthquakes above 4.0, the annual average number of earthquakes with magnitudes above 0 is $N(0) = 10^a = 10^{4.97}$. To find $N(4.0)$, the distribution function gives the probability of having earthquakes above 4.0 as

$$P[M \geqslant 4.0] = 1 - F(4.0) = \mathrm{e}^{-\beta 4.0} = \mathrm{e}^{-8.0}$$

The favourable fraction is therefore

$$N(4.0)/N(0) = \mathrm{e}^{-8.0} \qquad \text{so} \qquad N(4.0) = \mathrm{e}^{-8.0} \, 10^{4.97} = 31$$

which checks well with the observed magnitudes and numbers of earthquakes in California.

Lomnitz, [129], has also proposed to obtain the magnitude probability distribution by applying Kolmogorov's law of fragmentation (see Example 1.27) in a study of the fault area A, which is associated with the earthquake mechanism. From Eq. (8.8), the energy released by the earthquake E is directly related to the fault area through the average stress drop $\Delta\sigma$, whereby

$$E = A\,\Delta\sigma \tag{8.21}$$

The average stress drop during the fault break $\bar{\Delta}\sigma$ that causes the earthquake is about 100 bars for most rocks. It is now possible to imagine that A is produced by breaking up of the fault by successive stages into smaller and smaller areas. Consider a small stressed area of size y, which is tributary to the total fault area A, which in turn must be proportional to some power of y. If the break-up process is random and homogeneous, the following functional relation can be written for the stage i, which is reached after the area of state $i-1$ is further broken up into still smaller areas.

$$g_i(y) = T_i g_{i-1}(y) \tag{8.22}$$

where the T_i are positive and identically distributed random variables. By treating the above relation as a recurrence equation, the total fault break can therefore be written as

$$g_n(y) = g_0(y) \prod_{i=1}^{n} T_i \tag{8.23}$$

Taking the natural logarithm of both sides,

$$\log_e \left(\frac{g_n(y)}{g_0(y)} \right)^{1/n} \approx \log_e Y = Z = \frac{1}{n} \sum_{i=1}^{n} \log_e T_i$$

where $Z = \log_e Y$ is a random variable representing the tributary stressed area Y. Therefore, for large n, Lyapunov's theorem, Eq. (1.127), can be applied, whereby

$$P[Z \leqslant z] = F(z) = \frac{1}{\sigma_Z \sqrt{2\pi}} \int_0^z \mathrm{e}^{-(t-\mu_Z)^2/\sigma_Z{}^2}\,\mathrm{d}t \tag{8.24}$$

and Y has the lognormal distribution. Any power of a lognormal variable is also lognormal. Therefore, since A is related to the tributary areas by a power law, it is also a lognormal variable and hence by Eq. (8.21), the energy released in an earthquake is a lognormal variable. Now the energy released in an earthquake is also related to the magnitude M as originally proposed by Gutenberg and Richter, [178], whereby

$$E = 10^{a+bM}, \log_e E = a \log_e 10 + b \log_e 10\, M = \alpha + \beta M \tag{8.25}$$

where the constants a and b, on a world scale, have been assigned the values 11.8 and 1.5. Since E is a lognormal random variable, then by Eq. (8.25), the magnitude is a normal random variable. The exact relationship, in this context, is immaterial, since it is only used here to show that the magnitude distribution is the normal distribution.

It would now seem that the basic assumption, i.e. using the exponential distribution for the magnitudes, is not adequate. However, for the lack of earthquake data in the range from very small to very large earthquakes, it is difficult to fit the normal distribution to the often scarce data. The exponential distribution, however, fits the upper tail of the normal distribution very well and has also the advantage to have only one parameter instead of the two needed for the normal distribution. The main drawback of using the exponential distribution is that it requires the introduction of two artificial magnitude cutoffs at both ends to prevent divergent energy distribution.

As for the other basic assumption that the earthquake occurrence is a Poisson random variable, a few remarks on its validity are in order. In a recent discussion of recurrence times for earthquakes along the Mexican subduction zone, Nishenko and Singh, [156], point out that the fallacy of the Poisson distribution is that it does not take into account the elapsed time since the last major event. The interarrival times of a Poisson distributed random variable, in this case the occurrences of an earthquake, have been found to be exponentially distributed (cf. Examples 6.5 and 6.6) but there is no dependence on the last arrival time, that is, the time elapsed since the last event. Time-dependent hazard models that take into account both recurrence times, the inter-arrival times, and the time elapsed since a previous event occurred, are better compatible with the so-called seismic gap hypothesis. This hypothesis suggests that the potential for a future earthquake shock as a function of time is small immediately after a characteristic earthquake (a characteristic earthquake is an earthquake which repeatedly ruptures the same fault segment and whose source dimensions define that fault segment, [155]) and increases as time progresses. Analysing data from the catalogue of earthquakes along the Mexican subduction zone, Nishenko and Singh were able to show that the recurrence times, that is, the inter-arrival times T between characteristic earthquakes, fitted the lognormal distribution very well. The probability density used has the form proposed by Nishenko and Buland, [155], i.e.

$$f\left(\frac{T}{\bar{T}}\right)=\left(\frac{\bar{T}}{T\sqrt{2\pi}}\right)\exp\left[-\frac{(\log_e(T/\bar{T})-\mu_D)^2}{2\sigma_{D^2}}\right] \tag{8.26}$$

where $\bar{T}$ is the median of the recurrence interval. It is found that $T/\bar{T}$ gives a better fit to the distribution than the previously used T/T_{ave} data, where T_{ave} is the arithmetic mean value of the recurrence time intervals. The mean value of the variable $D=\log_e(T/T_{ave})$ is $\mu_D=-0.0099$ and its standard deviation $\sigma_D=0.215$ for Mexican subduction zone earthquakes. Now, $E[\log_e T]=\log_e T_{ave}+\mu_D=\log_e \bar{T}$, since the median of a lognormal variable $\bar{T}$ is $\exp(E[\log_e T])$, (see Example 1.27). From Eq. (1.129) and (8.26), the expected recurrence times are therefore given by

$$E[T]=\bar{T}\exp(\mu_D+\tfrac{1}{2}\sigma_D^2) \tag{8.27}$$

This result can be used for an earthquake prediction scheme based on a so-called forecast or prediction time window. Following Wallace *et al.*, [231], a time window for prediction can be defined as the time interval $[T_{pred}-\Delta, T_{pred}+\Delta]$ where the predicted time of occurrence of a characteristic earthquake $T_{pred}=t_0+T_{exp}$, that is, the time of occurrence of the last characteristic event t_0 plus the estimated expected recurrence time T_{exp}. The width of the time window 2Δ is defined as the difference between the 90% confidence limits for the estimate of $E[T]$ denoted by T_{exp}, that is,

$$P[t_0-\Delta<T_{exp}<t_0+\Delta]=0.90 \tag{8.28}$$

The confidence coefficient for a 90% confidence level is 1.645 (1.645 times the standard deviation) such that $\Delta=1.645\,(\mathrm{Var}[T_{exp}])^{1/2}$, [83], [214]. The variance of the sampling distribution for the mean is given by

$$\mathrm{Var}[T_{\mathrm{exp}}] = (\overline{\sigma^2}/\overline{T^2} + \sigma_D^2)\, T_{\mathrm{exp}}^2 \tag{8.29}$$

where $\bar{\sigma}$ describes how well the recurrence times are known and depends on the number of observations for each fault segment. All the above values can now be estimated from a catalogue of historical earthquakes.

The conditional probability P_C of a recurrence of a characteristic earthquake in a time interval $[t_1, t_2]$, conditional to the fact that by the current time t_C the event has not yet happened, can now be computed, [155]. If the date of the last event in a fault segment is $t_0 \pm \sigma_0$, where σ_0 is the variation in the determination of the date of the last event, and the value of $\bar{T} \pm \bar{\sigma}$ has been obtained, the result is

$$P_C = \frac{P[(t_1 - t_0) < T < (t_2 - t_0)]}{P[T > (t_C - t_0)]} = \frac{\displaystyle\int_{t_1 - t_0}^{t_2 - t_0} f(T/\bar{T})\mathrm{d}t}{\displaystyle\int_{t_C - t_0}^{\infty} f(T/\bar{T})\mathrm{d}t} = \frac{F((t_2 - t_0)/\bar{T}) - F((t_1 - t_0)/\bar{T})}{1 - F((t_C - t_0)/\bar{T})} \tag{8.30}$$

For most historical events, σ_0 is zero of course. Therefore, P_C gives the probability of having an earthquake happen within a time interval of say 20 years from the time the last earthquake occurred, given that the earthquake has not occurred yet. Nishenko and Singh [156], have computed these probabilities for the Pacific subduction zone of Mexico. The result is reproduced in Fig. 8.2.

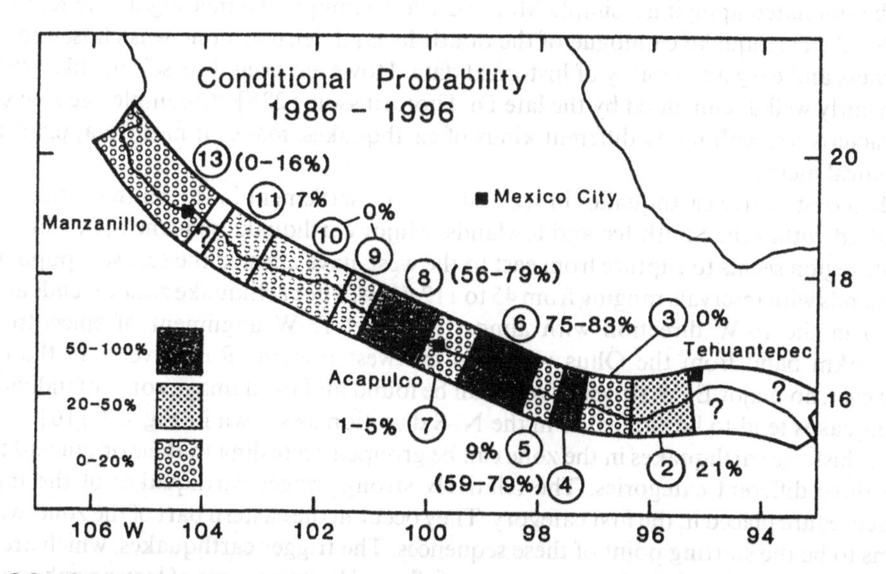

Fig. 8.2 *A map view of the conditional probability estimates along the Mexican subduction zone for the time interval 1986–1996, (after Nishenko and Singh, [156]), reproduced by permission of The Seismological Society of America*

8.1.3 A Statistical Model for Earthquake Events

Academic and practical studies in order to define the earthquake hazard and produce earthquake zoning maps abound in most earthquake-prone regions of the world. These studies have been both heuristic and based on classical statistical methods. In most cases, the existing earthquake catalogue has been the basis for the hazard assessment. The earthquake occurrence probabilities have been calculated using extreme value distributions, and adopting various decay formulas for the seismic waves, the intensities or accelerations at chosen grid points have been calculated, [7], [14], [62], [104], [107], [129], [130], [188], [197] and [223]. The main difficulty in this approach is the lack of instrumentally recorded earthquakes. Moreover, often the intensities are poorly documented and the geological research needed to amplify the magnitude and intensity assessments has not been completed. For this reason, a new approach has been proposed in which synthetic future catalogues entries are generated, based on the statistical information, which can be derived from the existing catalogues, Sólnes *et al.*, [212], and Sólnes and Halldórsson, [213] (reproduced with permission of Ouest Éditions, Presses Académiques). In this manner, the catalogue is extended towards the future and hazard maps based on the extended catalogue can be generated.

Earthquake events can be treated as singular or interrelated random events, which occur in space and time with a random magnitude. These characteristics can be described by a stochastic process of the Klondike type, which has been used to describe the location and magnitude of a gold strike as a random event, [193]. Let $X(\lambda, \phi, h, m, t)$ be a stochastic point process in which (Λ, Φ, H) are the random spatial coordinates of the origin of an earthquake, M is the random magnitude and T is the random year of occurrence. Each point in the process describes the occurrence of an earthquake, giving the location, the magnitude and the time. If the probabilistic law of the process can be derived, each point can be simulated applying a simple Monte Carlo technique. To investigate the Klondike model, the earthquake catalogue of the South Iceland Seismic Zone was chosen for the nearness and easy availability of historical data. However, even if its seismic history has been fairly well documented by the late Dr Thorarinsson, [225], its complex geophysical characteristics, with many different kinds of earthquakes, makes it hard to apply direct statistical methods.

Most destructive earthquakes in Iceland since its settlement in the ninth century have occurred within the South Iceland lowlands. Major earthquake sequences, in which the whole region seems to rupture from east to the west, have affected the sparsely populated farmlands with intervals ranging from 45 to 112 years. The earthquake zone extends about 70 km in the E–W direction with almost a perfect E–W alignment of epicentres in a 5–10 km band from the Ölfus region in the west towards Rangárvellir in the east. However, no major E–W striking faults can be found and the damage zones of individual earthquakes tend to be elongated in the N–S direction as shown in Fig. 8.3, [16].

The historic earthquakes in the zone can be grouped according to their origin and type into three different categories. The relatively strong trigger earthquakes of the major sequences are placed in the first category. They occur in the eastern part of the zone, which seems to be the starting point of these sequences. The trigger earthquakes, which are the largest magnitude earthquakes to occur, are followed by a sequence of lesser earthquakes, which take place all over the zone in a matter of days, weeks or months. Some sequences have lasted up to two even three years. The major sequences in this category have been

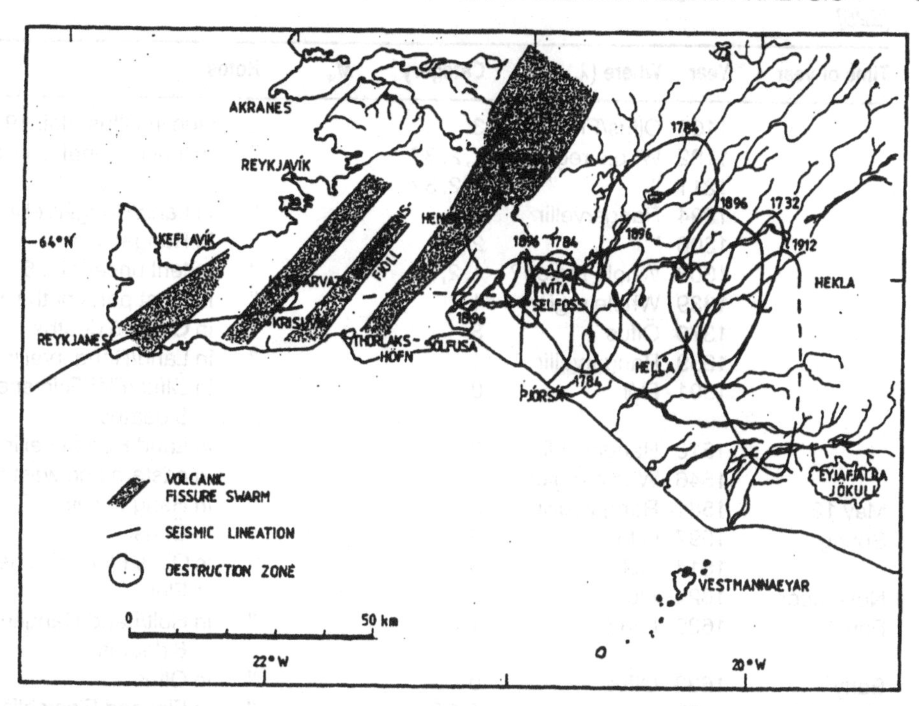

Fig. 8.3 *The Reykjanes Peninsula and South Iceland Seismic Zone. Damage areas of past earthquakes where more than 50% of all houses collapsed, [15]*

the 1294, 1339, 1389–91, 1630–33, 1732–34, 1784, and the 1896 earthquakes. The large interval including the fifteenth century may be due to poor historical records from that time of poverty and chaos in the country, the so-called historical gap, rather than scarcity of earthquakes. In all these episodes, severe damage to farmhouses in the entire region with several deaths is reported in the chronicles, [225].

The second category classifies single earthquakes, which have occurred without any noteworthy aftershocks in the eastern part of the Zone, and the third category classifies similar single events in the western part, which are generally smaller. The eastern part 1912 earthquake of magnitude 7, which is the last major earthquake to have occurred with the zone, is an example of this type of earthquake. It caused damage to a sparsely populated area close to the volcano Hekla in the Eastern Volcanic Zone with one death. Smaller earthquakes have occurred in the western part in more recent times. In Table 8.4, a revised list of all known major earthquakes in the South Iceland Zone is presented with assessed values for the maximum intensity and magnitude, [211].

The above classification of earthquakes can be related to the accumulation and subsequent release of the total seismic moment in the region $M_0 = \mu Au$. The average shear modulus μ in the zone has been estimated to be 34 GPa (see Stefánsson *et al.*, [215]). A is the area of the fracture zone, estimated to be 805 km^2 based on the E–N length of the zone, which is 70 km, the maximum focal depth, which is 8 km in the west part and 15 km in the east part, and the mean lateral displacement of the fault u in metres, [215]. The lateral displacement can be related to the average drift between the two continental plates, which is estimated to be 2 cm/year in South Iceland, [215]. It is further assumed that the

Table 8.4 *A catalogue of destructive and large magnitude earthquakes in South Iceland, [212], [225]*

Time of year	Year	Where (λ, ϕ)*	Category†	M_w	Notes
	1164	Ölfus/Flói	3		Damage in Ölfus/Flói, 19 deaths
	1182	Whole region?	1,2,3?		" extent uncertain, 11 deaths
	1211	" "	1,2,3?		" " " 18 "
	1294	Rangárvellir	1		" in Land/Rangárvellir
	1300	"	2		" in Rangárvellir
	1308	Whole region?	1,2,3?		" extent uncertain, 6 deaths
	1339	Whole region	1		" in most parts of the zone
	1370	Ölfus	3		" in Ölfus, 6 deaths
	1389	Rangárvellir	1		" in Land/Rangárvellir
	1391	Flói	0		" in Ölfus/Flói/Grímsnes, 5 deaths
	1510	Rangárvellir	2		" in Land/Rangárvellir
	1546	Whole region?	3		" in eastern and western parts
May 12	1581	Rangárvellir	2		" in Rangárvellir
Spring	1597	Ölfus	3		" in Ölfus
	1614	Flói	3		" in Ölfus/Flói/Grímsnes/Skeiö
November	1624	Flói	3		" in Flói
Feb 21	1630	Land	1		" in Holt/Land/Rangárvellir, 6 deaths
Early	1633	Ölfus	0		" in Ölfus
March 16	1657	Flói	1,3?		" in Flói and Fjóstshlíö?
Summer	1671	Ölfus	3		" in Ölfus/Grímnes, 5 deaths
Jan 28	1706	(64·N, 21·W)	3	(6–6.5)	" in Ölfus/Flói, 4 deaths
April 1	1725	Rangárvellir	2?		" in Rangárvellir
Summer	1726	"	2		" " "
September 7	1732	(64·, 20.2)·	1	(6.5	" in Hreppar/Land/Rangárvellir
March 21	1734	(64·, 20.5)·	0	(6.5)	" in Ölfus/Flói/Grímsnes, 9 deaths
	1749	Ölfus	3		" in Ölfus
	1752	"	3		" " "
	1757	Fljótshliö	2		" in Fijótshlíö
September 9	1766	Ölfus/Holt	3		" in Ölfus/Holt/Land
August 14	1784	(64.1·, 20.5)·	1	(7 +)	" in whole region, 3 deaths
June 10	1789	(64°, 21.5)°	3	(6–6.5)	" in Reykjanes and Ölfus
Feb 21	1829	(64°, 20°)	2	(6–6.5)	" in Land/Rangárvellir, 1 death
Aug 26	1896	(64°, 20.2)°	1	(7)	" in whole region, 4 deaths
May 6	1912	(64°, 19.9)°	2	(7)	" in Land/Ring./Fljótshlíö, 1 death
October 9	1935	West Ölfus	3	6	" Uninhabited area
March 29	1947	(64°, 19.7)°	(3)	5	No damage
April 1	1955	64.1°, 21.2°	(3)	5.5	Slight damage near Sog
July 2	1967	64°, 20.7°	(3)	5.0	Slight damage at Brúnastaöir
May 25	1987	63.91°, 19.78°	2	5.8	Uninhabited area

* Values within parentheses have been estimated or assessed.

† Earthquakes with the numbers 0 (sequential earthquakes), 1, 2 or 3, denoting the category, have been used in the analysis. Earthquakes with a question mark have not been used due to lack of information. The last three earthquakes in category 3 are not used, as the lowest magnitude level was set at 6 for the simulation process.

trigger earthquake together with the sequential earthquakes account for the release of about two thirds of the seismic moment, which has been accumulated due to the continental drift, whereas the single earthquakes of categories two and three account for one third of the seismic moment. Obviously, if a longer time elapses before the next release, the trigger earthquake can be expected to be larger than usual followed by a larger number of sequential earthquakes. This is used in the following analysis with the moment magnitude, Eq. (6.145), of the earthquakes defined as

$$M_w = \tfrac{2}{3}\log_{10}(M_0) - 5.7 \tag{8.31}$$

In an earlier study, [212], the simulation of earthquake occurrences for all three categories of earthquakes was based on the lognormal distribution of earthquake inter-arrival times, Eq. (8.26). Attempting to fit the Nishenko–Buland probability density to the scarce histogram data available for the South Iceland Earthquake Zone showed questionable results. The category 3 earthquakes are particularly difficult, with a very interesting anomaly in the inter-arrival times, i.e. with a gap between 52 years and 140 years. It was therefore decided to simply calculate the probability densities of the inter-arrival times using the kernel technique, [199].

The kernel technique is an extension of the classical density estimation using histograms of existing statistical data. Consider a histogram of statistical samples $x_i, i = 1, 2, \ldots,$ which are arranged in an histogram with the bin width $2h$. A so-called naive estimate of the probability density for the histogram bin $[x - h, x + h]$ is given by

$$f^*(x) = \frac{1}{n2h}[\text{number } n \text{ of samples } x_i \text{ within the bin } (x-h, x+h)] \tag{8.32}$$

The naive estimator can be defined in a more transparent manner using the weight function

$$w(x) = \begin{cases} \frac{1}{2} & \text{if } |x| < 1 \\ 0 & \text{otherwise} \end{cases} \tag{8.33}$$

The naive estimator is therefore written as

$$f^*(x) = \frac{1}{n}\sum_{i=1}^{n} \frac{1}{h} w\left(\frac{x - x_i}{h}\right) \tag{8.34}$$

It can be said that the naive estimate is constructed by placing a 'box' of width $2h$ and height $1/(2nh)$ around each sample and then summing to obtain the estimate. Just as in digital signal analysis, the box width (bin width) will be of crucial importance for the estimate. Whereas the naive estimator produces density functions of somewhat ragged character, it is possible to use more sophisticated kernels to overcome this difficulty. Replace the weight function $w(x)$ with a *kernel function* $K(x)$, which satisfies the condition

$$\int_{-\infty}^{\infty} K(x)\mathrm{d}x = 1 \tag{8.35}$$

Generally, a symmetric probability density like the Gaussian density function is selected for $K(x)$. By analogy with the naive estimator, the kernel estimator with a kernel $K(x)$ is defined by

$$f^*(x) = \frac{1}{nh} \sum_{i=1}^{n} K\left(\frac{x - x_i}{h}\right) \tag{8.36}$$

where h is the window width of the estimation. The actual estimation process can be described as a sum of 'bumps' placed over each sample. For a Gaussian density kernel function, the optimum window width h_{opt}, minimizing the approximate mean integrated square error $E[\int (f^*(x) - f(x))^2 dx]$ can be shown to have the value (see Silverman, [199])

$$h_{opt} = \left(\frac{4}{3}\right)^{1/5} \sigma_X n^{-1/5} \tag{8.37}$$

where σ_X is the standard deviation of the Gaussian distribution, and n is the number of samples.

The histograms for all categories are now obtained by using the kernel technique. The Gauss density was selected for the kernel function $K(x)$ with an optimal sample interval of $h_{opt} = 1.06\sigma_X n^{-1/5}$, where σ_X and n are the standard deviation and the dimension of the sample vector $\{X\}$ respectively. The relative recurrence interval length X between the earthquake events is formed by dividing by the average interval length based on the historical earthquake catalogue for all three categories, $X = T/T_{ave}$. For category 3 earthquakes, however, an $h < h_{opt}$ was selected to obtain a sufficiently distinctive double hump distribution (see Fig. 8.4). A polynomial function is then found as an approximation to the histograms thus obtained, and the probability densities for the relative recurrence periods are determined by scaling (the area under the polynomial function is put equal to one). In Fig. 8.4, the three densities are depicted, showing the peculiarity of the distribution of recurrence intervals for category 3 earthquakes. The difficulty of course is the scarcity of the data, making the histograms less reliable. The category 3 earthquakes are particularly difficult to interpret, and it is possible that earthquakes in this category are poorly documented. However, it is believed that, for the moment, this is the best estimate available. Finally, for the subsequent random generation of the inter-arrival times, the three distributions were cut at the high and low ends as shown in Fig. 8.4.

As for the sequential earthquakes, where no histogram data are available, the Gamma distribution was adopted. The Gamma distribution has the probability density

$$p(x) = \begin{cases} \dfrac{1}{\beta^{\alpha}\Gamma(\alpha)} x^{\alpha-1} e^{-x/\beta} & x \geqslant 0 \\ 0 & x < 0 \end{cases} \tag{8.38}$$

with two positive parameters α and β (note that the Gamma distribution turns into the exponential distributions for $\alpha = 1$). As for sequential earthquakes, the inter-arrival times may vary between a few hours and several months, according to the last major rupture episodes. For this reason, the Gamma distribution with its concentration of probability close to zero is deemed suitable. On a trial and error basis, the parameters were selected as $\alpha = 20$, and $\beta = 1.1$. This corresponds to the condition that there is a 50% probability of recurrence periods less than 30 weeks and a 73% probability of periods less than a year.

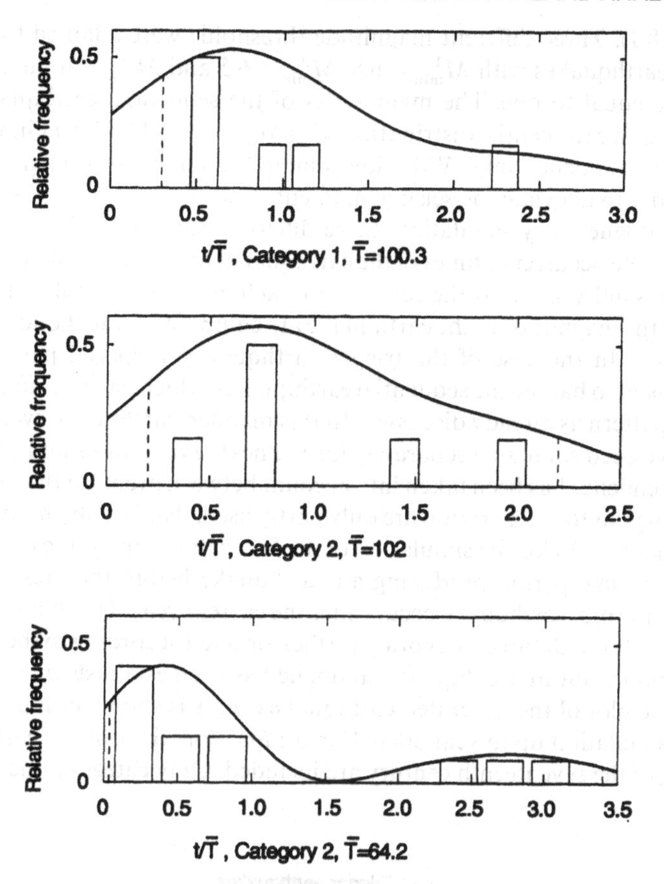

Fig. 8.4 *Histograms and probability densities of category 1–3 earthquakes*

The total number of sequential earthquakes following each trigger earthquake (category one), was estimated on the basis that the accumulated stress drop of the trigger earthquake and its sequential earthquakes would be less than the total accumulated stress since the last earthquake series or singular event, according to the seismic moment, Eq. (8.31).

The spatial coordinates of the earthquake origin were distributed as follows. The latitude λ was fixed in the middle of the seismic zone at 63.9° N. For the longitude Φ, a truncated normal distribution with the mean value in the east part of the zone was used for earthquakes in category one and three, and similarly a truncated normal distribution with its mean in the west part was used for earthquakes in category two, in both cases fitted by using a histogram of the historical data. The depth of the earthquakes is disregarded as it affects only the surface intensities and not the synthetic catalogue itself. The origins of the sequential earthquakes, following the trigger earthquakes (category one), are assumed to be uniformly distributed between 20° W and 21° W, as the E–W extension of the Zone is approximately from 19.9° W to 21.2° W.

The magnitudes of all earthquakes are assumed to be distributed according to the simple exponential model for the probability distribution of earthquakes (see Eq. (8.19) and Example 8.1). Three different magnitude thresholds were adapted for the different categories of earthquakes with $M^1_{min} = 6.8$, $M^2_{min} = 6.5$ and $M^3_{min} = 6$. In all cases, b was assumed to be equal to one. The magnitudes of the sequential earthquakes were also simulated using an exponential distribution with $M_{min} = 6.5$. This lower limit is kept large for the sake of convenience only. With a lower limit, the number of sequential earthquakes increases rapidly to alleviate the seismic moment.

For the catalogue entry simulation, three different random vectors are generated in order to obtain the occurrence times of the earthquakes in the three different categories up to several thousand years into the future. For each event, the simulated values for the longitude and the magnitude of the earthquake are computed using the adopted probability distributions. In the case of the trigger earthquakes, a special parallel computing routine is invoked to handle the sequential earthquakes, which have a completely different probabilistic pattern as already discussed. In this manner, earthquake events for all three categories have been simulated separately for the next several thousand years. However, as no interdependence has been taken into account between events of the three categories, the results of the random generation are only partly used, that is, only the events until year 3000 are considered. As for the simulation of the first catalog entry, the computer routine simply rejects a time period producing an earthquake before the present time, and is restarted until the first earthquake occurs after the current year. The thus generated future earthquake events for all three categories are then ordered according to the year or time of occurrence, and the future earthquake catalogue has been established.

A parametric plot of the generated earthquake events is shown in Fig. 8.5. This is the result of one simulation up to year 3000. For comparison, the historic earthquakes from the beginning of the seventeenth century are included. Only category one, two and three

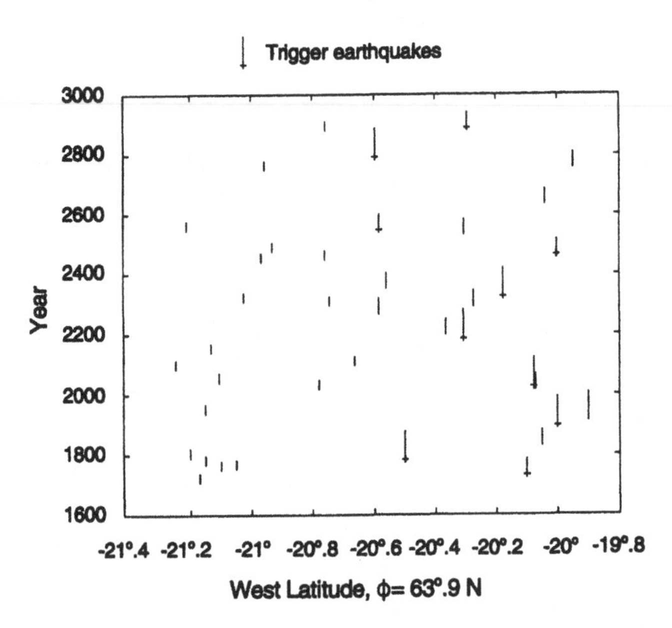

Fig. 8.5 *A parametric plot of generated earthquake events till year 3000 together with historic earthquakes from beginning of the 16th century*

Table 8.5 *A synthetic catalogue of future earthquakes in the South Iceland seismic zone*

Year	M_w	Where (ϕ)*	Category †	Year	M_w	Where (ϕ)*	Category †
2014	6.6	−20.07	2	2464	6.7	−20.96	0
2017	6.4	−20.78	3	2464	6.6	−20.25	0
2027	7.0	−20.08	1	2464	6.9	−20.01	0
2027	6.5	−20.56	0	2464	6.5	−20.37	0
2027	6.8	−20.84	0	2475	6.0	−20.30	3
2027	6.9	−20.31	0	2530	6.8	−20.30	2
2027	6.8	−20.86	0	2546	6.4	−21.21	3
2027	6.6	−20.65	0	2548	7.0	−20.58	1
2027	6.6	−20.38	0	2548	6.9	−20.24	0
2027	6.6	−20.76	0	2548	6.8	−20.68	0
2028	6.5	−20.24	0	2549	6.6	−20.35	0
2028	6.7	−20.06	0	2549	6.5	−20.20	0
2039	6.0	−21.11	3	2549	6.5	−20.34	0
2082	6.1	−21.25	3	2633	6.5	−20.04	2
2095	6.4	−20.66	3	2748	6.3	−20.96	3
2137	6.2	−21.13	3	2754	6.7	−19.95	2
2186	7.1	−20.31	1	2793	7.0	−20.60	1
2186	6.6	−20.66	0	2793	6.8	−20.27	0
2186	6.9	−20.52	0	2793	6.7	−20.29	0
2186	6.5	−20.78	0	2793	6.5	−20.29	0
2186	6.8	−20.81	0	2793	6.6	−20.61	0
2186	6.7	−20.70	0	2793	6.6	−20.91	0
2186	6.9	−20.56	0	2793	6.8	−20.51	0
2186	6.6	−20.10	0	2793	6.9	−20.21	0
2186	6.8	−20.50	0	2793	6.6	−20.19	0
2196	6.7	−20.37	2	2793	6.7	−20.77	0
2265	6.8	−20.58	2	2793	6.7	−20.13	0
2291	6.6	−20.77	2	2793	6.6	−20.27	0
2294	6.2	−20.74	3	2793	6.5	−20.45	0
2305	6.4	−21.02	3	2793	6.8	−20.70	0
2325	7.1	−20.18	1	2793	6.7	−20.89	0
2325	6.7	−20.46	0	2793	6.5	−20.62	0
2325	6.9	−20.33	0	2793	6.5	−20.50	0
2325	6.8	−20.08	0	2793	6.7	−20.02	0
2325	6.6	−20.95	0	2793	6.8	−20.46	0
2325	6.8	−20.98	0	2794	6.6	−20.66	0
2326	6.9	−20.45	0	2880	6.0	−20.76	3
2351	6.8	−20.56	2	2888	6.9	−20.29	1
2439	6.4	−20.97	3	2888	6.5	−20.90	0
2448	6.1	−20.76	3	2888	6.9	−20.49	0
2463	7.0	−20.00	1	2889	6.6	−20.24	0
2463	6.6	−20.13	0	2889	6.6	−20.82	0
2463	6.6	−20.61	0	2889	6.5	−20.30	0
2463	6.8	−20.69	0	2889	6.7	−20.20	0
2464	6.5	−20.87	0	2889	6.7	−20.57	0
2464	6.9	−20.38	0	2950	6.7	−20.08	2

* Latitude λ is fixed, 63.9°

† Category 1: trigger earthquakes
Category 0: sequential earthquakes
Category 2: east part earthquakes
Category 3: west part earthquakes

earthquakes are shown, that is, the sequential earthquakes are discarded. The magnitude size is indicated by a vertical bar. The trigger earthquakes, which set off the rupture sequences, are further marked with a horizontal bar. Apparently, the generated future catalogue compares well with the historical catalog. In Table 8.5, the earthquake events are listed with year of occurrence, the magnitude and location.

It has been shown that it is possible to generate synthetic earthquake catalogues, which show different scenarios for the future seismic development of a certain region with established geophysical properties based on its existing seismic history. The catalogues can be used to strengthen earthquake risk analysis for a region in the case of scarce earthquake data. Together with the existing historical catalogues, synthetic catalogues can be used to produce acceleration contours of equal risk and other types of zoning contours and aseismic design parameters. Also, by using different synthetic catalogues, the sensitivity of zoning contours can be investigated, giving an indication of the overall reliability of zoning maps based on the historical earthquake catalogues. It should be strongly emphasized that the synthetic catalogue has very little to do with earthquake prediction. Although it shows a possible scenario of future earthquake events, this must not be confused with the prediction of the occurrence of an awaited earthquake event, which has to be addressed in a different manner, as has already been discussed in connection with the Mexican subduction zone earthquakes. Thus in Table 8.5, the first trigger earthquake is 'occurring' in the year 2027. This is just a statistical possibility and does not mean anything else.

Finally, a MATLAB program was written to handle the entire simulation process. The program is available to interested readers, which can download the program by accessing the web site: //www.hi.is/ ~ solnes/.

8.2 ELEMENTS OF SEISMIC RISK

In the preceding sections, the earthquake hazard has been discussed, a few remarks on earthquake prediction have been made, and the basic relations giving the frequency and probability distributions of the main parameters have been given. Seismic risk analysis on the other hand involves the assessment of risk due to possible earthquake action on engineering structures in an earthquake-prone region. According to the UNDRO manual *Mitigating Natural Disasters*, [228], seismic risk is defined as the expected number of lives lost, persons injured, damage to property and disruption of economic activity due to a particular earthquake. In mathematical terms, seismic risk deals with expected maximum surface accelerations with prescribed return periods, expected average earthquake load spectra, vulnerability assessment of structures and damage expectation, to name but a few of the many topics of interest. There is perhaps no need to address the seismic risk in a different way from the above treatment of frequency distributions and prediction probabilities for earthquake hazard assessment. However, certain aspects of seismic risk analysis have developed in different directions and few examples will be shown in the following.

The magnitude–frequency relationship, Eq. (8.8), can be also be adapted to the ground intensities, [104]. In an exponential form this may be written as

$$N(I) = C \times 10^{-\gamma I} \tag{8.39}$$

where $N(I)$ is the average number of earthquakes in one year, which produce intensities in excess of I, and C and γ are two constants to be determined. As already mentioned, the best way of evaluating the seismic risk is to study the seismicity of each region in a historical and geological/geophysical context. Knowledge of the exact location of active faults in a region, to be avoided as building sites, is often a better earthquake protection policy than, say, knowledge of the maximum acceleration, intensities, magnitudes etc. Assuming such knowledge, however, the second most important parameter for the engineer is the maximum expected seismic intensity or acceleration at the site of a structure. In order to make the structure sufficiently earthquake resistant to withstand the potential earthquake exposure during its lifetime with minimum expense, it is necessary to determine the maximum expected intensity/acceleration at the site during its lifetime.

The following procedure has been proposed by Kawasumi, [104]. As evident from Tables 8.1 and 8.2, human perception of earthquakes will underestimate the acceleration level. It is a well-known phenomenon regarding the human senses that an increase by powers of a physical quantity (level of acceleration, intensity of light etc.) is perceived by the human mind as an increase by a constant difference (quotient series vs difference series). This phenomenon is sometimes referred to as the Weber–Fechners law of perception. In a mathematical sense this can be described as follows

$$\alpha_{max}(I) = A \times 10^{\beta I} \tag{8.40}$$

where A and β are constants determined by the intensity scale applied. The average annual number of earthquakes, which produce intensities at a certain locality in excess of I, is given by Eq. (8.39). Taking a period of T years, the average frequency of occurrence of the event that the intensity is above I is

$$p(I, T) = N(I)/T = C/T \times 10^{-\gamma I} \tag{8.41}$$

By Eq. (8.40), the maximum acceleration corresponding to the intensity I can be incorporated into the expression Eq. (8.41), whereby

$$p(\alpha_{max}, T) = \frac{C}{T}\left(\frac{A}{\alpha_{max}}\right)^{\delta} \tag{8.42}$$

For the Japanese intensity scale, β is approximately 0.5 (Table 8.2), and δ is close to unity for most localities in Japan. Assuming this to be the case,

$$1/p = T/(CA)\,\alpha_{max} = \bar{T} \tag{8.43}$$

that is, for an average interval of $\bar{T}$ years, a maximum acceleration at the locality considered will be in excess of α_{max}. Therefore, the expression

$$\bar{\alpha}(\bar{T}) = \left(\frac{CA}{T}\right)\bar{T} \tag{8.44}$$

indicates the maximum expected acceleration with a return period of $\bar{T}$ years. Now the risk U, that is, the probability of an earthquake with a maximum acceleration exceeding $\bar{\alpha}(\bar{T})$ at least once in n years is given by

$$U(\bar{\alpha}(\bar{T})) = 1 - (1 - 1/\bar{T})^{n} \tag{8.45}$$

The risk taken by adopting $\bar{\alpha}(\bar{T})$ as the design acceleration is therefore in the range

$$0.75 > U > 0.63$$

with U decreasing as $\bar{T}$ increases. Combining Eqs. (8.44) and (8.45), the maximum acceleration to be encountered in n years with the risk U (of still higher acceleration) is

$$\alpha(n, U) = \left(\frac{CA}{T}\right)\{1 - (1 - U)^{1/n}\}^{-1} = \bar{\alpha}(\bar{T})R \tag{8.46}$$

where the risk function R is given by

$$R = \frac{nU^{n-1}}{T}(1 - 0.5\,U - 0.1\,U^2) \tag{8.47}$$

neglecting a slight variation dependong on n, with U as the risk that an acceleration larger than $\alpha(n, U)$ will be encountered in n years.

The practical application of the above method is for example to draw contour lines for mean value accelerations $\bar{\alpha}(100)$, whereby $\alpha(n, U)$ can be computed for any structure with a projected lifetime and accepted risk. In Fig. 8.6, an example of such a contour map for Japan is shown, [104].

Example 8.2 Take a structure with a projected lifetime equal to $T = 150$ years, which is to be designed for a horizontal load equal to 10% gravity a return period of 50 years. That is, an acceleration level of 100 Gal is expected at least once in 50 years ($\bar{T} = 50$). Then by Eqs. (8.46) and (8.47),

$$\alpha(150, U) = 100\,R$$

$$R = (150/50)U^{-1}(1 - 0.5\,U - 0.1\,U^2)$$

or

$$\alpha(150, U) = 300\,U^{-1}(1 - 0.5\,U - 0.1\,U^2)$$

Accepting 50% risk,

$$\alpha(150, \tfrac{1}{2}) = 450\,\text{Gal}$$

which means that there is 50% risk that the structure will be exposed to an acceleration more than four times the design acceleration.

Another approach to the definition of seismic risk is to use extreme value distributions, Lomnitz, [129]. Starting with the exponential magnitude distribution Eq. (8.13), where $1/\beta$ is the average magnitude, and the magnitude–frequency relation Eq. (8.14) in the form

$$N(m) = \mathrm{e}^{\alpha - \beta m}, \qquad \alpha = a/\log_{10}\mathrm{e} \qquad \text{and} \qquad \beta = b/\log_{10}\mathrm{e} \tag{8.48}$$

the maximum annual earthquake magnitude M can then be considered to be an extreme distribution variable. Applying the Gumbel distribution, Eq. (1.149), in the form

$$\Phi(y) = \exp(-\mathrm{e}^{\alpha_1 - \beta_1 y}), \qquad -\infty < y < \infty \tag{8.49}$$

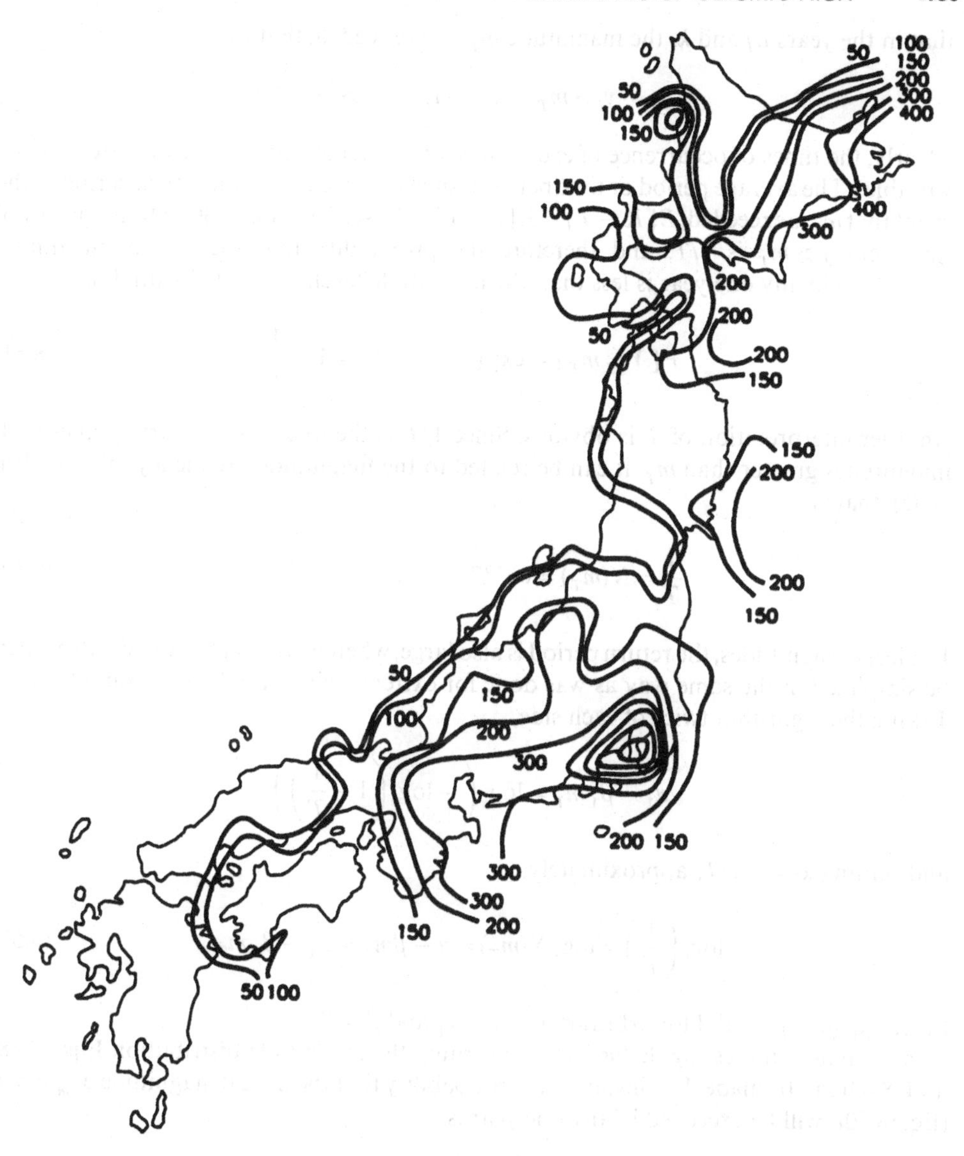

Fig. 8.6 *Distribution of $\bar{\alpha}(100)$ in Japan for a return period of 100 yrs, (after Kawasumi [104]). Reproduced with permission of the Earthquake Research Institute, University of Tokyo*

where the mean value and mode of the distribution are

$$E[Y] = \frac{\alpha_1 + 0.5772}{\beta_1}, \qquad y_{\text{mode}} = \frac{\alpha_1}{\beta_1} \tag{8.50}$$

It will be shown that $\alpha_1 = \alpha$ and $\beta_1 = \beta$. First define and find the return period T of large earthquakes with a magnitude larger than m_T. The largest earthquakes in a region, that is, the maximum annual magnitudes, are listed by the array $\{y_1, y_2, \ldots, y_N\}$. Now assume that in the years i, j and k, the magnitude m_T was exceeded, that is,

$$y_0 \cdots y_i > m_T \cdots y_j > m_T \cdots y_k > m_T \cdots$$

Clearly, the times of occurrence of exceedance of the magnitude m_T are discrete random variables. The average period or time between such occurrences or the return period of the event that m_T is exceeded, $E[K-J]=E[J-I]=T$ (see Example 1.30). The frequency of the event $y > m_T$ is $1/T$, and therefore the probability that the maximum annual magnitude in any one year is less than the magnitude level, associated with T, is

$$P[Y \leqslant m_T] = \exp(-e^{\alpha_1 - \beta_1 m_T}) = 1 - \frac{1}{T} \tag{8.51}$$

Another interpretation of T is obvious. Since $1/T$ is the frequency of earthquakes with magnitudes greater than m_T, it can be related to the magnitude–frequency relation, Eq. (8.48), that is,

$$\frac{1}{T} = N(m_T) = e^{\alpha - \beta m_T}, \qquad T = e^{-(\alpha - \beta m_T)} \tag{8.52}$$

For large magnitudes, the return period is also large, whereby the expression Eq. (8.51) can be simplified in the same way as was done for extreme wind speeds (see Example 1.30). Taking the logarithm twice of each side,

$$\alpha_1 - \beta_1 m_T = \log_e\left(-\log_e\left(1 - \frac{1}{T}\right)\right)$$

and putting $x = -1/T$, approximately,

$$\log_e\left(\frac{1}{T}\right) = \log_e N(m_T) = \alpha - \beta m_T = \alpha_1 - \beta_1 m_T \tag{8.53}$$

Since Eq. (8.53) is valid for arbitrary m_T, $\alpha = \alpha_1$ and $\beta = \beta_1$.

Now, many interesting deductions concerning the magnitude distribution Eqs. (8.48) and (8.49) can be made. For instance, the probability that the modal magnitude $M_m = \alpha/\beta$ (Eq. (8.50)) will be exceeded in any one year is

$$P[Y \geqslant M_m] = 1 - \exp[-e^{\alpha - \beta\alpha/\beta}] = 1 - e^{-1} = 0.633 \tag{8.54}$$

that is, the modal magnitude is most likely exceeded in an average year. If the probability that a certain magnitude level M_p is exceeded in any one year is called p, then by Eqs. (8.49) and (8.50)

$$M_p = M_m - (1/\beta)\log_e[-\log_e(1-p)] \tag{8.55}$$

Next take a period of C years. The value of a certain magnitude level $M_p(D)$ that is exceeded with probability p in D years is then obtained as follows:

$$P[(Y_1 \leqslant M_p(D)) \cap (Y_2 \leqslant M_p(D)) \cap \cdots \cap (Y_D \leqslant M_p(D))] = P[(Y \leqslant M_p(D)]^D = 1 - p$$

or

$$\exp[-e^{\alpha-\beta M_p(D)}]=1-p$$

and

$$M_p(D)=M_p+(1/\beta)\log_e D \tag{8.56}$$

An earthquake risk function R_D can now be defined as the probability of occurrence of an earthquake of magnitude M_D or more in a period of D years. From Eqs. (8.40) and (8.41),

$$M_D=\alpha/\beta-1(1/\beta)\log_e(-\log_e(1-R_D))+(1/\beta)\log_e D$$

or

$$R_D=1-\exp[-D\exp(\alpha-\beta M_D)] \tag{8.57}$$

Example 8.3 (after Lomnitz, [129])

In Table 8.6, the largest annual magnitudes exceeding $M=4.0$ for the period 1932–62, $D=31$, are listed, using a published catalogue of Californian earthquakes. The data has been ordered according to increasing magnitudes. The probability of occurrence of the maximum magnitudes has been calculated using the favourable fraction

$$P[Y\leqslant y_i]=\Phi(y_i)\simeq\frac{i}{n}$$

since the probability that the yearly maximum is less than or equal to, say, y_i is simply the number of maxima less than or equal to y_i divided by the total number of maxima. The coefficients of the extremal distribution α and β, which by Eq. (8.49) are given by the functional relationship $\log_e[-\log_e\Phi(y)]=\alpha-\beta y$ can therefore be obtained by a least squares fit, plotting the tabulated values on an extremal logarithmic paper. The thus

Table 8.6 *Maximum yearly magnitudes of Californian earthquakes 1932–62 (After Lomnitz, [129])*

y	$\Phi(y)$	y	$\Phi(y)$	y	$\Phi(y)$
4.9	0.0312	5.8	0.344	6.2	0.656
5.3	0.0625	5.8	0.375	6.2	0.687
5.3	0.0937	5.8	0.406	6.3	0.719
5.5	0.125	5.9	0.437	6.3	0.750
5.5	0.156	6.0	0.469	6.4	0.781
5.5	0.187	6.0	0.500	6.4	0.781
5.5	0.219	6.0	0.531	6.5	0.844
5.6	0.250	6.0	0.562	6.5	0.875
5.6	0.281	6.0	0.594	6.5	0.906
5.6	0.312	6.0	0.625	7.1	0.937
				7.7	0.696

obtained least squares line yields the values $\alpha = 11.43$ (the intercept with the vertical axis) and $\beta = 2$, (the slope), which corresponds very well with the Gutenberg and Richter values, Example 8.1.

By Eq. (8.52), the return period of large magnitude earthquakes can be computed. Thus for $M = 8$, $T = 100$ yr, that is, a magnitude 8 or greater earthquake is expected once every century on the average. The statistical basis for this statement is very weak, however, since the period range of data is only 31 years. However, if anything, this estimate of the return period for very large magnitude earthquakes is likely to be conservative as it neglects the observed downwards curvature of the magnitude–frequency relation, Fig. 8.1.

Using the above results, Eq. (8.57) can be utilized to produce a design acceleration map with contours of similar risk accelerations, assuming that within the region for which the earthquake risk map is to be drawn, the slope coefficient β is constant. The following procedure is due to Lomnitz, [129]. The design earthquake is specified as an earthquake that produces a maximum surface acceleration at a given site exceeding a certain acceleration limit, e.g. $\alpha_D = 150$ Gal. Over a period of T years ($T > D$, the 'design period'), all epicentres m in the region are localized. The region is now subdivided using a grid of mn points, covering the total area. For each grid point, compute the following expression:

$$p_{ij} = 1 - \exp(\beta y_j) \tag{8.58}$$

where y_j is the magnitude of an earthquake with epicentre j, which could produce the design acceleration at grid point i. Clearly (see Eq. (8.7)), p_{ij} is the probability that the design earthquake will not occur (magnitude is less than y_i for epicentre j) during the T years. The probability that no epicentral point will have an earthquake with sufficiently large magnitude to produce accelerations over the design acceleration at i is therefore the product of the probabilities. Therefore, the earthquake risk at point i, i.e. the probability $R_T(i)$ of having an acceleration larger than the design acceleration during the T years, is

$$R_T(i) = 1 - \prod_j p_{ij} \tag{8.59}$$

Finally, the earthquake risk at point i during the design period D can be obtained from $R_T(i)$ using Eq. (8.57), that is,

$$\frac{\log_e(1 - R_D)}{\log_e(1 - R_T)} = \frac{[-D\,e^{\alpha - \beta M_D}]}{[-T\,e^{\alpha - \beta M_D}]}$$

so

$$\log_e(1 - R_D) = (D/T)\log_e(1 - R_T) \tag{8.60}$$

Fairly extensive earthquake risk maps utilizing this procedure have been produced for California and Chile, [129].

Finally, a group of Mexican researchers, headed by Mario Ordaz, [163], have been working on a detailed seismicity and earthquake risk mapping of the whole of Mexico, employing highly sophisticated methods. In a recent preliminary report, [162], the methodology and preliminary results are presented. Primarily, the magnitude–frequency relations for the various parts or regions of the country have been studied and modified, using maximum likelihood estimation to obtain the a and b values for each region.

Deviation from the straight logarithmic Gutenberg–Richter line is quite evident in most cases. As for the distribution of arrival times, the Poisson model is adopted for earthquakes of magnitude less than 7, whereas for larger earthquakes, the distribution is modified using Bayesian statistical methods, assuming that all the parameters that define the distribution are themselves random variables which have to be estimated. This will naturally divide all Mexican earthquakes into those that have their origin in the subduction zone parallel to the Pacific coast and the lesser inland or intra-plate earthquakes, which follow the Poisson pattern more closely. For the inter arrival times of the subduction zone earthquakes, the lognormal distribution is adopted and fitted to the magnitude–frequency relations, which then assume the form

$$E[M|t] = \max\{77.5, 5.36 + 0.621 \log_e t\} \tag{8.61}$$

$$\sigma[M|t] = 0.27 \tag{8.62}$$

Here $E[M|t]$ and $\sigma[M|t]$ denote respectively the expectation and the standard deviation of the magnitude of the next event, given that $T = t$ years have passed since the last.

If it is assumed that the magnitude M, conditional of the time, has a normal distribution with the parameters given by Eqs. (8.61) and (8.62), the fact that the lognormal distribution has been adopted for the inter-arrival times will lead to a marginal distribution for the magnitudes, which is normal, with the parameters,

$$E[M] = \max\{7.3, 5.36 + 0.621 \log_e \mu_t\} \tag{8.63}$$

and

$$\sigma^2[M] = \sigma^2[M|t] + (0.621\,\sigma \log_e t)^2 \tag{8.64}$$

In the report, [162], the attenuation of seismic waves is taken into account when defining the earthquake risk at any particular site, together with local site effects, soil magnification, and resulting frequency modulation of the spectra. Compiling all these data together, design spectra for different locations and three different types of soil conditions, that is, type I for firm ground, and types II and III for soft ground, have been proposed. Thus an overall seismic risk analysis for the country as a whole has been effectuated, resulting in design spectra to meet the seismic risk with minimum cost and acceptable risk.

In a recent conference paper, Esteva *et al.*, [63], have described an extended Bayesian approach to the estimation of seismic risk in a multiple source region in the south-western part of Mexico. They divide the earthquakes into three categories instead of the two categories used in the above study by Ordaz *et al.*, [162]. The first two categories are the same, whereas a third category comprises normal-fault earthquakes generated at the lower end region of the subducting plate. An algorithm making using of a Monte Carlo simulation technique is formulated and used for the estimation.

8.3 CONCLUDING REMARKS

In the two preceding sections, a few topics concerning earthquake hazard and seismic risk analysis have been discussed. Only the tip of the iceberg has been viewed as this field has been subject to dramatic development, especially during the United Nations International Decade for Natural Disaster Reduction (IDNDR). The IDNDR effort has given rise to a multitude of meetings and conferences where an enormous quantity of papers on the various aspects of earthquake hazard and seismic risk have been presented. It would not serve a purpose to try to compact all this material into a short text contained in one chapter. Therefore, only a few classical examples have been given, and an emphasis has been placed on the fundamental aspects of an earthquake treated as a random event, following relatively simple probabilistic laws.

When the engineers started to show interest in earthquakes and seismic risk assessment in the early decades and around the middle of this century, many innovative ideas were put forward, which often lacked the correct geophysical and seismological background or basis. This often led to a controversial dialogue between the seismologists and the earthquake engineers with many misunderstandings. In later years, however, the two professions have learned to work together to greatly enhance the knowledge and analysis of earthquake hazard and seismic risk, [182]. Many sophisticated methods have been introduced, which take the earthquake mechanism into account, [14], [22], and use complicated probabilistic models, [63], [224]. The ultimate goal is to be able to set up well-established design parameters such as maximum expected surface accelerations, average maximum earthquake response spectra etc. at any given location within a seismically active area. The probability distribution and recurrence periods of these design parameters should be reasonably well defined in order to provide engineering design within established safety criteria. Another application is to be able to assess the vulnerability of engineered structures and expected property damage in a future earthquake for a particular seismic region with respect to insurance coverage and possible loss redemption through disaster assistances grants after the event. A particularly vivid reporting of an actual situation after the 6.7 magnitude Northridge earthquake of 17 January 1994, is presented by Tobin *et al.*, [221]. Within two days after the earthquake, damage estimates had been calculated, using geographical information systems (GIS) methodology, to set up an assistance grant programme for those in Los Angeles who had suffered severe damage or total property loss.

Seismic zoning and risk mapping is already a well-established field, and most of the seismic regions in the world have been subjected to precursory up to detailed analysis, [62]. The application of GIS methodology to facilitate the seismic zoning and risk mapping has proved to be a welcome addition to the probabilistic toolboxes being applied, [108], [140], [148] and [198]. The geographic information systems offer a clear and powerful method of compiling and registering various kinds of geographical data, which can be used to provide thematic maps containing almost any conceivable information. These kind of applications in seismic zoning and risk mapping can be said to be at the forefront of research efforts in this particular field at the time this book is written.

References

[1] Akaike, H., Fitting autoregressive models for prediction, *Annals of the Institute of Statistical Mathematics*, **21**, 1969, pp. 243–347.

[2] Akaike, H., A new look at the statistical model identification, *IEEE Transactions, Automatic Control*, **AC-19**, 1974, pp. 716–723.

[3] Aki, K. and Richards, P.G., *Quantitative Seismology, Theory and Methods*, Freeman, New York, 1980.

[4] Alanís, A., Sánchez Sesma, J. and López López, A. Determinación de velocidades del viento maximas producidas por huracanes. VIII Congreso Nacional de Ingeniera Estructural, Vol. II, Manzanillo, 1991.

[5] Alexandrov, A.D., Kolmogorov, A.N. and Lavrent'ev, M.A. *Mathematics, its Constants, Methods, and Meaning*, Moscow, 1956; English translation: MIT Press, Cambridge, 1963.

[6] ALGOR, Inc., *ALGOR Finite Element Analysis System*, Pittsburgh, 1993.

[7] Allin Cornell, C., Engineering seismic risk analysis, *Bulletin of the Seismological Society of America*, **58**, (5), 1968, pp. 1583–1606, San Francisco.

[8] Amin, M. and Ang, A.H-S., Non-stationary stochastic models of earthquake motions, *Proceedings ASCE, Journal of the Engneering Mechanics Division*, **94**, EM2, 1968.

[9] Ariaratnam, S.T., Schuëller, G.I.and Elishakoff, I., Stochastic Structural Dynamics, Progress in Theory and Applications, Elsevier, New York, 1988.

[10] Marcus Båth, *Introduction to Seismology, Birkhäuser Verlag, Basel, 1979*.

[11] Behan, M.J., *Statistical Continum Theories*, John Wiley and Sons, New York, 1968.

[12] Bendat, J., *Noise and Stochastic Processes*, John Wiley and Sons, New York, 1958.

[13] Benjamin, J. and Cornell, C.A. *Probability, Decision and Statistics for Civil Engineers*, McGraw-Hill, New York, 1967.

[14] Bessason, B., Assessment of earthquake loading and response of seismically isolated bridges, PHD Thesis, MTA-Report 1992:88, Institute of Marine Constructions, Technical University of Norway, Trondheim, 1992.

[15] Bharucha-Reid, A.T., *Elements of the Theory of Markov Processes and Their Applications*, McGraw-Hill, New York, 1960.

[16] Björnsson, S. and Einarsson, P., *Earthquakes in 'Náttúra Íslands'*, (in Icelandic) pp. 121–155, Almenna bókafélagið, Reykjavík, 1980.

[17] Blackman, R.B. and Tukey, T.W., *The Measurements of Power Spectra from the View of Communications Engineering*, Dover, New York, 1959.

[18] Bogdanoff, J.L., Goldberg, J.E. and Bernard, M.C., Response of simple structure to a random earthquake-like disturbance, *Bulletin of the Seismological Society of America*, **51**, (2), 1961, pp. 293–310.

[19] Boore, D., Stochastic Simulation of high frequency ground motions based on seismological models of the radiated spectra, *Bulletin of the Seismological Society of America*, **73**, (6), 1983, pp. 1865–1894, San Francisco.

[20] Borges, J.F., (Gen, Rapporteur), *CEB-FIP Model Code for Concrete Structures*, Vol. I, Common Unified Rules for Different Types of Construction and Materials, Comité Euro-International du Béton, 1978.

[21] Borgman, L.E., Ocean wave simulation for engineering design, *Proceedings, ASCE, Journal of the Waterways and Harbours Division*, **95**, (WW4), 1969.

[22] Box, G.E.P. and Jenkins, G.M., *Time Series Analysis: Forecasting and Control* (revised edn.), Holden Day, San Francisco, 1976.

[23] Boyer, C.B., *A History of Mathematics*, Wiley, New York, 1968.

[24] Breiman, L., *Probability and Stochastic Processes With a View Towards Applications* (2nd edn), The Scientific Press, Palo Alto, 1986.

[25] Brune, J.N., Tectonic stress and the spectra of seismic shear waves from earthquakes, *Journal of Geophysical Research*, **75**, 1970, pp. 4997–5009.

[26] Brune, J.N., Correction, *Journal of Geophysical Research*, **76**, 1971, pp. 5002.

[27] Bycroft, G.N., White noise representation of earthquakes, *Proceedings of the ASCE, Journal of the Engineering Mechanics Division*, **86**, (EM2), 1960, pp. 1–16.

[28] Cakmak, A.S., (ed.), *Structures and Stochastic Methods*, Elsevier, Amsterdam, 1987.

[29] Campell, N.R., The study of discontinuous phenomena, *Proceedings of the Cambridge Philosophical Society*, **15**, Cambridge, 1909.

[30] Cartwright, D.E. and Longuet-Higgins, M.S., The statistical distribution of the maxima of a random function, *Proceedings of the Royal Society*, Ser. A, **237**, 1956, London.

[31] Castro, R., Singh, S.K. Mena, E., An empirical model to predict Fourier ampitude spectra of horizontal ground motion, *Earthquake Spectra*, **4**, November, 1988.

[32] Caughey, T.H. and Stumpf, H.J., Transient response of a dynamic system under random excitation, *Journal Applied Mechanics, Trans. ASME*, **28**, 1961.

[33] Chang, M.K., Kwiatkowski, J.W., Nau, R.F., Oliver, R.M. and Pister, K.S., ARMA models for earthquake ground motions, *Earthquake Engineering and Structural Dynamics*, **10**, 1982, pp. 651–662.

[34] Childers, D. and Durling, A., *Digital Filtering and Signal processing*, West Publishing Co., New York, 1975.

[35] Chiu, A.N.L., Response of structures to time-varying wind loads, *Proceedings ASCE, Journal of the Structural Division*, **96**, (ST2), 1970.

[36] Choi, C.C.E., Correlation and Spectral Functions of Atmospheric Turbulence, *Proceedings of the 3rd Conference on Wind Effects of Buildings and Structures*, Tokyo, 1971.

[37] Clough, R.W. and Penzien, J., *Dynamics of Structures*, McGraw-Hill, New York, 1975.

[38] Collins, J.A., *Failure of Materials in Mechanical Design, Analysis, Prediction, Prevention*, (2nd edn), Wiley, New York, 1993.

[39] Cooley, J.W. and Tukey, J.W., An algorithm for the machine calculation of complex fourier series, *Mathematics of Computation*, **19**, 1965.

[40] Cox, D.R. and Miller, H.D., *The Theory of Stochastic Processes*, Methuen, London, 1965.

[41] Cramér, H., *Mathematical Methods of Statistics*, Princeton, 1946.

[42] Cramér, H. and Leadbetter, M.R., *Stationary and Related Stochastic Processes*, Wiley, New York, 1967.

[43] Crandall, S.H. and Mark, W.D., *Random Vibration in Mechanical Systems*, Academic Press, New York, 1963.

[44] Crandall, S.H., Measurement of random processes, Chapter 2 in S.H. Crandall (ed.), *Random vibration*, MIT Press, Cambridge, 1963.

[45] Crandall, S.H., Envelope distribution of a broad band process and other statistical measures of random response, *Journal of the Acoustic Society of America*, **35** (11), 1963.

[46] Crandall, S.H., Distribution of maxima of an oscillator to random excitation, *Journal of the Accoustic Society of America*, **47**, 1970.

[47] Crandall, S.H., First crossing probabilities of the linear oscillator, *Journal of Sound and Vibration*, **12** (3), 1970.

[48] Cruz-Reyna, S. De la, Poisson-distributed patterns of explosive eruptive activity, *Bulletin of Volcanology*, **54**, Springer-Verlag, Berlin, 1991.

[49] Cunningham, E.P., *Digital Filtering, An Introduction*, Houghton Mifflin, Boston, 1992.

[50] Daley, D.J. and Vere-Jones, D., *An Introduction to the Theory of Point Processes*, Springer Verlag, New York, 1988.

[51] Davenport, A.G., The application of statistical concepts to the wind loading of structures, *Proceedings of the Institution of Civil Engineers*, London, **19**, 1962.

[52] Davenport, A.G., Note on the distribution of the largest values of a random function with applications to gust loading, *Proceedings of the Institution of Civil Engineers*, London, **28**, 1964.

[53] Derman, C. and Klein, M., *Probability and Statistical Inference for Engineers, Oxford University Press*, New York, 1963.

[54] Ditlevsen, O., Approximate Extreme Value Distributions and First Passage Time Probabilities in Stationary and Related Stochastic Processes (doctoral dissertation), Copenhagen, 1969.

[55] Doob, J.L. *Stochastic Processes*, Wiley, New York, 1953.

[56] Dynkin, E.B., *Markov Processes*, Vols. 1 and 2, Academic Press, New York, 1965.

[57] Dyrbye, C., Gravesen, S., Lind, N.C. and Madsen, H.O., *Konstruktioners Sikkerhed*, den private Ingeniørfond, Danmarks tekniske Højskole, Copenhagen, 1979.

[58] Dyrbye, C and Hansen, S. Ole, Vindlast på bærende konstruktioner, SBI-Anvisning, Danish Building Research Institute, Report No. 158, Statens Byggeforskningsinstitut, Copenhagen, 1989.

[59] Einarsson, P., *et al.*, *Seismicity Pattern in the South Iceland Seismic Zone, Earthquake Prediction*, An International Review, Maurice Ewing Series 4, AGU, pp. 141–151, 1981.

[60] Elishakoff, I. and Lyon, R.H., (eds.), *Random Vibration–Status and Recent Developments*, The Stephen Harry Crandall Festschrift, Elsevier, Amsterdam, 1986.

[61] Epstein, B. and Lomnitz, C., A model for occurrences of large earthquakes, *Nature*, **211**, 1966, pp. 954–956.

[62] Escuela Politécnica Nacional *et al.*, *The Quito, Equador Earthquake Risk Management Project, An Overview*, GeoHazards International, San Francisco, 1994.

[63] Esteva, L., Diaz, O. and Garcia, J., Seismicity estimates of multiple-source regions, using uncertain earthquake locations, *Proceedings of the Fifth International Conference on Seismic Zonation*, Nice, France, Ouest Éditions, Presses Académiques, Vol. II, pp. 1538–1553, Nantes, 1995.

[64] Feller, W., *An Introduction to Probability Theory and its Applications* (2nd edn), Wiley, New York, 1957.

[65] Fertis, D.G., *Dynamics and Vibration of Structures*, Wiley, New York, 1973.

[66] Fisz, M., *Wahrscheinlichkeitsrechnung und matematische Statistik*, VEB Deustcher Verlag der Wissenschaften, Berlin Ost, 1970.

[67] Freedman, H.W., Estimating earthquake magnitude, *Bulletin of the Seismological Society of America*, San Francisco, **57**, 1967, pp. 747–760.

[68] Galambos, J., Lechner, J. and Simiu, E., (eds.), *Extreme Value Theory and Applications*, Kluwer, Dordrecht, 1993.

[69] Gardner, W.A., *Introduction to Random Processes with Application to Signals and Systems*, Macmillan, New York, 1986.

[70] Goldberg, J.E., Bogdanoff, J.L. and Sharp, D.R., The response of simple non-linear systems to random distribution of the earthquake type, *Bulletin of the Seismological Society of America*, **54**, 1964.

[71] Gnedenko, B.V., *The Theory of Probability* (4th edn), Chelsea, New York, 1968.

[72] Gröbner, W. and Hofreiter, N., *Integraltafeln*, Springer Verlag, Berlin, Vol. I, 1973, Vol. II, 1975.

[73] Gumbel, E.J., *Statistics of Extremes*, Columbia University Press, New York, 1958.

[74] Gutenberg, B., Amplitudes of surface waves and magnitudes of shallow earthquakes, *Bulletin of the Seismological Society of America*, January, **35**, 1945, pp. 3–12.

[75] Gutenberg, B. and Richter, C.F., *Seismicity of the Earth* (and Associated Phenomena), Princeton University Press, Princeton, N.J., 1954.

[76] Gutenberg, B. and Richter, C., Earthquake magnitude, intensity, energy, and acceleration (second paper), *Bulletin of the Seismological Society of America*, **46**, 1956, pp. 105–145.

[77] Hald, A., *Statistical Theory with Engineering Applications*, Wiley, New York, 1965.
[78] Halldórsson, P., Stefánsson, R., Einarsson, P. and Björnsson, S., Evaluation of Earthquake Risk, A Report to the Committee for Site Selection of Large Industries, 52. pp., Reykjavík, 1984.
[79] Hanks, T.C. and Kanamori, H., A moment magnitude scale, *Journal of Geophysical Research*, **84** (6), 1979, pp. 2348–2350.
[80] Hanks, T.C. and McGuire, R.K., The character of high frequency strong ground motion, *Bulletin of the Seismological Society of America*, San Francisco, **71** (6), 1981, pp. 154–174.
[81] Hart, G.C., *Uncertainty Analysis, Loads, and Safety in Structural Engineering*, Prentice Hall, New Jersey, 1982.
[82] Henry, D.L., Theory of fatique damage accumulation in steel, *Transactions of the ASME*, **77**, 1955, p. 913.
[83] Hogg, R.V. and Craig, A.T., *Introduction to Mathematical Statistics* (2nd edn), Collier-McMillan, London, 1966.
[84] Hong, H.P. and Rosenbulueth, E., The Mexico earthquake of Sept. 19, 1985—model for generation of subduction earthquakes, *Earthquake Spectra*, **4**, 1988.
[85] Housner, G.W. and Jennings, P.C., Generation of artificial earthquakes, *Proceedings ASCE, Journal of the Engineering Mechanics Division*, **90**, EM1, 1964.
[86] Hsu, H.-I. and Bernard, M.C., A random process for earthquake, *Earthquake Engineering and Structural Dynamics*, **6** (4), 1978.
[87] Hüsler, J. and Reiss, R.-D., (eds.), *Extreme Value Theory*, Lecture Notes in Statistics 51, Springer Verlag, New York, 1987.
[88] Icaza, M. de, Instituto de Física, UNAM, Personal communication, 1992.
[89] Itô, K., On stochastic differential equations, *Memoirs of the American Mathemetical Society*, **4**, 1951.
[90] Jacobsen, L.S. and Ayre, R.S., *Engineering Vibrations with Applications to Structures and Machinery*, McGraw-Hill, New York, 1958.
[91] James, H.M., Nichols, N.B. and Phillips, R.S., *Theory of Servomechanism*, MIT Radiation Laboratory Series, Vol. 25, McGraw-Hill, New York, 1947.
[92] Jara, J.M. and Rosenblueth, E., The Mexico earthquake of Sept. 19, 1985—probability distribution of times between subduction earthquakes, *Earthquake Spectra*, **4**, 1988.
[93] Jara, J.M. and Rosenblueth, E., Variación de los Coeficientes Sismicos de Diseño como Functión del Tiempo. *Boletín, Centro de Investigación Sísmica de la Fundación Javier Barros Sierra*, México, **2**, 1992.
[94] Jazwinski, A.H., *Stochastic Processes and Filtering Theory*, Academic Press, New York and London, 1970.
[95] Jennings, P.C., Housner, G.W. and Tsai, N.C., Simulated Earthquake Motions. Report, Earthquake Engineering Research Laboratory, California Institute of Technology, April 1968.
[96] Joyner, W.B. and Boore, D.M., *Measurement, Characterization, and Prediction of Strong Ground Motion*, Geotechnical Special Publication, No. 20, ASCE New York, 1988.
[97] Kachanov, L.M., *Introduction to Continuum Damage Mechanics*, Martinus Nijhoff, Dordrecht, 1986.
[98] Kanai, K., Semi-empirical formula for the seismic characteristics of the ground, *Bulletin of the Earthquake Research Institute*, University of Tokyo, **35**, Part 2, 1957, pp. 309–325.
[99] Kanai, K., An empirical formula for the spectrum of strong earthquake ground motion, *Proceedings of the 2nd World Conference on Earthquake Engineering*, Tokyo and Kyoto, pp. 1541–1551, 1960.
[100] Karlin, S. and Taylor, H.M., *A First Course in Stochastic Processes*, Academic Press, New York, 1975.
[101] Kárnik, V., *Seismicity of the European Area*, Part 1, Reidel, Dordrecht, 1969.
[102] Kárnik, V., *Seismicity of the European Area*, Part 2, Reidel, Dordrecht, 1971.

[103] Karnopp, D.C., Basic theory of random vibration, Chapter 1 in S.H. Crandall (ed.), *Random Vibration*, Vol. 2, MIT Press, Cambridge, 1963.

[104] Kawasumi, H., Measures of earthquake danger and expectancy of maximum intensity throughout Japan as inferred from the seismic activity in historical times, *Bulletin of the Earthquake Research Institute*, Tokyo University, **29**, 1951.

[105] Keulegan, G.H. and Carpenter, L.H., Forces on cylinders and plates in an oscilating fluid, *Journal of Research of the National Bureau of Standards*, **60**, 1958.

[106] Khinchine, A.J., Korrelationstheori der stationären stochastischen Prozesse, *Matematische Annalen*, Berlin, **109**, 1934, p. 604.

[107] Kiureghian, A.D. and Ang, A.H.-S., A fault-rupture model for seismic risk analysis, *Bulletin of the Seismological Society of America*, **67** (4), 1977, pp. 1173–1194.

[108] King, S.A., Kiremidjian, A.S., Law, K.H. and Basoz, N.I., Earthquake damage and loss estimation through geographic information systems, *Proceedings of the Fifth International Conference on Seismic Zonation*, Nice, France, Ouest Éditions, Presses Académiques, Vol. I, pp. 265–272, Nantes 1995.

[109] Klotter, K., *Technische Schwingungslehre*, Vols 1 and 2, Springer Verlag, Berlin, 1960.

[110] Knopoff, L., *Q, Rev. Geophysics*, **2**, 1964.

[111] Kolmogorov, A.N., Über das logarithmische normale Verteilungsgesetz der Dimensionen der Teilchen bei Zerstückelung, *Comptes Rendus (Doklady) de L'Académie des Science de l'URSS*, **31**(2), 1941.

[112] Kolmogorov, A.N., Local structure of turbulence in noncompressible fluid with high Reynolds number, *Dokl. Acad Sci.*, USSR, **30**, 1941.

[113] Kolmogorov, A.N., *Grundbegriffe der Wahrscheinlichkeitsrechung, Ergebnisse der Mathematik*, Springer Verlag, Berlin 1933. An English translation: *Foundations of the Theory of Probability*, Chelsea, New York, 1950.

[114] Kosko, B., *Neural Networks and Fuzzy Systems: A Dynamical Systems Approach to Machine Intelligence*, Prentice Hall, Englewood Cliffs, N.J., 1992.

[115] Kosko, B., *Neural Networks for Signal Processing*, Prentice Hall, Englewood Cliffs, N.J., 1992.

[116] Kosko, B., *Fuzzy Thinking: The New Science of Fuzzy Logic*, Prentice Hall, Englewood Cliffs, N.J., 1993.

[117] Kozin, F., Autoregressive moving average models of earthquake records, *Journal of Probabilistic Engineering Mechanics*, **3** (2), 1988.

[118] Lam, R.P., Dynamic response of a tall building to random wind loads, *Proceeding of the 3rd Conference on Wind Effects on Buildings and Structures*, Tokyo, 1971.

[119] Langen, I. and Sigbjörnsson, R., *Dynamisk Analyse av Konstruksjoner*, Tapir, Trondheim, 1979.

[120] Laning, J.H. and Battin, R.H., *Random Processes in Automatic Control*, McGraw-Hill, New York, 1956.

[121] Lee, Y.W., *Statistical Theory of Communications*, Wiley, New York, 1960.

[122] Lin, Y.K., *Probabilistic Theory of Structural Dynamics*, McGraw-Hill, New York, 1967.

[123] Lin, Y.K. and Cai, C.Q., *Probabilistic Structural Dynamics*, McGraw-hill, New York, 1995.

[124] Lindgren, B.W., *Statistical Theory*, Macmillan, New York, 1963.

[125] Liu, S.C. and Jhavery, D.P., Spectral simulation and earthquake site properties, *Proceedings ASCE, Journal of the Engineering Mechanics Division*, **95**, (EM5), 1969.

[126] Liu, S.C., Synthesis of stochastic representations of ground motions, *Bell Systems Technical Journal*, **49**, 1970, pp. 521–541.

[127] Ljung, L., *System Identification: Theory for the User*, Prentice Hall, Englewood Cliffs, N.J., 1987.

[128] Loeve, M.H., *Probability Theory*, Springer Verlag, New York, 1977–78.

[129] Lomnitz, C., Global tectonics and earthquake risk, *Developments in Geotectonics*, **5**, Elsevier, Amsterdam, 1974.

[130] Lomnitz, C. and Rosenblueth, E., (eds.), *Seismic Risk and Engineering Decisions*, Elsevier, Amsterdam, 1976.

[131] Lomnitz, C., Nonlinear dynamics of strong earthquake motion on soft ground, *International Symposium on the Effects of Surface Geology on Seismic Motion*, Vol. 1, Odawara, Tokyo, 1992.

[132] Lomnitz, C., *Earthquake Prediction*, Wiley, New York, 1994.

[133] Lomnitz-Adler, J., Automation, models of seismic fracture: constraints imposed by the magnitude–frequency relation, *Journal of Geophysical Research*, **98**, 1993, p. 17745.

[134] Loy, N.J., *An Engineer's Guide to FIR Digital Filters*, Prentice Hall, New Jersey, 1988.

[135] Lumley, J.L., *Stochastic Tools in Turbulence*, Academic Press, New York, 1970.

[136] Mann, N.R., Schafer, R.E. and Singpurwalla, N.D., *Methods for Statistical Analysis of Reliability and Life Data*, Wiley, New York, 1974.

[137] Marco, S.M. and Starkey, W.L., A concept of fatigue damage, *Transactions of the ASME*, **76**, 1954, p. 627.

[138] Marple, S.L., *Digital Spectral Analysis with Applications*, Prentice Hall, Englewood Cliffs, N.J., 1987.

[139] MathWorks, Inc., *MATLAB Reference Guide*, Natick, MA, 1995.

[140] Matsuoka, M. and Midorikawa, S., GIS-based integrated seismic hazard mapping for a large metropolitan area, *Proceedings of the Fifth International Conference on Seismic Zonation*, Nice France, Ouest Éditions, Presses Académiques, Vol. II, pp. 1334–1341, Nantes 1995.

[141] Medhi, J., *Stochastic Processes* (2nd edn), Wiley, New York, 1993.

[142] Medvedev, S.V., Sponheuer, W. and Karnik, Vit, *Seismic Intensity Scale*, MSK 1964, Academy of Sciency of the USSR, Soviet Geophysical Committee, 13 pp., Moscow 1965.

[143] Melbourne, W.H., Probability distributions associated with the wind loading of structure, *Civil Engineering Transactions*, Institution of Engineers in Australia, **19**, 1977.

[144] Piralla, R.M., *Diseño Estructural*, Editorial Limusa, México, 1985.

[145] Melsa, J.L. and Sage, A.P., *An Introduction to Probability and Stochastic Processes*, Prentice Hall, New Jersey, 1973.

[146] Mena, E. *et al.*, Acelerograma en el centro scop de la Secretaría de Comunicaciones y Transportes, Sismo del 19 de Septiembre de 1985, Informe IPS-10B, Instituto de Ingeniería, UNAM, 1985.

[147] Mendenhall, W., *Introduction to Probability and Statistics*, (7th edn), Duxbury Press, Boston, 1981.

[148] Meroni, F., Tomasoni, R., Grimaz, S., Petrini, V., Zonno, G. and Cella, F., Assessment of Seismic Effective Vulnerability using ARC/INFO connected to Nexpert, *Proceedings of the Fifth International Conference on Seismic Zonation*, Nice, France, Ouest Éditions, Presses, Académiques, Nantes, **I**, 1995, pp. 68–75.

[149] Miner, M.A., Cumulative damage in fatigue, *Journal of Applied Mechanics, Transactions ASME*, **12**, 1945.

[150] Narayanan, R. and Roberts, T.M., (eds.), *Structures Subjected to Repeating Loading, Stability and Strength*, Elsevier Applied Science, London, 1991.

[151] Nelson, R., *Probability, Stochastic Processes, and Queuing Theory*, Springer Verlag, New York 1995.

[152] Newland, D.E., *An Introduction to Random Vibration and Spectral Analysis*, Longman, London, 1975, (2nd edn, 1984).

[153] Newmark, N.M. and Rosenblueth, E., *Fundamentals of Earthquake Engineering*, Prentice-Hall, New Jersey, 1971.

[154] Nigam, N.C., *Introduction to Random Vibrations*, The MIT Press, Cambridge, MA, 1983.

[155] Nishenko, S.P. and Buland, R., A generic recurrence interval distibution for earthquake forecasting, *Bulletin of the Seismological Society of America*, **77**, 1987.

[156] Nishenko, S.P. and Singh, S.K., Conditional probabilities for the recurrence of large and great intraplate earthquakes along the Mexican subduction Zone, *Bulletin of the Seismological Society of America*, **77**, 1987.

[157] Norris, C., Hansen, R.J., Holley, M.J. Jr., Biggs, J.M. Namyet, S. and Minami, J.K., *Structural Design for Dynamic Loads*, McGraw-Hill, New York, 1959.
[158] Ólafsson, S., The use of ARMA models in strong motion modelling, *Proceedings of the Tenth World Conference on Earthquake Engineering*, Madrid, pp. 857–859, 1992.
[159] Ólafsson, S. and Sigbjörnsson, R., Application of ARMA models to estimate earthquake ground motion and structural response, *Earthquake Engineering and Structural Dynamics*, **24**, 1995, pp. 951–966.
[160] Oppenheim, A.V. and Schafer, W., *Discrete-time Signal Processing*, Prentice Hall, Englewood Cliffs, N.J., 1989
[161] Ordaz, M. and Singh, S.K., Source spectra and spectral attenuation of seismic waves from Mexican earthquakes and evidence of amplification in the hill zone of Mexico City, *Bulletin of the Seismological Society of America.*, **82**, 1992.
[162] Ordaz, M. *et al.*, Riesgo Sismigo y Espectros de Diseño en el Estado de Guerrero. Informe Conjunto del Instituto de Ingeniería, UNAM y el Centro de Investigación Sísmica AC de la Fundación Javier Barros Sierra, Enero 1992.
[163] Ordaz, M., Instituto de Ingenieria, UNAM, Personal Communication 1992.
[164] Otnes, R.K. and Enochsson, L., *Applied Time Series Analysis*, Wiley, New York, 1978.
[165] Palmgren, A., Die Lebensdauer von Kugellagern, *Berichte des Vereins Deutschen Ingenieuren*, Berlin, **68**, 1924.
[166] Papoulis, A., *Probability, Random Variables and Stochastic processes*, Int. Stud. Ed., McGraw-Hill Kogakusha, Tokyo, 1965.
[167] Papoulis, A., *Signal Analysis*, McGraw-Hill, New York, 1977.
[168] Papoulis, A., *The Fourier Integral and its Applications*, McGraw-Hill, New York, 1987.
[169] Parzen, E., *Modern Probability Theory and its Applications*, Wiley, New York, 1960.
[170] Parzen, E., *Stochastic Processes*, Holden Day, San Francisco, 1962.
[171] Petersen, C. Aerodynamische und seismische Einflüsse auf die Schwingungen insbesondere schlanker Bauwerke, *Fortshrittberichte der VDI Zeitschriften*, Düsseldorf, **11** (11), 1971.
[172] Petersen, G., *et al.*, A large Earthquake in South Iceland, A Report of a Working Group to the Civil Defence Authorities in Iceland, (in Icelandic), 54 pp., 1978.
[173] Polhemus, N.W. and Cakmak, A.S., Simulation of earthquake ground motions using autoregressive moving average (ARMA) models, *Earthquake Engineering and Structural Dynamics*, **9**, 1981, pp. 343–354.
[174] Powell, A., On the fatigue failure of structures due to vibrations excited by random pressure fields, *Journal of the Acoustical Society of America*, **30**, (12), 1958.
[175] Proakis, J.G. and Manolakis, D.G., *Digital Signal Processing, Principles, Algorithms, and Applications*, Macmillan, New York, 1992.
[176] Resnick, S.I., *Extreme Values, Regular Variation, and Point Processes*, Springer Verlag, New York, 1987.
[177] Rice, S.O., Mathematical analysis of random noise, *Bell Systems Technical Journal*, **23** and **24**. Reprinted in N. Wax (ed.), *Selected Papers on Noise and Stochastic Processes*, Dover, New York, 1954.
[178] Richter, C.F., *Elementary Seimsology*, Freeman, San Francisco, CA, 1958.
[179] Roberts, J.B. and Spanos, P.D., *Random Vibration and Statistical Linearization*, Wiley, Chichester, 1990.
[180] Robson, J.D., *An Introduction to Random Vibration*, Edinburgh University Press, Edinburgh, 1964.
[181] Rogers, G.L., *Dynamics of Framed Structures*, Wiley, New York, 1959.
[182] Safak, E., Optimal-adaptive filters for modelling spectral shape, site amplifications and source scaling, *Soil Dynamics and Earthquake Engineering*, **8**, 1989, pp. 75–95.
[183] Sauter, F., *Fundamentos de Ingeniería Sísmica I, Introducción a la Sismologia*, Editorial Tecnologica de Costa Rica, 1989.
[184] Schmetterer, L., *Introduction to Mathematical Statistics*, Springer Verlag, New York, 1974.

[185] Schuëller, G.I., and Shinozuka, M., (eds.), *Stochastic Methods in Structural Dynamics*, Martinus Nijhoff, Dordrecht, 1987.

[186] Schulz-DuBois, E.O. and Rehberg, I., Structure functions in lieu of correlation functions, *Journal of Applied Physics*, Springer Verlag, **24**, 1981.

[187] Scruton, C. and Rogers, E.W.E., Steady and unsteady wind loading of buildings and structures, *Philosophical Transactions of the Royal Society of London*, **A269**, 1971.

[188] Shah, H.C., *Earthquake Engineering and Seismic Risk Analysis*, CE 282B, The John, A. Blume Earthquake Engineering Center, Stanford University, Stanford, 1985.

[189] Shinosuka, M. and Sato, Y., Simulation of non-stationary random processes, *Proceedings ASCE, Journal of the Engineering Mechanics Division*, **93**, (EM1), 1967.

[190] Shinosuka, M. and Yang, J.N., On the bound of first excursion probability, *Proceedings ASCE, Journal of the Engineering Mechanics Division*, **95**, (EM2), 1969.

[191] Shinosuka, M., Random processes with evolutionary power, *Proceedings ASCE, Journal of the Engineering Mechanics Division*, **96**, (EM4), 1970.

[192] Shinoshuka, M., Maximum structural response to seismic excitations, *Proceedings ASCE, Journal of the Engineering Mechanics Division*, **96**, (EM5), 1970.

[193] Shiotani, M. and Iwatani, Y., Correlation of wind velocities in relation to gust loadings, *Proceedings 3rd Conference on Wind Effects on Buildings and Structures*, Tokyo, 1971.

[194] Sigbjörnsson, R., On the Theory of Structural Vibration due to Natural Wind, (Ph.D. thesis), Report Nr. R 59, Structural Research Laboratory, Technical University of Denmark, Copenhagen 1974.

[195] Sigbjörnsson, R., Ólafsson, S. and Bessason, B., Jarðskjalfti 25. maí 1987 (in Icelandic), Report No. 88005, Engineering Research Institute, University of Iceland, 1988.

[196] Sigbjörnsson, R. and Ólafsson, S., Application of parametric time series models in earthquake engineering, *European Earthquake Engineering*, **3**, 1992, pp. 43–49.

[197] Sigbjörnsson, R., *et al.*, On seismic hazard in Iceland—A stochastic simulation approach, *Proceedings of the 10th European Conference on Earthquake Engineering*, Vienna, August 1994.

[198] Ragnar Sigbjörnsson *et al.*, The mapping of seismic hazard using stochastic simulation and geographic information systems, *Proceedings of the 11th World Conference on Earthquake Engineering*, Acapulco, Mexico, June 1996.

[199] Silverman, B.W., *Density Estimationfor Statistics and Data Analysis*, Monograph of Statistics and Probability, Chapman Hall, New York, 1986.

[200] Singh, S.K., Astiz, L. and Havskov, J., Seismic gaps and recurrence periods of large earthquakes along the Mexican subduction zone: a reexamination, *Bulletin of the Seismological Society of America*, **71**, June 1981.

[201] Snæbjörnsson, J.P. and Sigbjörnsson, R., A study of earthquake and wind induced structural response, *Proceedings of the 2nd European Conference on Structural Dynamics: EURODYN '93*, Trondheim, Norway, A.A. Balkema, Rotterdam, 1993.

[202] Snæbjörnsson, J.P. and Sigbjörnsson, R. Estimation of structural parameters from full scale wind induced response, *Proceedings of the 9th International Conference on Wind Engineering*, Wiley Eastern, New Delhi, 1995.

[203] Snæbjörnsson, J.P., Hansen, E.H. and Sigbjörnsson, R., Variability of natural frequency and damping ratio of a concrete building, case study in system identification, *Proceedings of the 3rd European Conference on Structural Dynamics: EURODYN '96*, Florence, Italy, A.A. Balkema, Rotterdam 1996.

[204] Sobzcyk, K. and Spencer, B.F. Jr., *Random Fatigue: From Data to Theory*, Academic Press, Boston, 1992.

[205] Söderström, T. and Stoica, P. *System Identification*, Prentice Hall, New York, 1989.

[206] Sólnes, E.J., Structural Vibration induced by Earthquakes. Ph.D. thesis, Reykjavik, Iceland 1966.

[207] Sólnes, J., The spectral character of earthquake motions, *Proceedings of the 3rd European Conference on Earthquake Engineering*, Sofia, 1970.

[208] Sólnes, J. and Sigbjörnsson, R., Along-wind response of large bluff buildings, *Proceedings ASCE, Journal of the Structural Division*, **99**, (ST3), 1973.

[209] Sólnes, J., (ed.), *Engineering Seismology and Earthquake Engineering*, NATO ASI series, Noordhoff, Leiden, 1974.

[210] Sólnes, J., Interpretation, and simulation of earthquake ground motions as non-stationary stochastic process. Safety of structures under dynamic loading. *Proceedings of the International Research Seminar*, June 1977, Trondheim, Norway, Vol. 1, Tapir, 1978.

[211] Sólnes, J., *The seismicity and earthquake hazard in Iceland*, Engineering Research Institute, University of Iceland, October, 1985.

[212] Sólnes, J., Halldórsson, B. and Sigbjörnsson, R., Assessment of seismic risk based on synthetic and upgraded earthquake catalogs of Iceland, *Proceedings of the 9th International Seminar on Earthquake Prognostics*, San José, Costa Rica, 1994.

[213] Sólnes, J. and Halldórsson, B., Generation of synthetic earthquake catalogs: application in earthquake hazard and seismic risk assessment, *Proceedings of the 5th international conference on seismic zonation*, Nice, France, pp. 1131–1140, Ouest Éditions, Presses Académiques, Nantes, 1995.

[214] Spiegel, M.R., *Theory and Problems of Statistics*, Schaum Outline Series, McGraw-Hill, New York, 1961.

[215] Stefánsson, R. *et al.*, Earthquake prediction research in the South Iceland seismic zone and the SIL project, *Bulletin of the Seismological Society of America*, **83**, 1993.

[216] Stein, S. Introduction to Seismology, Earthquakes and Earth Structure, Unpublished Manuscript, Northwestern University, Chicago, 1991.

[217] Svesnikov, A.A., *Problems in Probability Theory, Mathematical Statistics and Theory of Random Functions*, Dover, New York, 1979.

[218] Tajimi, H., A statistical method of determining the maximum response of a building structure during an earthquake, *Proceedings 2nd World Conference on Earthquake Engineering*, Tokyo and Kyoto, 1960.

[219] Thoft-Christensen, P. and Baker, M.J., *Structural Reliability Theory and its Applications*, Springer-Verlag, Berlin, 1982.

[220] Thomson, W.T., *Theory of Vibration with Applications*, Chapman & Hall, London, 1993.

[221] Tobin, L.T., Davis, J.F., Eguchi, R.T. and Nathe, S.K., Northridge earthquake case study, *Proceedings of the 5th International Conference on Seismic Zonation*, Nice, France, Ouest Éditions, Presses Académiques, Vol. III, pp. 1805–1825, Nantes 1995.

[222] Todorovic, P., *An Introduction to Stochastic Processes and Their Applications*, Springer-Verlag, New York, 1992.

[223] Toro, G.R. and McGuire, R.K., Calculational procedures for seismic hazard analysis and its uncertainty in the eastern United States, *Proceedings of the 3rd International Conference on Soil Dynamics and Earthquake Engineering*, Princeton, 1987.

[224] Toro, G.R., Probabilistic seismic-hazard analysis: a review of the state of the art, *Proceedings of the 5th International Conference on Seismic Zonation*, Nice, France, Ouest Éditions, Presses Académiques, Vol. III, pp. 1829–1857, Nantes, 1995.

[225] Tryggvason, E.S., Thoroddsen, S. and Thorarinsson, S., Report on earthquake risk in Iceland, (in Icelandic), *Bulletin of the Association of Chartered Engineers*, Reykjavík, **43**, 1959, pp. 1–9.

[226] Tsuboi, C., Determination of the Gutenberg–Richter magnitude of earthquakes occurring in and near Japan, *Journal of the Seismological Society of Japan*, Tokyo, **7** (3), 1954.

[227] Uhlenbeck, G.E. and Ornstein, L.S., On the theory of Brownian motion, *Physical Review*, **36**, 1930, pp. 93–112, Reprinted in N. Wax (ed.), *Selected Papers on Noise and Stochastic Processes*, Dover, New York, 1954.

[228] UNDRO, *Mitigating Natural Disasters: Phenomena, Effects and Options*, A Manual For Policy Makers and Planners, United Nations, New York 1991.

[229] Vickery, B.J., A model of atmospheric turbulence for studies of wind loads on buildings, *Proceedings of the 2nd Australian Conference on Hydraulics and Fluid Mechanics*, Auckland, 1965.

[230] Vilar Rojas, J.I., Muños Black, C. and Sánchez Sesma, J., Velocidades regionales para diseño eolico en la republica Mexicana. *Proceedings of the 8th National Congress on Structural Engineering*, Vol. II, Manzanillo 1991.

[231] Wallace, R.E., Davis, J.F. and McNally, K.C., Terms for expressing earthquake potential, predictions and probability, *Bulletin of the Seismological Society of America*, **74**, 1984.

[232] Weast, R.C. and Selby, S.M., (eds.), *Handbook of Tables for Mathematics*, (4th edn), The Chemical Rubber Co., Cleveland, Ohio, 1970.

[233] Weaver, H.J., *Applications of Discrete and Continuous Fourier Analysis*, Wiley, New York, 1983.

[234] Wiener, N., Generalized harmonic analysis, *Acta Mathematica*, **55**, 1930, p. 117.

[235] Wöhler, A., Über die Festigkeitsversuche mit Eisen und Stahl, *Zeitschrift für Bauwesen*, **8**, 1858.

[236] Wood, H.O. and Neumann, F., Modified Mercally intensity scale of 1931, *Bulletin of the Seismological Society of America*, **21** 1931, pp. 277–283.

[237] Wooton, L.R. and Scruton, C., Aerodynamic stability, *Proceedings of the Seminar on Modern Design of Wind-Sensitive Structures*, CIRIA, London, 1970.

[238] Yaglom, A.H., *An Introduction to the Theory of Random Functions*, Prentice Hall, New Jersey, 1962.

[239] Yang, J.N., Lin, Y.K. and Sae-Ung, S., Tall building response to earthquake excitations, *Proceedings ASCE, Journal of the Engineering Mechanics Division*, **106**, (EM4), 1980.

[240] Þrainsson, H. and Sigbjörnsson, R., Álag og öryggi burðarvirkja (Action and Reliability of Structures, Lecture Notes in Icelandic), Engineering Research Institute Report No. 94005, University of Iceland, Reykjavík 1994.

Index

9 780471 971924